2020 27th International Conference on Mixed Design of Integrated Circuits and Systems (MIXDES 2020)

Lodz, Poland
25 – 27 June 2020

IEEE Catalog Number: CFP20MIX-POD
ISBN: 978-1-7281-9781-4

Copyright © 2020, Department of Microelectronics and Computer Science, Lodz University of Technology
All Rights Reserved

*** *This is a print representation of what appears in the IEEE Digital Library. Some format issues inherent in the e-media version may also appear in this print version.*

IEEE Catalog Number: CFP20MIX-POD
ISBN (Print-On-Demand):   978-1-7281-9781-4
ISBN (Online):   978-83-63578-18-3

Additional Copies of This Publication Are Available From:

Curran Associates, Inc
57 Morehouse Lane
Red Hook, NY 12571 USA
Phone: (845) 758-0400
Fax: (845) 758-2633
E-mail: curran@proceedings.com
Web: www.proceedings.com

Proceedings of the 27th International Conference "Mixed Design of Integrated Circuits and Systems"

25-27 June 2020, Łódź, POLAND

MIXDES 2020

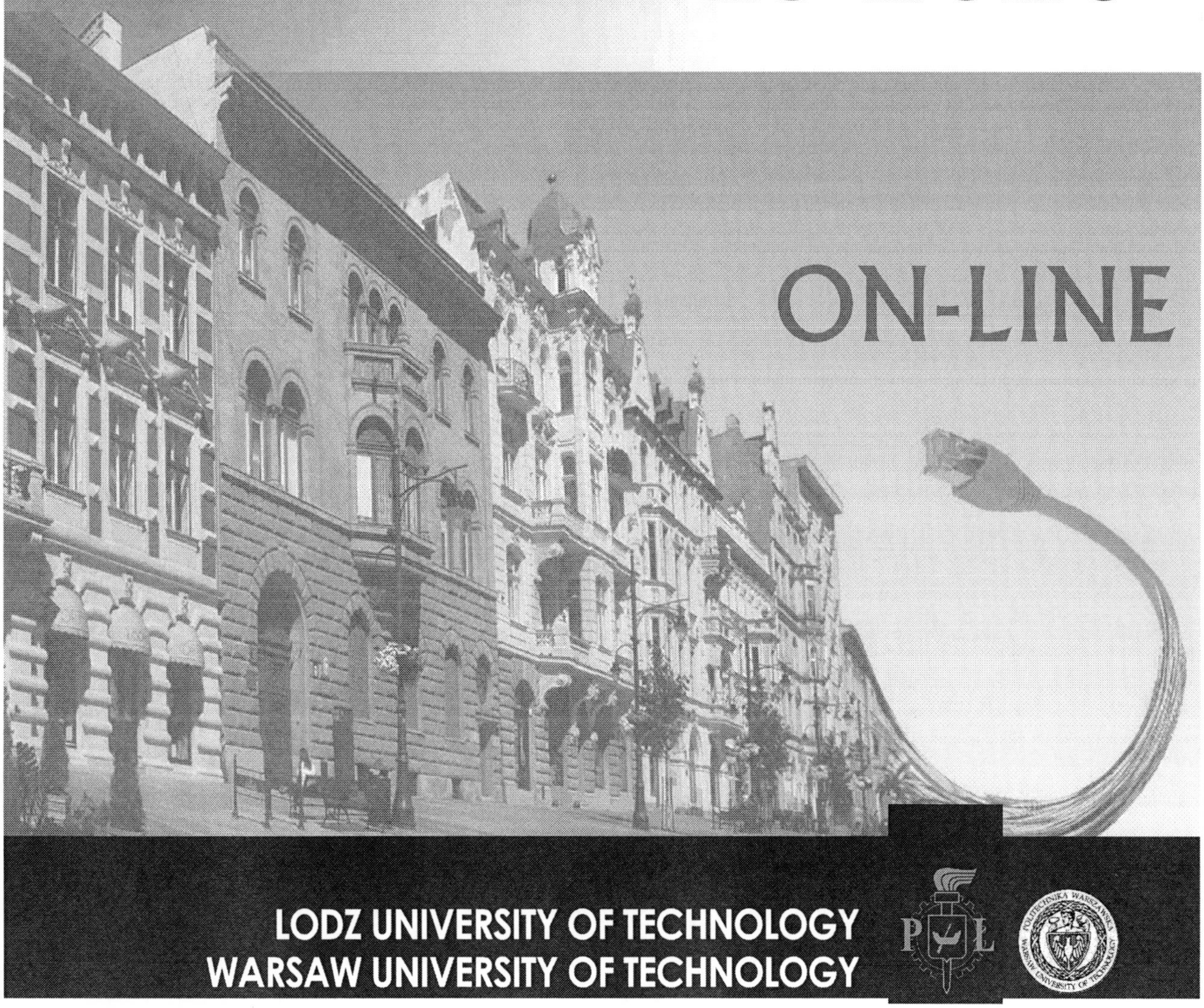

ON-LINE

LODZ UNIVERSITY OF TECHNOLOGY
WARSAW UNIVERSITY OF TECHNOLOGY

Proceedings of 27th International Conference

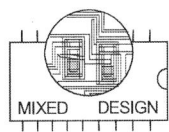

MIXED DESIGN OF INTEGRATED CIRCUITS AND SYSTEMS
MIXDES 2020

Łódź, Poland
June 25 – 27, 2020

Organised by:
**Department of Microelectronics and Computer Science,
Lodz University of Technology, Poland**

**Institute of Microelectronics and Optoelectronics,
Warsaw University of Technology, Poland**

in co-operation with:
Poland Section IEEE - ED & CAS Chapters

**Section of Microelectronics & Electron Technology
and Section of Signals, Electronic Circuits & Systems
of the Committee of Electronics and Telecommunication
of the Polish Academy of Sciences**

**Commission of Electronics and Photonics
of Polish National Committee
of International Union of Radio Science – URSI**

supported by:
Ministry of Science and Higher Education

Edited by
Andrzej Napieralski

Lodz University of Technology
Department of Microelectronics and Computer Science
ul. Wólczańska 221/223
90-924 Łódź, Poland

Phone: +48 42 631 27 27
Fax: +48 42 638 03 27
E-mail: napier@dmcs.pl
Web: http://www.mixdes.org

Preface

Since 26 years the MIXDES Conference is a forum devoted to recent advances in micro- and nanoelectronics design methods, modelling, simulation, testing and manufacturing technology in diverse areas including embedded systems, MEMS, sensors, actuators, power devices and biomedical applications.

Due to the coronavirus (COVID-19) pandemic, the Scientific and Organising Committees have decided this year to hold a virtual conference. We are extremely disappointed that is was not possible to meet in person but despite this, the presentations and discussions were efficient.

The program of the conference consisted of three days of sessions starting each day with invited talks. The following invited talks were presented:

- *Maximizing the Efficiency of CMOS Front-illuminated Solar cell for Self-powered IoT Sensor Applications*
  Poki Chen (National Taiwan University of Science and Technology, Taiwan)

- *Miniaturized Sensors for Planetary Applications*
  Mina Rais-Zadeh (NASA Jet Propulsion Laboratory, California Institute of Technology, USA)

- *Semiconducting Oxide Electronics for Newly Emerging Applications*
  Arokia Nathan (Cambridge Touch Technologies, Cambridge, UK)

- *Sensor Design Made by Bosch*
  Mike Schwarz (Robert Bosch GmbH, Germany)

- *The Qucs/QucsStudio and Qucs-S Graphical User Interface: An Evolving "White-Board" for Compact Device Modelling and Circuit Simulation in the Current Era*
  Mike Brinson (London Metropolitan University, UK)

The program of MIXDES 2020 also included two special sessions:

- *Compact Modeling Support for Micro and Nanoelectronic System Development*
  organised by D. Tomaszewski (Institute of Electron Technology, Poland) and W. Grabiński (GMC, Switzerland)

- *Fusion Diagnostics I&C Workshop*
  organised by S. Simrock (ITER, France) and D. Makowski (Lodz University of Technology, Poland)

In addition to the technical sessions, opportunities for the conference attendees was (free of charge) EDS Distinguished Lecturer Mini-Colloquium: "Semiconductor-based sensors - technology, modeling, applications", organized by ED Poland Chapter with collaboration of Institute of Electron Technology, Warsaw, Poland, which took place June 27, 2020.

Number of presented papers and authors by country

Country	Number of		Country	Number of		Country	Number of	
	papers	co-authors		papers	co-authors		papers	co-authors
Austria	2	7	Iran	1	3	Slovakia	1	2
Brazil	2	12	Italy	1	2	Switzerland	0	1
Czech Republic	0	1	Lithuania	1	2	Taiwan	1	1
Egypt	0	1	Peru	1	3	UK	3	2
France	2	12	Poland	27	58	USA	1	3
Germany	4	12	Portugal	1	2	Total:	50	133
Greece	1	7	Saudi Arabia	1	2			

Number of authors by country

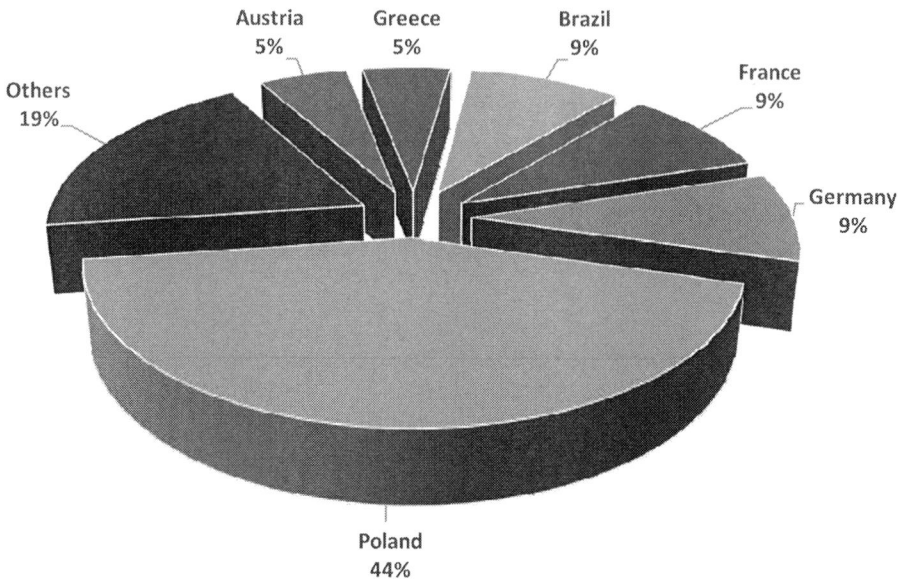

All regular papers were reviewed and selected from submissions from 19 countries. The organisers would like to thank all the distinguished scientists who have supported the conference by taking part in the International Programme Committee and reviewing contributed papers.

We hope that you are safe and healthy and remain so, and we will meet together next year in Wrocław, June 24-26, 2021.

Łódź, June 2020

Andrzej NAPIERALSKI
Department of Microelectronics and Computer Science
Lodz University of Technology, Poland
General Chairman of MIXDES 2020

International Programme Committee

Prof.	M. Bucher	Technical University of Crete, Greece
Prof.	J. Cabestany	Universitat Politecnica de Catalunya, Spain
Prof.	J. Collet	LAAS - CNRS, Toulouse, France
Prof.	A. Dąbrowski	Poznan University of Technology, Poland
Prof.	G. De Mey	University of Ghent, Belgium **(Vice-Chairman)**
Prof.	J. Deen	McMaster University, Canada
Prof.	M.H. Fino	Universidade Nova de Lisboa, Portugal
Dr.	D. Foty	Gilgamesh Associates, USA
Prof.	L. Golonka	Wrocław University of Science and Technology, Poland
Prof.	K. Górecki	pGdynia Maritime University, Poland
Dr.	W. Grabiński	GMC, Switzerland
Prof.	P. Gryboś	AGH University of Science and Technology, Poland
Prof.	V. Hahanov	Kharkiv National University of Radioelectronics, Ukraine
Prof.	A. Handkiewicz	Poznan University of Technology, Poland
Prof.	A. Hatzopoulos	Aristotle University of Thessaloniki, Greece
Prof.	S. Hausman	Lodz University of Technology, Poland
Dr.	G. Jabłoński	Lodz University of Technology, Poland
Prof.	A. Jakubowski	Warsaw University of Technology, Poland
Prof.	A. Kos	AGH University of Science and Technology, Poland
Prof.	W. Kuźmicz	Warsaw University of Technology, Poland
Prof.	C. Lallement	Strasbourg University, France
Prof.	M. Lobur	State University Lviv, Ukraine
Dr.	M.M. Louerat	Université Pierre et Marie Curie, Paris, France
Prof.	T. Łuba	Warsaw University of Technology, Poland
Prof.	B. Macukow	Warsaw University of Technology, Poland
Prof.	J. Madrenas	Universitat Politecnica de Catalunya, Spain
Dr.	D. Makowski	Lodz University of Technology, Poland
Prof.	A. Martinez	LAAS - CNRS, Toulouse, France **(Honorary Chairman)**
Prof.	A. Materka	Lodz University of Technology, Poland
Prof.	W. Mathis	Leibniz University of Hannover, Germany
Prof.	J.M. Moreno	Universitat Politecnica de Catalunya, Spain
Dr.	M. Napieralska	Lodz University of Technology, Poland
Prof.	A. Napieralski	Lodz University of Technology, Poland **(General Chairman)**
Prof.	J. Nishizawa	Semiconductor Research Institute, Japan
Dr.	J.L. Noullet	INSA de Toulouse, France
Prof.	L. Opalski	Warsaw University of Technology, Poland
Prof.	A. Pfitzner	Warsaw University of Technology, Poland **(Programme Chairman)**
Prof.	E. Piętka	Silesian University of Technology, Poland
Prof.	W. Pleskacz	Warsaw University of Technology, Poland
Dr.	B.F. Romanowicz	Nano Science and Technology Institute, USA
Prof.	J.A. Rubio	Universitat Politecnica de Catalunya, Spain
Prof.	A. Rybarczyk	Poznan University of Technology, Poland
Dr.	J.-M. Sallese	Swiss Federal Institute of Technology, Switzerland
Prof.	D. Sankowski	Lodz University of Technology, Poland
Dr.	M. Schwarz	Robert Bosch GmbH, Germany
Prof.	N. Stojadinović	University of Niš, Serbia
Prof.	T. Szmuc	AGH University of Science and Technology, Poland
Dr.	P. Śniatała	Poznań University of Technology, Poland
Prof.	M. Tadeusiewicz	Lodz University of Technology, Poland
Dr.	D. Tomaszewski	Institute of Electron Technology, Warsaw, Poland
Dr.	P. Tounsi	INSA de Toulouse, France
Dr.	M. Turowski	Alphacore, Inc.,USA
Prof.	R. Ubar	Tallinn Technical University, Estonia
Prof.	G. Wachutka	Technische Universitaet Muenchen, Germany
Prof.	K. Wawryn	Technical University of Koszalin, Poland
Prof.	B. Więcek	Lodz University of Technology, Poland
Prof.	S. Yoshitomi	Toshiba Corporation, Japan
Prof.	J. Zarębski	Gdynia Maritime University, Poland
Prof.	M. Zubert	Lodz University of Technology, Poland

Organising Committee

Prof.	A. Napieralski	(Chairman)
Dr.	M. Orlikowski	(Secretary)
Dr.	M. Napieralska	(Vice-Chairman)
Dr.	G. Jabłoński	Department of Microelectronics and Computer Science, Lodz University of Technology, Poland
Prof.	W. Kuźmicz	Institute of Micro- and Optoelectronics, Warsaw University of Technology, Poland

Table of Contents

Preface 3

Table of Contents 7

I General Invited Papers

Sensor Design Made by Bosch 13
M. Schwarz (Robert Bosch GmbH, Germany)

The Qucs/QucsStudio and Qucs-S Graphical User Interface: An Evolving "White-Board" for Compact Device Modeling and Circuit
Simulation in the Current Era 23
M. Brinson (London Metropolitan Univ., UK)

S1 Compact Modeling Support for Micro and Nanoelectronic System Development

Compact Model for Continuous Microfluidic Mixer 35
A. Bonament, A. Prel (ICube, France), J.-M. Sallese (EPFL, Switzerland), M. Madec, C. Lallement (ICube, France)

Impact of Dynamic Trapping on High Frequency Organic Field-Effect Transistors 40
M. Mueller, S. Donnhäuser, S. Mothes, A. Pacheco-Sanchez, K. Haase, S. C. B Mannsfeld, M. Claus (Tech. Univ. Dresden, Germany)

Parameter Extraction for a Simplified EKV-model in a 28nm FDSOI Technology 45
K. Bajer, S. Paul, D. Peters-Drolshagen (Univ. Bremen, Germany)

Qucs-S/QucsStudio/Octave Schematic Synthesis Tools for Device and Circuit Parameter Extraction from Measured Characteristics 50
M. Brinson (London Metropolitan Univ., UK)

1 Design of Integrated Circuits and Microsystems

A Capacitive Feedback 80 dBOhm 1.1 GHz CMOS Transimpedance Amplifier with Improved Biasing 59
A. Romanova, V. Barzdenas (Vilnius Gediminas Tech. Univ., Lithuania)

A New Architecture of Thermometer to Binary Decoder in a Low-Power 6-bit 1.5GS/s Flash ADC 65
M. Keyhanazar, A. Kalami (Islamic Azad Univ., Urmia Branch, Iran), A. Amini (Sina Bioelectronics Company, Iran)

A Survey on the Application of Parametric Amplification in Next Generation Digital RF Transceivers 69
L.M. Pires, J.P. Oliveira (Univ. Nova de Lisboa, Portugal)

A W-band SiGe BiCMOS Transmitter Based on K-band Wideband VCO for Radar Applications 74
M. Kucharski, M. Widlok, R. Piesiewicz (SIRC Sp. z o.o., Poland)

Active Feedbacks Comparative Analysis for Charge Sensitive Amplifiers Designed in CMOS 40 nm 78
G. Węgrzyn, R. Kłeczek, P. Kmon (AGH Univ. of Science and Techn., Poland)

ASIC Architecture and Implementation of RED Scheduler for Mixed-Criticality Real-Time Systems 83
L. Kohutka, V. Stopjaková (Slovak Univ. of Techn. in Bratislava, Slovakia)

CMOS Interface for Capacitive Sensors with Custom Fully-Differential Amplifiers 89
M. Jankowski, P. Zając, P. Amrozik, M. Szermer (Lodz Univ. of Techn., Poland)

Comparative Analysis of Power Consumption of Parallel Prefix Adders 94
I. Brzozowski (AGH Univ. of Science and Techn., Poland)

Low Hardware Complexity Filters for On-Chip Algorithm Used in Air Pollution Sensors for Dense Urban Areas in Smart Cities 101
*Z. Długosz, M. Rajewski (UTP Univ. of Science and Techn., Poland), M. Banach (Poznan Univ. of Techn., Poland), T. Talaśka,
R. Długosz (UTP Univ. of Science and Techn., Poland)*

Low Power Preamplifier for Biomedical Signal Digitization 107
N. ALjehani (King Saud Univ., Saudi Arabia), M. Abbas (King Saud Univ., Saudi Arabia and Assiut University, Egypt)

Multichannel Programmable Readout IC for Photodiodes Array 112
P. Pieńczuk, C. Kołaciński, A. Szymański, P. Janus, K. Kucharski, D. Obrębski, M. Zbieć, M. Jakubowski (Łukasiewicz Research Network - Institute of Electron Techn., Poland)

Relocatable Partial Bitstreams for Overlay Architectures atop FPGAs 117
Z. Mudza (Lodz Univ. of Techn., Poland)

2   Thermal Issues in Microelectronics

Comparison of Set-ups Dedicated to Measure Thermal Parameters of Power LEDs 127
K. Górecki, P. Ptak (Gdynia Maritime Univ., Poland), M. Janicki (Lodz Univ. of Techn., Poland)

Investigations Properties of Selected Methods of Measurements of Thermal Parameters of the IGBT 133
K. Górecki, P. Górecki (Gdynia Maritime Univ., Poland)

Thermal Characterization of Electronic Components Using Single-detector IR Measurement and 3D Heat Transfer Modelling 139
M. Kopeć, B. Więcek (Lodz Univ. of Techn., Poland)

3   Analysis and Modelling of ICs and Microsystems

A Process, Voltage and Temperature Dependent Modeling Methodology for Industrial Requirements 147
I. Sejc, R. Kappel (ams AG, Austria)

Capacitance Deviation Caused by Mechanical Deformation of MEMS Inertial Structure 151
J. Nazdrowicz, A. Stawiński, A. Napieralski (Lodz Univ. of Techn., Poland)

Noise Resistance Estimation for a GaN JFET Using Small Signal Measurements for an X-band LNA 156
E. Karagianni (Hellenic Naval Academy, Greece), C. Lessi (National Tech. Univ. Athens, Greece), C. Vazouras (Hellenic Naval Academy, Greece), A. Panagopoulos (National Tech. Univ. Athens, Greece), G. Deligeorgis, G. Stavrinidis, A. Kostopoulos (Foundation for Research and Techn. Hellas, Greece)

On Applications of Fractional Derivatives in Circuit Theory 160
J. Gulgowski (Univ. Gdansk, Poland), T. Stefanski (Gdansk Univ. of Techn., Poland), D. Trofimowicz (SpaceForest Ltd., Poland)

Simulation of Signal Propagation Along Fractional Order Transmission Lines 164
T. Stefanski (Gdansk Univ. of Techn., Poland), D. Trofimowicz (SpaceForest Ltd., Poland), J. Gulgowski (Univ. Gdansk, Poland)

Subban Structure and Ballistic Conductance of a Molybdenum Disulfide Nanoribbon in Topological 1T' Phase: A k·p Study 168
V. Sverdlov, A.-M. El-Sayed, S. Selberherr (Tech. Univ. Wien, Austria)

4   Microelectronics Technology and Packaging

Challenges in Performance Improvement of Silicon Systems on Chip in Advanced Nanoelectronics Technology Nodes 175
A. Malinowski (Globalfoundries, Germany), S.K. Mishra (Globalfoundries, USA)

Recessed and P-GaN Regrowth Gate Development for Normally-off AlGaN/GaN HEMTs 181
C. Haloui (LAAS-CNRS and CEA-Tech, France), G. Toulon (EXAGAN, France), J. Tasselli (LAAS-CNRS, France), Y. Cordier, E. Frayssinet (CRHEA-CNRS, France), K. Isoird, F. Morancho (LAAS-CNRS, France), M. Gavelle (CEA-Tech, France)

5   Testing and Reliability

A Human Immunity Inspired Intrusion Detection System to Search for Infections in an Operating System 187
P. Widulinski, K. Wawryn (Koszalin Univ. of Techn., Poland)

The Application of NIR Spectrometer for Average Temperature Measurement in Optical Fibers Based on Spontaneous Raman Scattering for DTS Applications 192
I.S.M. Shatarah, B. Więcek (Lodz Univ. of Techn., Poland)

6   Power Electronics

1MHz Gate Driver in Power Technology for Fast Switching Applications 199
R. Di Lorenzo, A. Baschirotto (Univ. Milan - Bicocca, Italy), A. Pidutti, P. Del Croce (Infineon, Austria)

An Influence of the Operation Mode of a LED Lamp of the HUE Type on Its Electrical and Optical Parameters 204
P. Ptak, K. Górecki, J. Heleniak (Gdynia Maritime Univ., Poland)

7   Signal Processing

Combining Epsilon-similar Fuzzy Rules for Efficient Classification of Cardiotocographic Signals 213
M. Jezewski, R. Czabanski, J.M. Leski (Silesian Univ. of Techn., Gliwice, Poland), A. Matonia (Łukasiewicz Research Network - Institute of Med. Techn. and Equipm., Poland), R. Martinek (VBS - Tech. Univ. of Ostrava, Czech Republic)

Fusion of Position Adjustment from Vision System and Wheels Odometry for Mobile Robot in Autonomous Driving 218
J. Zwierzchowski (Lodz Univ. of Techn., Poland), D. Pietrala (Kielce Univ. of Techn., Poland), J. Napieralski (Lodz Univ. of Techn., Poland)

Marker Detection Algorithm for the Navigation of a Mobile Robot 223
A. Annusewicz (Kielce Univ. of Techn., Poland), J. Zwierzchowski (Lodz Univ. of Techn., Poland)

Modified Particle Swarm Optimization Algorithm Facilitating Its Hardware Implementation 227
M. Rajewski, Z. Długosz, R. Długosz, T. Talaśka (UTP Univ. of Science and Techn., Poland)

Testing Stability of Digital Filters Using Multimodal Particle Swarm Optimization with Phase Analysis 232
D. Trofimowicz (SpaceForest Ltd., Poland), T. Stefanski (Gdansk Univ. of Techn., Poland)

8   Embedded Systems

A Database Proposal for an Application Involving Industrial Networks for Industry 4.0 Concepts 239
A. Lugli, E. Neto, J.P. Henriques, M.T. de Carvalho Silva, N. Dias Pereira (Inatel, Brazil), T.C. Pimenta (UNIFEI, Brazil)

Consistency Preserving Development of Embedded Systems Using AADL 245
T. Szmuc, W. Szmuc (AGH Univ. of Science and Techn., Poland)

Indoor Precise Infrared Navigation 249
P. Marzec, A. Kos (AGH Univ. of Science and Techn., Poland)

Linux Kernel Driver for External Analog-to-Digital and Digital-to-Analog Converters 255
P. Skrzypiec, Z. Marszałek (AGH Univ. of Science and Techn., Poland)

Multipoint Wireless Humidity and Temperature Monitoring Network for HVAC Systems Validation 260
M. Zbieć, D. Obrębski (Sieć Badawcza Łukasiewicz - Instytut Technologii Elektronowej, Poland)

Performance Constraints of Machine Learning on Embedded Devices 266
Ł. Grzymkowski, T. Stefanski (Gdansk Univ. of Techn., Poland)

Rigorous Development of Embedded Systems Supported by Formal Tools 272
T. Szmuc, W. Szmuc (AGH Univ. of Science and Techn., Poland)

Sensor Fusion Algorithm Implementation on Microchip PIC Microcontroller 277
S. Salas, C. Valdez, K. Lau (Univ. Peruana de Ciencias Aplicadas, Peru), M.H. Amini (Florida International University, USA), M. Kropidłowski, P. Śniatała (Poznan Univ. of Techn., Poland)

Virtualization of an Aluminum Cans Production Line Using Virtual Reality 282
L. Sales de Oliveira Almeida, A. Baratella Lugli (Inatel, Brazil), T. Cleber Pimenta (UNIFEI, Brazil), M.V. Cirino e Silva, J.P. Carvalho Henriques, R. Paranaiba Mesquita (Inatel, Brazil)

Index of Authors 289

10

General Invited Papers

Proceedings of the 27th International Conference *"Mixed Design of Integrated Circuits and Systems"*
June 25-27, 2020, Łódź, Poland

Sensor Design Made by Bosch

Invited Paper

Mike Schwarz
Robert Bosch GmbH
Email: mike.schwarz@de.bosch.com

Abstract—In this paper the sensor design made by Bosch, one of the world's largest supplier of micromechanical sensors in automotive and consumer applications, is briefly introduced. The design is one of the key elements to bridge the different domains between process technology, electronics, system and customers needs. The here presented flow and methodology ensures an integration of all perspectives from prototyping to series production. Examples of typical Sensor/MEMS design including various mechanical and electronical constraints are given.

Keywords—MEMS, micromechnical sensors, design methodology, automotive applications, consumer applications

I. INTRODUCTION

The design is one of the key parts within the development process of new sensors. It bridges the physics, e.g. the physical process, process development, electronic circuit(s), ASIC interface and the customer (market) requirements. The need of design, simulation and modeling methodologies is therefore an essential part in the product development cycle. They enable to concentrate on the Key Performance Index (KPI) of sensor and beyond future performance improvements. Furthermore, knowing the design and interaction between the various system components allows for a significant sample phase reduction and the "1st time right" approach. This ensures finally criteria as high functionality, yield and reliability of micromechanical devices in mass production.

To emphasize the reader for the demand of sensor design and parts of its methodologies one notice the penetration of sensors and sensor systems. Sensors are almost everywhere, dominating humans life every minute within electronics and/or sensor interfaces, e.g. for automotive and consumer applications. To give some rough numbers, a car typically contains more than 50 sensors made of Micro Electro Mechanical Systems (MEMS) and a mobile phone between 20-25 sensors [1]. Today the MEMS and sensor market represents more than 10% of the total Integrated Circuit (IC) market, as more and more MEMS devices and sensors are integrated in products, such as MEMS, image sensors, RF filters, etc. in consumer and automotive [1]. In a few years the global amount of MEMS sensors will reach more than 80 billion [2]. For improving sensor performance combined with new features and smaller development cycles, enhanced design flows and methodologies are required taking criteria into account such as quality, cost effectiveness, reliability and time to market.

The paper is organized as follows: Section II gives an overview in the design flow and methodologies as influence strength analysis, models, process- and device simulations, multi-domain and mechanical structure simulations, artificial intelligence and big data. Within section III some specific sensor design examples are highlighted using the presented approaches. Finally, the paper is closed by a conclusion in section IV.

II. DESIGN FLOW AND METHODOLOGY

New sensor development for industrial mass production requires various recursions of design, process verification and optimization, and testing of system specification. Each sample phase requires a characterization of the samples with a benchmark of the design, layout and wafer processing. Especially, the challenges and topics of micromechanical sensors, i.e. dedicated processes and equipment, processes contamination risks to ICs, combined design and control of mechanical and electrical parameters, controlling the influence of mechanical stress of package, e.g. temperature influence, mold, etc., testing of physical parameters, and robustness and media compatibility (e.g. for pressure, flow sensors) need to be addressed during each sample phase. If required additional loops of all these development steps have to be fulfilled to achieve a high functionality and reliability of the micromechanical sensor device. The typical design flow at a certain sample phase is given in Figure 1.

It is obvious that process parameters as doping, temperature, pressure, and lithography (denoted as $P_1 \ldots P_n$) always vary to some rate by its nature of fluctuations. These results in deviations of geometry ($G_1 \ldots G_n$) as layer thickness variations or structure loss due to mask underetching. Besides, material ($M_1 \ldots M_n$) as the elastic modulus or thermal expansion of the micromechanical structure vary by natural fluctuations. During the sensor design one has taken into account each of these aspects. The results of these spreads are variations of the functional parameters, e.g. sensitivity, frequency, temperature coefficient, higher order parameters, etc. [3].

The design itself is established by complex models, which represent the physical and system behavior of the sensor and sensor system. The model G_i, M_i and system parameters S_i, e.g. operating voltage, temperature range, etc. act as input for the physical models to describe the functional parameters of the sensor element. The distributions of the model and system parameters including their c_{pk}-values have to be known from process monitoring. Design parameters D_i, e.g. structural length and width, also taken as input of the physical models.

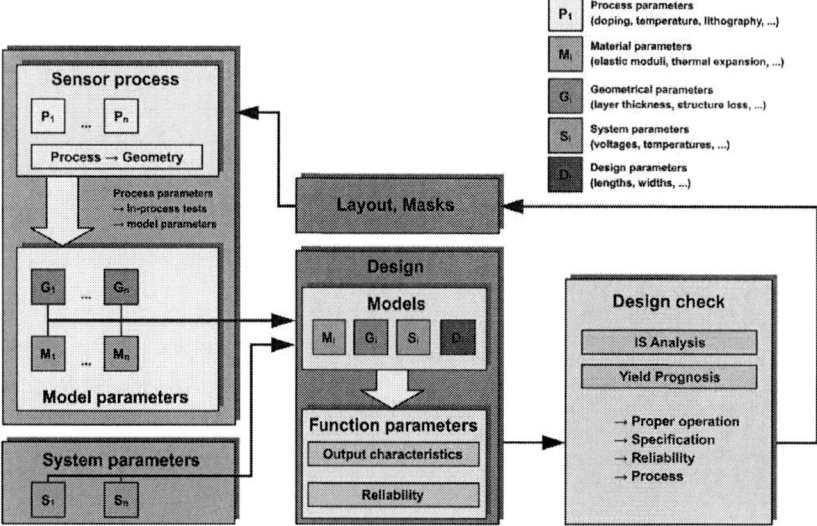

Fig. 1. Schematic design flow for micro-electro-mechanical sensors.

The relationship of functional parameters to model parameters is

$$F_j = f\left(G_i, M_i, S_i, D_i\right). \qquad (1)$$

Every function parameter of the sensor, e.g. sensitivity, or temperature coefficient of sensitivity, etc. requires an analytical expression, numerical Finite Element Method (FEM) simulations or empirical studies (multi-chip designs) for F_j.

A design check of various modifications requires a Monte Carlo based yield prognosis and an Influence Strength Analysis (ISA) to identify optimization parameters with the most signifcant impact on performance improvements.

A. Influence Strength Analysis & Monte Carlo

The ISA, a methodology to identify the most critical parameters of a design identifies the influence and sensitivity of single parameters G_i, M_i or S_i on the function parameters F_j. This allows for the optimization of the most critical parameter in the design and/or process.

The basic idea is to represent a model parameter by its nominal (actual operation point), and lower and upper limits, $\pm 3\sigma$ of a probability function. The influence of these model parameters onto function parameter(s) should be within the specification (lower and upper limit). The influence is then defined as the distance between variation and nominal divided by the distance between specification and nominal value [3]–[7].

The representation of the actual operation point of a probability function is made via Monte Carlo simulations. This method takes into account process tolerances in the mass production of sensor devices. These can be derived from simple functions, e.g. Gaussian or more complex functions, e.g. Rayleigh or Weibull or from process monitoring. The

Monte Carlo analysis itself selects randomly from a various set of these parameters.

To obtain a sufficient statement of the distributions for the functional parameters depending on the input parameters, the random analysis has to be repeated many times, e.g. approximate 1000 shots, which is a typical value for the amount of Monte Carlo runs.

The number of model parameters can vary between 50 and more depending on the present information. The number of the functional parameters is between 10 to 20 depending on the sensor type and customer requirements. The strength analysis for a functional parameters F_j is obtained by the root of the square sum of the single model parameter influences [3]–[7].

The proof of the sensor design with the ISA is a very useful method to identify potential risks of yield losses. Futhermore, the value of the strength analysis define the required action. Is the value 0, no influence is observed by any model parameter. A value below "one" states an allowed scattering of all model parameters and an expectation of high yield. Values above "one" for the strength analysis offer a potential risk and yield losses, even for an allowed scattering of a model parameter but its scattering impact is significant on the function parameter limits.

Of interest are values above "one" for a design at its sample phase, because these identify the parameters to be improved. Further details of the methodology and details in calculation can be found in [3]–[7].

B. Models

The accuracy of the prediction depends on the model description. In general one can correlate the complexity of a model with its accuracy and effects to be included.

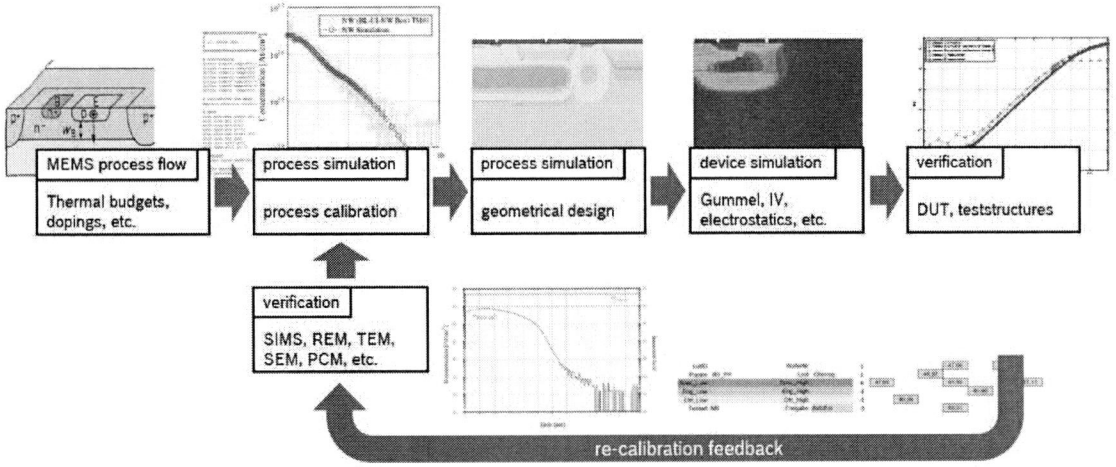

Fig. 2. Process and Device simulation methodology and flow presenting the different aspects.

However, every type of model has its validity during the various sample phases. In an early development phase an analytical model description including first and maybe second order effects is resonable to concentrate on the key performance parameters.

With the development progress the demand for more precise and powerful models can require the shift from analytical description to Finite Element Method (FEM) models. This depends on the demand of information required and the experience of the used technology. Furthermore, the point of view defines the needs of such simulation. In an early development phase the sensor element itself is of interest, while in the later development complete system simulations and the interaction might be in the focus. However, such simulation require more computation power and limit therefore the space for the Design of Experiments (DoE). Here, in general corners and worst case predictions are made and as first rough assumption gaussian distributions assumed if no further information is available. They are fed backwards into more simple sensor model descriptions.

Nevertheless, the experience of the design and development progress can allow for the Reduced Order Models (ROM) methodology. Such methodologies can improve the speed during the design and samples phases and allow for larger DoEs in contrast to FEM simulation models.

C. Process and Device Simulation

At a certain phase during the design, the need of more adavanced simulation support is required, because the complexity in interactions of semiconductor physics cannot be modeled by pure analytics.

On top, additional boundary conditions have to be taken into account. The most critical and contradictory one is the simultaneous development of the sensor/MEMS process, due to the interaction of mechanics and electronics within the design. The consideration of "floating" boundary conditions requires special design flows.

Here, the process and device simulation enter the game and bridge the interface between technology and design. Furthermore, it acts as part of the model verfication between sensor process and model parameters as shown in Figure 1. Depending on the demands and needs, those simulation can represent a part of a process or a complete geometrical design in a process simulation with applied physics from device simulation towards the extraction of classicical current voltage characteristics. As one may see, from the design flow (Fig. 1), such simulation methodologies can target the technology pillar or the design pillar with the modeling itself.

To emphasize one in the need of such methodologies, a brief example is given. As a support for technology in an early design phase, a process simulation can help in extraction of first distributions of specific process parameters. Huge DoE can be applied without processing lots of wafers. If one is interested in the distribution of the spread of implant dose or energy and its impact on a resistivity parameter, setup the DoE and extract the distributions for the models and design. This saves cost and time of huge statistics in an early design phase. Finally, some corners are required to benchmark simulation vs reality.

An extension of this is the full simulation of a geometrical design with a calibrated flow as shown in [8]. Here, identical options as for a single process step are available. The possibility to setup a DoE on each single process steps may give some first guess of the impact and spread on specification parameters. Nevertheless, one should consider the amount of time required to simulate a DoE with such a complexity. Here, a trade off between complexity and benefit has to be choosen and in comparison to the ISA methodology, in here no Monte Carlo approach is available. This limits the capabilty. However, by combining intelligent corners and simulations, it is possbile to capture the most significant parameters and spreads.

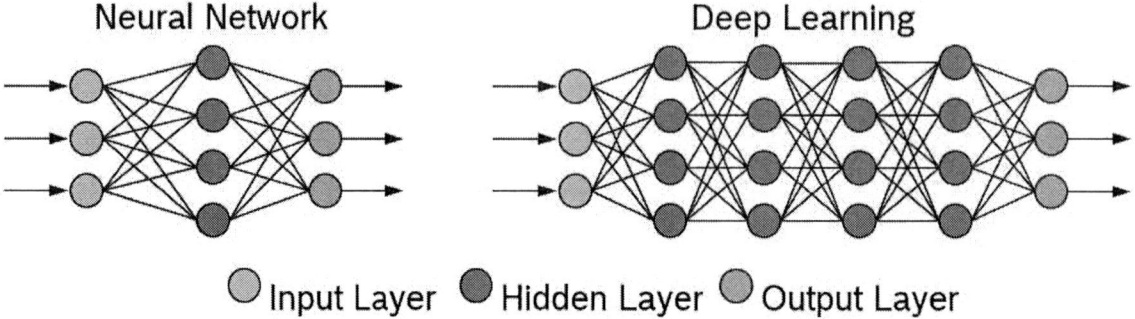

Fig. 3. Neural network or deep learning networks used to identify with big data hidden correlations.

To introduce one the methodology behind, here it is briefly introduced. More detail can be found from Schwarz et al. in [9]. The methodology flow including different aspects is given in Figure 2.

The methodology starts with process, its comic and detailed descriptions of the parameters, e.g. the implants of the species, dose, energy and tilt, etc. Afterwards, a calibration of the process is usually required to adjust the models to reality, to adopt the difference in each process and the physical behavior. The need for calibration by e.g. Secondary-Ion Mass Spectrometry (SIMS), Spreading Resistance (SPR), and/or Scanning Spreading Resistance Microscopy (SSRM) is required, because the process differs from fab to fab and even tool to tool.

Afterwards, device simulation with well-known physical models of solid state physics can be performed, including various investigations, e.g. IV curve analysis, extraction of geometrical effects, etc. Furthermore, the behavior over temperature and charge influences on the device performance can be investigated by DoE of process simulation and device simulation. The possibility for combinations is almost unlimited, except the required time for large DoE.

Finally, the verification with structures, e.g. SIMS boxes, Process Control Monitor (PCM) structures and device characterizations are done to proof and benchmark the simulation. The gained knowledge is fed back to the beginning of the design-flow to re-calibrate for instance deviations or effects, which have not been considered initially.

D. Multi-Domain and Mechanical Structure Simulations

Besides the process and device simulation possibilities, which allow for multi-domain simulation in a limited domain, multi-domain simulation i.e. mechanical, thermo-mechanical, fluid-dynamics, and magnetic domains are of interest during the chip design with almost full functionality. Here, the simulation domain generally include full sensor/chip dimensions, packaging and/or ASIC.

Especially, the interaction and cross coupling of the components and multi-domains are in the focus. Here, one wants to figure out the various impacts on the sensor, i.e. housing/-mounting influences in terms of stress and thermal impacts

or stability and modal analyses of current and/or new sensor designs.

Of importance to such simulation enviroments is the interface between design and layout. To simulate and predict the expected influences, it is recommended to included the structures of interest as precise as possible. Therefore, parts of the sensor reasonably generated by the design and included into the layout as designed with all the required layer information.

Finally, this procedure/flow include predicted process tolerances from the design perspective and consider the limitations given by the layout rules at all.

E. Artifical Intelligence and Big Data

A new methodology used during the design phases is the Artificial Intelligence (AI) and Big Data (BD) approach. The combination of both enables the enhancement of models in a way, which was not present a few years ago.

AI and BD are used to feed and enhance simulation models, physical models, etc., which allows for prediction of critical components and enhanced physical understanding for the demands and boundaries.

To make that happen AI is one of the potential future fields to make the most of technology and release the maximum potential. Topics such as BD make it possible to bundle existing data, which arise during production, to detect hidden correlation, influences, etc., which until now have been unknown and/or played a subordinate role. To emphasize one, such hidden correlations could be identified by the use of Neural Networks (NN), Deep Learning (DL) or similar statistical methods.

In terms of NN/DL, as depicted in Figure 3, one may train the network with the amount of data of PCM, and characterization data and extract those weights which are the most signifcant. This way, one may be able to identify hidden correlations. The importance here lays in the data preparation and knowledge one invests during the setup of such an enviroment.

In addition, this method allows estimates of future restrictions, recognize them early in the process or design and thus influence the design or roadmap. Furthermore, it allows for decision of trade offs and various decision trees. An example might be a pressure sensor. Especially, during the interaction between electronics and mechanics, the mechanics plays a decisive role with attention to the overall system. By significantly improving parameters such as resolution, sensitivity and accuracy with respect to mechanics, new design capabilities are provided, e.g. in the electronics, or vice versa.

Furthermore, the subject of AI is also associated with Industry 4.0 (I4.0), where topics such as predictive maintenance and the interaction on sensor performance parameters are a significant part. For example, drifting, aging of tools like implanter, etc. and its influence on specification parameters of sensors may give the engineer the required knowledge for robust future designs and the freedom to extend the design space.

III. SENSOR DESIGN EXAMPLES

A. Diode Simulation and Modeling

A typical example of the bench of design methodologies and their application is given during this example. The integrated diode, acting as temperature sensing element is modeled, calibrated and simulated as in [3], [8], [9] in detail explained. Here, a brief review is given from the global design perspective.

First, as in section II-C presented, the process and device methodology is accomplished with all its implications of calibration and verification. Once this is done detailed studies of device functionality and evaluation of various process corners and their impact onto device performance are taken into account and verified against measurements of the process variants. Both, PCM data and cross sections of process and device simulation are used to model the diode device for the inhouse Monte Carlo simulator of section II-A.

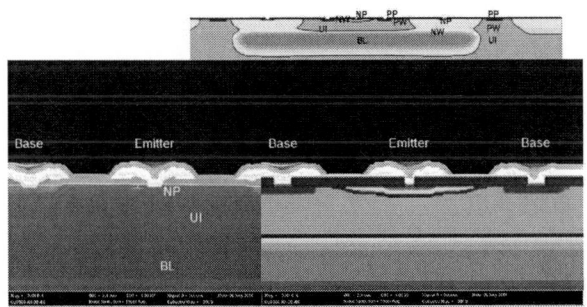

Fig. 4. Comparison of process simulation vs SEM image of chamfered edges.

Here, it is the aim to create a Monte Carlo compact model of the diode [8], which enables to consider process distributions and tolerances of dominating process, geometry and design parameters. This allows for analysis of influence strength and yield prognosis by analyzation of such tolerances onto the diode performance parameters i.e. diffusion voltage (offset) or temperature coefficients. Furthermore, it enables the possibility in a reasonable amount of time for e.g. 10000 Monte Carlo shots in a few minutes compared to FEM simulation with several hours each.

TABLE I
EXEMPLARY DoE FOR BOTTOM ISOLATION.

Process (M_1)	Dose	Buffer	Etch rate (M_3)
Bottom Isolation Implant (BI)	+10%	smaller	target
	-10%	larger	
	-10%	larger	provoke
	-60%		

1) Diode Influence Processes: As an initial example a study of the key process parameters with focus on the possible maximum reductions is presented. Here, various corners were simulated and processed in a DoE to verify the limitations given by different critical process steps (M_i), e.g. doping concentrations. Impacts of these on device performance parameters, such as offset (F_1), temperature coefficients (F_2) and breakdown voltage (F_3) were extracted. From that a behavioral model with a potential design parameters (D_i) were developed to judge the limitations of critical process steps.

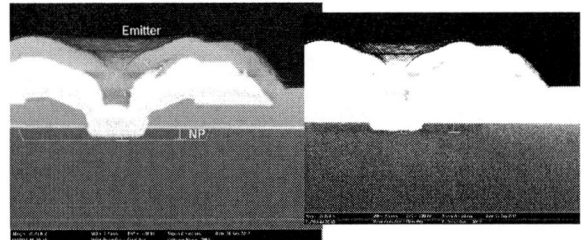

Fig. 5. Chamfered edges of critical emitter component.

In the example a special focus was taken on the backend of the diode, e.g. contact hole etch process and its influence on device functionality. Here, the from the process simulation (Figure 4) the critical component relies at the emitter contact. If the etch rate is too high, a influence on the device performance is expected. This is caused by the extrinsical induced influence of the etch process on the space charge region of the diode, which finally influence the offset, temperature coefficients and breakdown voltage. This is visualized in detail in Figure 5 for the emitter contact.

As explained above, various influences were analyzed by simulation and silicon. Etch rates as well as process steps in the front end with expected significant impacts were considered in the DoE. Simulation support was considered for significant diffusion process steps, as well as for variations in implant doses and thickness variations of the epitaxial layer. Silicon was processed for implant variations, epitaxial thickness variations and various etch rates and cross checked with simulation expectations.

A buffer from the etch edge towards the space charge region as a potential design parameter (D_1) was extracted. It defines the etch rate to provoke a failure and significant influence on the diode performance parameters (Figure 6).

Furthermore, one provoke a single process step and improved his counterpart to extract potential improvement possibilities and judge on the significance of each process step.

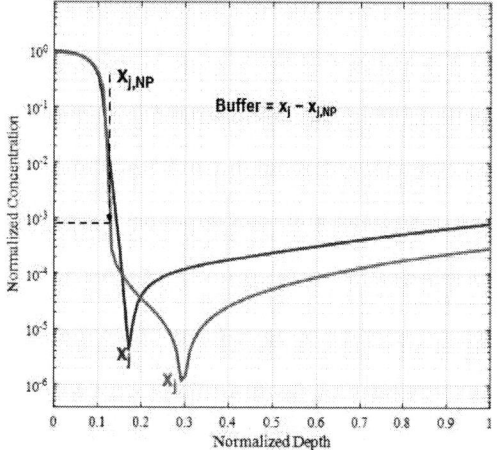

Fig. 6. Definition of the buffer (D_1) to validate the process influences.

In the following two process step examples and their influences for two etch rates are benchmarked against each other by the impact on the performance parameters as offset and breakdown voltage.

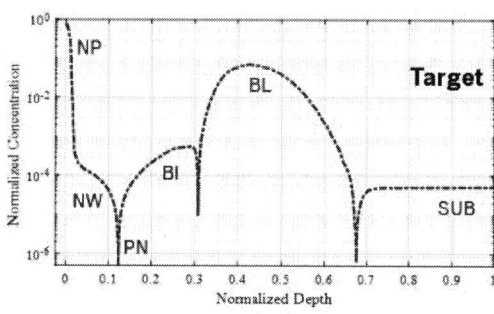

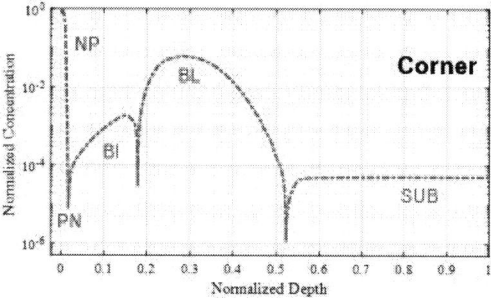

Fig. 7. 1D simulation of the the well formations for a target and corner process with normalized values.

The first process step considers the bottom isolation (BI) and the second the nwell (NW). Both processes act contradictonary and define the performance of the bipolar transistor

TABLE II
EXEMPLARY DoE FOR NWELL.

Process (M_2)	Dose	Buffer	Etch rate (M_3)
Nwell Implant (NW)	+10%	larger	target
	-10%	smaller	
	+10%	larger	provoke
	+1400%		

and/or transdiode. Futhermore, as one may observe in Figure 7 from the backend point of view they define the buffer towards the pn-junction and space charge region.

A part of the DoE for the bottom isolation is presented in Table I including the expectations in terms of the buffer and finally on the device performance parameters. The corresponding measurements at a certain operation point are visualized in Figure 8, where certain simulation results marked as well. Here, one clearly see the dependencies as expected.

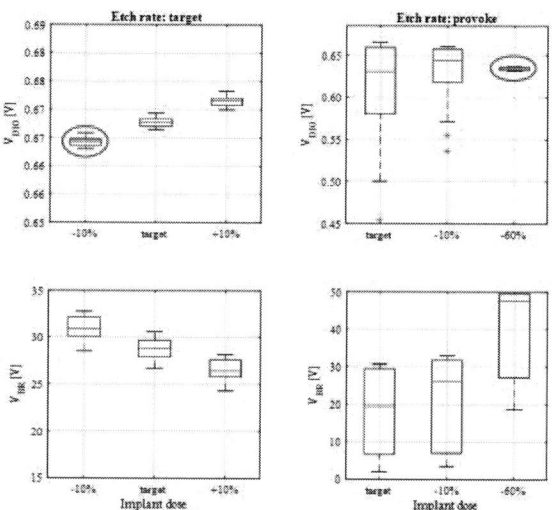

Fig. 8. Exemplary results for a bottom isoloation (M_1) provoke for two different echt rates (M_3) on offset V_{DIO} (F_1) and breakdown voltage V_{BR} (F_3).

The nominal/target etch rate in combination with the bottom isolation variances offer a variation of the operation point due to the shift of the pn-junction by the overcompensation of the counterpart of the nwell. This also influences and lowers the breakdown voltage. By decreasing the doping the buffer enhances and the opposite behavior is observed, see Figure 8.

If one move forward to the provoke of etch rate in combination with variation of the bottom isolation, the picture as expected and visualized in Figure 8 result. The target dose in combination with the provoke of the etch rate result in a huge spread for the offset and breakdown voltage. This results from the impact of the etch process onto the space charge region by extrinsic Shockley-Read-Hall (SRH) recombination phenomenon. The lower the dose, the higher the buffer, the lower the spread of the offset. However, the impact is still present, because the spread off the breakdown voltage is almost unaffected by the variation.

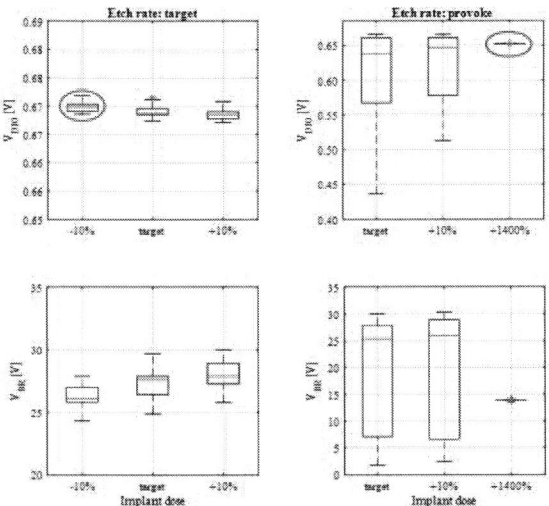

Fig. 9. Exemplary results for a nwell (M_2) provoke for two different echt rates (M_3) on offset V_{DIO} (F_1) and breakdown voltage V_{BR} (F_3).

The second part of the DoE deals with the nwell as presented in Table II. The corresponding measurements at a certain operation point are visualized in Figure 9.

The nominal/target etch rate and nwell variations show the typical dependency as from physics expected. The higher the dose, the larger the overcompensation of the bottom isolation and a shift towards the buried layer results, which drops the offset by a certain amount. Furthermore, the breakdown voltage enhance due to more available space of the space charge region within the nwell as first order approximation. The opposite occurs when decreasing the dose.

The more interesting part and results are shown for the etch rate provoke. The target dose shows a huge spread for offset and breakdown voltage. Increasing the dose by +10% offers a small improvement, where the dose increase of +1400% to generate a surface concentration increase of one order shows significant performance boost in both parameters, offset and breakdown voltage. The effect obviously result from the enormous pn-junction shift into the bulk and furthermore offer spread improvements. However, cross coupling effects on others devices were not taken into account and need further investigations.

The methodology took into account the systematic introduction of successively exceeding the parameters to extract the critical parameter of the buffer, before the device fails in terms of specification limits.

As a result of the ISA the etch rate (M_3) shows the most significant impact on the presented functional parameters (F_j). The methodology identified the different impacts as follows: etch rate (M_3) > epitaxial thickness (not presented, M_4) > NP implant (not presented, M_5) > BI implant (M_1) > NW implant (M_2). Depending on the parameter, the provoke offered yield losses between 5% and 100% for the defined specification limits of the functional parameters.

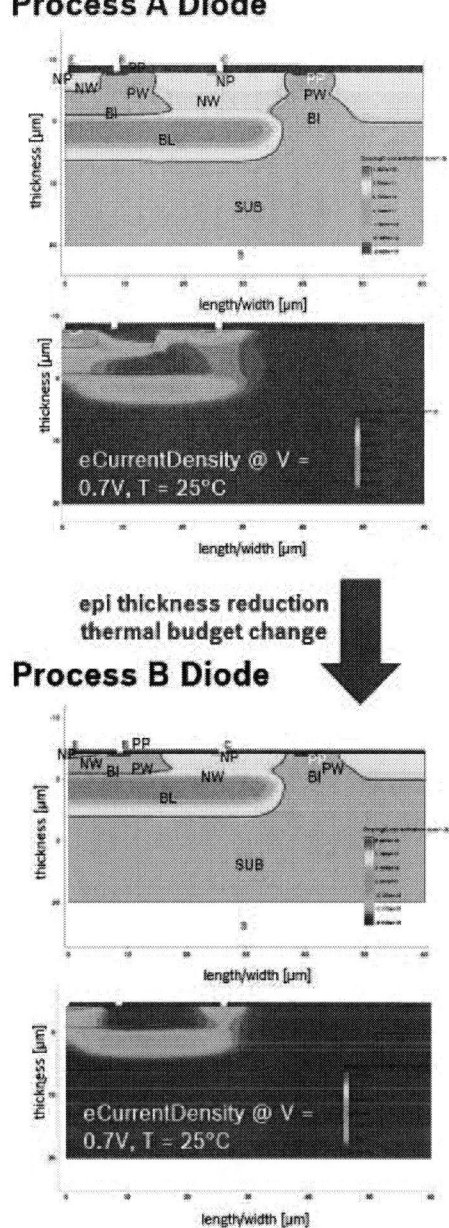

Fig. 10. Analysis of process change A to B influencing device physics, e.g. current density.

2) Diode Noise: The above brief explained methodology for process and device simulation is now used for further investigations, i.e. possible noise sources. The intrinsic noise contribution in sensor systems is always a hot topic. To improve it, various strategies are common, i.e. enhancement of surfaces of pn-junctions for active semiconductor devices, change of operation conditions, etc.

Some of the noise source can be improved by changing the operation point of the diode, i.e. Johnson/Shot noise. Other noise source can be a challenging task, i.e. the 1/f noise which depend on the process.

The 1/f noise problem in various electronic devices has been investigated over 80 years, and over 60 years in solids, but the origin of the 1/f noise is still open on discussions.

The studies of the physical origin of flicker noise for over 60 years in semiconductors devices led to the two major theories how 1/f noise is generated. The carrier number fluctuation theory related to the McWhorter charge trapping and de-trapping model [10], and the mobility fluctuation theory based on Hooge's hypothesis [11].

Nowadays, the discussion on the origin of the 1/f noise is still open. Various studies have been widely discussed for different materials in many works [12]–[19]. To emphasize one regarding this topic, the following example of the temperature sensor is explained in more detail.

As in [8] explained, the example and its process change was forced by the performance improvement of the sensitivity of the pressure sensor in combination with a reduction of the MEMS footprint. Figure 10 shows the impact of this change resulting in a current density change due to the process change from A to B. This is caused by the epitaxy reduction (M_4) and thermal budget change (M_6), which influence the well formation of the semiconductor diode device. While in process A the current flow (electron conduction current) is split and dominated in lateral and vertical direction, process B forces the current flow (electron conduction current) into the vertical direction towards the buried layer.

spectrum potentially occur due to the process change and its resulting current flow change as shown in Figure 10.

The shift/drop of slope in the low frequency behavior potentially result in the first order from the current flow change of process A towards process B. In process B, the epitaxial thickness reduction change the well formation and conducting mechanism within the diode. Less surface current and a forced current towards the buried layer (BL) is observed in comparison to process A. By changing the current flow towards the bulk, less scattering in charge motion or trapping in states occur and result in reduced noise contribution.

If the slope change result from surface charges which occur in a trapping mechanism requires further investigations. Measurements and simulation need to be setup to identify the impact, which are currently in progress.

The difference between red (process B) and pink (process B with different ASIC) occur due to a change in der operation condition. The change of the noise floor is reasonable due to the change of the operation point which results in less Shot noise but higher Johnson noise contribution.

Out of the measurements Spice models were extracted for further system simulations to judge the impact on the sensor system performance for particular electronic circuits and filters.

B. Hall Sensor and Flux Guide

Fig. 12. Bosch BMC150 magentic field sensor incorperating hall sensor elements for out of plane field detection.

A further design example for the need of the above discussed design methodologies and simulation tools is presented by the following study for hall sensors. In Figure 12 the Bosch BMC150 magnetic field sensor is shown. The image shows the integrated sensor with its Flipcore sensing elements for x and y magnetic field and hall sensor elements for z magnetic field detection. This configuration allows for 3D magnetic field detection and enables compass applications for low power application.

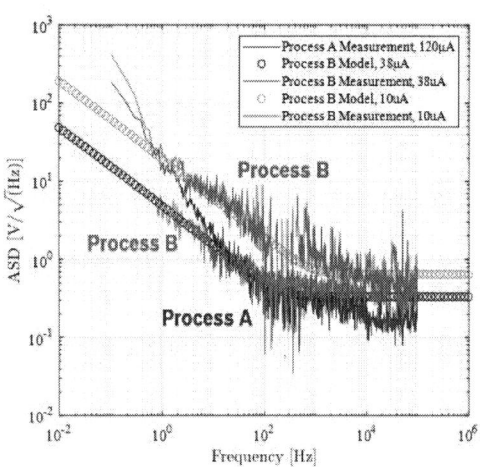

Fig. 11. Normalized amplitude noise spectrum density for different processes and operation points.

The corresponding noise measurements of the amplitude spectrum are shown in Figure 11. Here, one can observe differences of the noise behavior (F_4) by the process impact (the geometrical designs of the sensor elements are identical). The difference between blue (process A) and red (process B)

Next generation requirements figured out, that one of the limiting KPI of such a configuration is the current consumption (F_1) of the hall sensor elements. To improve this parameter for a next sensor generation various studies took place, i.e. a flux guide simulation study to identify potential improvements and implementation options, e.g. placement (M_1), thickness (M_2) and overlap (M_3). Besides, general concept questions also limited tolerances were taken into account for the simulations study.

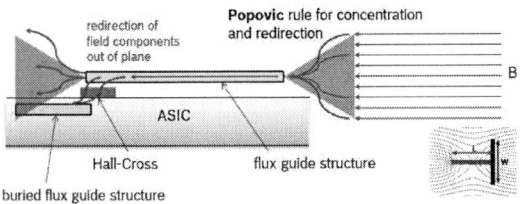

Fig. 13. Popovic rule: A magnetic concentrator (MC) harvest magnetic field of a width W, which corresponds to the lenght L of the MC [20].

The flux guide principle is given in the schematic of Figure 13 and offers a magnetic concentrator (MC) which harvest magnetic field and redirects it by a certain amount when leaving the concentrator [20]. As one may observe by the schematics, a magnetic concentrator (MC) harvest magnetic field of a width W, which corresponds to the lenght L of the MC [20].

This configuration allows for improving the sensitivity of a hall sensor, which finally leads to a current consumption reduction to tackle the identical sensitivity (F_2) as a sensor element without a magnetic concentrator. Furthermore, it defines the design space of the sensor system in terms of length, width and thickness of the flux guide.

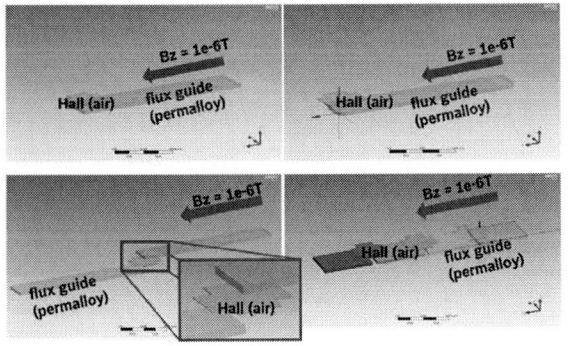

Fig. 14. Exemplary DoE with various geometries, aspect ratios and overlaps for Hall sensor and flux guide.

In Figure 14 an exemplary DoE as part of a broad study with various geometries, aspect ratios, overlaps, etc. is presented. In here, the dimensions of the sensor element from Figure 12 were implemented plus a variety of flux guides in different configurations using various model parameters. The simulations were performed in ANSYS Magnetostatic

simulation environment, where a z magnetic field was applied and the redirection into the x direction is extracted.

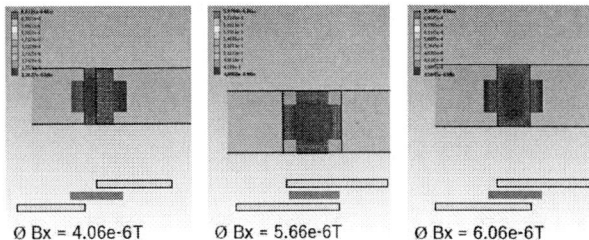

Fig. 15. Exemplary results of the Hall sensor and flux guide simulation DoE.

A part of the results of this study are presented for different overlap conditions in Figure 15. Here, one may observe that the maximum gain (F_3) for a two flux guide configuration results for an overlap of approx. 75%. The overlap of the top and bottom flux guide as function of the gain are presented in Figure 16. As one can observe, the gain varies between factors of two and six, which states that an improvement is to be expected by implementing such magnetic concentrator structures. Furthermore, possible process variations were taken into account in this study (not presented) and offered a robust gain improvement. Such results were taken into account in the discussion of possible implementation options to be realized.

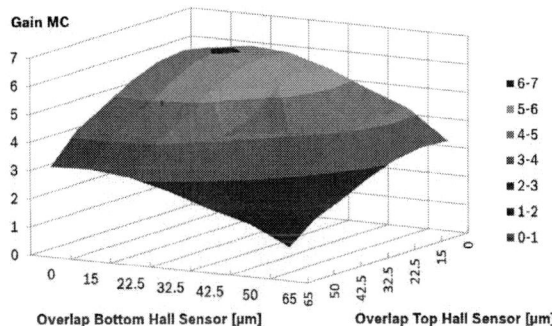

Fig. 16. Study results of the Hall sensor and flux guide simulation DoE.

IV. CONCLUSION

The sensor design methodology and flow ensures an integration of all perspectives from prototyping to series production. It is one of the key processes to bridge technology, mechanics, electronics and system and considers various aspects required during the development process.

The methodology is not limited to specific parts of the sensor design. Quite the reverse! The interfaces between the different parts allow for an abstract description which enables the transfer to various domains. It does not matter, if the point of view relies on process, sensor and/or system development or all of them in once.

The here presented sensor design examples incorperating the dicussed and suggested methods, enable and show the potential of the flow advantages. They emphasize the different abstraction level taken into account. Besides, they minimize the development cycles and allow for considering multi dimensional engineering questions.

Besides, the shown examples offer insights into the different aspects and questions during prototyping and/or series production. Depending on the sample phase various methodologies were applied with different abstraction level of the question. This allows for answering the required information in an efficient manner, without taking to much effort in complex simulations and/or models during an early development.

ACKNOWLEDGMENT

This project was funded by the Federal Ministry for Economic Affairs and Energy by resolution of the German Bundestag under IPCEI framework (Pluto_S and Pluto_P).

Gefördert durch:

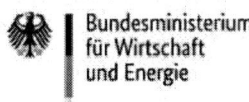 

on Microelectronics

aufgrund eines Beschlusses
des Deutschen Bundestages

FUNDING PROVIDED BY THE FEDERAL MINISTRY
FOR ECONOMIC AFFAIRS AND ENERGY

The author thank all colleagues at Robert Bosch Automotive Electronics, Engineering Sensor Technology, for their valuable contributions to this work. Special thanks to my colleagues Peter Engelhart, Pascal Gieschke, Stefan Krause, Christian Banzhaf, Bassem Baffoun, Arne Dannenberg, Isolde Simon, Frank Rettig, Friedjof Heuck, Wolfgang Feiler, Alexander Mann, Bastian Schmitt, Matthias Boehringer, and Jochen Franz.

REFERENCES

[1] E. Mounier and D. Damianos, *"Status of the MEMS Industry 2019 - Market and Technology Report 2019"*, http://www.yole.fr, Yole Développement, Lyon, France, 2019.

[2] E. Mounier, *"The MEMS market is showing a 17.5% CAGR between 2018 and 2023, strongly supported by current mega trends"*, in Press Releases http://www.yole.fr, Yole Développement, Lyon, France, 2018.

[3] M. Schwarz, J. Franz, and M. Reimann, *"The future is MEMS - Design considerations of microelectromechanical systems at Bosch"*, in Proc. MIXDES, Torun, Poland, 2015.

[4] J. Classen, J. Franz, O. A. Prütz, and A. Kretschmann, *"Design methodology for micromechanical sensors in automotive applications"*, in Eurosensors, 2004.

[5] H.-P. Trah, J. Franz, and J. Marek *"Physics of semiconductor sensors"*, in Advances in Solid State Physics Volume 39, pp 25-36, 1999.

[6] W. Romes, J. Muchow, S. Finkbeiner, J. Franz, O. Schatz and H.-P. Trah *"Simulation of nonideal behavior in integrated piezoresistive silicon pressure sensors"*, in Proc. MSM99, San Juan, Puerto Rico, 1999.

[7] S. Finkbeiner, J. Franz, S. Hein, A. Junger, J. Muchow, B. Opitz, W. Romes, O. Schatz and H.-P. Trah *"Simulation of nonideal behavior in integrated piezoresistive silicon pressure sensors"*, in Proc. SPIE 3680, Design, Test, and Microfabrication of MEMS and MOEMS, 188, 1999.

[8] M. Schwarz, *"The Need of Simulation Methodologies for Active Semiconductor Devices in MEMS : Invited Paper"*, in Proc. MIXDES, Rszesow, Poland, 2019.

[9] M. Schwarz, V. Senz, A. Dannenberg, W. Feiler, F. Heuck, T. Friedrich, C. Sorger, and J. Franz, *"Simulation methodology for active semiconductor devices in MEMS"*, EuroSimE Congress on Modelling and Simulation, Hannover, Germany, 2019.

[10] A.L. McWhorter, *"1/F Noise and Germanium Surface Properties"*, in Semiconductor Surface Physics. ed. by Kingstone R. H., Univ. of Pensylv. Press, 207–228, (1957).

[11] F. Hooge, *"1/f noise is no surface effect"*, in Physics Letters A 29 (3), 139–140 (1969).

[12] V.P. Palenskis, G.E. Leont'ev, H.S. Mykolaitis, *"On the origin of the 1/f noise in linear resistors and p-n junctions"*, in Radiotekhnika i Elektronika 21, 2433–2434, (1976).

[13] F. Hooge, *"Discussion of resent experiment on 1/f noise"*, in Physica 60, 130–144 (1972).

[14] F. Hooge, T.G.M. Kleinpenning, L.K.J. Vandamme, *"Experimental studies on 1/f noise"*, in Rep. Progr. Phys. 44, 479–532 (1981).

[15] P. Dutta, P.M. Horn, *"Low-frequency fluctuations in solids: 1/f noise"*, in Rev. of Modern Phys. 53, 497–516 (1981).

[16] B. Pelegrini, *"One model of flicker, burst, and generation-recombination noises"*, in Phys. Rev. B 24, 7071–7083 (1981).

[17] B. Pelegrini, *"Electric charge motion, induced current, energy balance, and noise"*, in Phys. Rev. B 34, 5921–5924 (1986).

[18] Martin Sandén, Ognian Marinov, M. Jamal Deen, and Mikael Östling, *"New Model for the Low-Frequency Noise and theNoise Level Variation in Polysilicon Emitter BJTs"*, in IEEE Transactions on Electron Devices 49 (3), 514–520 (2002).

[19] V. Palenskis and K. Maknys, *"Nature of low-frequency noise in homogeneous semiconductors"*,Sci. Rep. 5, 18305, (2015).

[20] R.S. Popovic, Z. Randjelovic, D. Manic, *"Integrated Hall-effect magnetic sensors"*,Sensors and Actuators A 91,46–50, 2001.

Proceedings of the 27th International Conference *"Mixed Design of Integrated Circuits and Systems"*
June 25-27, 2020, Łódź, Poland

The Qucs/QucsStudio and Qucs-S Graphical User Interface: An Evolving "White-Board" for Compact Device Modeling and Circuit Simulation in the Current Era

Invited Paper

Mike Brinson
Centre for Communications Technology
London Metropolitan University
UK
Email: mbrin72043@yahoo.co.uk

Abstract—The Qucs/QucsStudio and Qucs-S simulators share a common graphical user interface which has slowly evolved into an interactive platform for drawing circuit schematics, controlling simulation and displaying simulation output data and measured device/circuit parameters/properties. This interface acts as a window for accessing circuit simulation software and is in many ways similar to the "White-Boards" that are popular among scientists and engineers for recording ideas when "brainstorming" circuit design or analysis problems. This paper outlines the evolution of the Qucs device modeling and simulation "white-Board" from concept to working media over the fifteen year period that Qucs, QucsStudio and Qucs-S have been under development. The operation of a number of the "White-Board" features are introduced with a compact tunnel diode model and the simulation data obtained from tests using the QucsStudio and Qucs-S software packages.

Keywords—Qucs/QucsStudio, Qucs-S, compact device modeling, Verilog-A, Graphical User Interface, data visualization, "White-Board" display,

I. INTRODUCTION

A. Background

It is over fifty years since the industrial standard SPICE 2g6 [1] and 3f5 simulators [2] were first released as tools for integrated circuit design. Originally, these were developed as applications for main frame computers. Today the high performance Personal Computer (PC) has become the work horse for compact modeling and circuit simulation, which in turn has encouraged the development of a range of new commercial and Free Open Source Software (FOSS) circuit simulators similar to, or derived from, the Berkeley SPICE FORTRAN (SPICE 2g6) or C (SPICE 3f5) code. The move from centralized main frame computers to individual PC work station has had a profound effect on the analysis/design capabilities and simulation output data processing tools available to the compact modeling and circuit simulation communities. This paper outlines the evolution, over a period of roughly fifteen years, of the "Quite universal circuit simulator" (Qucs) modeling and simulation facilities [3], placing particular emphasis on the evolution

from "SPICE text-in netlist input and simulation text-out output data" to highly interactive PC "White-Board" controlled compact device modeling, circuit schematic drawing, circuit simulation and output data visualization. Illustrated in Fig. 1 is a block diagram that shows pictorially the links between Qucs and the "forked" QucsStudio [4] and Qucs-S [5] circuit simulators. In this figure the vertical arrows signify modeling and simulation information flow, culminating in entries displayed on a PC Graphical User Interface (GUI) "White-Board" window. Other GUI background information also indicates additional features, like for example where SPICE netlists apply. Throughout this paper a series of compact device modeling, circuit design/simulation and data visualization examples based on a tunnel diode compact model are presented. These have been chosen to demonstrate the QucsStudio and Qucs-S PC "White-Board" features that exemplify, without complex detail, the application of this innovative approach to compact modeling and circuit simulation.

B. The Qucs Graphical User Interface (GUI)

The Qucs project was started by German engineer Michael Margraf [3] as a universal simulation tool with a Graphical User Interface (GUI) for drawing Qucs/Qucs-S and QucsStudio schematics, developing compact device models with Equation-Defined Devices (EDD) and Verilog-A module synthesis, launching multi-engine circuit simulation, undertaking parameter sweeps and device/component optimization, plus post-simulation data processing and visualization. The GUI acts as a wrapper for circuit schematic entry and post simulation data processing. It also gives access to a color highlighted text editor, 2D and 3D graph plotting, the Octave numerical analysis package [6] (for advanced output data processing and visualization), plus a group of drop down menus for launching simulations and undertaking other modeling and design tasks. The GUI is the foundation for a powerful user platform that functions as a "White-Board" on a high resolution PC display window, allowing users to freely experiment with compact

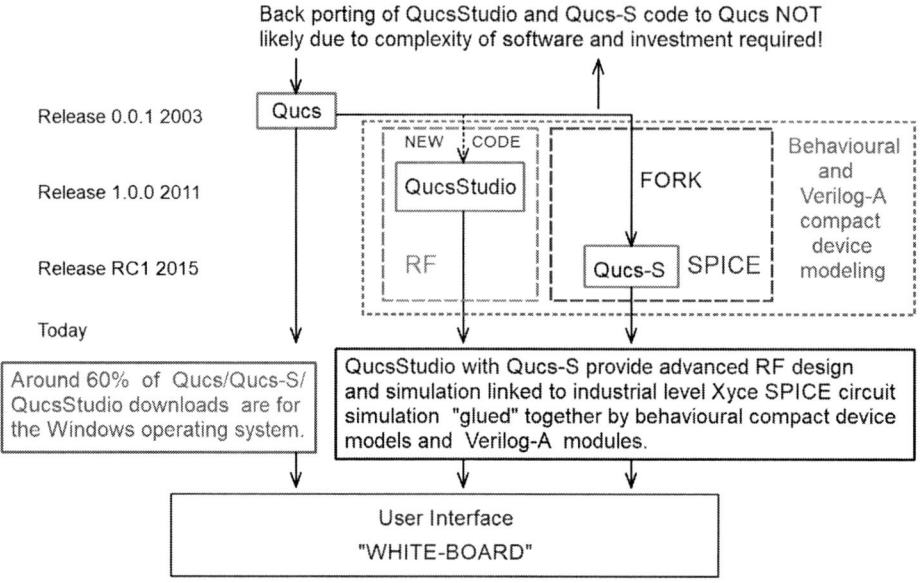

Fig. 1. A block diagram illustrating links between Qucs, QucsStudio, Qucs-S and a common GUI "White-Board".

modeling, circuit simulation and output data analysis on a single graphics screen. The PC "White-Board" is interpreted by the GUI software and its use is largely limited by a users imagination.

C. The relationship between Qucs, Qucs-S and QucsStudio

The QucsStudio and Qucs-S versions of the current Qucs compact modeling and circuit simulation software both derive from recent releases of the Qucs package. However, since their original release there has been inevitable divergence between the capabilities of the three software packages. This is largely the result of their target use; Qucs and QucsStudio are aimed squarely at RF circuit analysis and design, while Qucs-S links the Qucs GUI to different versions of SPICE. By combining the different facilities provided by individual packages a more versatile group of design and analysis tools has resulted. The diagram in Fig. 2 shows the relationship between QucsStudio and Qucs-S, highlighting the use of standardized simulator independent Verilog-A module code [7] as a model interchange vehicle between the two packages, and indeed other circuit simulators, effectively minimizing the effects of differences in model netlist formats.

II. INTEGRATED ANALYSIS AND DESIGN BLOCKS IN SIMULATION SCHEMATICS

Subcircuits are essential building blocks when sectioning parts of a complex circuit schematic into smaller manageable units, such as circuit blocks that represent specific electronic functions or integrated circuits. Unfortunately, in the original SPICE 2g6 and 3f5 netlist format subcircuits were defined without the ability to pass parameters via an argument list.

Later commercial and FOSS simulators corrected this omission, allowing both subcircuit parameters and, in the case of the Qucs series of simulators, blocks of algebraic and numeric equations for calculating component values from physical parameter values coupled to circuit design routines [8]. Similar capabilities are also available for circuit macromodels. Fig.3 shows a typical device model with component values derived from subcircuit parameters calculated by a Qucs style Equation block [9]. Each simulation schematic drawn on the PC "White-Board" window is allowed one or more Equation blocks at the highest hierarchy design level, where the left hand equation variables may only be defined once across the blocks. Equation blocks are also allowed within subcircuits. With Qucs and QucsStudio ordering of the equations in a set of Equation blocks is not important. With Qucs-S however, when simulating with a SPICE engine, it is. The diagram sections labeled (a), (b) and (c) displayed in Fig. 3, are examples of fundamental items that can be combined with other component symbols to form a model schematic and placed on a Qucs style "White-Board". Allowed items include, simulation icons, data tables, data plots, text blocks, library models, predefined passive and active component symbols and subcircuit symbols.

III. THE QUCS EQUATION DEFINED-DEVICE

At the center of the Qucs/Qucs-S and QucsStudio compact device modeling capabilities is a new highly innovative component called an Equation-Defined Device (EDD) [10] [11]. The primary task of this multi-terminal device is to define a static or dynamic nonlinear component that represents, at electrical level, a physical process, where the nonlinear component is built from controlled current generators who's properties are

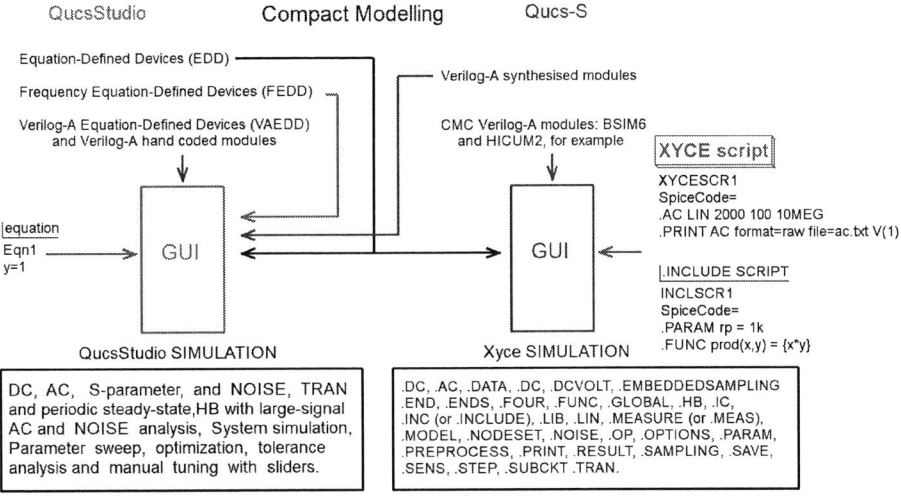

Fig. 2. A compact device modeling and simulation tool set derived from QucsStudio and Qucs-S.

expressed as explicit algebraic functions of one or more EDD component terminal voltages, $I(Vn)$, or a differential time dependent function of branch stored charge $I(Q(Vn, In))$. Unfortunately, both SPICE 2g6 and 3f5 are only equipped with limited capabilities of this type (polynomial current source in SPICE 2g6 and B type controlled current source in SPICE 3f5), making dynamic current, $I = d/dt(Q(Vn, In))$, particularly difficult to model.

A. Explicit Equation-Defined Device (EDD) models

The component drawn in Fig. 4 represents a generalized EDD. Qucs and QucsStudio have a maximum of eight two terminal ports per EDD. Qucs-S has this number increased to twenty. All three packages allow more than one EDD per PC "White-Board" schematic. This new nonlinear component allows interpretive modeling of both static and dynamic device properties, where terminal current In can be an algebraic function of branch voltages Vn plus a dynamic current component expressed as $d/dt(Qn(Vn, In))$, where $Qn(Vn, In)$ is the charge associated with EDD branch n. The explicit equations listed in Fig. 4 give a more complete specification. EDD can be combined with conventional component models, subcircuits, and *Equation Eqn* blocks to construct compact device subcircuits or circuit macromodels. EDD is an advanced component that allows users to build prototype nonlinear compact device models, based on sets of physical properties defined as algebraic equations, attached to a schematic, and placed on a PC "White-Board". The d/dt operator is automatically evaluated by the circuit simulation software. Fig. 5 illustrates an example of this powerful interactive form of compact modeling applied to a tunnel diode. The tunnel diode static Id/Vd EDD equations are

$$I1 = I_p \cdot exp\left(\frac{-V_{pp}}{V1}\right) \cdot \left(exp\left(\frac{V1}{VTH}\right) - 1.0\right) \quad (1)$$

$$I2 = I_p \cdot \frac{V1}{Vp} \cdot exp\left(1.0 - \frac{V1}{Vp}\right) \quad (2)$$

$$I3 = I_p \cdot exp(V1 - V_v) \quad (3)$$

$$Id = I1 + I2 + I3 \quad (4)$$

where I_p and Vp are the diode peak current and voltage, Iv and Vv are the diode valley current and voltage respectively, V_{pp} is the protected peak voltage and VTH is the thermal voltage at $TempK$ Kelvin. The tunnel diode test bench in Fig. 5 shows the minimum set of items common to most PC "White-Boards", namely a test circuit (including components with lists of parameters), simulation Icons (*dc simulation* and *Parameter sweep* in the tunnel diode example), and output data (a d.c. 2D plot of $Itd(A)$ against $Vtd(V)$). The order and placement position of these items on a PC "White-Board" is quite arbitrary. Notice also that the charge associated with a fixed capacitor Cp, defined as $Cp *\_ v1$ and stored in branch one of EDD X1 Fig. 3, is given in (a), where $\_v1$ is the QucsStudio format for the voltage across branch one.

B. Debugging compact device models and circuits during development and simulation

During the development of complex compact device models, or simulation test circuits, it is likely that errors in a schematic drawing, parameter list or an equation block will occur. Finding and debugging such errors can often be difficult and indeed very time consuming, particularly in those case where a PC "White-Board" includes many different elements. To minimize such problems complex compact modeling and simulation tasks are normally split into multiple self-contained sections and tested, when possible, separately. The use of subcircuits is particularly helpful when finding and eliminating bugs. The Qucs, QucsStudio and Qucs-S GUI have an additional aid for debugging problems during the development process. Fig. 6

25

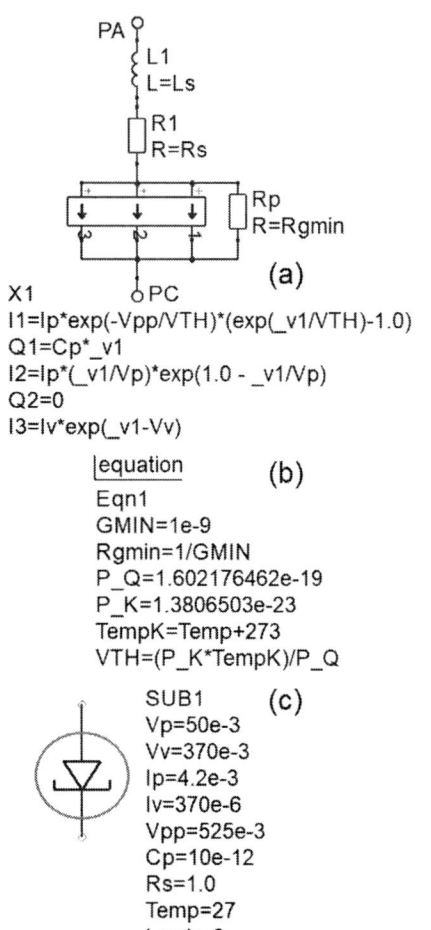

```
PA ○
    L1
    L=Ls

    R1
    R=Rs
```

```
                    Rp
                    R=Rgmin
```

(a)

```
X1    ○ PC
I1=Ip*exp(-Vpp/VTH)*(exp(_v1/VTH)-1.0)
Q1=Cp*_v1
I2=Ip*(_v1/Vp)*exp(1.0 - _v1/Vp)
Q2=0
I3=Iv*exp(_v1-Vv)
```

```
|equation      (b)
Eqn1
GMIN=1e-9
Rgmin=1/GMIN
P_Q=1.602176462e-19
P_K=1.3806503e-23
TempK=Temp+273
VTH=(P_K*TempK)/P_Q
```

```
SUB1          (c)
Vp=50e-3
Vv=370e-3
Ip=4.2e-3
Iv=370e-6
Vpp=525e-3
Cp=10e-12
Rs=1.0
Temp=27
Ls=1e-9
```

Fig. 3. Qucs-S subcircuit model of a tunnel diode: (a) subcircuit body, (b) design equation block, and (c) subcircuit schematic symbol with attached parameter list.

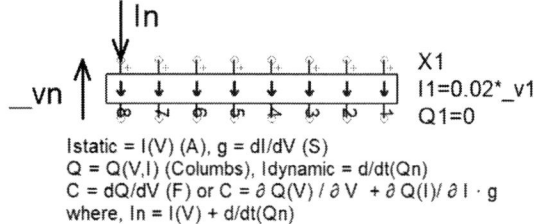

```
              X1
              I1=0.02*_v1
              Q1=0
```

Istatic = I(V) (A), g = dI/dV (S)
Q = Q(V,I) (Columbs), Idynamic = d/dt(Qn)
C = dQ/dV (F) or C = ∂ Q(V) / ∂ V + ∂ Q(I)/ ∂ I · g
where, In = I(V) + d/dt(Qn)

Fig. 4. Specification of Qucs-S and QucsStudio EDD models with explicit equations.

shows the body of the tunnel diode model previously drawn in Fig. 3. In this diagram component Rp has a red cross superimposed on it's symbol and component Ls has a green cross superimposed on it's symbol, indicating open circuit (red) and short circuit (green) respectively. Hence, it becomes possible to build and test subcircuits by removing single or

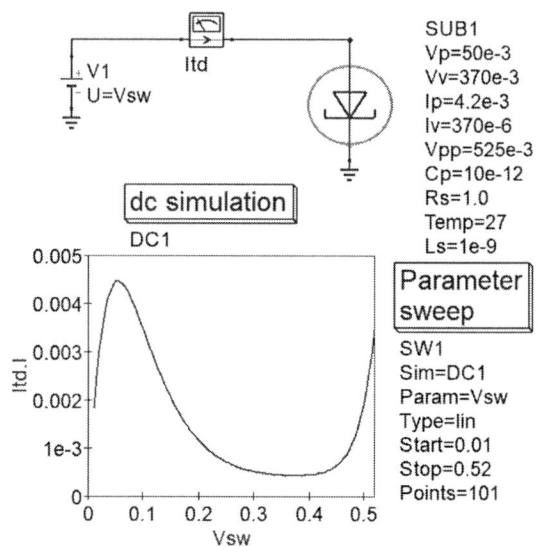

```
                         SUB1
                         Vp=50e-3
                         Vv=370e-3
                         Ip=4.2e-3
                         Iv=370e-6
                         Vpp=525e-3
                         Cp=10e-12
                         Rs=1.0
                         Temp=27
                         Ls=1e-9
```

```
  dc simulation
              DC1
```

```
                         Parameter
                         sweep

                         SW1
                         Sim=DC1
                         Param=Vsw
                         Type=lin
                         Start=0.01
                         Stop=0.52
                         Points=101
```

Fig. 5. Tunnel diode d.c. test bench and $Itd(A)$ plotted against $Vtd = Vsw$.

groups of components from a compact device model. The short circuit and open circuit features are available at all hierarchical levels of a schematic placed on a PC "White-Board". They can also be used to remove complete subcircuits, icons or other valid items. Removing, the red and green crosses reconnects components in a device model or circuit. However, remember this feature is an aid for finding and eliminate drawing and textual syntax errors, having little or no direct effect when tracking down model physical property errors or omissions.

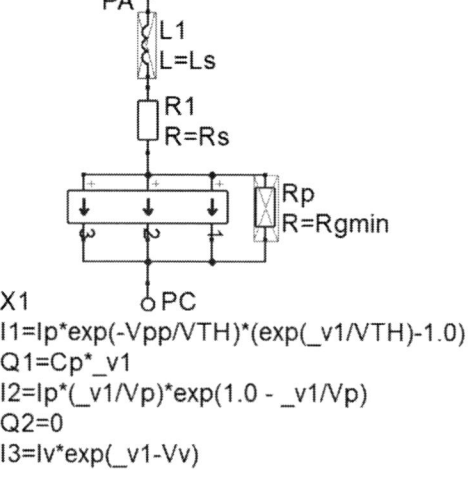

```
PA ○
    L1
    L=Ls

    R1
    R=Rs
```

```
                    Rp
                    R=Rgmin
```

```
X1    ○ PC
I1=Ip*exp(-Vpp/VTH)*(exp(_v1/VTH)-1.0)
Q1=Cp*_v1
I2=Ip*(_v1/Vp)*exp(1.0 - _v1/Vp)
Q2=0
I3=Iv*exp(_v1-Vv)
```

Fig. 6. The tunnel diode compact device model showing Ls shorted (green crossed box) and Rp open circuit (red crossed box) for debugging purposes.

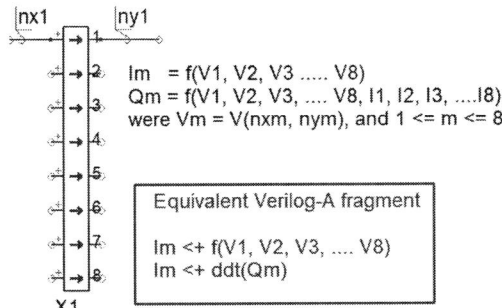

Fig. 7. An explicit EDD with eight two terminal ports and equivalent Verilog-A code fragment.

C. The Qucs Radio Frequency Equation-Defined Device (FEDD) and equivalent QucsStudio component

In some compact modeling instances the physical attributes of a device can be a nonlinear function of a.c. signal frequency f where a model element is expressed as an algebraic function of f. To handle this form of modeling an extended form of the EDD has been implemented in Qucs, called FEDD, and as a "Frequency Domain" component in QucsStudio. These devices add frequency domain two port and multi-port nonlinear modeling to the Qucs group of simulators, including S, Y, Z, H, G, A, T and VCVS types. Examples of the use of these elements can be found in reference [12].

IV. COMPACT MODELING WITH VERILOG-A

The previously described process for constructing compact device models with the built in components is essentially one of placing all the required elements on a PC "White-Board", simulating and making any required changes in an interactive fashion. Such an approach provides an ideal environment for prototyping and testing device models and circuits. Unfortunately, compact device models or circuits that include significant numbers of EDD and FEDD tend to simulate slowly. Moreover, their simulation performance also deteriorates as the number of nonlinear components increases. To overcome this limitation, and indeed other factors, the compact modeling community has adopted the Verilog-A [14] analog hardware description language for defining and constructing C++ machine code models. Qucs and QucsStudio employ the Analog Device Model Synthesizer (ADMS) [15] to translate Verilog-A module code to C++ code. Finally, the C++ version of a model is compiled to dynamic machine code and linked to the main body of simulator code. A schematic symbol has been specially developed which allows compiled Verilog-A models to be combined with EDD and other conventional component symbols. The function and details of this advanced modeling technique are introduced in the following sections.

A. The relationship between EDD and Verilog-A modules

The design of the EDD has been tailored to match the functions provided by the Verilog-A static and dynamic cur-

rent contribution statements. The generalized form of these relationships are defined in Fig. 7. In Fig. 8 a complete tunnel diode subcircuit is drawn along side its equivalent Verilog-A module code with each of the different sections indicated by horizontal arrow pointers. Note the structure of the EDDTD Verilog-A module. This is typical for a large number of compact models, allowing the module code to be written by hand without difficulty, as is the case in Fig. 8. In QucsStudio the PC "White-Board" concept has been extended to add a second window which acts as a color highlighted text editor similar to the popular Windows "notepad++" text editor package. Simply simulating the Verilog-A hardware description language listed in the second window causes QucsStudio to convert the module code from Verilog-A to C++, followed by an automatic C++ compile and link sequence, attaching the machine code model to the verilog-A file name.

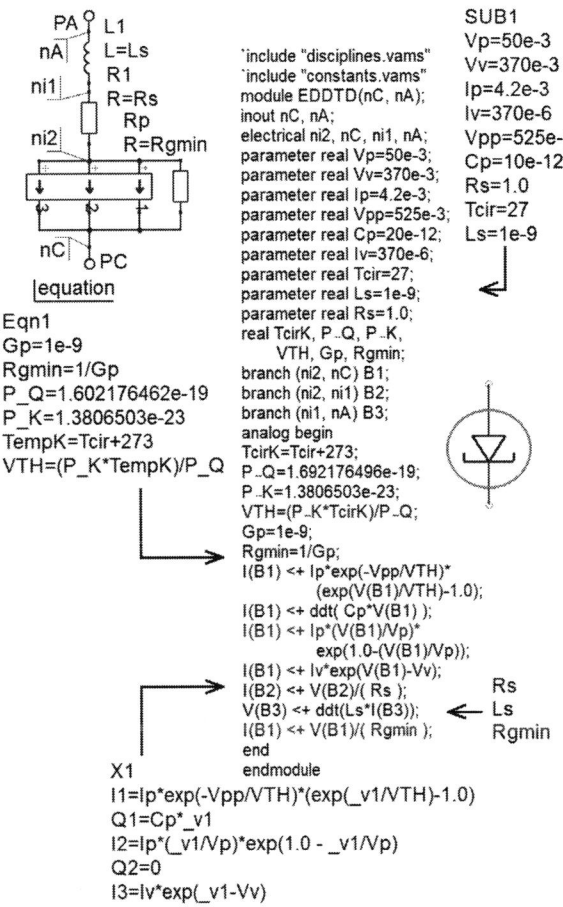

Fig. 8. A QucsStudio tunnel diode compact model drawn at both symbol and internal body hierarchy level, showing Verilog-A module structure and positioning of the Equation:Eqn1 block, subcircuit parameters, components Rs, Ls, and $Rgmin$ within the module code. Note that parameter $Temp$ has been renamed $Tcir$ in the Verilog-A code to remove a clash of names between the EDD and Verilog-A naming conventions.

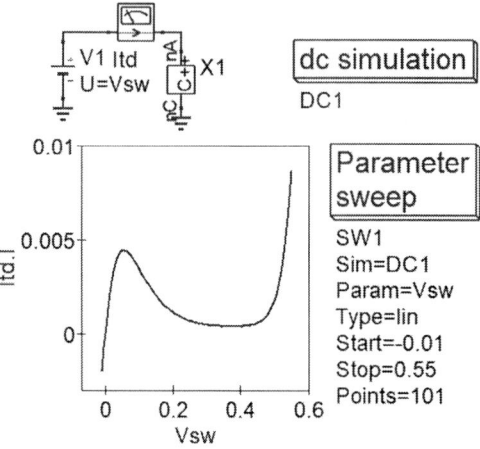

Fig. 9. Tunnel diode d.c. test bench showing compiled C++ compact model symbol and Itd plotted against $Vtd = Vsw$.

Indeed any time a change occurs in the Verilog-A module code QucsStudio automatically recompiles and links the new version of the model to the main body of the simulator C++ code. Fig. 9 shows the simple tunnel diode test set with the EDD subcircuit model replaced by the compiled Verilog-A equivalent.

B. Qucs-S Verilog-A module synthesis

Qucs-S has a number of important extensions when compared to Qucs. One of these is a built in synthesizer for generating Verilog-A module code from a subcircuit schematic [5]. In this process the Qucs-S subcircuit must only be constructed with the following Qucs components: R, C, L, $VCCS$, $CCCS$, $VCVS$, $CCVS$, a *subcircuit* wrapper, a set of *subcircuit parameters*, EDD, $Equation \quad Eqn$, *pinspx* and the SPICE B style nonlinear current source. Fig. 10 lists the tunnel diode Verilog-A module code generated by the Qucs-S synthesizer. Notice that there appears to be a significant number of differences between the hand crafted Verilog-A code listed in Fig. 8 and the synthesized code, for example in Fig. 10 the current contributions are written in terms of node names, rather than branches, and the ADMS Verilog-A statement @(*initial_model) begin.....end;* is included. Notice also that in Fig. 10 the inductor is synthesized by three Verilog-A statements, based on the electrical equivalent circuits given in Fig.11, rather than the single statement in Fig.8. However, in reality both sets of code provide the same overall function. The synthesized Verilog-A code is in a format from which it is possible to hand adjust the statements to meet the capabilities of different Verilog-A to C++ translators. Synthesized Verilog-A modules can be added to both Qucs and QucsStudio projects, compiled and used like standard built-in components. The Qucs/QucsStudio circuit simulators also provide facilities that allow Verilog-A modules to be stored in libraries of new components.

C. The Verilog-A Equation-Defined Device (VAEDD)

Compact modeling of nonlinear devices with EDD and Verilog-A modules represent two techniques at opposite ends of a scale going from purely interpreted models to compiled C++ models. From a practical point of view what is often required is an approach to modeling that has the convenience of interpreted EDD models coupled with the high simulation speed obtained with compiled C++ models synthesized from Verilog-A code. It is also worth noting that in a large number of models, often with more than one EDD, simulation time is largely determined by the complexity of the $In(Vn)$ and $Qn(Vn, In)$ equations for each of the EDD branches. Hence, significant improvement in simulation speed can be obtained by replacing one or more of the most complex EDD branches with an equivalent Verilog-A module called a VAEDD [13]. The tunnel diode compact model shown in Fig. 12 has branch

```
`include "disciplines.vams"
`include "constants.vams"
module EDDTD(nC, nA);
inout nC, nA;
ele[ctrical ni2, nC, ni1, nA, _netOL1;
parameter real Vp=50e-3;
parameter real Vv=370e-3;
parameter real Ip=4.2e-3;
parameter real Vpp=525e-3;
parameter real Cp=20e-12;
parameter real Iv=370e-6;
parameter real Tcir=27;
parameter real Ls=1e-9;
parameter real Rs=1.0;
real TcirK, P_Q, P_K, VTH, Gp, Rgmin;
analog begin
@(initial_model)
begin
TcirK=Tcir+273;
P_Q=1.692176496e-19;
P_K=1.3806503e-23;
VTH=(P_K*TcirK)/P_Q;
Gp=1e-9;
Rgmin=1/Gp;
end
I(ni2,nC) <+ Ip*exp(-Vpp/VTH)*(exp(V(ni2,nC)/VTH)-1.0);
I(ni2,nC) <+ ddt( Cp*V(ni2,nC) );
I(ni2,nC) <+ Ip*(V(ni2,nC)/Vp)*exp(1.0-(V(ni2,nC)/Vp));
I(ni2,nC) <+ Iv*exp(V(ni2,nC)-Vv);
I(ni2,ni1) <+ V(ni2,ni1)/( Rs );
I(ni2,ni1) <+ white_noise( 4.0*`P_K*( 26.85 + 273.15) /
            ( Rs ), "thermal" );
I(_netOL1) <+ ddt(V(_netOL1));
I(_netOL1) <+ V(ni1,nA);
I(ni1,nA) <+ V(_netOL1)/(Ls+1e-20)];
I(nC,ni2) <+ V(nC,ni2)/( Rgmin );
I(nC,ni2) <+ white_noise( 4.0*`P_K*( 26.85 + 273.15) /
            ( Rgmin ), "thermal" );
end
endmodule
```

Fig. 10. Qucs-S synthesized Verilog-A module code for the EDD tunnel diode model.

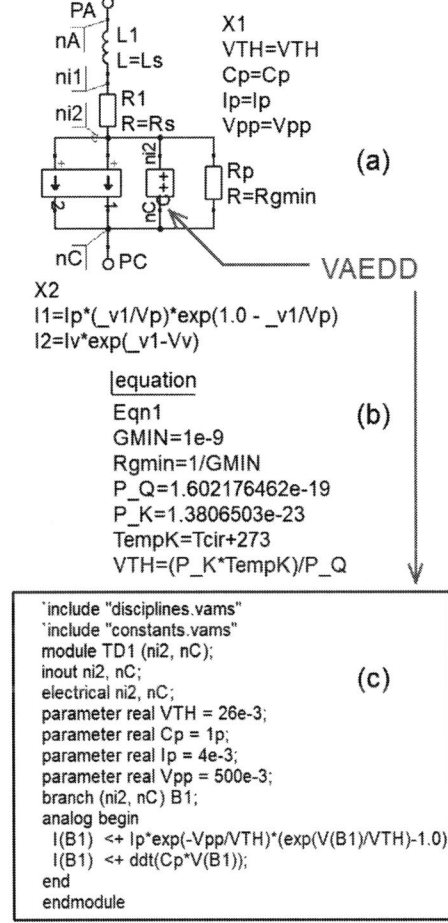

Fig. 11. Synthesized nonlinear inductor and capacitor models: (a) inductor and (b) capacitor.

$U = Ls \cdot dI(\tau)/dt \qquad C = dQ_T/d\tau = dQ_T/d\tau \cdot d\tau/d\tau$

But $dQ_T/d\tau = U$ and $C = 1$

Hence, $d\tau/dt = U$ and $I(\tau) = \tau/Ls$

$I(\tau) = dQ(U)/dt = d(Cp \cdot U)/dt \qquad \tau = L \cdot dI(L)/dt$

But $I(L) = Q(U)$ and $L = 1$

Hence, $\tau = dQ(U)/dt = d(Cp \dot U)/dt$ and $I(\tau) = \tau$

one of the original EDD, see Fig. 8, replaced by a VAEDD branch, see Fig. 12 (a) and (c). This is in reality a tiny Verilog-A module with a structure that is simple to construct, follows a standard template and simulates at speeds significantly faster than its EDD equivalent. Further gains in simulation speed can be achieved by replacing more than one EDD branch with equivalent Verilog-A VAEDD.

V. SPICE MULTI-SIMULATOR SNGINE COMPACT MODELING

Qucs-S is a ground-breaking circuit simulation package in that it allows users to select a simulation engine from (1) the Qucs built-in Qucsator simulator, (2) the SPICE 3f5 compatible SPICE OPUS simulator, (3) the next generation SPICE Ngspice circuit simulator, and (4) the new SPICE compatible Xyce circuit simulator. As these SPICE 3f5 related simulation engines all have some form of extensions when compared to the original Berkeley software, Qucs-S selects which built-in models and library models are allowed with each package. In order for users to be made aware of which model works with each package schematic symbols are color coded; dark blue denoting a legacy Qucs item, red/brown signifying a Ngspice/SPICE OPUS item and dark green an Xyce item. This makes identification of components/devices placed on the PC "White-Board" straight forward. Qucs-S automatically synthesizes the different netlist formats for each of the SPICE engines, taking into account individual package extensions. This process is not simply a one-to-one translation of the Qucs-S symbols to a single line SPICE statement but is a more complex procedure that takes into account component function, often resulting in more than one SPICE statement per symbol. A typical Xyce translation output

netlist is given in Fig. 13 for the tunnel diode test circuit shown in Fig. 5. Note the use of Qucs-S "*dc simulation*" and "*Parameter sweep*" icons in both Fig. 5 and Fig. 9. Qucs-S icons are implemented for all the fundamental SPICE simulation types. However, with advance mature packages like Xyce continuous development is underway, regularly adding a range of new simulation types, for example Harmonic Balance analysis, and output data manipulation statements like .MEASURE, making it difficult to keep up with the volume of changes. With Qucs-S this situation was anticipated and a scripting icon developed that allows users to incorporate into Qucs-S future changes made by the Xyce development team. Fig. 14 (a) indicates how scripts of Xyce simulation and output data processing are added to a Qucs-S PC "White-Board". Fig. 14 (b) introduces a number of SPICE related icons, who's names identify their function, for adding and manipulating Xyce netlist elements.

Fig. 12. The tunnel diode compact model with branch one of the EDD replaced with a VAEDD: (a) and (b) the revised model and (c) the VAEDD Verilog-A code.

```
* Qucs 0.0.22 EDDTD.sch
.SUBCKT EDDTD nA nC Vp=50e-3 Vv=370e-3
+ Ip=4.2e-3 Vpp=525e-3 Cp=20e-12 Iv=370e-6
+ Tcir=27 Ls=1e-9 Rs=1.0
.PARAM TcirK={Tcir+273}
.PARAM P_Q=1.692176496e-19
.PARAM P_K=1.3806503e-23
.PARAM VTH={(P_K*TcirK)/P_Q}
.PARAM Gp=1e-9
.PARAM Rgmin={1/Gp}
BD1I0 ni2 nC I=Ip*exp(-Vpp/VTH)*(exp(V(ni2,nC)/VTH)-1.0)
GD1Q0 ni2 nC nD1Q0 nC 1.0
LD1Q0 nD1Q0 nC 1.0
BD1Q0 nD1Q0 nC I=-(Cp*V(ni2,nC))
BD1I1 ni2 nC I=Ip*(V(ni2,nC)/Vp)*exp(1.0-(V(ni2,nC)/Vp))
BD1I2 ni2 nC I=Iv*exp(V(ni2,nC)-Vv)
R1 ni2 ni1 {RS}
L1 ni1 nA {LS}
R2 nC ni2 {RGMIN}
.ENDS
V1 ncir1  0 dc 0.1
VITD ncir1 ncir2 DC 0
XSUB1 ncir2 0 EDDTD Vp=50E-3 Vv=370e-3
+ Ip=4.2E-3 Vpp=525E-3 Cp=1E-12 Iv=370E-6
+ Tcir=27 Ls=1E-9 Rs=1.0
.dc v1 -0.01 0.55 0.00278607
.PRINT dc format=raw file=testTDEDDdc.txt
+ I(VITD) v(ncir1) v(ncir2)
.END
```

Fig. 13. Xyce SPICE netlist for the tunnel diode test bench given in Fig. 5.

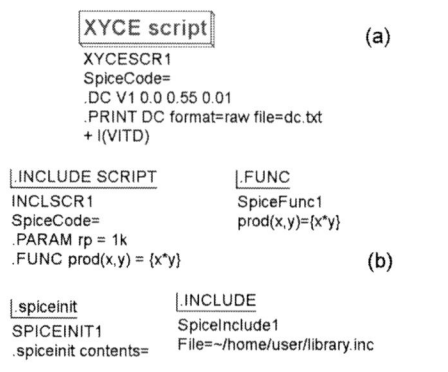

Fig. 14. Qucs-S SPICE netlist manipulation icons: (a) $XYCE\ script$, and (b) other related SPICE netlist handling scripts.

VI. CURRENT AND FUTURE DEVELOPMENTS

Both Qucs-S and QucsStudio are active projects with new features, improvements and bug fixes regularly released by their Development Teams. The same is true for the Ngspice and Xyce circuit simulators. SPICE OPUS has been include for use with Qucs-S because it provides a base line simulator who's properties are essentially the same as the published Berkeley SPICE 3f5 software. The following sections introduce a number of current and proposed developments that aim at improving the Qucs-S/Xyce and Qucsstudio compact modeling and simulation capabilities, and the PC "White-Board" platform.

A. Xyce

Since the release of Xyce version 6.0, as open source software, under the General Public License (GPL) 3.0, new versions of the software have been appeared roughly every six months. As of March 2020 Version 7.0 is the current stable package. For the Qucs-S Development Team this is both a good and a bad feature of the Xyce software. Good in the sense that the Xyce package offers new features at every release plus bug fixes. Bad in the sense that continuous changes and adaptions of the Qucs-S code are needed to keep in step with Xyce. It was largely for this reason that the $XYCE\ script$, $.INCLUDE$, $.FUNC$, $.INCLUDE$, and $spiceinit$ control icons have been implemented. These allow Xyce SPICE netlists to be placed on a PC "White-Board" and interpreted at simulation run time. Qucs-S has in fact a two level GUI system; items common to SPICE 3f5, and other equivalent simulators, operate via built-in Icons or a $XYCE$ $script$, while the less used or recently added features, can only be accessed via a $XYCE\ script$. For example, since 2018 approximately 20 important additions to Xyce functionality have been implemented, including Monte Carlo analysis and Lattice hypercube sampling via a new $.SAMPLING$ feature. Transient simulation direct sensitivity analysis that supports $.FOUR$, $.LIN$ for S parameter multiport analysis with Y and Z output data in Touchstone level 1 and level 2 format, and a new charge expression variant for capacitors that is similar to the EDD branch charge implementation. The tunnel diode model test circuit shown in Fig. 15 illustrates both the use of the $XYCE\ script$ icon and a number of new Xyce

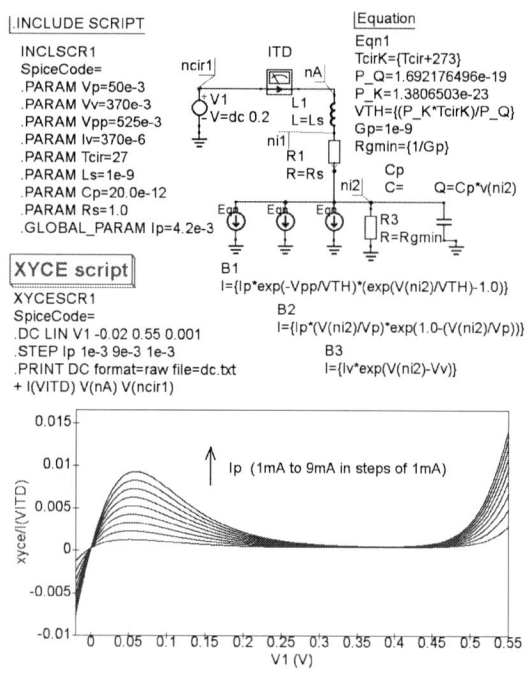

Fig. 15. A Xyce SPICE style tunnel diode compact model with a test bench for investigating the effects of stepped device parameters.

features: firstly the diode compact model has been built from SPICE B style current sources with the capacitor Cp modeled with a Q style component ($Q = Cp \cdot V(ni2)$), secondly note the extensive use of {...} round equations, thirdly the use of $.GLOBAL\_PARAM$ to identify parameters to be stepped, and finally the combination of SPICE directives $.DC, .STEP$ and $.PRINT$ to direct simulation and output data.

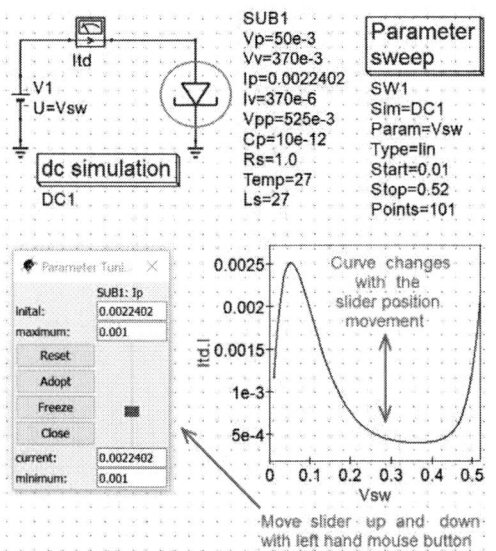

Fig. 16. The QucsStudio EDD tunnel diode model test bench with an interactive slider for investigating the effects of changing device parameter Ip over range of values: in this example parameter Ip is changed over the range 1mA (minimum = slider down) to 10mA (maximum = slider up).

B. QucsStudio

The current version of QucsStudio is 2.57. This package has reached an advanced stage of development in that it offers an almost complete set of circuit simulation routines covering the d.c. to transient domains with significant additions beyond SPICE 3f5 like multi-tone Harmonic Balance analysis, Monte Carlo analysis, parameter sweep, multi-port S parameter and noise simulation, optimization and system simulation. Full "turn-key" Verilog-A compact modelling is also offered via the ADMS software. In terms of White-Board" development QucsStudio is particularly interesting in that it is the first of the Qucs series of circuit simulators to introduce interactive animation as a tool for advanced circuit simulation. Illustrated in Figure 16 is the basic tunnel diode d.c. test circuit, introduced previously, where the value of current peak parameter I_p can be set by changing the position of a slider with the left hand mouse button. QucsStudio allows one or more parameter values to be simultaneously controlled by sliders. With the computational power of a modern PC changes in simulation output data can be observed as movements in plotted curves as the sliders are moved. The parameter slider technique is particularly useful as a process for obtaining (good guess) starting values in parameter sets prior to full computer controlled optimization.

C. Merging simulation and measurements

Illustrated in Fig. 17 is an example of a comprehensive "White-Board" for simulating the a.c. performance of a simple passive first order low pass RC filter. The latest extensions to the Qucs "White-Board" repertoire includes pictures (in Fig. 17 a picture of an "Analog Discovery 2" transfer function measurement system [16]), information pointer directed data flow diagrams and associated text blocks (in Fig. 17 a note explaining the use of an Octave script for converting measured CSV formatted output data to simulation control icons with data lists (*AC simulation AC*1, *Equation Eq*2 and *Equation Equ*3)), a QucsStudio simulation schematic plus plotted measured and simulated output data controlled by parameter tuning (of values R1 and C1). In this example two points are worth noting, firstly the AC simulation frequency range is determined by the "Points" list, synchronizing the measured and simulation frequency values, and secondly the picture, text and arrow pointers shown on the "White-Board" are transparent during simulation and play no part in computing the RC voltage transfer function. Moreover, their primary role is to provide a clear indication of the processes involved in the function of the test bench, making the concept of a self-documenting "White-Board" an important step in the development of the next generation compact modeling and circuit simulation tools.

VII. SUMMARY

Low cost high performance PC engineering work stations have encouraged the development of compact device modeling and circuit simulation tools centered on high resolution graphics interfaces for schematic drawing, simulation and output data visualization. This paper outlines the structure and capabilities of a "White-Board" display system developed for the Qucs series of circuit simulators and modeling tools. These tools allow interactive prototyping of compact device models and their testing using the Qucs/QucsStudio and Qucs-S "White-Board" environment as a central platform in the construction of production level Verilog-A device models. The "White-Board" environment offers significant interactive circuit entry and data display improvements when compared to those implemented in previous generations of circuit simulator, namely high resolution schematic drawing, simulation control Icons, output data visualization, attached design equations and self documenting text and flow diagram features. Each of these, when coupled with established, or new, compact modeling techniques, like non-linear EDD, FEDD, mixed Equation-Defined Device and Verilog-A models (VAEDD) plus verilog-A modules, make the evolving "White-Board" a highly flexible and innovative platform for compact modeling and circuit simulation in the current era. Future expansion of the "White-Board" concept indicates that by merging device parameter measurements with established circuit simulation will significantly extend the scope of traditional circuit simulation.

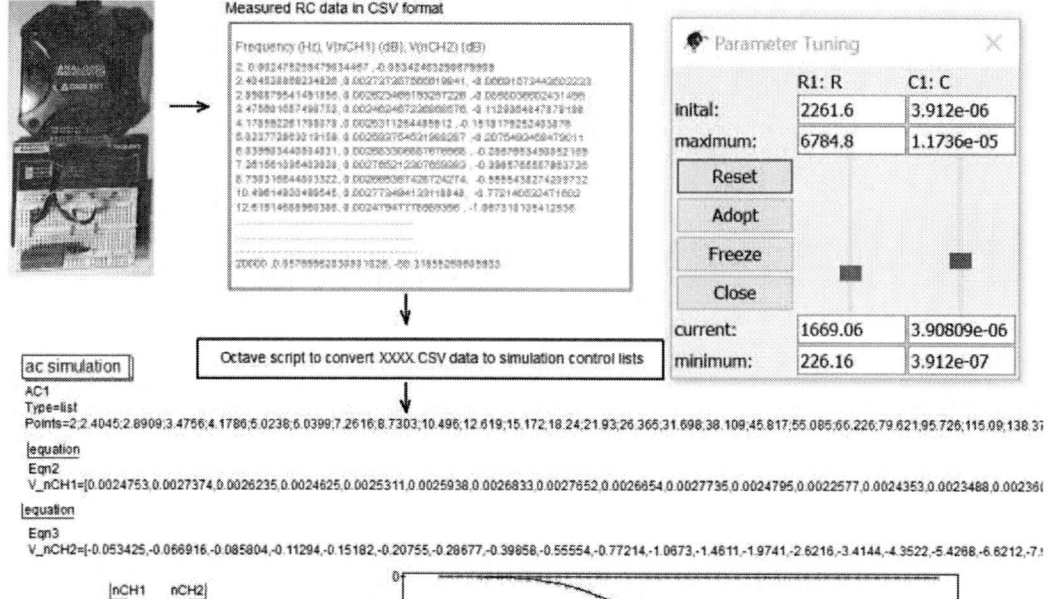

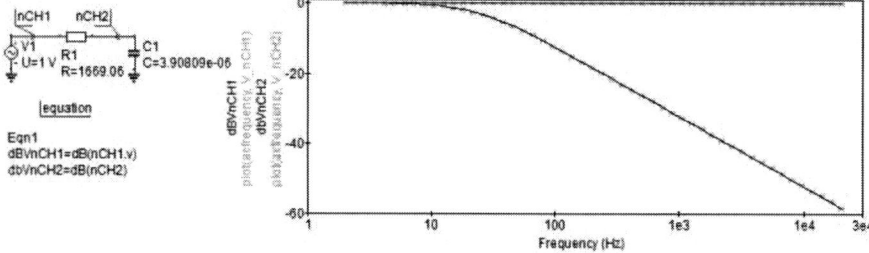

Fig. 17. A simple RC low pass passive filter example showing an advanced "White-Board" with a computer controlled transfer function measurement system where the measured output data is converted from CSV format to Qucs simulation control ICONS. (Note the right hand side of the *AC simulation* and the two *Equation* Icons have been truncated.)

REFERENCES

[1] A.R. Newton, D. O. Pederson and A Sangiovanni-Vincentelli. "SPICE Version 2g User's Guide", Department of Electrical Engineering and Computer Science, University of California, Berkeley, CA, 1981.

[2] B. Johnson, T. Quarles, A.R. Newton, D. O. Pederson and A Sangiovanni-Vincentelli. "SPICE3 Version 3f User's Manual", Department of Electrical Engineering and Computer Science, University of California, Berkeley, CA, 1992.

[3] M. Margraf et al. Qucs (Quite Universal Circuit Simulator). Version 0.0.19, 2020. Available from http://qucs.sourceforge.net/index.html. [Accessed May 2020].

[4] M. Margraf, "QucsStudio Version 2.5.7. Available from http://dd6um.darc.de/ QucsStudio/qucsstudio.html, 2018. [Accessed May 2020].

[5] V. Kusnetsov and M. Brinson, "Qucs-S: Qucs with SPICE". Version 0.0.22. Available from https://ra3xdh.github.io/. 2020. [Accessed January 2020].

[6] John W. Eaton, David Bateman, Søren Hauberg and Rik Wehbring. "GNU Octave version 5.1.0 manual: a high-level interactive language for numerical computations". 2019, CreateSpace Independent Publishing Platform. ISBN 144141300. Available from http://www.gnu.org/software/octave/doc/interpreter/, [Accessed January 2020].

[7] S. Jahn, M. Brinson, H. Parruitte, B. Ardouin, P. Nenzi and L. Lemaitre, "GNU simulators supporting Verilog-A compact model standardization", MOS-AK meeting, Premstaetten 2007. Available from http://www.mos-ak.org/premstaetten/papers/MOSAK_QUCS_ngspice_ADMS.pdf. [Accessed May 2020].

[8] M. Brinson and H. Nabijou, "Adaptive subcircuits and compact Verilog-A macromodels as integrated design and analysis blocks in Qucs circuit simulation", International Journal of Electronics, 2011, 98(5), pp. 631-645.

[9] M. Brinson, "Qucs:Component, compact device and circuit modelling using symbolic equations", 2007. Available from http://qucs.sourceforge.net/docs/tutorial/equations.pdf. [Accessed May 2020].

[10] S. Jahn and M. Brinson, "Interactive compact device modelling using Qucs equation-defined devices", International Journal of Numerical Modelling: Electronic Networks, Devices and Fields. 2008, 21(5), pp 335-349.

[11] M. Brinson and S. Jahn, Qucs: "A GPL software package for circuit simulation, compact modelling and circuit macromodelling from DC to RF and beyond", International Journal of Numerical Modelling: Electronic Networks, Devices and Fields, 2009, 22(5), pp 297-319.

[12] M. Brinson and S. Jahn, "Modelling of high-frequency inductance with Qucs non-linear radio frequency equation defined devices", International Journal of Electronics, 2009, 96(3), pp. 307-321.

[13] M. Brinson, "FOSS Compact Model Prototyping with Verilog-A Equation-Defined Devices (VAEDD)", 2019 MIXDES - 26th International Conference "Mixed Design of Integrated Circuits and Systems", Rzeszów, Poland, 2019, pp. 92-97.

[14] Accellera, "Verilog-AMS Language Reference Manual, version 2.2". Available from: http://www.accellera.org, [Accessed May 2020].

[15] L. Lemaitre, C. McAndrew and S. Hamm, "ADMS- Automatic device model synthesizer", IEEE Custom Integrated Circuits Conference, 2002, pp. 27-30.

[16] Digilent (A National Instruments Company), "Analog Discovery 2: 100MS/s USB Oscilloscope, Logic Analyzer and Variable Power Supply", Digilent Inc. 1300 Henley Ct 3, Pullman, WA 99163, 2020.

Compact Modeling Support for Micro and Nanoelectronic System Development

Proceedings of the 27th International Conference *"Mixed Design of Integrated Circuits and Systems"*
June 25-27, 2020, Łódź, Poland

Compact Model
for Continuous Microfluidic Mixer

Alexi Bonament[1], Alexis Prel[1], Jean-Michel Sallese[2], Morgan Madec[1], Christophe Lallement[1]

[1]Laboratory of Engineer Sciences, Computer Science and Imagine (ICube),
UMR 7357 (Université de Strasbourg / Centre National de Recherche Scientifique), Strasbourg, France
[2]STI-IEL-Electronics Laboratory,
Ecole Polytechnique Fédérale de Lausanne (EPFL), Lausanne, Switzerland

Abstract— **While the development of lab-on-chip is increasing, the lack of dedicated computer-aided design tools appears as a bottleneck preventing the emergence of large-scale industrial applications. One of the answer relied on 50 years of CAD experience in microelectronic. Based on this fact, and using this environment, multi-domain libraries (fluidic, biological, chemical) are to be designed. Among other, the development of efficient compact model for microfluidic devices is a first step toward such design tool. This paper deals with a continuous microfluidic mixer. Our model takes as inputs the flow rates and the concentrations of each fluid to mix and returns the flow rate and the concentration profiles across the channel at its output. The model depends also on some physical parameters (e.g. diffusion coefficient of fluids) and mixing channel geometry. The model is validated by comparison with finite-element simulation performed with COMSOL Multiphysics. Comparisons are made on several cases. We demonstrated that the model gives a good estimation of the concentration profile, with an error of less than 2% compared to the finite element simulator.**

Keywords—**compact modeling; lab-on-chip; microfluidics; continuous mixers;**

I. INTRODUCTION

Over the past decade, lab-on-chips have established themselves as a reference technology for high-quality high-throughput analysis in several fields like environment [1] and healthcare [2], [3]. These lab-on-chips are mainly composed of a microfluidic system to drive fluids, bio-chemical reactions, electronic transducers and driving and processing electronic circuits. The design of such a system is a tricky challenge, especially if we aim at merging all these domains early in the design process. Up to now, lab-on-chip designers lack a computer-aided design tool encompassing microfluidic, biochemistry and electronics. Designers might use multi-physical simulators to study a small part of the system but such an approach cannot be used at the scale of the complete device due to computation power limitation.

Regarding biochemistry, its integration in standard electronic design automation (EDA) environment has already been demonstrated. It is already possible to co-simulate bio-chemical reactions and electronic transducer in SPICE environment. With this work we already simulated a simple lab-on-chip device for which molecular transport is only driven by diffusion phenomenon [4]. However, most of the time, lab-on-chips include a microfluidics system to manage fluids.

Generally speaking, driving fluids with microfluidics chips can be performed in two ways. On the one hand, digital microfluidic chips are composed of a glass plate and an array of electrodes and manipulate droplets containing reactants. They can be moved by electrical fields created by the electrodes and/or mixed to induce reactions and analyzed [5]. For such devices, compact models and associated design automation tools already exists [6]. On the other hand, in continuous microfluidics, displacements and mixing of reactants are made by continuous flow driven channels, valves and pumps. The physics beyond such circuits is mostly based on Navier-Stokes equations [7] that are much more complicated to manipulate for the purpose of writing compact models.

Previous work has already shown analogies that can be exploited to simulate microfluidic channels as electronic equivalent circuits [8]. This approach provides a nice approximation as long as fluids are homogeneous, which is not the case for passive mixers [9]. Due to the small dimensions of the channels, fluids flows are laminar which means that the mixture between the two fluids is often imperfect. Equivalent circuit approach is only valid if it can be assumed that channels are long enough to have homogeneous fluids at their output.

In this paper, a new modeling approach based on an analytic compact model is described. The technique consists of a reformulation of partial differential equations into time-dependent ordinary differential equations by using Fourier series and a time-displacement change of variable. First, the theoretical concepts required to understand the physics of passive microfluidic mixer are introduced. Then, the model as well as the assumptions on which the model relied are presented. Finally, simulation results of the compact model and comparison with simulations using the COMSOL Multiphysics software are given and discussed.

II. PHYSICS OF MICROFLUIDIC PASSIVE MIXER

In fluid mechanics, behavior of fluid is characterized by a dimensionless quantity, namely the Reynolds number. In microfluidics, this Reynold number is low (typ. <1), meaning that the fluid flows laminarly [7]. Thus, fluid dynamic is specific. In the following, physics required to understand laminar flows is briefly described.

A. Poiseuille Flow

Consider a long cylindrical channel of radius R. Let the y-axis be the direction along the channel. At steady state, the velocity field $\mathbf{u}$ is unidirectional and laminar and there is no acceleration of the fluid. Due to the boundary condition ($\mathbf{u} = 0$ at $r = R$), the pressure-driven motion in the channel, so-called Poiseuille flow [7], has a parabolic profile across the channel. Thus, the x-component u_x of $\mathbf{u}$ can be modeled as a function of the radial distance r by the following equation:

$$u_y(r) = u_{max} * \left(1 - \frac{r^2}{R^2}\right) \quad (1)$$

where u_{max} (m/s) is the maximum velocity at the center of the channel which is proportional to the ratio of the flow rate Q (m^3.s^{-1}) and the channel section S (m^2):

$$u_{max} \propto \frac{Q}{S} \quad (2)$$

B. Passive mixer

The structure of a passive mixer is described in Figure 1. The mixer is a straight channel in which both fluids flow laminarly. The mixing between both fluid is entirely driven by diffusion. The concentration of a molecule in the mixing channel is described by the advection-diffusion equation that mass conservation law for an elementary volume. This leads to the Fick's law [10]:

$$\frac{\partial C}{\partial t} = D\Delta C - \mathbf{u}(x, y) \cdot \nabla C + s(x, y, t) \quad (3)$$

where C is the concentration of a given molecule (mol.L^{-1}), D is the diffusion coefficient of that molecule (m^2.s^{-1}), $\mathbf{u}$(x,y) is the velocity field of the fluid (m.s^{-1}), $s(x, y, t)$ is a local source (or a sink, depending on its sign) of molecule (e.g. due to a chemical reaction occuring in the channel), Δ is the Laplacian operator and ∇ is the gradient operator.

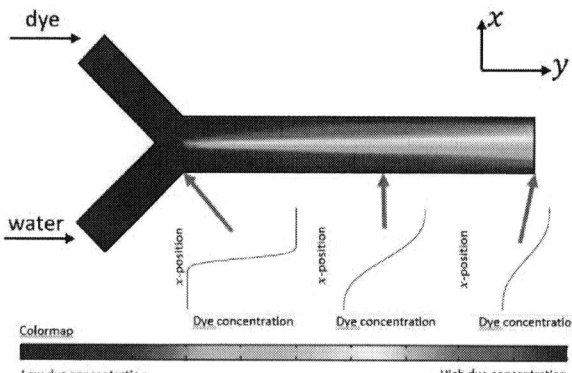

Figure 1. Schematic of a typical concentration distribution within a microchannel without a specific mixing element. The graphs below show the concentration distribution as a function of the coordinate normal to the channel wall, y, at several channel sections.

C. Working hypothesis

In a straight channel, due to laminar flow, the velocity field does not have any transversal component, thus $u_x = 0$. Moreover, if there is no source of molecule (e.g. no chemical reaction occurs in the channel), $s(x, y, t) = 0$.

$$\frac{\partial C}{\partial t} = D \cdot \left(\frac{\partial^2 C}{\partial x^2} + \frac{\partial^2 C}{\partial y^2}\right) - u_y(x) \cdot \frac{\partial C}{\partial y} \quad (4)$$

Finally, we assume the diffusion along the canal to be negligible with respect to the transport due to advection. The dye molecule experience an average speed $v_y = <u_y(x)>$. Under these assumptions, equation (3) becomes:

$$\frac{\partial C}{\partial t} = D \cdot \frac{\partial^2 C}{\partial x^2} - v_y \cdot \frac{\partial C}{\partial y} \quad (5)$$

Only the diffusion along the x direction is considered. In steady-state conditions:

$$0 = D \cdot \frac{\partial^2 C}{\partial x^2} - v_y \cdot \frac{\partial C}{\partial y} \quad (6)$$

III. MODEL CONSTRUCTION

A. Reference frame change

We simplify equation (6) by changing the reference frame. First, we decide to change the reference frame. The new reference frame is centered on a point that follow the diffusion front. The coordinates in the new reference frame (x', y') are related to the terrestrial one (x, y) through equations (5).

$$\begin{cases} x' = x \\ y' = y + v_y \cdot t \end{cases}$$
$$\Rightarrow 0 = D \frac{\partial^2 C}{\partial x^2} - v_y \cdot \frac{\partial y'}{\partial y} \cdot \frac{\partial t}{\partial y'} \cdot \frac{\partial C}{\partial t} \quad (7)$$

Let's drop the apostrophes for clarity

$$\frac{\partial C}{\partial t} = D \cdot \frac{\partial^2 C}{\partial x^2} \quad (8)$$

B. Resolution of diffusion equation in time

The method of separation of variables is used to solve this equation. It is assumed that the function $C(x, t)$ can be written as the product of a function $C_X(x)$ that depends only on x and a function $C_T(t)$ that depends only on time.

$$C(x, t) = C_X(x) \cdot C_T(t) \quad (9)$$

Equation (4) becomes

$$C_X(x) \cdot \frac{dC_T(t)}{dt} - D \cdot C_T(t) \cdot \frac{d^2 C_X(x)}{dx^2} = 0 \quad (10)$$

This equation can be rewritten

$$\frac{1}{D \cdot C_T(t)} \cdot \frac{dC_T(t)}{dt} = \frac{1}{C_X(x)} \cdot \frac{d^2 C_X(x)}{dx^2} \quad (11)$$

This equation must hold for all x and all t. But the left-hand side does not depend on x and the right-hand side does not depend on t. Hence, each side must be a constant α.

The left side is a 1st order differential equation of solution:

$$C_T(t) = \exp(\alpha \cdot D \cdot t) \quad (12)$$

Clearly, $\alpha \geq 0$ can not validate the mass conservation law, hence $\alpha < 0$ and we can write $\alpha = -\lambda^2$ with $\lambda \in \mathbb{R}$.

The right-hand writes as:

$$\frac{d^2 C_X(x)}{dx^2} = -\lambda^2 \cdot C_X(x) \quad (13)$$

This 2nd order differential equation has solution:

$$C_{X,n}(x) = A_n \cdot \sin(\lambda_n \cdot x) + B_n \cdot \cos(\lambda_n \cdot x) \qquad (14)$$

with $C_{X,n}(x)$ the n-th term of the Fourier series of $C_X(x)$, A_n and B_n the associated n-th sine and cosine Fourier coefficient and λ_n the constant λ associated with the n-th term and defined as a function of the channel width W as following.

If $\lambda = 0$:

$$C_{X,n}(x) = A_n \cdot + B_n \cdot x \qquad (15)$$

Considering the boundary conditions, one gets $A_n = B_n = 0$, so for $\lambda = 0$ only the trivial solution exists.

If $\lambda > 0$:

Substituting of the boundary conditions of equation 11 leads to the following equations for the constants A_n and B_n:

$$\begin{cases} C_{X,n}(0) = B_n = 0 \\ C_{X,n}(W) = A_n \cdot \sin(\lambda_n \cdot W) \end{cases} \qquad (16)$$

$$\sin(\lambda_n \cdot W) = 0 \Rightarrow \lambda_n = \frac{n \cdot \pi}{W} \qquad (17)$$

The left-hand side of equation (9) is a simple first-order equation. Its solution is a exponential decay of λ_n. Putting everything together, we obtain the final expression for $C(x,t)$:

$$C(x,t) = \sum_0^n \left(A_n \cdot \sin\left(\frac{n \cdot \pi \cdot x}{W}\right) + B_n \right. \\ \left. \cdot \cos\left(\frac{n \cdot \pi \cdot x}{W}\right) \right) \cdot e^{-D*t*\left(\frac{n*\pi}{W}\right)^2} \qquad (18)$$

The last step is to compute the values of A_n and B_n. For that purpose, we can use the concentration profile at $t = 0$:

$$A_n = \frac{1}{W} \int_{-W}^{W} C(x,0) \sin\left(\pi * n * \frac{x}{W}\right) dx \qquad (19)$$

$$B_n = \frac{1}{W} \int_{-W}^{W} C(x,0) \cos\left(\pi * n * \frac{x}{W}\right) dx \qquad (20)$$

The initial profile $C(x,0)$ is computed by COMSOL simulations.

C. COMSOL Multiphysics simulation

COMSOL Multiphysics is a partial differential equation solver based on the finite element method [12]. The geometry used for the simulation is depicted in Figure 2.

We simulate the mix between a pure solvent (input 1) and a solution with the molecule of interest (input 2). For both fluids, the flow rate is fixed to $Q_1 = Q_2 = 5 \cdot 10^{-4}$ µl/s. The concentration C_1 is equal to zero (pure solvent) while $C_2 = 1$mM. The diffusion coefficient of the molecule of interest is $D = 10^{-9}$ m²/s. the boundary condition at the exit of the mixer is a free flow.

This configuration is used both for the determination of $C(x,0)$, and as a reference for the evaluation of the quality of the compact model in the next section.

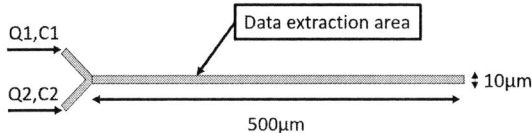

Figure 2. COMSOL simulation scheme of a Y microfluidic channel mixer

D. Sigmoid fit

According to the COMSOL simulations , $C(x,0)$ can be approximated by a sigmoid profile (figure 3)

$$C(x,0) = \frac{a}{1 + e^{-b(x-c)}} \qquad (21)$$

with a, b and c three fitting parameters: a is directly proportional to the concentration input, b gives the slope of the sigmoid at the inflection point and is proportional to the input flow rates and c is the position on the x-axis of the inflection points. COMSOL simulations with different couples of input flow rates show that a, b and c can be written as follow:

$$\begin{cases} a = C_2 \\ b = k_1 \cdot (Q_1 + Q_2) + k_2 \\ c = \frac{Q_1}{Q_1 + Q_2} \cdot W \end{cases} \qquad (22)$$

where C2 is the concentration of the molecule of interest, k1 and k2 are two fitting parameters obtained from COMSOL simulation, they depend on the Y junction geometry and fluid homogeneity. In our case k1=605s/µm⁴ and k2=1.3µm⁻¹.

The maximal deviation of the sigmoid fit from the true profile is 0.01 mM, which represent a relative error of about 2%.

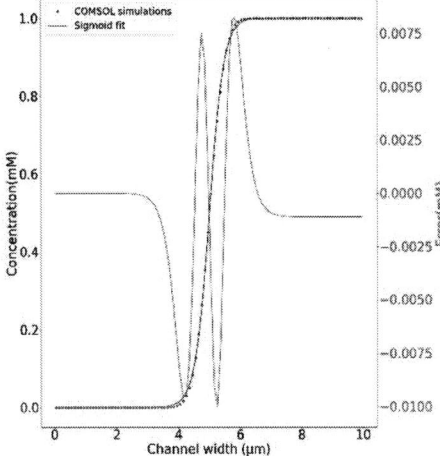

Figure 3. Results of COMSOL simulation vs sigmoid equation: concentration distribution at the entrance of the microfluidic mixer channel.

E. Computation of Fourier coefficients

Fourier decomposition applied to a periodic function, which is not the case of a sigmoid. To overcome this, we "*periodize*" $C(x,0)$ by considering a function $\tilde{C}(x,0)$ of period 2 W and is defined in the range $[-W;W]$ by the following equation:

$$\tilde{C}(x,0) = \frac{a}{1 + e^{-b(|x|-c)}} \qquad (23)$$

The comparison between the sigmoid and its Fourier series is given in figure 4.

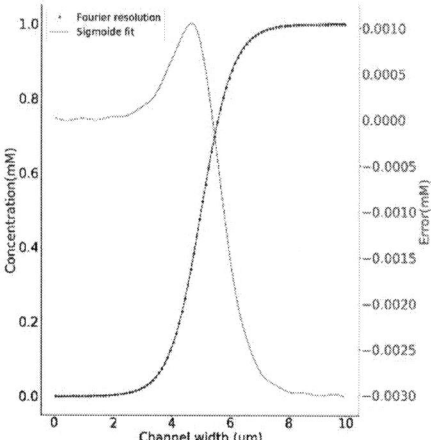

Figure 4. Results of Python Fourier resolution vs sigmoid equation: concentration distribution at the entrance of the microfluidic mixer channel

We finally test optimization of the number of basis functions n in the Fourier series (Table I). We made several tests with different n values and we compute the quadratic error in all the channel between COMSOL results, Fourier resolution and sigmoid fit. We can see there is no interest to go beyond n = 5 because the error is dominated by the sigmoid approximation dominated by the sigmoid approximation.

TABLE I.
QUADRATIC ERROR IN THE CHANNEL FOR N-ORDER FOURIER SERIES

N-order	1	3	5	7	9+
MSE COMSOL/Fourier (%)	2.70	0.418	0.330	0.317	>0.311
MSE Sigmoid/Fourier (%)	2.76	0.835	0.303	0.116	>0.05

IV. RESULTS AND DISCUSSION

A. Hypothesis validation

COMSOL simulations will be used to validate the hypothesis explained in subsection III-A. We compare the results obtained with an isotropic diffusion (standard case) and with an anisotropic diffusion (no diffusion along the channel axis). Results are given in Figure 5 for different couples of input flow rates. The maximum error over all the tests is around 1.2%, which small enough to validate our hypothesis.

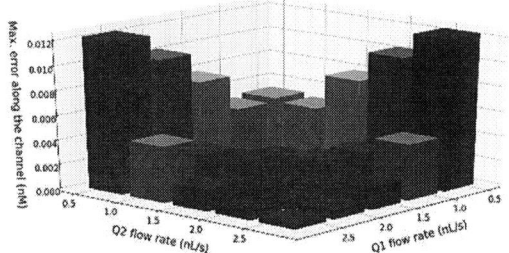

Figure 5. Hypothesis validation: maximal error for different flow rates.

B. Compact model

The compact model was first written in Python. Results are compared to COMSOL simulations for several setups described in Table II. These experiments are made with different flow rates Q1 and Q2 to challenge our numerical solving.

These experiments show a concentration distribution close to results find in the literature [13](Figure 6).

TABLE II.
SIMULATIONS SETUP

Experience #	Flow rate Q1 (nL/s)	Flow rate Q2 (nL/s)	Ratio R=Q1/Q2	Channel Flow Q = Q1+Q2 (nL/s)
1	0.5	0.5	R = 1	Q = 1
2	2.5	2.5	R = 1	Q = 5
3	1	0.5	R = 0.5	Q = 1.5
4	0.5	1	R = 2	Q = 1.5
5	2.5	0.5	R = 0.2	Q = 3
6	0.5	2.5	R = 5	Q = 3

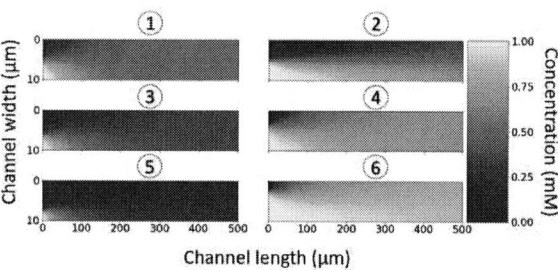

Figure 6. Results of experiences 1-6 from python Fourier series resolution. Y axis channel width is stretch 1:10 to improve visibility. The ladder shows the concentration in the channel.

The absolute difference between our model and COMSOL simulation are given in Figure 7. The absolute maximal error is about 0.07 mM, which represent 7% of concentration range inside the channel. It should also be noticed that the maximal error occurs at the beginning of the channel. This can be explained by the perturbation of laminar flow due to the Y-junction, which is computed by COMSOL but not integrated in our model. Error decrease along the channel.

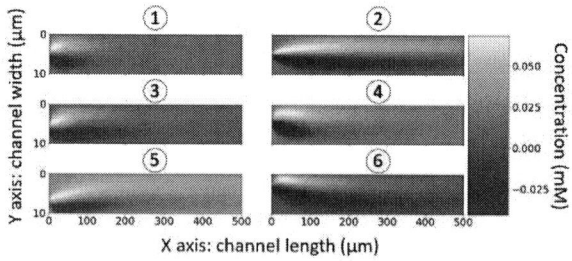

Figure 7. Results of experiences 1-6. Data python Fourier series – Data COMSOL simulation. Y axis channel width is stretch 1:10 to improve visibility. The ladder shows relative difference concentration in the channel.

Figure 8 show a comparison between the output concentration profile computed by COMSOL and the one estimated by our compact model. The maximal error is 0.02 mM, which is 4% of the average concentration at the output of the channel. We also compute the quadratic error for different couple of flow rates inside the channel and at its output (Figure 9). The quadratic errors are very weak (< 2% on the whole channel and <0.3% at the output).

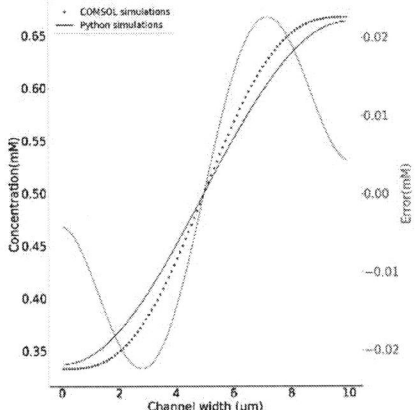

Figure 8. Transversal slice at the end of the channel experience 2. Differences obtain with our python simulation vs COMSOL simulation

(1)

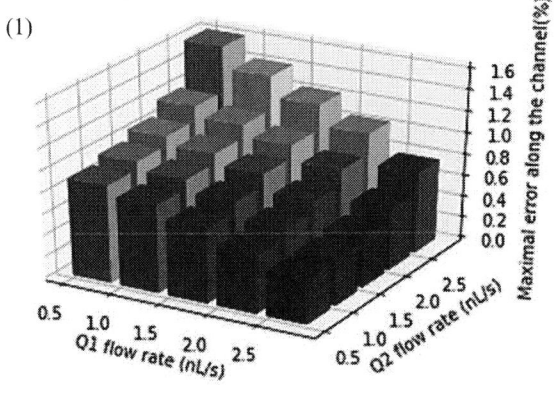

(2)

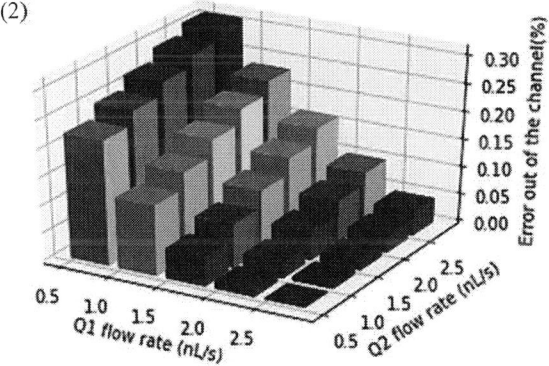

Figure 9. Quadratic Error (1) maximum along the channel and (2) at the end of the channel for different flow rate combinations

CONCLUSION

In this paper we presented a compact model for a passive microfluidic channel mixer. The model is based on a time-to-space transformation, a sigmoidal approximation of the concentration profile after the Y-junction and Fourier series to resolve diffusion equation in the microfluidic channel. This compact model has been written in Python but is only composed of analytics equations, which make it compatible with VHDL-AMS or Verilog-A implementation. Thus, it can be added to our database of compact models and interfaced with other microfluidic, biochemical or electrical models for the virtual prototyping of lab-on-chip.

In our model we only present molecule concentration from one side of the Y channel. But we can consider a mix of two different molecules C1 and C2. In this case we need two data set for each molecule and due to our symmetric situation we reverse the x axis to obtain the other concentration field.

In parallel, investigations will go on in order to improve the model. We plan to integrate in the model the possibility for both fluids to react inside the mixer.

REFERENCES

[1] S. Guiton, A. Rezgui, M. Madec, C. Lallement, F. Rufi, and J. Haiech, 'Modeling and simulation of a Lab-On-Chip for micropollutants detection', in *2014 Proceedings of the 21st International Conference Mixed Design of Integrated Circuits and Systems (MIXDES)*, 2014, pp. 256–261.

[2] A. Weltin *et al.*, 'Cell culture monitoring for drug screening and cancer research: a transparent{,} microfluidic{,} multi-sensor microsystem', *Lab Chip*, vol. 14, no. 1, pp. 138–146, 2014.

[3] P. Xu, X. Li, H. Yu, and T. Xu, 'Advanced Nanoporous Materials for Micro-Gravimetric Sensing to Trace-Level Bio/Chemical Molecules', pp. 19023–19056, 2014.

[4] M. Madec, A. Bonament, E. Rosati, L. Hebrard, and C. Lallement, 'Virtual prototyping of biosensors involving reaction- diffusion phenomena', in *2018 16th IEEE International New Circuits and Systems Conference (NEWCAS)*, 2018, no. Umr 7357, pp. 3–6.

[5] C. M. Collier, J. Nichols, and J. F. Holzman. '4 - Digital microfluidics technologies for biomedical devices', in *Woodhead Publishing Series in Biomaterials*, X. (James) Li and Y. B. T.-M. D. for B. A. Zhou, Eds. Woodhead Publishing, 2013, pp. 139–164.

[6] D. Grissom and P. Brisk, 'Path Scheduling on Digital Microfluidic Biochips', pp. 26–35, 2012.

[7] H. Bruus, 'Theoretical microfluidics', pp. 31–32, 2006.

[8] K. W. Oh, K. Lee, B. Ahn, and E. P. Furlani, 'Design of pressure-driven microfluidic networks using electric circuit analogy', *Lab Chip*, vol. 12, no. 3, pp. 515–545, 2012.

[9] L. Renaud, D. Selloum, and S. Tingry, 'Xurography for 2D and multi - level glucose / O 2 microfluidic biofuel cell', no. January 2016, 2015.

[10] P. Tabeling, *Introduction to Microfluidics*. Oxford University press, 2005.

[11] A. E. Kamholz and P. Yager, 'Theoretical Analysis of Molecular Diffusion in Pressure-Driven Laminar Flow in Microfluidic Channels', *Biophys. J.*, vol. 80, no. 1, pp. 155–160, 2001.

[12] 'COMSOL Multiphysics® v. 5.4.'

[13] Y. K. Suh and S. Kang, 'A review on mixing in microfluidics', *Micromachines*, vol. 1, no. 3, pp. 82–111, 2010.

Proceedings of the 27th International Conference *"Mixed Design of Integrated Circuits and Systems"*
June 25-27, 2020, Łódź, Poland

Impact of Dynamic Trapping on High Frequency Organic Field-Effect Transistors

Markus Mueller[1], Shabnam Donnhäuser[1], Sven Mothes[1], Anibal Pacheco-Sanchez[1],
Katherina Haase[2], Stefan C. B Mannsfeld[2], Martin Claus[2]

[1]Chair for Electron Devices and Integrated Circuits, Technische Universität Dresden, Germany
[2]Center for Advancing Electronics Dresden, Technische Universität Dresden, Germany
Email: markus.mueller3@tu-dresden.de, Shabnam.Donnhaeuser@tu-dresden.de

Abstract—Many emerging organic semiconductor devices suffer from hysteresis effects caused by traps. The impact of such traps on AC device performance has not been investigated yet. In this paper, high-frequency non-quasi-static effects related to dynamic trapping based on a theoretical framework, TCAD simulations and experimental data have been studied. In contrast to previous studies, the focus of this paper is the OFET's high-frequency performance and not the characterization of the traps.

Keywords—Dynamic Trapping, non-quasi-static (NQS), high frequency(HF), Organic field effect transistor (OFET).

I. INTRODUCTION

The industrial, scientific, and medical radio band (ISM band) with center frequency 13.56 MHz has been identified as a target frequency that would enable first applications of organic field effect transistors (OFETs) in commercial applications [1]. Although the organic semiconductor technology has made huge progress in the last years, their high frequency (HF) performance is still not sufficient for applications such as Radio-frequency identification (RFID) tags [1]. Hence, a further increase of the transit frequency f_t is required.

The limitation of the device performance can be attributed to, among other things, high contact resistance, low mobility and traps. Trap related hysteresis effects have been reported in many fabricated OFETs [2] and other emerging device technologies [3]. Traps are known to shift the internal bias point of a transistor but - depending on the trap related time constants- they can also affect the HF performance. Thus, a deep understanding in characterization and modeling of trap-related phenomena is important. However, hysteresis effects are challenging to be characterized because established characterization methods [4], [5] require either a bulk contact or a MOS capacitor. Furthermore, to the best of our knowledge, the effects of such traps on OFETs high frequency performance for circuit design purposes have not been investigated so far.

We explain how one can exploit the trap dynamics of OFETs for circuit design at high frequency. In principle our results can be applied to a wide range of emerging FET devices such as CNTFETs or 2D FETs, however in this work we focus on OFETs.

II. THEORY

A. Carrier Density, Drift-Diffusion Equation System

The mobile hole carrier density p is calculated using the conventional equation,

$$p = N_V \exp\left(\frac{E_V - E_{Fp}}{k_B T}\right), \qquad (1)$$

where N_v is the effective density of states in the valence band, k_B is Boltzmann's constant, T is the temperature, E_V is the valence band energy and E_{Fp} is the quasi Fermi energy of holes. E_V is shifted by the electrostatic potential Ψ and E_{Fp} is shifted by the quasi Fermi potential of holes φ_p. Eq. 1 can be applied to semiconductors with a parabolic band-structure and to semiconductors with a Gaussian band-structure [6].

The density of the trapped carriers are calculated based on the assumption that trapped holes populate a single trap energy level E_t with a density of states N_{tt}. Therefore, the trapped hole density is given by:

$$p_t = \int_{-\infty}^{\infty} N_{tt} F_t(E)\,\mathrm{d}E = \frac{N_{tt}}{1 + \exp\left(\frac{E_{Ft} - E_t}{k_B T}\right)}, \qquad (2)$$

where F_t is the Fermi-Dirac distribution function of the trapped carriers and E_{Ft} is the trapped carrier's Fermi level. E_t and E_{Ft} are shifted by Ψ and by the quasi Fermi potential of traps φ_t, respectively.

If a trap is not occupied by a hole, hence, it is occupied by an electron. Therefore the density of trapped electrons n_t is given by

$$n_t = N_{tt} - p_t. \qquad (3)$$

Inside the semiconductor, the conventional drift-diffusion (DD) equation system is solved for holes, extended by the continuity equation of trapped holes. The dynamics of the trapped carrier is calculated based on the trapped carrier continuity equation [7],

$$\frac{\partial p_t}{\partial t} = \tau_c p n_t - \tau_e p_t. \qquad (4)$$

Here τ_c is the capture rate of holes and τ_e is the emission rate of trapped holes. Equations 1-4 are implemented into our in-house multi-material TCAD simulator COOS capturing semi-classical [8] as well as quantum transport [9]. Note that only traps whose trapping and release times are slow compared to

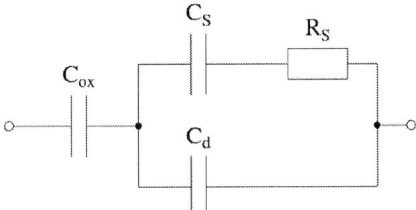

Fig. 1. Equivalent circuit of a MOS capacitor capturing the impact of slow traps.

the channel transit time (inverse of the transit frequency f_t) are considered in this study. This excludes all "fast" traps where $\varphi_t = \varphi_p$ which have already been investigated in detail [10]. Furthermore we would like to stress that only slow traps cause the hysteresis effects observed in experiments [11].

It has been shown in [7] that the relation between τ_e and τ_c is given by:

$$\frac{\tau_e}{\tau_c} \equiv p_1 = N_V \exp\left(\frac{E_V - E_t}{k_B T}\right) \qquad (5)$$

which reduces Eq. 4 to

$$\frac{\partial p_t}{\partial t} = \tau_c \left(p n_t - p_1 p_t\right) . \qquad (6)$$

B. DC and AC Analysis of the Equation System

In the DC case, the solution of Eq. 6 is given by $\varphi_t = \varphi_p$. The impact of traps on the DC characteristics has been summarized in [12]. The impact of slow and fast traps on the DC characteristics is exactly the same, theoretically. However, only the slow traps cause a hysteresis in the electrical characteristics since in conventional staircase DC measurements the DC condition $\frac{\partial}{\partial t} \approx 0$ is never perfectly satisfied.

The impact of slow traps on a MOS capacitor can be modeled with the equivalent circuit shown in Fig. 1 [5]. C_{ox} and C_d are the oxide capacitance and the depletion layer capacitance per unit area, respectively. C_s is the trap capacitance per unit area and R_s the hole capture resistance. In the case of a single level trap energy, the traps act like a series RC-network with the series capacitance,

$$C_S = \frac{q}{V_t} N_{tt} F_t \left(1 - F_t\right) \qquad (7)$$

and the time constant

$$\tau = \frac{F_t}{\tau_c p} . \qquad (8)$$

III. TCAD Study of High Frequency Effects Caused by Slow Traps

Next we qualitatively analyze the high frequency effects caused by traps. For this study we have simulated a TIPS-Pentacene based OFET. The parameters and the device structure are shown in table I and in the inset of Fig. 2, respectively.

The simulation results in Fig. 3 demonstrate that H_{21} does not drop with the usual 20dB/dec due to a distinct non-quasi

TABLE I
TCAD SIMULATION PARAMETERS

Parameters	Values and units
l_{ch}	$25\,\mu\text{m}$
t_{ox}	$30\,\text{nm}$
t_{semi}	$30\,\text{nm}$
$\varepsilon_{r,semi}$	3
$\varepsilon_{r,SiO2}$	3.9
E_g	$2\,\text{eV}$
m_{eff}	$11.7 * m_e$
μ_{const}	$1\,\frac{\text{cm}^2}{\text{Vs}}$
E_t	$-0.7\,\text{eV}$
N_D	$-1e22\,\frac{1}{\text{m}^3}$
c_p	$1e-22\,\frac{\text{m}^2\text{eV}}{\text{s}}$
contact boundary condition	ohmic

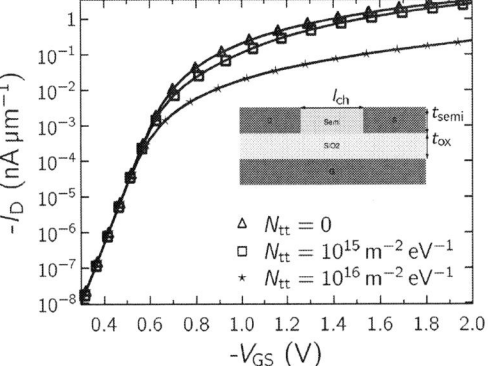

Fig. 2. Transfer characteristics in logarithmic scale for different trap concentration (N_{tt}) at $V_{DS} = -1\,\text{V}$ (The inset shows the structure of the investigated devices).

static (NQS) effect in the trap affected devices. Therefore f_t can not, as usually done, be calculated by extrapolating H_{21} at an arbitrary frequency for trap affected devices. Particularly, this means that the extrapolated transit frequency,

$$f_{t,extr} = f_{meas} H_{21}\left(f_{meas}\right) \qquad (9)$$

underestimates the actual f_t. At low frequencies, $f_{t,extr}$ is equal to

$$f_{t,extr} = \frac{g_{m,DC}}{2\pi C_{gg}}, \qquad (10)$$

which is usually employed to predict the high-frequency performance of FETs.

In order to understand why f_t is higher than $f_{t,DC}$ we analyze the AC transconductance,

$$g_{m,AC} = \Re\{Y_{21} - Y_{12}\}, \qquad (11)$$

the total gate capacitance,

$$C_{gg} = \Im\left\{\frac{Y_{11}}{\omega}\right\} . \qquad (12)$$

and the output conductance Y_{22}. The typical double plateau behavior for $g_{m,AC}$ (Fig. 4) that has been predicted in [13], [14] can be observed before $g_{m,AC}$ starts to decrease (not

41

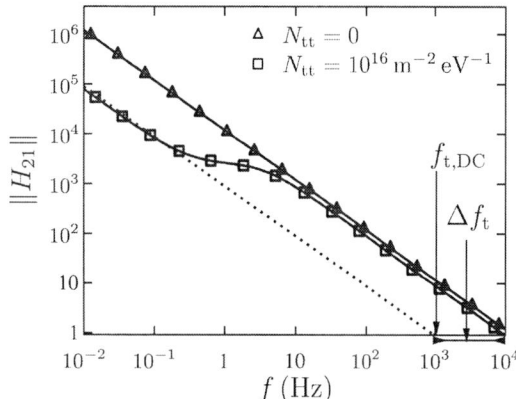

Fig. 3. Current gain H_{21} for a trap affected and a trap free OFET at $V_{GS} = -1.5\,V$ and $V_{DS} = -1\,V$.

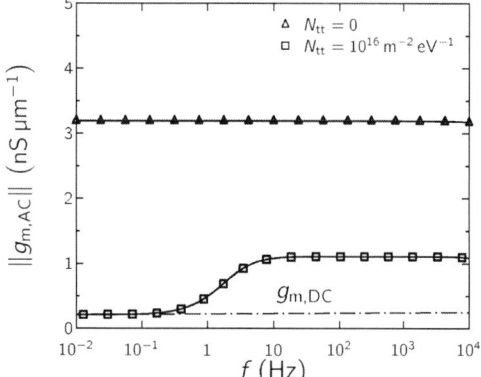

Fig. 4. Frequency dependent transconductance $g_{m,AC}$ for a trap affected and a trap free OFET at $V_{GS} = -1.5\,V$ and $V_{DS} = -1\,V$.

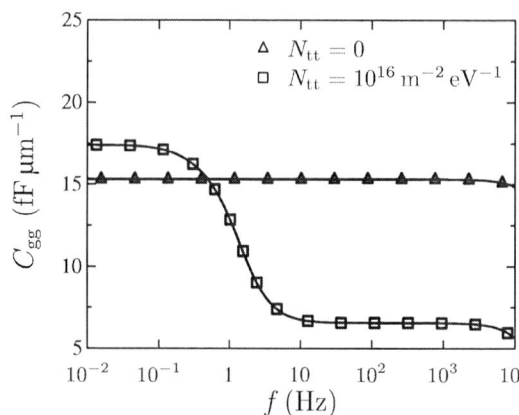

Fig. 5. Input capacitance from Y_{11} for different N_{tt} at $V_{GS} = -1.5\,V$ and $V_{DS} = -1\,V$.

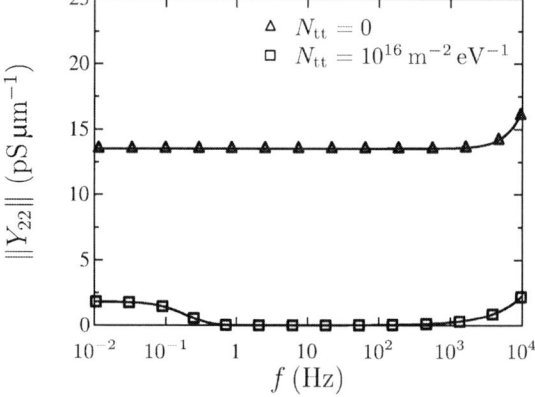

Fig. 6. Output conductance for different N_{tt} at $V_{GS} = -1.5\,V$ and $V_{DS} = -1\,V$.

shown here) at higher frequencies. As soon as the slow traps can not follow the surface potential anymore, $g_{m,AC}$ rises, whereas C_{gg} decreases (Fig. 5). The combination of these effects causes the current gain to rise and therefore f_t increases. The output conductance decreases as the traps stop following the AC signal, as can be seen in Fig. 6.

In summary our simulations predict the following trap-related effects:

- g_m at high frequencies is higher than one would expect from DC measurements
- C'_{gg} at high frequencies is lower than one would expect from DC measurements
- f_t is higher than $f_{t,DC}$
- Y_{22} at high frequencies is lower than one would expect from DC measurements

IV. OBSERVATION BY EXPERIMENTS

According to the simulations, the influence of slow traps on the AC behavior of trap affected devices is observed at extremely low frequencies. Hence, the characterization of the effect is challenging using conventional AC measurement equipment, e.g. VNAs and bias tees. Furthermore , the required voltage for operation of OFETs is rather large which can be another limiting factor when using commercial measuring equipments.

In order to characterize the low frequency small signal behavior of OFETs we employed a Bode 100 from Omicron, together with an in-house built interface circuit. The used OFETs were prepared on Si wafers with a 300 nm thick SiO2 gate dielectric. The cleaned wafer substrates were treated with phenyltrichlorosilane (PTS) to provide sufficient wettability for the subsequent solution deposition process. The organic semiconductor 2,7-Dioctyl[1]benzothieno[3,2-b][1]benzothiophene (C8-BTBT) was used as active channel material and deposited via the solution-shearing method. Top-contacts were prepared by thermal evaporation through a shadow mask.

Fig. 7 shows the measured frequency dependent transconductance $g_{m,AC}$ for two different fabricated devices. As predicted, the transconductance increases with frequency as shown in Fig. 4.

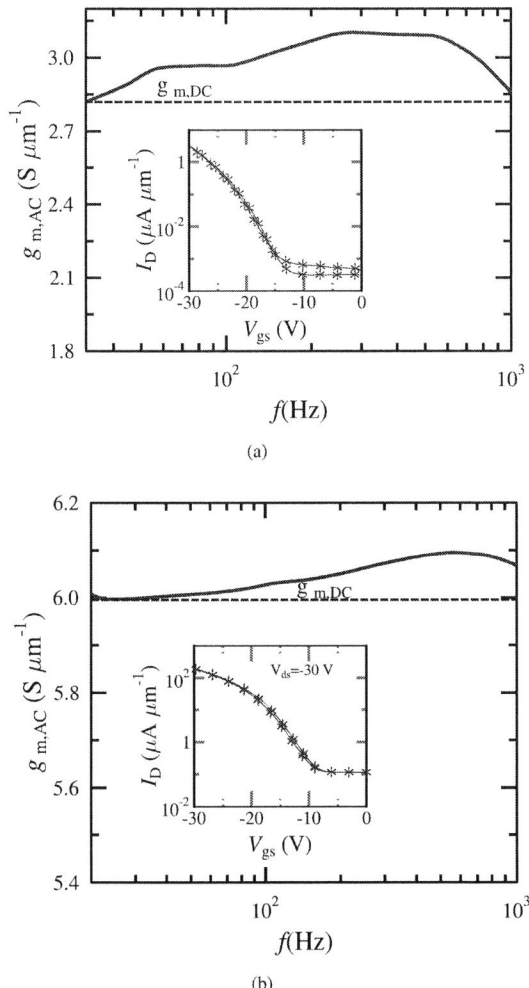

(a)

(b)

Fig. 7. Frequency dependent transconductance $g_{m,AC}$ of two different fabricated OFETs (a) Device with low gain (b) Device with high gain at $V_{DS} = -50\,\text{V}$ and $V_{GS} = -50\,\text{V}$. The inset figures show transfer characteristics of devices with forward-backward sweep.

The increase of $g_{m,AC}$ in the experimental results is clearly visible, however the magnitude strongly depends on the operating point and the device fabrication process, which affects the density of slow traps. The small hysteresis of transfer characteristics (inset of Fig. 7) confirms the presence of less slow traps in our fabricated devices.

V. TOWARD PHYSICS BASED COMPACT MODELING TRAP RELATED HIGH FREQUENCY EFFECTS

While the equivalent circuit in Fig.1 gives basic insights into trap related effects, for compact models the approach presented in [3], [15] is better. Fig. 8a shows an equivalent circuit for modeling the shift of the threshold voltage by the voltage drop V_T due to the presence of traps. This model besides existing FET-compact models can be used to describe the impact of traps. Since V_T in Fig. 8a is time-dependent, a proper selection

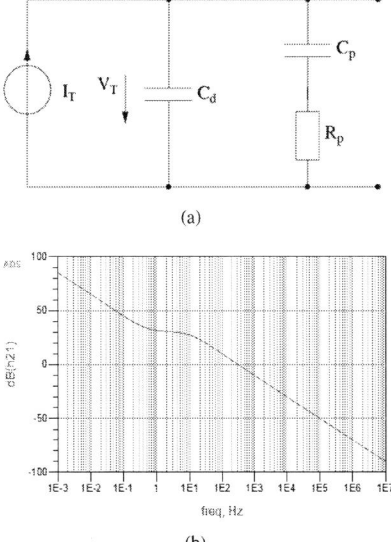

(a)

(b)

Fig. 8. (a) Equivalent circuit to capture trap related effects [15] (b) Current gain H_{21} of an OFET compact model extending by (a) at $V_{DS} = -12\,\text{V}$ and $V_{GS} = -15\,\text{V}$.

of C_p and R_p allows to capture fast as well as slow traps. The trap dependencies of the threshold voltage and the drain current are described by

$$I_T = \alpha V_{gs} + \beta V_{ds} V_{gs} + \gamma \tag{13}$$

where α, β and γ are fitting parameters. The qualitative agreement of circuit simulation results for the frequency dependence of the current gain h_{21} in Fig. 8b and the TCAD simulation results in Fig. 3 confirms that the described compact modeling approach could be used to study the impact of slow traps at the circuit level by adjusting the RC-elements and the parameters in Eq. 13 properly.

VI. CONCLUSION

This study indicates that one can not predict the AC behavior from DC measurements of trap affected devices. While the present results reveal trap-related effects in a narrow frequency range, it is expected to see these effects for a larger frequency range for traps with a broad energy distribution.

The results so far have implications for high frequency circuit design. The most important finding is that f_t is higher than one might predict from DC measurements. Furthermore $g_{m,AC}$ is higher than $g_{m,DC}$. Another conclusion of our work is, that C_{gg} decreases at high frequencies which is highly beneficial for input matching.

For low bandwidth applications such as RFID tags, it might be interesting to turn on the transistor only for a short time to prevent the charging of the traps and thus, to get access to the almost trap-free high frequency capability of the OFET.

ACKNOWLEDGMENT

The authors would like to thank the German Research Foundation (Deutsche Forschungsgemeinschaft-DFG CL384/5 and MA 3342/2-1 SPP Fflexcom).

REFERENCES

[1] R. Rotzoll, S. Mohapatra, V. Olariu, R. Wenz, M. Grigas, K. Dimmler, O. Shchekin, and A. Dodabalapur, "Radio frequency rectifiers based on organic thin-film transistors," *Applied Physics Letters*, vol. 88, no. 12, p. 123502, mar 2006.

[2] S. H. Kim, W. M. Yun, O.-K. Kwon, K. Hong, C. Yang, W.-S. Choi, and C. E. Park, "Hysteresis behaviour of low-voltage organic field-effect transistors employing high dielectric constant polymer gate dielectrics," *Journal of Physics D: Applied Physics*, vol. 43, no. 46, p. 465102, nov 2010.

[3] M. Haferlach, M. Claus, T. Nardmann, A. Pacheco-Sanchez, P. Sakalas, and M. Schröter, "Trap-induced apparent linearity of cntfets," in *Nanotech, Workshop on compact modeling*, 2014.

[4] D. V. Lang, "Deep-level transient spectroscopy: A new method to characterize traps in semiconductors," *Journal of Applied Physics*, vol. 45, no. 7, pp. 3023–3032, jul 1974.

[5] E. H. Nicollian and A. Goetzberger, "The si-SiO2interface - electrical properties as determined by the metal-insulator-silicon conductance technique," *Bell System Technical Journal*, vol. 46, no. 6, pp. 1055–1133, jul 1967.

[6] G. Paasch and S. Scheinert, "Charge carrier density of organics with gaussian density of states: Analytical approximation for the gauss–fermi integral," *Journal of Applied Physics*, vol. 107, no. 10, p. 104501, may 2010.

[7] C.-T. Sah, "The equivalent circuit model in solid-state electronics—part i: The single energy level defect centers," *Proceedings of the IEEE*, vol. 55, no. 5, pp. 654–671, 1967.

[8] S. Mothes and M. Schröter, "Three-dimensional transport simulations and modeling of densely packed cntfets," *IEEE Transactions on Nanotechnology*, vol. 17, no. 6, pp. 1282–1287, 2018.

[9] M. Claus, S. Mothes, S. Blawid, and M. Schröter, "COOS: a wavefunction based schrödinger–poisson solver for ballistic nanotube transistors," *Journal of Computational Electronics*, vol. 13, no. 3, pp. 689–700, jun 2014.

[10] W. L. Kalb and B. Batlogg, "Calculating the trap density of states in organic field-effect transistors from experiment: A comparison of different methods," *Physical Review B*, vol. 81, no. 3, jan 2010.

[11] D. Fleetwood, "Fast and slow border traps in mos devices," pp. 1–8, 1995.

[12] J. H. Schön and B. Batlogg, "Trapping in organic field-effect transistors," *Journal of Applied Physics*, vol. 89, no. 1, pp. 336–342, jan 2001.

[13] U. Sharma, R. Booth, and M. White, "Static and dynamic transconductance of MOSFETs," *IEEE Transactions on Electron Devices*, vol. 36, no. 5, pp. 954–962, may 1989.

[14] L. Schrader, "The influence of the interface states on the dynamic transconductance of mis-fets," *Solid-State Electronics*, vol. 20, no. 8, pp. 671–674, aug 1977.

[15] M. Haferlach, A. Pacheco, P. Sakalas, M. Alexandru, S. Hermann, T. Nardmann, M. Schroter, and M. Claus, "Electrical characterization of emerging transistor technologies: Issues and challenges," *IEEE Transactions on Nanotechnology*, vol. 15, no. 4, pp. 619–626, jul 2016.

Proceedings of the 27th International Conference *"Mixed Design of Integrated Circuits and Systems"*
June 25-27, 2020, Łódź, Poland

Parameter Extraction for a Simplified EKV-model in a 28nm FDSOI Technology

Konstantin Bajer, Steffen Paul, Dagmar Peters-Drolshagen
Institute of Electrodynamics and Microelectronics (ITEM.me)
University of Bremen, Bremen, Germany, +49(0)421/218-62539
Email: {bajer, peters, steffen.paul}@me.uni-bremen.de

Abstract—**The g_m/I_D methodology is applicable for the circuit design in advanced nanometer technologies. This work proposes a systematic parameter extraction process for a simplified EKV-model with only three model parameters which is applicable to all CMOS technologies. The extraction procedure relies only on the drain current for a sweep of the gate voltage without the need of additional extraction simulations in SPICE or parameters from the model card. Therefore, it is independent from the applied technology or the compact model of the SPICE simulation. For devices with short channel lengths, three variations of the EKV model were evaluated which consider velocity saturation. The resulting model provides good results compared to the SPICE simulation over the complete operation region of the technology for long and short channel devices while keeping simplicity for fast tool-based circuit design procedures and hand calculations.**

Keywords—**Transistor Modeling. Parameter Extraction, Design Process, EKV, MOSFET, FDSOI, Nanometer Technology.**

I. INTRODUCTION

The g_m/I_D design methodology is gaining popularity with smaller technologies, especially for low power designs for amplifiers [1]. While g_m is a small signal parameter determining the gain of an amplifier, I_D is an important parameter for the operation point. Both construct the transconductance efficiency g_m/I_D. In addition, the g_m/I_D characteristic for a transistor covers all possible operation regions in one consistent and smooth model that is easy to use for hand calculations. This methodology can be applied in deep submicron or nanoscale bulk, single- and double-gate FDSOI and FinFET CMOS technologies [2], [3]. However, different approaches for the use of the g_m/I_D curves do exist. Most often the semi-empirical design procedure based on lookup tables is referred to as the g_m/I_D methodology [4]. The inversion coefficient is defined and used in most approaches [5], [6], [7] while not mandatory and not always used [4]. Likewise, the parameter extraction is handled differently. While [5] shows a variety of possible definitions for the subthreshold slope of n for different operation regions, it is usually assumed to be constant over the inversion coefficients (IC) in many cases for simplified calculations [8], [9]. In order to work with the g_m/I_D design procedure, the model parameters have to be extracted. This can be performed using the charge equations as in [4]. However, the authors decided for a lookup table based design approach with small signal parameters without the use of the inversion coefficient rather than analytical model calculations. A Verilog-A model is given in [10] with a detailed parameter extraction procedure.

This model is suitable for SPICE simulations and can model the MOS behaviour precisely. However, it is an iterative simulation model and is not suitable for hand calculations. A non-iterative simplified EKV model is shown in [9], which covers the complete operation region of the EKV model. Some model parameters can be extracted from simulations while most of them are taken from the BSIM model card. This makes the acquirement of the parameters very process dependent and manual modifications have to be performed for the parameter extraction from different SPICE models. Other simplified EKV models are presented in [6]. These models vary in complexity and are approximations for a certain operation region or cover several regions of the MOS device. While they are very useful for hand calculations, no exact parameter extraction procedure is proposed. [11] shows a parameter extraction for a 28nm bulk technology for a simplified EKV-model while the exact extraction points are not provided. In this work, a simplified EKV model is used which is suitable for hand calculations. This model uses only the subthreshold slope n, the technology specific current I_{spec} and λ_c to model velocity saturation.

For a tool-based design procedure as presented in [4], [12], [13] the g_m/I_D curves are already available. The approach in this work uses only this database to extract the model parameter without the need for an additional SPICE simulation or parameters from the SPICE model card and is based on the slope of the g_m/I_D curve. This makes it possible to adapt this method to other technologies and to already existing databases.

II. EKV MODEL

The Enz-Krummenacher-Vittoz (EKV) model is a charge-based transistor model suitable for advanced technology nodes [5], [8]. Different versions for SPICE simulations do exist (EKV2.6 and EKV3.0). Nevertheless, its core is based on a few parameters and it is also suitable for hand calculations. Furthermore, very simple expressions do exist which cover certain operation regions or cases, i.e. weak inversion (WI), strong inversion (SI) or velocity saturation (VSAT). For WI the transconductance efficiency g_m/I_D can considered as a constant:

$$\frac{g_m}{I_D} = \frac{1}{nV_T} \qquad (1)$$

where n is the subthreshold factor and V_T is the thermal voltage. For SI, the g_m/I_D is proportional to $1/\sqrt{IC}$ and can be expressed as follows:

$$\frac{g_m}{I_D} = \frac{1}{nV_T}\frac{1}{\sqrt{IC}} \qquad (2)$$

with $IC = \dfrac{I_D}{I_{spec}}$ and $I_{spec} = 2n\mu C_{OX}V_T^2\dfrac{W}{L} = I_{sp}\dfrac{W}{L}$. (3)

This approximation describes the negative slope of the g_m/I_D in Fig. 1 and 2 for large inversion coefficients and long channel devices without significant velocity saturation. Taking care of both operation regions for devices without velocity saturation, the WI+SI model from [6] is as follows:

$$\frac{g_m}{I_D} = \frac{1}{nV_T}\frac{1 - \exp^{-\sqrt{IC}}}{\sqrt{IC}} . \qquad (4)$$

With increasing V_{GS} the electric field approaches a critical point where it reaches its maximum, resulting in a velocity saturation of the charge carriers in the channel. For VSAT the g_m/I_D is given by

$$\frac{g_m}{I_D} = \frac{1}{nV_T}\frac{1}{\lambda_c IC} . \qquad (5)$$

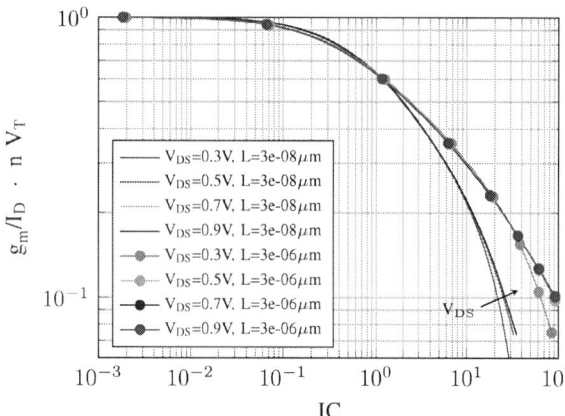

Fig. 1. V_{DS} and L variation of the g_m/I_D curves of the SPICE model.

The effect of velocity saturation is more pronounced in short channel devices. A comprehensive summary of these simplified models for certain operation regions is presented in [6], i.e. a model which covers weak inversion and velocity saturation (WI+VSAT)

$$\frac{g_m}{I_D} = \frac{1}{nV_T}\frac{1}{1 + \lambda_c IC} \qquad (6)$$

and a model which includes strong inversion as well, denoted here as (WI+SI+VSAT)

$$\frac{g_m}{I_D} = \frac{1}{nV_T}\frac{1}{\sqrt{IC}}\frac{1 - \exp^{-\sqrt{IC}}}{1 + \lambda_c\sqrt{IC}} . \qquad (7)$$

In [8] a simplified model is presented which covers weak inversion and velocity saturation (WI+VSAT Enz)

$$\frac{g_m}{I_D} = \frac{1}{nV_T}\frac{\sqrt{(\lambda_c IC + 1)^2 + 4IC} - 1}{IC\left[\lambda_c\left(\lambda_c IC + 1\right) + 2\right]} . \qquad (8)$$

These simplified models are evaluated and compared in the next section of this work.

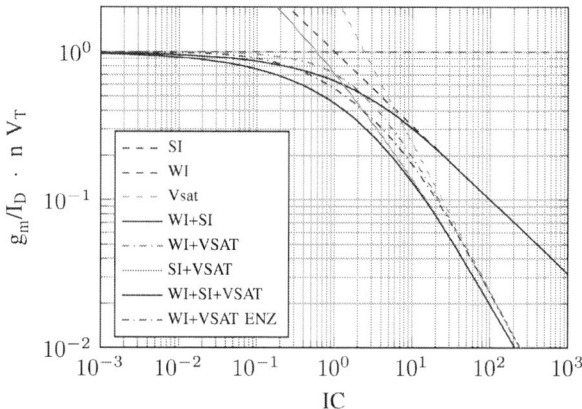

Fig. 2. Comparison of different EKV models considering different operation regions of a MOSFET device.

III. PARAMETER EXTRACTION

The extraction procedure is based on g_m/I_D curves from SPICE simulations, where V_{DS} is swept for each curve, as shown in Fig. 3 and 5.

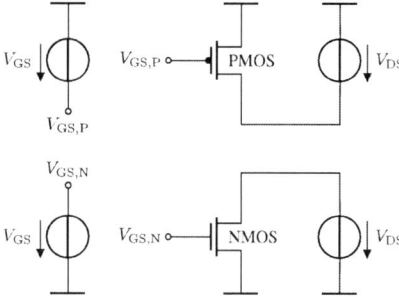

Fig. 3. Measurement circuit in SPICE for the extraction of the g_m/I_D curves.

The threshold slope n is calculated using the feature that g_m/I_D converges for $IC \to 0$ in eq. (6) to the WI solution in eq. (1). This is used by calculating the maximum of g_m/I_D from the SPICE simulation in order to extract n:

$$n = \frac{1}{V_T \max\left(\frac{g_m}{I_D}\right)} \qquad (9)$$

Fig. 1 demonstrates that the $g_\mathrm{m}/I_\mathrm{D}$ curves variate over the gate length and V_DS, which is also reported for other technologies [5], [9]. Thus, the model is extracted for each of these data tuples of the gate length L and V_DS. From equations (2) and (5), I_spec and λ_c can be determined. In order to find an extraction point for these parameters, the logarithmic function is used for SI

$$\log\left(\frac{g_\mathrm{m}}{I_\mathrm{D}}nV_\mathrm{T}\right) = \log\left(\frac{1}{\sqrt{\mathrm{IC}}}\right) = \frac{1}{2}\left[\log\left(I_\mathrm{spec}\right) - \log\left(I_\mathrm{D}\right)\right] \tag{10}$$

and for VSAT:

$$\log\left(\frac{g_\mathrm{m}}{I_\mathrm{D}}nV_\mathrm{T}\right) = \log\left(I_\mathrm{spec}\right) - \log\left(\lambda_c\right) - \log\left(I_\mathrm{D}\right) \tag{11}$$

with the slope denoted as

$$\xi = \frac{\partial \log\left(\frac{g_\mathrm{m}}{I_\mathrm{D}}nV_\mathrm{T}\right)}{\partial \log I_\mathrm{D}} \tag{12}$$

Eq. (10) and (11) show that $\log\frac{g_\mathrm{m}}{I_\mathrm{D}}$ has a slope of -1/2 and -1 with respect to I_D for SI and VSAT, respectively. Fig. 4 shows the slope of $\log\frac{g_\mathrm{m}}{I_\mathrm{D}}$ from the SPICE simulation for the long and the short channel device. The slope changes gradually from 0 to approx. 1.1 for the short channel device. However, the slope for the long channel device does not reach -1. In order to extract a model for long as well as short channel devices, I_spec is derived evaluating eq. (10) from a long channel device with negligible velocity saturation at the I_D where the slope of $\log\frac{g_\mathrm{m}}{I_\mathrm{D}}$ is -1/2:

$$I_\mathrm{sp} = \frac{L}{W} I_{D,\xi=-1/2}\, 10^{2\left[\log\left(\frac{g_\mathrm{m}}{I_\mathrm{D}}\right)\right]_{\xi=-1/2}} \tag{13}$$

The same procedure is performed for the short channel device using eq. (11) and evaluating $\log\frac{g_\mathrm{m}}{I_\mathrm{D}}$ at a slope of -1 in order to calculate λ_c:

$$\lambda_c = \frac{I_\mathrm{spec}}{10^{\left[\log\left(\frac{g_\mathrm{m}}{I_\mathrm{D}}\right)\right]_{\xi=-1}} I_{D,\xi=-1}} \tag{14}$$

This extraction is performed for each V_DS from the SPICE database. V_DS and V_GS are swept in a range from 0 to 1 V which covers the complete operation region of the technology and L from 30 nm to 3 μm. If the slope of $\log\frac{g_\mathrm{m}}{I_\mathrm{D}}nV_\mathrm{T}$ does not reach -1, only the long channel model without the velocity saturation is used.

The complete extraction algorithm is shown in Fig.5 and can be easily automated for a typical tool-based $g_\mathrm{m}/I_\mathrm{D}$ design procedure. The resulting model covers the V_DS dependence of the $g_\mathrm{m}/I_\mathrm{D}$ curves as well as the velocity saturation, if possible, with only three model parameters.

Fig. 6 and 7 show the results for the extracted parameters as well as the mean absolute relative error (MARE) for the long channel model compared to the SPICE simulations

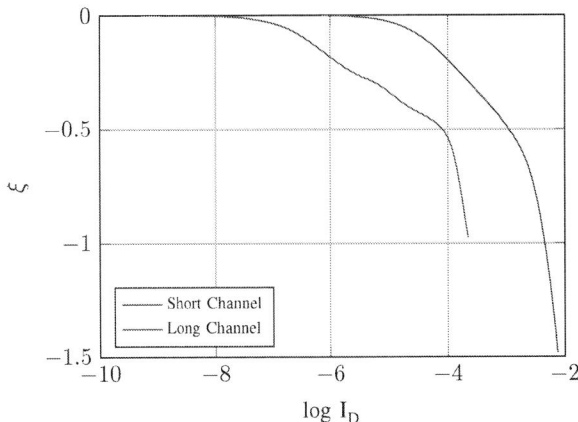

Fig. 4. Slope of $\log\frac{g_\mathrm{m}}{I_\mathrm{D}}nV_\mathrm{T}$ for the long and short channel device.

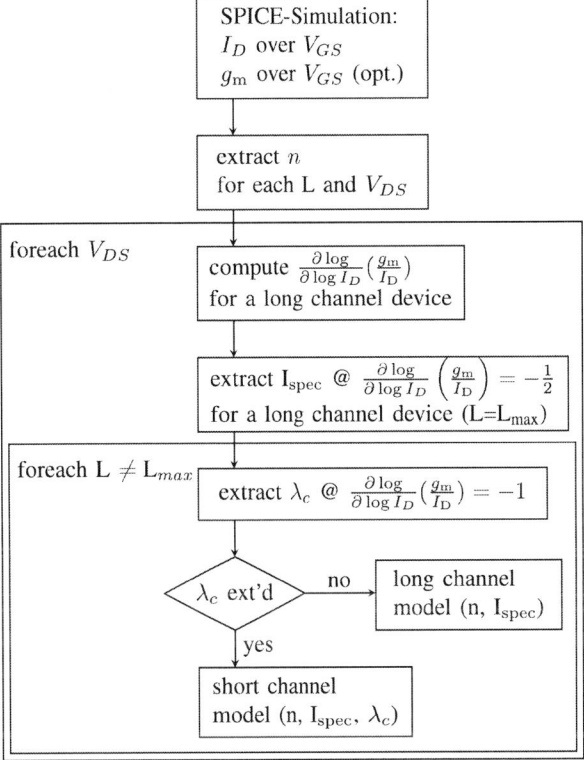

Fig. 5. Parameter extraction procedure from the SPICE database.

$$\mathrm{MARE} = \frac{1}{n_\mathrm{IC}}\sum_{\mathrm{IC}_i}\left|\frac{\left(\frac{g_\mathrm{m}}{I_\mathrm{D}}\right)_\mathrm{SPICE} - \left(\frac{g_\mathrm{m}}{I_\mathrm{D}}\right)_\mathrm{model}}{\left(\frac{g_\mathrm{m}}{I_\mathrm{D}}\right)_\mathrm{SPICE}}\right| \tag{15}$$

where n_IC is the number of simulation points over all inversion coefficients over a V_GS sweep.

As expected, n is almost constant except for very short devices and λ_c shows a L_{sat}/L dependence as [2] indicates. I_{sp} is constant over L and therefore only shown for the V_{DS} sweep. The MARE for the long channel model shows a good agreement of the model for large L and high V_{DS} with only two model parameters and a MARE below 0.04. However, for small device lengths and/or low V_{DS} the model error increases drastically up to a MARE above 1.

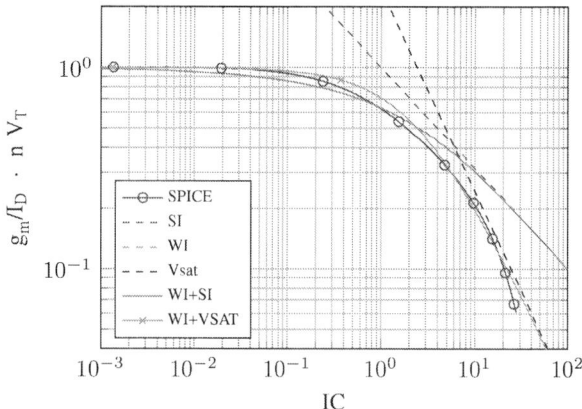

Fig. 8. $g_{\mathrm{m}}/I_{\mathrm{D}}$ from SPICE simulation and from the EKV model for $V_{\mathrm{DS}} = 0.4\,\mathrm{V}$, $L = 300\,\mathrm{nm}$ $n = 1.4572$ and $\lambda_c = 0.41218$.

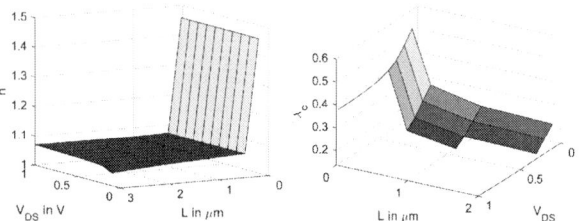

Fig. 6. Left: Extracted n. Right: Extracted λ_c

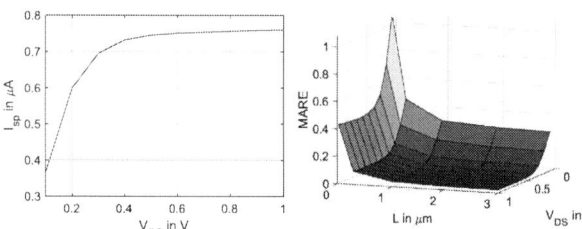

Fig. 7. Left: Extracted I_{sp} Right: MARE of the long channel model

The choice of the appropriate model for the short channel devices is not obvious. Fig. 2 shows the different $g_{\mathrm{m}}/I_{\mathrm{D}}$ curves for the models used in this work and some additional from [6], which share the same values for n, I_{spec} and λ_c and are capable of modelling short channel devices. It can be clearly seen that for very small and very large inversion coefficients in WI and VSAT the models converge to the same $g_{\mathrm{m}}/I_{\mathrm{D}}$ value and the same VSAT slope, respectively. For moderate inversion ($0.1 \leq IC \leq 10$), the $g_{\mathrm{m}}/I_{\mathrm{D}}$ values deviate for each model. Nevertheless, none of the models can describe the behavior of the slope in Fig. 4 adequately which goes below -1 for large I_{D} and IC, respectively. Therefore, the three models from eq. (6) to (8) are evaluated and compared to the SPICE simulations.

In order to evaluate the parameter extraction for the short channel models, the MARE for these models are compared in Fig. 9 to the MARE of the long channel model, which is lightly overlayed in (a) to (c). The view angle is different than in Fig. 7 in order to show the improvement of WI+VSAT ENZ, WI+SI+VSAT and WI+VSAT using λ_c over the WI+SI.

The short channel models are only defined for certain L and V_{DS} values where λ_c can be extracted. Nevertheless, this is the very operation region where the MARE for the long channel model increases drastically. WI+VSAT ENZ and WI+VSAT have very similar error characteristics with WI+VSAT performing slightly better. The MARE of WI+SI

reaches only a peak of slightly over 0.2 and is around 0.1 to 0.15 for the most L and V_{DS} values with a minimum of 0.05. Still, WI+VSAT and WI+VSAT ENZ support significant improvement compared to the long channel model. However, WI+SI+VSAT performs poorly compared to the other short channel models and gives a moderate improvement over the long channel model. The advantage of the WI+SI model over the other models, which take VSAT into account, is not in general and is very technology dependent. Therefore, one of the other two models presented here may be the better choice for other CMOS technologies. Fig. 8 shows the $g_{\mathrm{m}}/I_{\mathrm{D}}$ curves from the SPICE simulation for a short channel device compared to the resulting long channel model and a short channel model. It can be clearly seen, that the long channel model cannot model the steep slope of the $g_{\mathrm{m}}/I_{\mathrm{D}}$ curve for strong inversion, whereas the short channel model (WI+VSAT) recreates the characteristic also for the strong inversion.

IV. CONCLUSION

This paper proposes a systematic parameter extraction for a simplified EKV-model with only three model parameters and applies it to a 28nm FDSOI technology as an example. Compared to other approaches, this work evaluates the slope of the $g_{\mathrm{m}}/I_{\mathrm{D}}$ curve. The algorithm determines the extraction points automatically from the database and whether the velocity saturation has to be considered or the long channel model with only two model parameters is sufficient. In case of the 28nm FDSOI technology for moderate and high V_{DS} and large L, the technology current I_{sp} and the subthreshold slope n are sufficient for the model. For the velocity saturation λ_c three different model approaches were compared in order to improve the model accuracy. The FDSOI can be modelled with the WI+VSAT model with a mean absolute relative error slightly above 0.2 down to 0.05, depending on the transistor dimension and the drain-to-source voltage. In addition to that, two other models are presented, which are capable to model the same operation and design space with less performance. Nevertheless, this extraction procedure can easily be applied

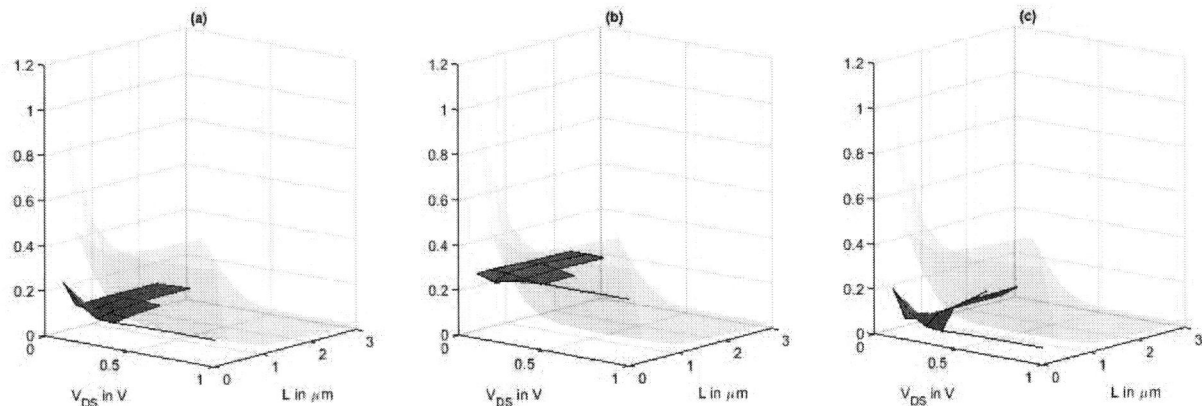

Fig. 9. Mean absolute relative error (MARE) for (a) WI+VSAT ENZ (b) WI+SI+VSAT (c) WI+VSAT with lightly overlayed MARE for long channel model.

to other CMOS technologies (i.e. bulk, FinFET) and is independent from the used SPICE simulation model. Furthermore, it can be easily automated and applied to already existing databases from tool based g_m/I_D design procedures without the need of additional SPICE simulation for measurements. While this resulting model is very simple and can be used for hand calculations, it is fairly accurate over the complete design space of the technology for a first design approach, even for nanometre scale feature sizes of the MOSFET and low voltage designs.

REFERENCES

[1] W. Sansen, "Minimum power in analog amplifying blocks: Presenting a design procedure," *IEEE Solid-State Circuits Magazine*, vol. 7, no. 4, pp. 83–89, 2015.

[2] C. Enz, F. Chicco, and A. Pezzotta, "Nanoscale mosfet modeling: Part 2: Using the inversion coefficient as the primary design parameter," *IEEE Solid-State Circuits Magazine*, vol. 9, no. 4, pp. 73–81, 2017.

[3] S. El Ghouli, D. Rideau, F. Monsieur, P. Scheer, G. Gouget, A. Juge, T. Poiroux, J.-M. Sallese, and C. Lallement, "Experimental gm/id invariance assessment for asymmetric double-gate fdsoi mosfet," *IEEE Transactions on Electron Devices*, vol. 65, no. 1, pp. 11–18, 2017.

[4] P. G. Jespers and B. Murmann, *Systematic Design of Analog CMOS Circuits*. Cambridge University Press, 2017.

[5] C. C. Enz and E. A. Vittoz, *Charge-based MOS transistor modeling: the EKV model for low-power and RF IC design*. John Wiley & Sons, 2006.

[6] W. Sansen, "Biasing for zero distortion: Using the ekv\/bsim6 expressions," *IEEE Solid-State Circuits Magazine*, vol. 10, no. 3, pp. 48–53, 2018.

[7] D. M. Binkley, M. Bucher, and D. Foty, "Design-oriented characterization of cmos over the continuum of inversion level and channel length," in *ICECS 2000. 7th IEEE International Conference on Electronics, Circuits and Systems (Cat. No. 00EX445)*, vol. 1. IEEE, 2000, pp. 161–164.

[8] C. Enz, F. Chicco, and A. Pezzotta, "Nanoscale mosfet modeling: Part 1: The simplified ekv model for the design of low-power analog circuits," *IEEE Solid-State Circuits Magazine*, vol. 9, no. 3, pp. 26–35, 2017.

[9] D. M. Binkley, "Tradeoffs and optimization in analog cmos design," in *2007 14th International Conference on Mixed Design of Integrated Circuits and Systems*. IEEE, 2007, pp. 47–60.

[10] W. Grabinski, D. Tomaszewski, F. Jazaeri, A. Mangla, J.-M. Sallese, M.-A. Chalkiadaki, A. Bazigos, and M. Bucher, "Foss ekv 2.6 parameter extractor," in *2015 22nd International Conference Mixed Design of Integrated Circuits & Systems (MIXDES)*. Icec, 2015, pp. 181–186.

[11] C.-M. Zhang, F. Jazaeri, A. Pezzotta, C. Bruschini, G. Borghello, S. Mattiazzo, A. Baschirotto, and C. Enz, "Total ionizing dose effects on analog performance of 28 nm bulk mosfets," in *2017 47th European Solid-State Device Research Conference (ESSDERC)*. IEEE, 2017, pp. 30–33.

[12] T. Schäfer, T. Hillebrand, N. Hellwege, M. Erstling, D. Peters-Drolshagen, and S. Paul, "Design and verification of analog cmos circuits using the gm/i d-method with age-dependent degradation effects," *Journal of Low Power Electronics*, vol. 13, no. 1, pp. 135–147, 2017.

[13] N. Hellwege, N. Heidmann, M. Erstling, D. Peters-Drolshagen, and S. Paul, "An aging-aware transistor sizing tool regarding bti and hcd degradation modes," in *2015 22nd International Conference Mixed Design of Integrated Circuits & Systems (MIXDES)*. IEEE, 2015, pp. 272–277.

Proceedings of the 27th International Conference *"Mixed Design of Integrated Circuits and Systems"*
June 25-27, 2020, Łódź, Poland

Qucs-S/QucsStudio/Octave Schematic Synthesis Tools for Device and Circuit Parameter Extraction from Measured Characteristics

Mike Brinson
Centre for Communications Technology
London Metropolitan University
UK
Email: mbrin72043@yahoo.co.uk

Abstract—**A universal technique for the extraction of device parameters or circuit component values from measured performance data is presented. The proposed method can be used by any circuit simulator that implements parameter sweep features and allows user defined tabulated data with independent voltage, or current, sources. A key feature of the reported extraction process is the use of schematic capture simulation icons, with their sweep parameters tabulated as a list of data points, synthesized from CSV measured data. By overlaying simulation output data on top of measured values, then varying user selected parameter/component values and re-simulating repeatable until the two data sets converge, it becomes possible to extract parameter/component values to within a specified error limit. In this paper FOSS circuit simulators Qucs-S/QucsStudio and the numerical analysis package Octave are used to demonstrate the application of the proposed schematic capture synthesis procedure in the investigate of diode inductance at high forward d.c. bias currents and a.c. signal band width.**

Keywords—**Qucs-S, QucsStudio, Octave. compact device modeling, parameter extraction, tuning, optimization**

I. INTRODUCTION

The extraction of device parameters and circuit component values from d.c., a.c., and transient measurements are significant steps in establishing the validity of device and circuit simulation models. One of the most important practical techniques used for this purpose relies on the comparison of measured and simulated output data where each dependent data set has a common independent axis selected from signal frequency (in the a.c., S, Z and Y domains) or time (in the transient domain) or swept parameter values (in all domains). The extraction process proceeds by overlaying simulation output data on top of measured data, varying user selected device parameters, or component values, then re-simulating the device/circuit under test until the two data sets line-up within a specified error limit. The Qucs-S/QucsStudio Free Open Source Software (FOSS) circuit simulators [1][2] allow individual or groups of parameters to be varied by "manual slider tuning" (QucsStudio) or by computer controlled optimization employing objective target functions (Qucs-S and QucsStudio). This paper introduces a groundbreaking parameter extraction technique which links measured and simulated output data via Qucs-S/QucsStudio

test bench schematics. To ensure that the independent X axis of the measured and simulated data have the same range and number of data points simulation is controlled by icons synthesized from the measured X scale data. The primary task of these icons is to set up and instantiate simulation while simultaneously ensuring that the independent axis of the measured and simulation output data are aligned automatically during parameter extraction. An overview of this process is shown diagrammatically in Fig.1. To illustrate the procedure data from a study of the admittance of a forward biased semiconductor pn junction diode is introduced, it's model parameters extracted, analyzed and commented on. In this example the pn junction is represented by a non-linear Verilog-A module that models diode inductance generated by conductivity modulation and frequency dependent minority carrier lifetime at high forward d.c. bias currents [3][4][5].

II. OCTAVE SYNTHESIS FUNCTIONS FOR THE EXTRACTION OF DEVICE AND CIRCUIT PARAMETERS FROM MEASUREMENTS

A key feature of the reported parameter extraction method is the use of Qucs-S/QucsStudio simulation icons with their independent sweep parameter tabulated as a list of data points, for example in the case of a.c. simulation variable "Sweep parameter" is set to acfrequency, variable "type" is set to list and vector "values" is set to a semicolon separated list of signal frequencies. The listed data points must be in ascending or descending order of magnitude. In those instances where only a few measured data points are needed, or indeed are available, they can be simply entered manually on a schematic diagram. In most cases however, this is impracticable due to the large number of data points, and it is better to convert CSV formatted data to a character separated number list using Octave [6]. Finally, after building the numerical lists, an Octave m function is used to synthesize one or more Qucs-S/QucsStudio simulation control icons. Although the icons shown in Fig.1 have different data formats, which largely depending on the simulation domain, the use of Octave to convert CSV tabulated measurements to Qucs-S/QucsStudio

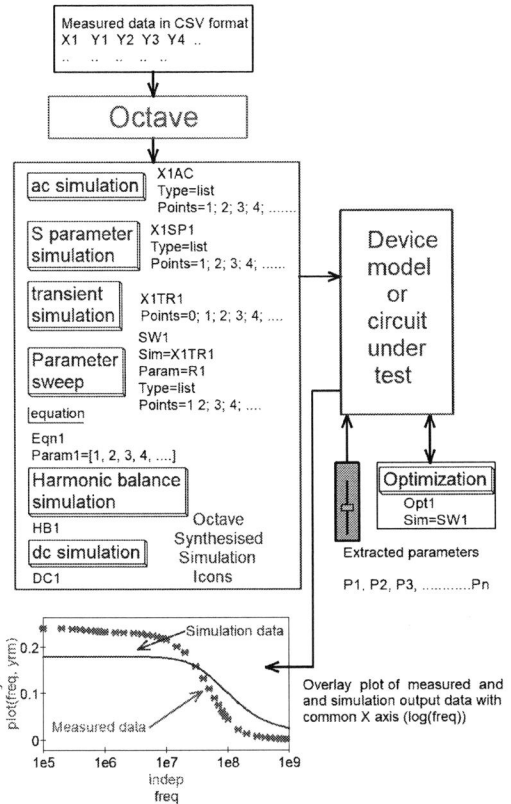

Fig. 1. A block diagram illustrating the fundamental stages for reading measured data, building extended independent and dependent variable lists and synthesizing sets of simulation control icons.

icon lists is in most cases similar. The example described in the text outlines the detail steps in the synthesis process. These can also be easily modified to change, for example, simulation type or add a higher number of dependent variables. The Octave m function given in Fig.2 lists the steps for synthesizing a set of d.c. simulation control icons applicable to extracting device parameters from measured diode Id/Vd characteristics. Sample sections of a typical diodeDCIV.csv data file and the initial dctemplate.sch template are also given in Fig.3.

III. THE EXTRACTION OF DIODE D.C. CIRCUIT PARAMETERS FROM MEASURED DATA

Illustrated in Fig. 4 is a QucsStudio test bench schematic for extracting diode best fit d.c. parameters Is and Rs, with $N = 1.0$, from the Id/Vd overlay graphs plotted in Fig.4 (b). These parameters are defined in Table I and equation 5. The parameter tuning sliders drawn in Fig.4 (c) can be adjusted manually to obtain the best fit by visual comparison of the measured and simulated output data plots. Synthesized simulation icons formed from a tabulated list of measured data are represented in Fig.3 (c) by the normal Qucs-S/QucsStudio schematic symbols with attached horizontal lists of data points separated by a semicolon or comer deliminator character. The

synthesized icons are generated, and stored in file DCsimTemplate.sch, by simulating Octave m function dcextractxy. Note that the simulation icon lists shown in Fig.3(c) have, for convenience, been truncated at the right hand side of the schematic diagram. Note also that the schematic illustrated in Fig.4 is different from the classical SPICE 3f5 diode model [7] in that it has an additional terminal F that senses signal frequency during simulation. In d.c. simulation the voltage on terminal F is set to 0 V d.c. to represent 0 Hz. Similarly, during other types of simulation it is set to a real voltage that represents the signal frequency in Hertz. Fig.5 in contrast to Fig.4 gives the details of an optimization simulation icon, its control settings and typical best fit output data, where in most instances, the slider extracted parameters act as initial values for a computer optimization refined fit.

```
function dcextractxy();
% An Octave function to extract measured x values (independent) and
% y (dependent) values from an n row, 2 column csv table. Extracted x
% and y lists are synthesized into a set of d.c. simulation icons.
% (C) 2020 Mike Brinson: Published under GNU General Public License
% Function dcextractxy() was developed from:
% fillmeas.m 1.0 (C) Z. Huszka 09-July-2019, and
% fillmeasurement1dep.m (C) Mike Brinson 11 October 2019.
% ==== Initialize variables =
measureddcdata="diodeDCIV.csv"; dctemplate="DCtemplate.sch";
dcsimtemplate="DCsimTemplate.sch";  csvdelim=";";
listdcx=[]; listdcy=[];
fidread=fopen(dctemplate,"r"); fidwrite=fopen(dcsimtemplate,"w");
% === Extract measured d.c. data ===
[dcx, dcy ]=textread(measureddcdata,"%f%f","delimiter",csvdelim);
% === Build modified x and y lists ===
for k=1:length(dcx)
  listdcx=[listdcx,num2str(dcx(k)),";"];
end
listdcx(end)=[];
for k=1:length(dcy)
  listdcy=[listdcy,num2str(dcy(k)),","];
end
listdcy(end)=[];
% === Synthesize modified simulation control icons  ===
while 1
  if findstr(line,"Eqn")
    found1=findstr(line,"["); found2=findstr(line,"]");
    line=[line(1:found1),listdcy,line(found2:end)];
    fprintf(fidwrite,"%s\n",line);
  elseif findstr(line,".SW")
    found=findstr(line,"");
    line=[line(1:found(end-1)),listdcx,line(found(end):end)];
    fprintf(fidwrite,"%s\n",line);
  else
    fprintf(fidwrite,"%s\n",line);
  end
  line=fgetl(fidread);
  if feof(fidread)
    if ischar(line)
      fprintf(fidwrite,"%s\n",line);
    end
    break
  end
end
fclose(fidread); fclose(fidwrite);
display("dcextractxy finished without error.\n");
return
```

Fig. 2. Octave function dcextractxy.m code: Works with QucsStudio netlist syntax. Modify as necessary for Qucs-S netlists.

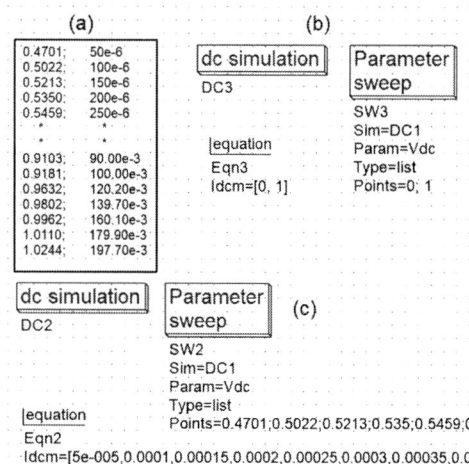

Fig. 3. Input and output data for Octave function dcextractxy(): (a) Measured input data, (b) dctemplate file "diodeDCIV.csv"; (c) dcsimtemplate file ;"DCsimTemplate;"

IV. MODELING SEMICONDUCTOR PN JUNCTION DIODES OF HIGH FORWARD D.C. BIAS OVER WIDE FREQUENCY BANDWIDTH

Conventional semiconductor diodes are constructed with metal contact terminals connected to short lengths of bulk or doped semiconductor material on either side of a pn junction. These add a finite amount of electrical resistance in series with the pn junction. The basic SPICE diode model represents this series resistance as a fixed parasitic resistance Rs. In reality however, at high d.c. currents, minority carrier charges accumulate in the series semiconductor material causing changes in the minority carrier density. Hence, as the diode d.c. current increases or decreases the resistance of the bulk/doped semiconductor varies. This process is often referred to as conductivity modulation [8] where R_0 is the intrinsic bulk semiconductor material resistance with no extrinsic charge, K_r is a real number that changes for different types semiconductor material, doping level and device geometry, I_d is the diode d.c. bias current, and the ratio K_r/R_0 is called the coefficient of conductivity modulation. One of the consequences of conductivity modulation in semiconductor diodes at high d.c. bias currents and high a.c. signal frequencies is the generation of an inductive component in the diode admittance $Y_d = Y_r + j \cdot Y_i(\omega)$ [9], where Y_r and Y_i are the real and imaginary components of the diode admittance respectively. The high frequency $Ldiode$ compact model in Fig.6 illustrates the structure and hierarchy of an experimental model that includes the basic conventional SPICE diode elements R_s, I_d and capacitance represented by nonlinear depletion and diffusion charges $Qdep$ and $Qdiff$ respectively. In the frequency domain resistor R_s models conductivity modulation and is taken to be a function of I_d. Similar to the conventional SPICE diode model it is considered to be a fixed value in the nonlinear d.c. domain. Experimental compact model $Ldiode$ models the

device current phase shift at high frequencies by a simple algebraic function of the diode minority carrier lifetime (Tt) and a.c. signal frequency F. The physical properties of the experimental $Ldiode$ model are represented by the following equations

$$Id = Is \cdot \left(exp\left(\frac{V_d}{n \cdot VTH} \right) - 1.0 \right) \quad (1)$$

$$IRx = V_d/RGMIN \quad (2)$$

$$Idep = C0 \cdot \frac{Vj}{1-M} \cdot \frac{d}{dt}\left[\left(1 - \frac{Vd}{Vj}\right)^{-M} \right], \quad (3)$$

$$\forall \quad Vd <= 0V$$

$$Idep = C0 \cdot \frac{d}{dt}\left(Vd + \frac{M \cdot Vd^2}{2.0 \cdot VJ} \right), \quad (4)$$

$$\forall \quad Vd > 0V$$

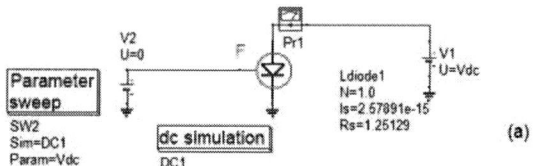

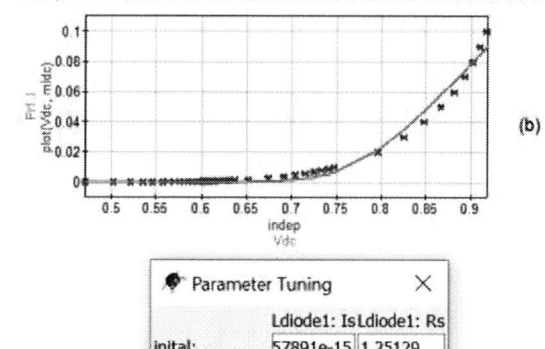

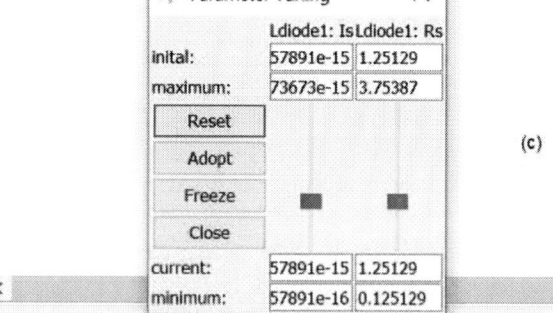

Fig. 4. Model Ldiode d.c. parameters $N = 1.0$, Is and Rs extracted using Octave m function dcextractxy() and manual "tuning sliders": (a) QucsStudio test bench schematic, (b) measured Id/Vd data (* plot) and simulation output data (solid line) overlay plots, and (c) slider settings displayed along side schematic.

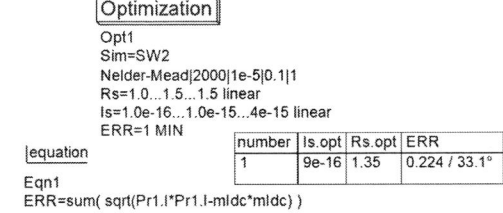

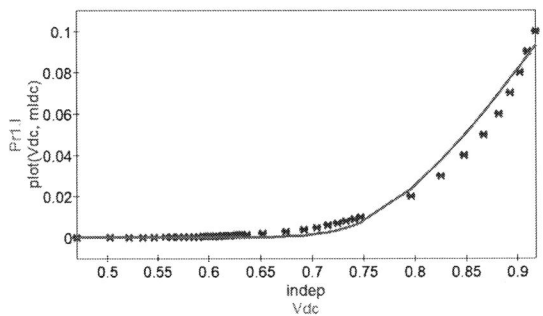

Fig. 5. Model Ldiode d.c. parameters N, Is and Rs extracted using Octave function dcextractxy(): QucsStudio optimization analysis using Nelder-Mead algorithm and objective minimized goal using target function $sum(sqrt((PrI1.I * PrI1.I) - (mIdc * mIdc)))$; measured Id/Vd and simulation output data overlay plots; best fit parameters $Is = 9e\text{-}16A$ and $Rs = 1.25|omega$.

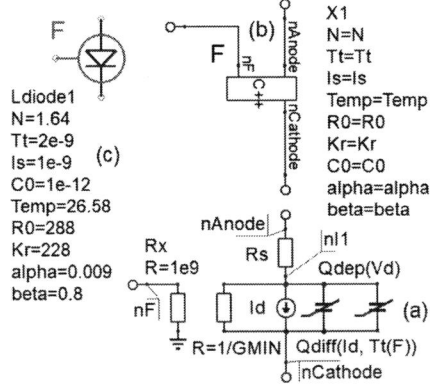

Fig. 6. Ldiode compact model: (a) electrical equivalent circuit, (b) Verilog-A synthesized C++ code block and (c) schematic symbol.

$$R_s = \frac{R_0}{(1 + K_r . I_d)} \quad (5)$$

$$Idiff = \frac{Id}{(N \cdot VTH)} \cdot Tt(F) \cdot \frac{dVd}{dt} \quad (6)$$

Where $Tt(F) = Tt \cdot \left(1.0 + \alpha \cdot F^\beta\right)$ and F is the a.c. signal frequency in Hz and $GMIN$ is the minimum conductance supported by Qucs-S/QucsStudio (normally in the range 1e-12S to 1e-9S). Other symbols are defined in Fig.6 or Table I. Fig. 7 lists the Verilog-A module code [10][11] for the experimental *ldiode* module introduced in Fig. 6. This module is structured around a subset of the basic SPICE diode model

with extensions in the forward .d.c. bias region to account for conductivity modulation at high currents. To ensure correct a.c. performance of the device in the reverse bias region a simplified depletion capacitance model is included in the *ldiode* module Verilog-A code.

TABLE I
LDIODE COMPACT MODEL PARAMETER VALUES

Name	Description	Unit	Default
N	Emission coefficient		1.64
Is	Saturation current at Temp.	A	1e-9
Tt	Minority carrier lifetime at Temp.	s	2e-9
$R0$	Bulk semiconductor resistance at Temp.	Ω	288
Kr	Value depends on doping and geometry	$1/A$	226
α	Conductivity modulation coeff.		0.009
β	Conductivity modulation power coeff.		0.8
M	Grading coefficient		0.5
Vj	Junction potential	V	0.7
$Temp$	Diode temperature	Celsius	27

```
`include "disciplines.vams"
`include "constants.vams"
module DiodeModel9(nAnode, nCathode, nfreq);
inout nAnode, nCathode, nfreq;
// (C) 2020 Mike Brinson: Published under GNU
//  General Public License V2 or later.
electrical nAnode, nCathode, nfreq, ni0, ni1, ni2;
parameter real N = 1.44; parameter real Tt = 2e-9;
parameter real Is = 5.6e-10; parameter real Temp=26.58;
parameter real R0 = 22.5; parameter real Kr = 180;
parameter real C0 = 1e-12; parameter real M = 0.5;
parameter real Vj = 0.7; parameter real alpha = 0.01;
parameter real beta = 2.0; parameter real Rs = 0.5;
branch (nAnode, ni1) B1; branch (ni1, nCathode) B2;
branch (nfreq)  B3;
real TempK, VTH, Id, GMIN;
analog begin
GMIN = 1e-12; TempK = Temp+273.15; VTH = (`P_K/`P_Q)*TempK;
Id =  Is*(limexp(V(B2)/(VTH*N))-1.0) + GMIN*V(B2);
if ( V(B3) <= 0.0)
  begin
  I(B1)  <+ V(B1)/(Rs+1e-9); I(B2)  <+ Id; I(B2)  <+ ddt(Tt*Id);
  I(B2)  <+ ddt(C0*(Vj/(1.0-M))*pow( (1.0-V(B2)/Vj), -M));
  end
else
begin
  I(B1)  <+ V(B1)*(1.0/(R0/(1+Kr*Id))); I(B2) <+ Id;
  I(B2)  <+ ddt(V(B2)*Id*Tt*(1.0+alpha*pow(V(B3), beta))/(N*VTH));
  I(B2)  <+ ddt(C0*(V(B2)+M*V(B2)*V(B2)/(2.0*Vj)));
end
I(B3)  <+ V(B3)*1e-9;
end
endmodule
```

Fig. 7. Verilog-A module DiodeModel9 for the experimental diode model *Ldiode*: with F=0.0 Hz the model reverts to a basic SPICE level 1 diode model. Noise and reverse bias breakdown effects are not modeled

V. MEASUREMENT OF DIODE ADMITTANCE, OPERATING AT HIGH D.C. FORWARD BIAS, OVER THE FREQUENCY RANGE 50KHZ TO 1GHZ.

A fundamental test bench for measuring or simulating the admittance of a forward biased diode is given in Fig. 9. Central to the test set up is the forward biased diode with separate d.c. and a.c. signal supplies. These signals are isolated by the d.c. blocking capacitor Cb and resistor Rm. Voltage source

Vm supplies d.c. power to the test circuit to set the diode d.c. bias current in the range 4mA to 100mA. The value of Rm is set at each of the specified measurement/simulation d.c. current level to be at least twenty times the diode d.c. forward resistance, yielding a high degree of isolation between the a.c. and d.c. signals. This simple form of bias tee network has been chosen to allow measurements of a satisfactory accuracy over the measurement/frequency range 50kHz to 1GHz. Values for the real and imaginary parts of diode admittance were obtained by measuring a wide band diode S11 parameters, at each of the different d.c. forward bias states with a vector network analyzer [14], followed by conversion of the measured data to CS tables composing n rows by 3 columns (F, Y_r, Y_i). The test bench in Fig. 9 gives details of the corresponding simulation test bench and the conversion of S11 values to admittance, see equation eqn2.

TABLE II
LDIODE EXTRACTED R, Kr AND α PARAMETER VALUES FOR DIFFERENT D.C. BIAS CURRENTS

name	100mA	60mA	30mA	10mA	5mA	4mA
kr(1/A)	927.2	1374.62	2130.6	5341.6	9828.5	10674
α(1/Hz)	6.8e-5	5.5e-5	3.9e-5	2.9e-5	2.79e-5	2.5e-5

TABLE III
LDIODE EXTRACTED F_r AND Y_i VALUES AT DIFFERENT D.C. BIAS CURRENTS

name	100mA	60mA	30mA	10mA	5mA	4mA
F_r(Hz)	1e7	1.5e7	3.5e7	8e7	12.9e7	17.4e7
$Y_i(\omega)$	-0.302	-0.229	-0.138	-0.076	-0.049	-0.041

current. Plots of a typical set of measured/simulation data are illustrated in Fig.8. The simulated values for Y_r and Y_i are shown fitted to the measured data. These were obtained with $R0 = 1.2e4$ and QucsStudio "tuning sliders" varying parameters Kr and α. The other parameters were either extracted from d.c. measurements (N=1.0, Is= 9.0e-16 A) or assumed to be typical physical values for a broad band diode (Tt=6.9e-12s, $C0$=1.0e-12F, M=0.5, Vj=0.7). In the case of parameter β a value of 0.87, approaching a linear function of F, was found to provide a reasonable fit to the measured data, particularly at frequencies in the mega Hertz region. At d.c. forward bias currents above 3 mA the imaginary part of the diode admittance Y_i shows inductive properties. Although, the sections of the $ldiode$ model that represent conductivity modulation and frequency dependent carrier lifetime are very simple the fit between measured and simulated data is good up to frequencies in the 100

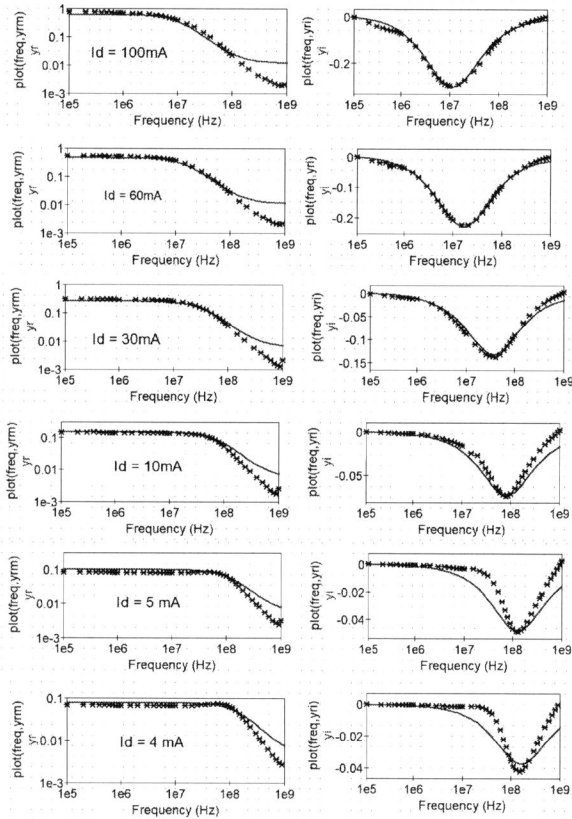

Fig. 8. Forward biased diode admittance plots for 4mA d.c. $<=$ Id $<=$ 100mA d.c.: plots in the left column are for Y_r (Ω) versus frequency (Hz) and plots in the right column are for Y_i (Ω) versus frequency (Hz): d.c. forward bias currents are shown on each row of plots. Solid lines represent simulation output and "*" measured data.

VI. DISCUSSION

Extraction of the diode parameters from measured F, Y_r and Y_i data requires one test bench per d.c. forward bias

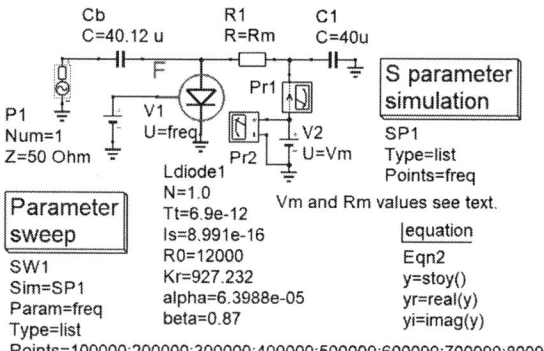

Fig. 9. A test bench for measuring or simulating the admittance of a forward biased diode in the range 4mA to 100mA over the frequency range 50kHz to 1GHz.

MHz region. Above 100 Mhz significant deviations occur, particularly beyond the clearly visible resonance in the Y_i plots. Values for the extracted parameters Kr and α are listed in Table II. Similarly, Table III gives values for the observed resonance frequencies (F_r) and the associated Y_i data. Further improvement in the data fit, if required, are potentially possible with the optimization extension to parameter extraction based on the initial parameter values extracted with "slider tuning", see Fig. 6.

VII. CONCLUSION

The ability to extract device or circuit parameter values from measured data is an important attribute of both circuit simulation tools and compact models. This paper introduces a new technique for parameter extraction from measured data using synthesized schematic capture icons that control circuit simulation and play a central role in the parameter extraction process. Provided a circuit simulator implements swept parameter/component value facilities and allows user defined data for voltage, or current, sources the proposed technique can be applied universally across simulators. When combined with Octave the development of more versatile compact modeling tools becomes a definite possibility, opening up a number of significant routes for future research. To demonstrate the proposed parameter extraction technique an example is introduced in the text that illustrates its use to investigate the inductive properties of d.c. forward biased semiconductor diodes.

REFERENCES

[1] V. Kusnetsov and M. Brinson, "Qucs-S: Qucs with SPICE". Version 0.0.22, https://ra3xdh.github.io/, 2020. [Accessed January 2020].

[2] M. Margraf, "QucsStudio Version 2.5.7, http://dd6um.darc.de/ QucsStudio/qucsstudio.html, 2018, [Accessed January 2020].

[3] W. Shockley, "The Theory of p-n Junctions in Semiconductors and p-n Junction Transistors", 1949, Bell System Technical Journal, pp. 435-489.

[4] A.A Barna and D. Horelick, "A Simple Diode Model Including Conductivity Modulation", IEEE Transactions on Circuit Theory, Vol. CT-18, No. 2, 1971, pp. 233-240.

[5] L. Ladany, " An analysis of inertial inductance in a junction diode", IRE Trans. Electron Devices, Vol. ED-7, 1960, pp. 303-310.

[6] John W. Eaton, David Bateman, Søren Hauberg and Rik Wehbring. GNU Octave version 5.1.0 manual: a high-level interactive language for numerical computations. 2019, CreateSpace Independent Publishing Platform. ISBN 1441413006, URL http://www.gnu.org/software/octave/doc/interpreter/, [Accessed January 2020].

[7] B. Johnson, T. Quarles, A.R. Newton, D. O. Pederson and A Sangiovanni-Vincentelli, "SPICE3 Version 3f User's Manual", Department of Electrical Engineering and Computer Science, University of California, Berkeley, CA, 1992.

[8] W.H Ko, "The forward transient behaviour of semiconductor junction diodes", Solid-State Electron., Vol. 3, 1961, pp. 59-69.

[9] Y. Kanai, "On the inductive part in the ac characteristics of the semiconductor diodes", J. ///phys. Soc. Japan, vol. 10, 1955, pp. 718-720.

[10] Accellera, "Verilog-AMS Language Reference Manual, version 2.2", available from: http://www.accellera.org, [Accessed February 2019].

[11] L. Lemaitre. C. McAndrew and S. Hamm, "ADMS- Automatic device model synthesizer", IEEE Custom Integrated Circuits Conference, 2002, pp. 27-30.

[12] DG8SAQ, "VNWA 3E Vector Network Analyzer", SDR-Kits Ltd. 2013, Trowbridge Wilts., UK. Available from https://www.sdr-kits.net.

```
function acyextract();
% (C) 2020 Mike Brinson: Published under GNU GPI V2.0 or later.
% Developed from: fillmeas.m 1.0 (C) Z. Huszka 09-July-2019, and
% fillmeasurement1dep.m (C) Mike Brinson 11 October 2019.
% == Initialize variables ==
measuredacdata="ywb4m.csv";
actemplateyr="actemplateyr.sch"; actemplateyi="actemplateyi.sch";
acsimtemplateyr="acsimtemplateyr.sch";
acsimtemplateyi="acsimtemplateyi.sch";
csvdelim=";"; listacx=[]; listacyr=[]; listacyi=[];
% == Extract measured d.c. data ==
[acx, acyr, acyi ]=textread(measuredacdata,"%f%f%f",
    "delimiter",csvdelim);
% == Build modified x and y lists ==
for k=1:length(acx) listacx=[listacx,num2str(acx(k)),";"]; end
listacx(end)=[];
for k=1:length(acyr) listacyr=[listacyr,num2str(acyr(k)),","]; end
listacyr(end)=[];
for k=1:length(acyi) listacyi=[listacyi,num2str(acyi(k)),","]; end
listacyi(end)=[];
% == Synthesize AC Yr Eqn and frequency simulation icons
fidreadyr=fopen(actemplateyr,"r");
fidwrite=fopen(acsimtemplateyr,"w");
line=fgetl(fidreadyr);
while 1
  if findstr(line,"Eqn") found1=findstr(line,"[");
    found2=findstr(line,"]");
    line=[line(1:found1),listacyr,line(found2:end)];
    fprintf(fidwrite,"%s\n",line);
  elseif findstr(line,".SW")
    found=findstr(line,"");
    line=[line(1:found(end-1)),listacx,line(found(end):end)];
    fprintf(fidwrite,"%s\n",line);
  else
    fprintf(fidwrite,"%s\n",line);
  end
  line=fgetl(fidreadyr);
  if feof(fidreadyr)
    if ischar(line) fprintf(fidwrite,"%s\n",line);
    end
    break
  end
end
fclose(fidreadyr); fclose(fidwrite);
display("acextractyr finished.\n");
% == Synthesize AC Yi Eqn simulation control icon  ==
fidreadyi=fopen(actemplateyi,"r");
fidwrite=fopen(acsimtemplateyi,"w");
line=fgetl(fidreadyi);
while 1
  if findstr(line,"Eqn") found1=findstr(line,"[");
    found2=findstr(line,"]");
    line=[line(1:found1),listacyi,line(found2:end)];
    fprintf(fidwrite,"%s\n",line);
  else
    fprintf(fidwrite,"%s\n",line);
  end
  line=fgetl(fidreadyi);
  if feof(fidreadyi)
    if ischar(line) fprintf(fidwrite,"%s\n",line);
    end
    break
  end
end
fclose(fidreadyi); fclose(fidwrite);
display("acextractyi finished.\n");
return
```

Fig. 10. An Octave function to extract measured frequency values (independent) and Yr and Yi (dependent) values from an n row, 3 column CSV table. Extracted x and y lists are synthesized into a set of a.c. simulation icons.

Design of Integrated Circuits and Microsystems

Proceedings of the 27th International Conference *"Mixed Design of Integrated Circuits and Systems"*
June 25-27, 2020, Łódź, Poland

A Capacitive Feedback 80 dBΩ 1.1 GHz CMOS Transimpedance Amplifier with Improved Biasing

Agata Romanova[1], Vaidotas Barzdenas[1,2]

[1]Department of Computer Science and Communications Technologies
Vilnius Gediminas Technical University, Vilnius, Lithuania
[2]Micro and Nanoelectronic Systems Design and Research Laboratory
Vilnius Gediminas Technical University, Vilnius, Lithuania
{agata.romanova,vaidotas.barzdenas}@vgtu.lt

Abstract—**The work presents the design of an area-efficient low-noise high-performance CMOS transimpedance amplifier for optical time-domain reflectometers. The proposed solution is based on a low-noise capacitive feedback structure and shows a gain of 83/80 dBΩ with the bandwidth reaching 1.1 GHz and average input-referred noise current density below 1.8 pA/$\sqrt{\text{Hz}}$ in the presence of a 0.7 pF total input capacitance. The noise-efficient feedback structure allows addressing noise problem of conventional feed-forward or resistive feedback devices with the total power consumption around 21 mW while running at 1.8 V power supply. A more accurate design methodology is proposed based on explicit modeling of the biasing circuits and decoupling capacitor and modifications to the reference design are suggested including circuits for PMOS-based biasing and DC current elimination.**

Keywords—**Analog integrated circuits, broadband amplifiers, CMOS integrated circuits, optical time-domain reflectometry, transimpedance amplifier.**

I. INTRODUCTION

Constantly growing data rates in communication systems had led to a wide adoption of optical fibers as the core technology for data transmission. As a result, the popularity of optical communication systems also triggered an increase in demand for the related instrumentation and maintenance equipment for systems' monitoring and repairs including the instruments such as Optical Time-Domain Reflectometers (OTDR) [1], [2]. These instruments are the well-established tools which are typically employed for fault detection and characterization in optical fiber links. The technique allows finding a precise location of the fault along with its nature, where the required information is deduced from the time-domain analysis of the reflected signals after a set of optical pulses are injected into the fiber. The measurement principle permits the user to estimate the fiber's loss characteristics as a function of the fiber length as well as to identify the possible causes of the problems in fiber.

The structure of the front-end of a typical OTDR device strongly resembles the one of the optical receiver. Here the front-end Transimpedance Amplifier (TIA) can be considered as the most critical part of the instrument as its performance often limits the overall sensitivity and the noise level of the measurement equipment. Although a number of CMOS TIAs have been already proposed for optical communication systems (see [3] for an overview), the OTDR context puts a set of additional constraints which have to be taken into account when compared to classical TIA topologies often employed for data transmission in high-speed optical communication systems. As the latter have been historically the major driving force for TIA development towards higher speeds, the noise requirements as well as those for the flatness of the pass-band and device linearity, have been eventually set for a lower priority in classical data transmission applications. However, exactly those are considered as important requirements for OTDR and shall be properly emphasized while selecting a candidate TIA topology for product implementation. Furthermore, while targeting low-cost moderate performance equipment one shall also consider a maximally area-efficient design under the rest of the performance constraints.

The TIA is typically a fast current measurement device which is responsible for converting relatively weak input current (e.g. from a diode or, in general, from any sensor with current output) to the output voltage of the amplitude which is sufficient for subsequent signal processing and analysis. Although the term TIA typically evokes an image of a so-called resistive shunt-feedback (SFB) TIA [3], this is only one realization of the general TIA concept. This design, however, gained almost a universal adoption and is, therefore, often considered as a default choice due to its reasonable balance of the most important performance characteristics such as transimpedance gain, bandwidth and noise [4]. Nevertheless, several OTDR-specific requirements are likely to make the design of a suitable higher-performance TIA being a challenging task when compared to a problem of designing a CMOS TIA for optical data transmission. Here, the requirement of a relatively low-noise eliminates straightforward high-speed feed-forward TIA circuits, while opting for an area-efficient solution with flat frequency response (less than 0.5 dB) is likely to limit or, at least, make less attractive the applicability of the intensive bandwidth enhancement methods such as those based on inductive peaking [5] or similar strategies. On the other hand, the classical bandwidth extension techniques may be not even relevant anymore as, in general, the performance of modern commercially available CMOS is more than enough for intended OTDR applications with requirements of around 1 GHz. Finally, an additional constraint comes from the requirement to employ the off-chip photodetectors. As a result, the circuit has to operate with the input capacitance C_T larger than 0.7 pF including not only the photodiode's capacitance itself, but also the rest of the input parasitics such as those due to ESD circuits, bond pads, etc. The presence of a large input capacitance at the input of the classical TIA circuits may result in inherent bandwidth limitations due to the dominant pole in TIA's transfer function and this issue may need to be addressed separately to avoid significant bandwidth constraints in classical TIA designs [6]. A relatively large gain of 10 kΩ must be delivered within the bandwidth of 1 GHz and input-referred noise density below 5 pA/$\sqrt{\text{Hz}}$ shall be ensured within the bandwidth of interest.

Historically, higher performance and low-noise TIAs have been developed using non-CMOS technologies such as GaAs, HEMT, HBT and SiGe BiCMOS. Although those have been often chosen due to their excellent noise, bandwidth and gain performance [7], here the decision have been made to employ a commercial deep submicron CMOS both due to its cost advantages and an inherent support for high integration. The latter is an important feature as it enables the performance-enhancing functionality including complex biasing circuits, advanced ESD designs etc., something which may be not commercially feasible for available III-V technologies. Even though CMOS is an excellent candidate for cost-saving purposes, the technological constraints may result in certain challenges which make a design of a low-cost CMOS analog circuit fairly non-trivial. Such challenges may include severe parasitic capacitances, limitations in supply voltage, restrictions on available transimpedance and poorer noise performance [8].

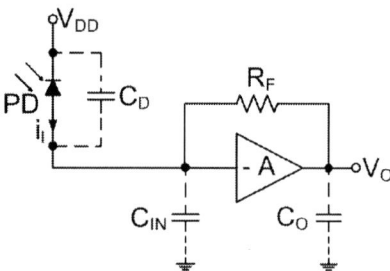

Fig. 1.   Basic configuration of resistive SFB TIA.

Any discussion on CMOS TIAs shall start with a brief comment on a classical SFB TIA as depicted in Fig. 1. The design results in low input impedance, as is beneficial for a current sensing block and, at the same time, low output impedance as is important for voltage output amplifiers [9]. Numerous works have reported modifications on this classical architecture such an additional feedback capacitor in parallel with R_F for flat frequency response, different configurations of core voltage amplifier, etc. Unfortunately, not all inherent limitations of SFB TIA could be effectively addressed with such simple means (e.g. sensitivity to input parasitic capacitance, stability issues) and alternative architectures have been also suggested. Among them, one shall mention a feed-forward common-gate

(CG) or more advanced regulated cascade (RGC) approaches. Even though they have solved some of the problems typical for SFB TIA, they introduced drawbacks on their own such as an increase in noise levels which may fundamentally forbid usage of these architectures in OTDR instruments with significant noise level constraints. Finally, although monolithic inductors have been often used for performance improvement in TIA, they may not only increase the area of the chip, but can also result in practical challenges of maintaining their inductive characteristics within the intended bandwidth and, due to substrate coupling, may also cause higher cross-talks when compared to inductor-less designs.

Due to a relatively small market volume for OTDR instruments, the amplifiers used in them were historically implemented using discrete components with the bandwidth often limited by 70 MHz. The latter was often caused by large parasitic capacitances and resulted in relatively poor performance of such low-cost instruments with custom TIA designs [2]. Obviously, here an integrated TIA will result in significantly reduced parasitic capacitances and may bring to a noticeable improvement in OTDR performance due to increase in bandwidth when compared to customized discrete solutions and an ability to deliver an optimal amplifier configuration. Few TIA designs have been reported specifically addressing OTDR requirements. The authors in [2] proposed a fully differential SFB TIA with variable gain implemented using feedback resistors connected in series. Unfortunately, the authors did not go beyond this classical architecture and typical issues of this reference topology including low noise performance and sensitivity to C_T have remained unaddressed. Several works reported on integrating of TIA into a complete lower performance OTDR ASICs [7], [10]. In these cases the TIA was only a minor part of the complete chip and either no specifics on TIA design were reported [10] or a classical SFB TIA with additional feedback capacitor was used [7]. Both works addressed low-cost single-chip OTDR applications with, apparently, no special emphasis on the noise performance of the front-end TIA.

II. CIRCUIT DESIGN

The classical SFB TIA has an inherent problem with the noise performance as the noise current of the feedback resistor R_F is directly added to the input-referred noise current. As the feedback approach itself has important structural benefits, one may want to preserve it while replacing the noisy R_F with a noise-free element such as a capacitor. Here, instead of using direct capacitive feedback, which results in a phase shift and, thus, needs a phase correction [3], we follow an approach first demonstrated by Razavi in [11] and shown in Fig. 2.

Here the voltage across C_1 is sensed by C_2 and is returned as a proportional current to the input. Assuming high gain $A \gg 1$, the current gain becomes approximately:

$$\frac{I_{out}}{I_{in}} \approx 1 + \frac{C_1}{C_2}. \tag{1}$$

The circuit operates as a current amplifier and, with the resistor R_2 connected to the drain of the output transistor M_2 (with

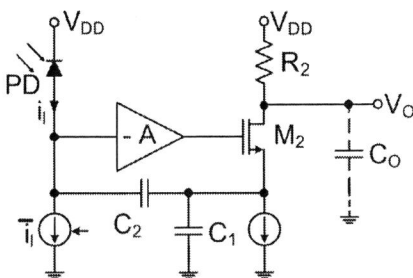

Fig. 2.   Capacitive feedback TIA.

transconductance $g_{m,2}$), the gain for low frequencies can be approximated as [9]:

$$R_T = \left(1 + \frac{C_1}{C_2}\right) R_2. \qquad (2)$$

This simple expression as derived by the feedback circuit assumes an infinite forward gain A with the overall gain determined solely by the feedback circuit. Furthermore, the approximation also does not consider possible coupling via the biasing circuit at the input and results in significant overestimate of the gain when compared to realistic circuit implementation under CMOS voltage headroom constraints. Although the expression above provides an important insight into the underlying principles of the proposed design, it can hardly serve as the basis for the circuit design methodology and a more elaborated model is definitely necessary.

The overall voltage transfer function V_T of the circuit can be derived from the combined forward gain $G_1 G_2 G_3$ and the feedback gain F_1 in addition to the output feed-out gain G_{out}:

$$V_T = \frac{V_{out}}{V_{in}} = \frac{G_1 G_2 G_3}{1 + G_1 G_2 G_3 F_1} G_{out}, \qquad (3)$$

where using the complex variable $s = j\omega$ one defines the gain for the input CS stage with M_1 and R_1 as:

$$G_1 = \frac{V_{d4}}{V_{in}} = g_{m,1} R_1, \qquad (4)$$

and:

$$G_2 = \frac{V_{g2}}{V_{d4}} = \frac{s R_{bias,2} C_c}{1 + s R_{bias,2} C_c}. \qquad (5)$$

The latter factor is responsible for the high-pass behavior due to coupling capacitor C_c and second biasing resistor $R_{bias,2}$. The last gain factor for the source-follower M_2 with:

$$G_3 = \frac{V_{s2}}{V_{g2}} = \frac{g_{m,2} r_{DS5}}{1 + g_{m,2} r_{DS5}} \cdot \frac{1}{1 + s \frac{r_{DS5} C_1}{1 + g_{m,2} r_{DS5}}}$$

$$\approx \frac{1}{1 + s \frac{C_1}{g_{m,2}}}, \qquad (6)$$

and possible further simplifications due to $g_{m,2} r_{DS5} \gg 1$. The feedback component becomes approximately:

$$F_1 = \frac{V_{in}}{V_{s2}} = \frac{s R_{bias,1} C_2}{1 + s R_{bias,1} (C_{IN} + C_2)}, \qquad (7)$$

where C_{IN} is the total input capacitance and $C_2 \ll C_1$ so that its path can be neglected. Finally one gets:

$$G_{out} = R_2 \frac{1 + s r_{DS5} C_1}{r_{DS5}}. \qquad (8)$$

For the transimpedance gain R_T the current-voltage transformation at input impedance shall be considered. For $C_{IN} \gg C_2$ this results in:

$$R_T = V_T \frac{V_{in}}{I_{in}} = V_T \frac{X_{C_{IN}} \cdot R_{bias,1}}{X_{C_{IN}} + R_{bias,1}} \qquad (9)$$

$$= V_T \frac{R_{bias,1}}{1 + s R_{bias,1} C_{IN}}.$$

With the assumption of large r_{DS5}, large bias resistances, small C_2, high forward gain g_{m1}, etc we get a simplified expression:

$$R_T \approx R_2 \frac{C_1}{C_2}, \qquad (10)$$

which matches the original approximate expression in (2) for high ratio C_1/C_2. Differently from [9], [11], the model above also considers the effects due to both biasing resistances $R_{bias,1}$ and $R_{bias,2}$ and those caused by C_c. The newly developed expression for R_T forms a basis for an improved design methodology when compared to simplified modeling demonstrated originally in [9], [11]. An operating point for M_1 shall be established for the expressions above to hold and for the given supply voltage this may be achieved by controlling the current source which shall be adjusted for particular R_1. This fact is of primary importance for the envisioned programmable-gain version of the proposed design where the changes in circuit parameters shall be aligned with the adjustment of the current source formed by M_3.

The presented design is believed to have several major advantages over other well-established TIA designs [11]. First of all, the major part of the gain definition network is still mainly formed by C_1 and C_2 and, therefore, shall contribute far less noise when compared to resistive feedback. Furthermore, the total noise current contributed by the first stage shall be significantly lower compared to that of SFB TIA for the given bandwidth. Finally, the capacitance as seen at the input node shall not degrade the stability of the design and shall only lower the DC loop gain.

Even though the noise of the current source and M_2 are also of some concern, with proper design the contribution of M_2 may be made negligible [11], thus, resembling the generic behavior of topologies such as CG [3]. At the same time, when compared to simple feed-forward designs, the noise current of R_2 and that of the current source is divided by capacitive ratio with the corresponding increase in the gain.

The complete implementation of the amplifier including the biasing circuits and the buffer is shown in Fig. 3. The circuit employs a single-stage CS as a core amplifier as one is opted for a simpler voltage amplifier due to stringent requirements for noise performance and, hence, less number of active components. Obviously, one shall ensure that sufficient gain is achieved with the provided CS configuration [9]. The

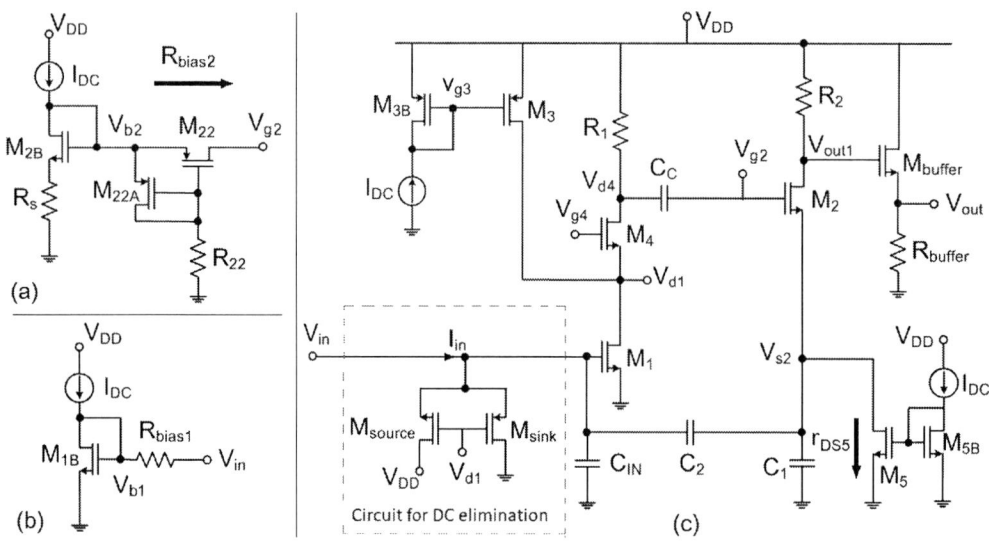

Fig. 3. Basic implementation of the single-gain amplifier circuit (c) with the output buffer and associated bias circuits (a,b).

Miller effect of the gate-drain capacitance in M_1 is mitigated by stacking an additional NMOS (M_4) on the top. The later is biased to 1.4 V via V_{g4} and shall also improve the overall noise efficiency of the design.

As the gain of the core amplifier strongly affects the accuracy of the feedback, one may opt to increase the value of R_1. However, such a straightforward attempt may not succeed due to the voltage headroom problem for the given 1.8 V supply. Therefore, in order to increase R_1 without sacrificing the current density in M_1, a PMOS transistor M_3 is added in parallel to handle the current which is fed to M_1. Note that by increasing R_1 one also decreases its noise current, while the transistor M_3 is itself biased for minimization of the noise current. At the end, the circuit is configured with only 20% of the current passing through R_1, while the rest of the current is fed via M_3. The approach not only reduces the noise contribution of the load, but also a relatively high gain of this simple core amplifier may be reached. A dedicated bias circuit is designed to minimize the noise contribution of M_3.

The issue of DC dark current, which may lead to instability and saturation problems in TIA, is addressed following an approach similar to the one suggested in [9], [11]. A pair of transistors is placed at the gate of M_1 (see DC eliminator block in Fig. 3). In the excess of DC current sourcing to the gate of M_1, the diode-connected transistor will sink the extra current to the ground while limiting the maximum voltage at the gate of M_1. In case of extra DC current-sinking from the input node, the M_{source} has to turn on and will provide an extra current. This transistor is off in the absence of the extra sinking current and will turn on only if simultaneously the voltage at the gate of M_1 drops and the voltage at the drain increases. Both transistors are optimally designed with minimum widths to reduce the input noise current of the circuit.

A special challenge in this TIA design comes from the requirement to have a low cut-off frequency of 100 kHz. A classical implementation would require relatively large resistors to be connected to the gates of both M_1 and M_2 while ensuring a minimum possible parasitic capacitance. An attempt to address a problem with relatively large resistors had been already demonstrated in the original work of [9], where the biasing solution using three NMOS transistors was also suggested. However, the approach is hardly possible in the present circuit due to voltage headroom problem (the author in [9] employed 2.2 V power supply). A straightforward approach is to employ a classical circuit with biasing resistor as shown in Fig. 3b. The approach may be considered feasible for the first bias voltage V_{in} as the area issue for $R_{bias,1}$ is of a lesser problem (16 kΩ). Unfortunately, such an approach, when applied to V_{g2} tends to occupy around 70% of the chip area for the bias resistor and is hardly acceptable for a practical design (see Fig. 4 (left)).

To mitigate the problem with $R_{bias,2}$ we suggest an alternative approach with two PMOS devices (M_{22} and M_{22A}) as shown in Fig. 3a formed by corresponding $L_{22} = L_{22A}$ and $W_{22} = 1/N \cdot W_{22A}$ constrained with $V_{eff,22} = V_{eff,22A}$. The equivalent bias resistance for the configuration shown becomes:

$$R_{bias,2} = R_{DS,22} = \frac{1}{\sqrt{2\beta_P I_{22A}}} \frac{L_{22}\sqrt{W_{22A}}}{W_{22}\sqrt{L_{22A}}} \quad (11)$$
$$= \frac{N}{\sqrt{2\beta_P I_{22A}}} \frac{\sqrt{L_{22}}}{\sqrt{W_{22}}},$$

where L_{22}, L_{22A} and W_{22}, W_{22A} are the corresponding dimensions of the M_{22} and M_{22A} and β_P is the transconductance parameter for PMOS device. In the proposed design the transistor M_{22A} is set to be 10 times bigger than M_{22}. While the M_{22} operates in linear mode, the M_{22A} ensures a stable voltage V_{g2} with resistor R_{22} used to limit the current I_{22A}

to just few µA. Note that if the current is added via R_{22}, the same current has to be removed from I_{DC} for M_{2B} transistor. The design results in 0.8 V at the gate of both transistors with V_{g2} stable around 1.35 V. Finally, an additional resistor R_s is employed to add an extra voltage. The proposed solution is based on PMOS as the $V_{DD} - V_{g2}$ is too small for operation of the NMOS devices, while $V_{g2} - V_{GND}$ is still sufficient for a PMOS-based approach. Finally, the current source at V_{s2} is implemented with a carefully designed current mirror and the output matching to the 50Ω load is realized using a standard buffer configuration.

III. RESULTS

The circuit was designed and optimized for standard TSMC 0.18 µm CMOS process. While the original solution with a pure resistive biasing for both M_1 and M_2 occupies 150 µm × 200 µm due to MΩ resistance in the second bias circuit, the transistor-based solution for M_2 reduces the TIA size down to 130 µm × 70 µm (reduction in 70% of total area) as can be seen in Fig. 4. The results of the simulation for R_T are shown in Fig. 5 for both designs. The approach with an optimized bias circuit reaches gain of 80 dBΩ after the output buffer (2.5 dB gain drop) which is also marginally higher than the original design with pure resistive biasing circuits. This comes at no penalty for the bandwidth which remains the same (around 1.1 GHz) with the cut-off frequency being still close to the OTDR requirement of 100 kHz. The flatness of the response in the pass-band satisfies the OTDR flatness requirement to be below 0.5 dB in both implementations.

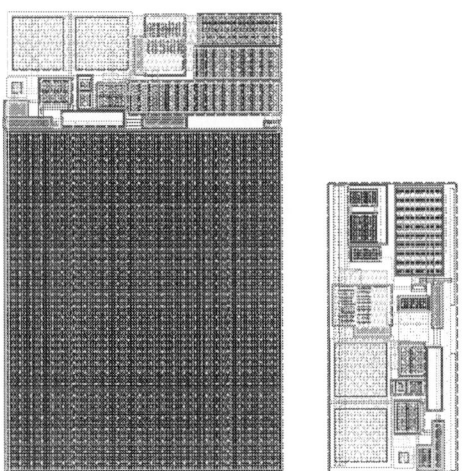

Fig. 4. Layout of the implemented TIA: with original resistive (left) and optimized biasing circuit (right).

The results of the input-referred current noises for both designs are shown in Fig. 6. The designs reach similar noise level below 1.8 pA/$\sqrt{\text{Hz}}$ within the bandwidth of interest. Although the circuit provides a significant noise margin with respect to the OTDR requirement to be below 5 pA/$\sqrt{\text{Hz}}$, the final design of the TIA for OTDR may likely employ

a differential configuration and the ultimate noise level may be higher when compared to the a single-ended approach demonstrated here.

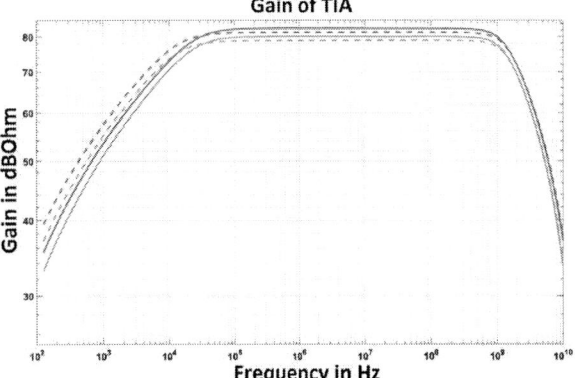

Fig. 5. Simulation results for the transimpedance gain before (red line) and after the output buffer (green line): dashed line - original design with resistive biasing, solid line - optimized bias circuit.

The results in Fig. 4 and Fig. 5 are given for the input capacitance C_{IN} = 0.7 pF, C_1 = 2.4 pF, R_1= 300Ω, and R_2= 450Ω. The value of C_2 is taken five times the minimum capacitance for the given technology and a classical source follower is added for 50Ω output matching. Within the input stage, the current through M_1 is around 10 mA. The total current is just slightly more than 11.4 mA with the overall power consumption being 21 mW at 1.8 V power supply. The design provides a sufficient power margin if a differential approach would be chosen for a final solution with the OTDR requirement opting for below 50 mW.

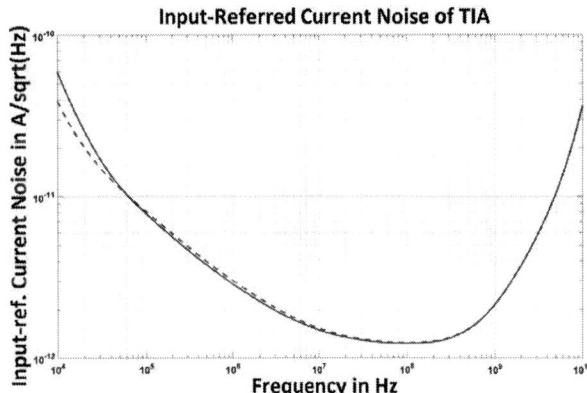

Fig. 6. Simulation results for input-referred current noise (dashed line - original design with resistive bias, solid line - optimized bias circuit).

IV. DISCUSSION

Rapid growth of the market for optical communications resulted in numerous CMOS TIA designs proposed in recent years. Although the CMOS TIA targeting OTDR applications

63

must adhere to rather specific requirements such as high linearity and constraints on the ripple in the pass-band, some of the TIA designs ideas or optimizations, originally suggested for data transmission application, may appear also useful for the discussed application.

In order to position the present design among similar works, the following figure-of-merit (FoM) is introduced:

$$FoM = \frac{\sqrt{BW\,[GHz]}\,R_T\,[\Omega]\,C_{IN}\,[pF]}{Noise\,\left[pA/\sqrt{Hz}\right]\,P[mW]}. \quad (12)$$

In the expression above P is the power in mW, C_{IN} is the total input parasitic capacitance, BW is the bandwidth of the TIA and $Noise$ is the input-referred noise current density. Although the proposed FoM strongly emphasizes the gain and penalizes TIA designs intended to operate with smaller C_{IN} (i.e. modern CMOS TIA for high-speed optical transmission), the ability of the design to handle higher input capacitance is an important OTDR requirement which may eliminate a number of otherwise promising topologies. The proposed FoM does not include explicitly penalty terms due to nonlinearity or ripple in the pass-band leaving these important OTDR parameters beyond the suggested performance indicator. This is caused by the fact that most of the recent works do not provide quantitative measures on these parameters and only qualitative claims can be done with some caution based on published results.

In terms of this FoM, the circuit reaches the performance close to 200 units and is close to those which can computed for [12] and [9]. However, the design of [12] is based on a relatively advanced 40 nm CMOS, while in [9] no details on total C_{IN} have been specified and a default value of 0.5 pF have been assumed by us for FoM calculation. Although much better gain and bandwidth results can be obtained for smaller values of C_{IN}, an ability to handle large input capacitance is an important OTDR requirement which cannot be ignored.

Recall that in ideal capacitive feedback TIA the gain is defined by the ratio of two main capacitors C_1 and C_2 and this effect was claimed by [9], [11] to be one of the major advantages for an area-efficient implementation using poly-capacitors and shall result in significantly reduced vulnerability to process variations. The revisited design methodology for practical TIA, as described in this work, shows that the original expression may be too simplistic. As the expression for the gain seems to be more intricate, this promising advantage may be not fully realizable in practical TIA configuration. The simulation results also reveal the approach not to result in an extremely low-power design when compared to some advanced feed-forward TIAs reported elsewhere. Despite of a large number of works addressing CMOS TIA design, only a few discuss an impact of process variation on the circuit's performance (see, e.g. [13]) and a systematic overview of the issue for different TIA designs seems to be still missing.

V. CONCLUSION

The work revisits the design methodology for low-noise and high-gain CMOS TIA. The proposed inductor-less design targets the OTDR application and is based on a capacitive-feedback topology. The circuit was implemented in 0.18 μm 1.8 V CMOS process and the simulation confirms the gain 83/80 dBΩ and the bandwidth of 1.1 GHz. An area-efficient biasing circuit was suggested to reduce the chip area for 70% while preserving all the advantages of the reference capacitive feedback structure. The circuit demonstrates the average input-referred current noise density below 1.8 pA/$\sqrt{\text{Hz}}$ and the power consumption of 21 mW. A more accurate design procedure is proposed while considering the the influence of the biasing circuits and decoupling capacitance. The suggested TIA design demonstrates no ripple in the pass-band and is a viable candidate for low-cost OTDR instruments, where a significant performance margin makes the approach also suitable for other low-noise applications. Further work is planned on the variable-gain version of the proposed design and on a comparative study of different CMOS TIA topologies in terms of sensitivity of their major performance characteristics to variations in CMOS processes.

REFERENCES

[1] M. Tateda and T. Horiguchi, "Advances in optical time-domain reflectometry," *Journal of Lightwave Technology*, vol. 7, no. 8, pp. 1217–1224, 1989.

[2] J. Charlamov and R. Navickas, "Design of CMOS Differential Transimpedance Amplifier," *Elektronika ir Elektrotechnika*, vol. 21, no. 1, pp. 38–41, 2015. [Online]. Available: http://eejournal.ktu.lt/index.php/elt/article/view/4548

[3] E. Sackinger, *Analysis and Design of Transimpedance Amplifiers for Optical Receivers.* Wiley, 2017.

[4] H. Escid, S. Salhi, and A. Slimane, "Bandwidth enhancement for 0.18 um cmos transimpedance amplifier circuit," in *2013 25th International Conference on Microelectronics (ICM)*, Dec 2013, pp. 1–4.

[5] S. S. Mohan, M. D. M. Hershenson, S. P. Boyd, and T. H. Lee, "Bandwidth extension in cmos with optimized on-chip inductors," *IEEE Journal of Solid-State Circuits*, vol. 35, no. 3, pp. 346–355, March 2000.

[6] O. Momeni, H. Hashemi, and E. Afshari, "A 10-Gb/s Inductorless Transimpedance Amplifier," *IEEE Transactions on Circuits and Systems II: Express Briefs*, vol. 57, no. 12, pp. 926–930, Dec 2010.

[7] J. H. Yeom, K. Park, J. Choi, M. Song, and S. Y. Kim, "Low-cost and high-integration optical time domain reflectometer using CMOS technology," in *2019 15th Conference on Ph.D Research in Microelectronics and Electronics (PRIME)*, July 2019, pp. 145–148.

[8] J. d. Jin and S. S. h. Hsu, "40-Gb/s Transimpedance Amplifier in 0.18-um CMOS Technology," in *2006 Proceedings of the 32nd European Solid-State Circuits Conference*, Sept 2006, pp. 520–523.

[9] S. Shahdoost, A. Medi, and N. Saniei, "Design of low-noise transimpedance amplifiers with capacitive feedback," *Analog Integrated Circuits and Signal Processing*, vol. 86, no. 2, pp. 233–240, 2016.

[10] C. Kuznia, J. Ahadian, D. Pommer, R. Hagan, P. Bachta, M. Wong, K. Kusumoto, S. Skendzic, C. Tabbert, and M. W. Beranek, "Novel high-resolution OTDR technology for multi-Gbps transceivers," in *OFC 2014*, March 2014, pp. 1–3.

[11] B. Razavi, "A 622 Mb/s 4.5 pA/$\sqrt{(Hz)}$ CMOS transimpedance amplifier [for optical receiver front-end]," in *2000 IEEE International Solid-State Circuits Conference. Digest of Technical Papers (Cat. No.00CH37056)*, Feb 2000, pp. 162–163.

[12] M. Atef and H. Zimmermann, "2.5Gbit/s transimpedance amplifier using noise cancelling for optical receivers," in *2012 IEEE International Symposium on Circuits and Systems*, May 2012, pp. 1740–1743.

[13] M. Atef and D. Abd-elrahman, "2.5 Gbit/s compact transimpedance amplifier using active inductor in 130nm CMOS technology," in *2014 Proceedings of the 21st International Conference Mixed Design of Integrated Circuits and Systems (MIXDES)*, June 2014, pp. 103–107.

Proceedings of the 27th International Conference *"Mixed Design of Integrated Circuits and Systems"*
June 25-27, 2020, Łódź, Poland

A New Architecture of Thermometer to Binary Decoder in a Low-Power 6-bit 1.5GS/s Flash ADC

Mohammad Keyhanazar, Arash Kalami
Department of Electrical Engineering
Urmia Branch, Islamic Azad University
Urmia, Iran
m.keyhanazar@gmail.com, a.kalami@iaurmia.ac.ir

Abdollah Amini
Sina Bioelectronics Company (SBC)
Roshd Center, Urmia Branch, Islamic Azad University
Urmia, Iran
amini7908@gmail.com

Abstract—This paper introduces a new structure of the thermometer to binary decoder utilizing combination of two conventional approach and modify them in a low-power 6-bit flash analog to digital converter (ADC). Considering advantages of each method to form the presented decoder can lead to minimum possible power consumption which is a critical parameter in all converters especially in flash ADCs. Moreover, in the high-resolution flash ADCs, the decoder structure will be simpler compared to conventional ones and decreases the amount of power dissipation as well. Simulation results through HSPICE software level 49 parameters in 0.18μm standard CMOS technology parameters, prove the precise operation and the great improvements. The 6-bit converter achieves sampling rate of 1.5 GS/s, and precision of 5.10 effective number of bits (ENOB). The proposed ADC works with 1.8V power supply and it has the power consumption of 4.57mW and the figure of merit (FOM) is 0.047 pJ/conversion-step. Hence, this architecture would be dedicated to communication transceivers and data acquisition systems where area and energy efficiency are paramount.

Keywords—Flash ADC, High speed, Low power, Simple Architecture, Thermometer to Binary Decoder.

I. INTRODUCTION

Flash analog to digital converters (ADCs) find wide application both as stand-alone components and as building blocks of more complex systems such as data acquisition systems and also systems on chip (SoC) applied to high-speed communication interfaces. Hence, it would be a circuit for all seasons [1], [2]. Although, a Flash ADC always exhibits a trade-off between resolution and power consumption, but one of highly preferred ADC architectures in high-speed conversion applications is flash ADC [3]. As shown in Fig. 1, a typical N-bit flash ADC employs 2^N-1 comparators along with a resistor ladder consisting of 2^N equal segments. Finally, a decoder block extracts the output binary codes from the thermometer code produced by the comparison block [1], [4]. On the other hand, design and implementation of a low-power and more compact integrated circuits (ICs) is an important issue in modern applications [5]. There are many ADC architectures including successive approximation and pipelined structures, etc., that each of which can reduce the power consumption but they cannot achieve high-speed operations due to their architectural limitations [6]. Flash ADC is well known for its simplicity, elegance and very fast operation instead of mentioned methods. The major drawback of the flash ADC is the exponential growth of its cost as a function of resolution [1].

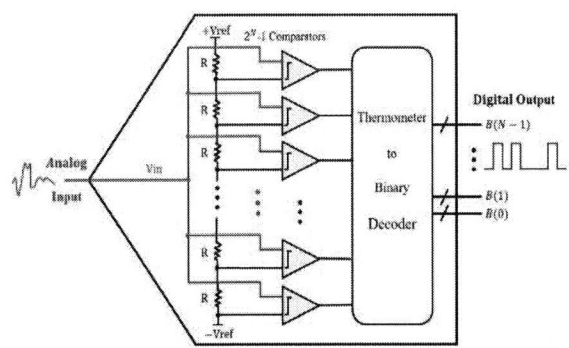

Fig. 1. Block diagram of Flash ADC

Therefore, IC designers have to focus on the essential factors such as power consumption and area occupation particularly in flash ADCs which suffer from a great amount of hardware and power in higher resolutions (more than 6 bits). To solve this problem, many techniques have been introduced. For instance, in order to decrease the power consumption amount and also minimize the size of required area, some approaches are used including folding and interpolation techniques [7-9]. An online method has recently introduced for flash ADCs in order to eliminate the offset of the comparators and also reduction in power and area utilizing simple circuitry, but it only focuses on comparator structures and does not consider other power hungry blocks as well as thermometer to binary decoder [4]. Averaging technique is also used to reduce the differential nonlinearity (DNL), offset and mismatches without increasing transistors sizes. Hence, it leads to the high resolution and high speed, without increasing area and power consumption [9-13]. Almost all the mentioned works have been considered the analog input part of the flash ADCs, but other parts should be put in center of attention because they can help to improve the critical parameters as well as input parts of the body. In flash ADCs, with increment in resolution, the encoder structure will increase exponentially and it would be leads to bigger chip area and more power consumption as well as circuit complexity. In addition, eradication of the bubble error effects and metastability will become hard to achieve. Although some encoding algorithms have been developed to suppress metastability and bubble errors [14, 15], but they haven't intended to solve other mentioned problems. Other structures, those are specifically developed for the Flash type ADC, also have been reported that each of them has some advantages and drawbacks [16-21]. For example, a replacement approach has reduced the power consumption in

[21], but because of using transmission gates, the undesirable disadvantages such as complexity, latency, and mismatches are unavoidable and make the structure inappropriate for high speed applications. Almost all the mentioned problems could be tackled utilizing a combination of two conventional work and simplifying them to reach the minimum amount of power consumption as well as possible. Hence, in this paper we applied the same idea in a typical flash ADC to show the proper results. The rest of the paper is arranged as follows: the design of the circuits, including the architecture and the circuit details of comparator and decoder, are discussed in Sect II. Simulation results and comparison are presented in Sect III. Concluding remarks are provided in Sect IV.

II. CIRCUIT DESIGN

The presented 6-bit flash ADC consists of three main blocks: a resistor string, comparison block and thermometer to binary decoder. The simple resistor string generates 2^6-1 couples of reference voltages. The reference voltages are applied to the inputs of the comparators and compared with the analog input signal.

A. Comparator [4]

The schematic of the single-stage comparator structure is depicted in Fig. 2 where the input differential pairs (M_3-M_4) and (M_5-M_6) with their current sources (M_1 and M_2) are responsible for converting the differential input voltage into the currents required in the main current comparator unit (M_{13} and M_{14}) [4]. In order to suppress the kickback noise and clock feed-through effect, cascade devices (M_7, M_8, M_9 and M_{10}) are employed as well. Consequently, in this comparator the reset process is executed by shorting the output nodes to the ground through the devices (M11 and M12). In this system a simple readout circuitry (M15-M21) is utilized as illustrated in Fig. 2 to capture and hold the output digital bits.

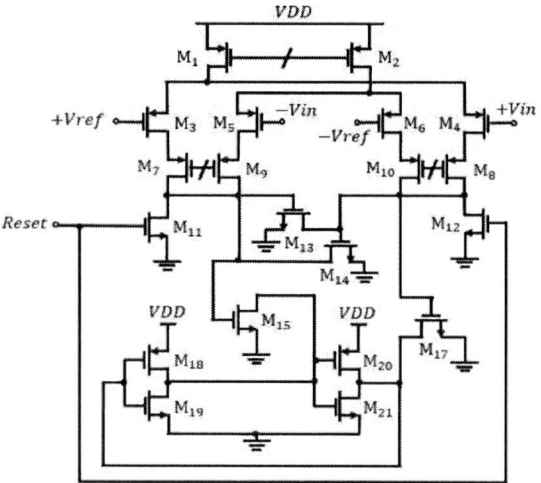

Fig. 2. The comparator circuit

In the readout circuit, the PMOS devices of the inverter gates (M18 and M20) are designed in minimum size. Hence, changing the output bits is easy for the toggling transistors (M15 and M17).

B. Thermometer to Binary Decoder

Any flash ADC decoder involves with the two key design factor including speed and handling capability. In addition, the bubble error which originated from voltage offset, can significantly affects the linearity and signal-to-noise ratio (SNDR) of the ADC. A proper solution against the mentioned error is to use of two conversion steps: at first, convert the thermometer code into a Gray code and then extract the binary code in the next stage using very simple circuitry. When bubbles appear in the thermometer code, only one bit changes between adjacent Gray codes and it is more efficient because the accuracy will be more gradually decreased [18]. The following equations show the relation between the digits of the thermometer-coded data (T_n), the Gray-coded data (G_n), and the binary-coded data (B_n) for a 4-bit decoder:

$$G_3 = T_8 ,$$
$$G_2 = T_4 \, \overline{T_{12}} ,$$
$$G_1 = T_2 \, \overline{T_6} + T_{10} \, \overline{T_{14}} ,$$
$$G_0 = T_1 \, \overline{T_3} + T_5 \, \overline{T_7} + T_9 \, \overline{T_{11}} + T_{13} \, \overline{T_{15}} \qquad (1)$$

$$B_3 = G_3 ,$$
$$B_2 = G_2 \oplus B_3 ,$$
$$B_1 = G_1 \oplus B_2 ,$$
$$B_0 = G_0 \oplus B_1 \qquad (2)$$

To simplify the implementation, the following equivalent expressions can be used for G_1 and G_0:

$$\overline{G_1} = \overline{T_2 \, \overline{T_6}} \cdot \overline{T_{10} \, \overline{T_{14}}} ,$$
$$\overline{G_0} = \overline{T_1 \, \overline{T_3}} \cdot \overline{T_5 \, \overline{T_7}} \cdot \overline{T_9 \, \overline{T_{11}}} \cdot \overline{T_{13} \, \overline{T_{15}}} \qquad (3)$$

Fig. 3 shows the schematic of a simple 4-bit decoder. Note that no gate is needed for inversion of the input thermometer code (e.g., T_{12} in Fig. 3), since the decoder inputs (i.e., the outputs of the comparators) are differential signals; that is, both the inverted and non-inverted inputs are available [18].

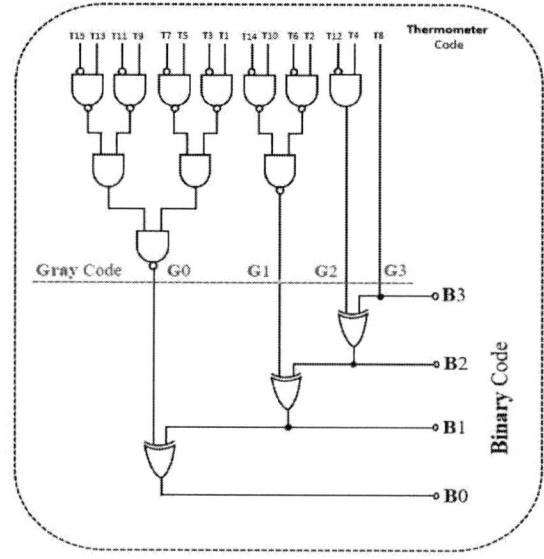

Fig. 3. Schematic of a simple 4-bit decoder in [18]

The proposed decoder:

A switching decoder was proposed in [19] which makes use of low resolution decoder as the core unit. In [19] four-bit decoder is designed using single 3:2 bit decoder, to the input of which fifteen signals are multiplexed to generate four output bits. The proposed 5-bit decoder is based on similar concept however, but here we used four-bits decoder which is presented in [18] and also we employed pseudo-NMOS logic to implement the internal gates, instead. Extending the concept of pattern matching, a five-bit decoder and then final six-bit decoder using 31:5 bit decoder are designed as shown in Figure 4(a) and (b), respectively.

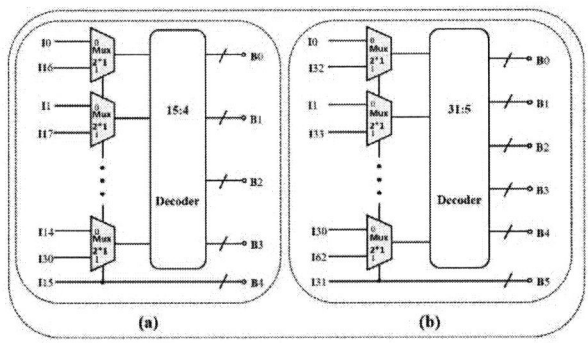

Fig. 4. a) Schematic of a simple 5-bit decoder, b) The proposed 6-bit decoder

The complete design is accomplished utilizing one 15:4 decoder and forty five multiplexers which can be easily reconfigured to higher number of bits by further multiplexing the inputs and also choosing the right select signal. This architecture is implemented in standard 0.18μm CMOS technology.

III. SIMULATION RESULTS AND COMPARISON

The simulation results using HSPICE software level 49 parameters in a typical 0.18 μm CMOS technology, indicates the proper performance of the proposed Flash ADC. The simulation result of the system is illustrated in Fig. 5. Figure 6 shows the total power consumption of the proposed ADC which the proposed ADC consumes about 4.57 mW from a 1.8 V supply voltage. Precisely, the resistor string, comparators, decoder and the other parts of ADC consume 28% (1mW), 61% (2.1mW), 10% (0.34mW) and 1% (0.03mW) of the overall power budget, respectively. So, the proposed decoder dissipates minimum possible power. Finally, applying a sinusoidal input signal and taking the Fast Fourier Transformation (FFT) from the analog version of the digital output of the proposed ADC, the results are depicted in Fig. 7 (in 128 points) which shows 5.10 ENOB and 32.50dB SNDR. In order to better compare the proposed ADC with other woks, the operating speed, dissipated power and the resolution should be simultaneously considered. The lesser the consumption power, the lower the FoM. Hence a figure of merit (FOM) is introduced as:

$$FoM = \frac{P}{f_s * 2^{ENOB}} \qquad (4)$$

where P is the total power dissipation, f_s the sampling frequency and ENOB the effective number of bits. Comparison with

similar previous works is tabulated in Table I. The performance specifications of the designed ADC are compared with the similar previous works in Table 2 which shows the obvious superiority of the proposed flash ADC.

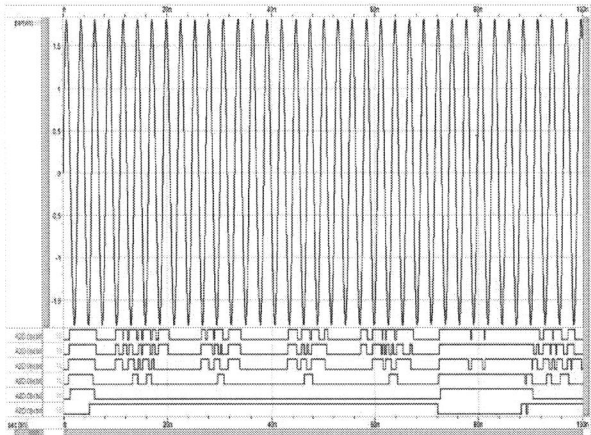

Fig. 5. Digital outputs of ADC for sinus output

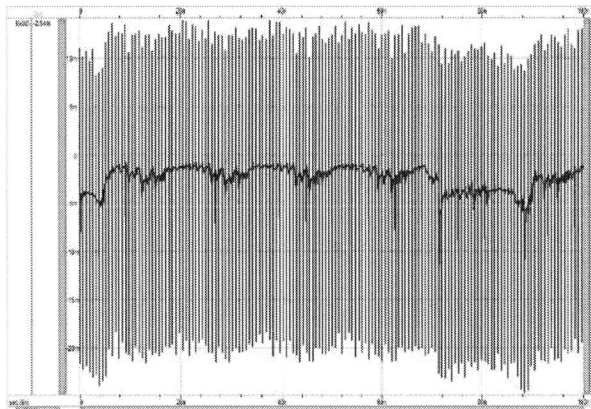

Fig. 6. Total current drawn from power supply

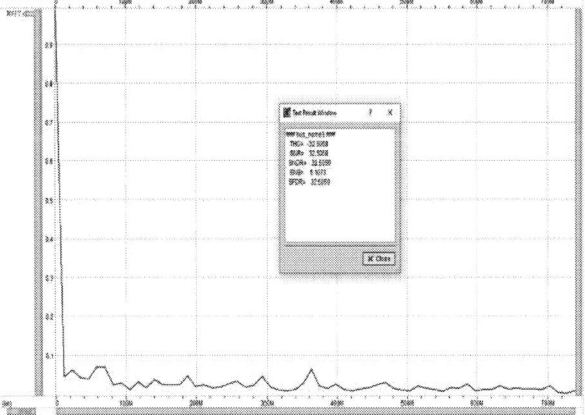

Fig. 7. The resulted parameters from Fast Fourier Transformation (FFT)

TABLE I.
COMPARISON

Ref	Tech (μm)	Resolution (Bits)	Fs (GS/s)	Power (mW)	FoM (pJ/step*Hz)
[3]	0.18	6	1.5	21	0.356
[4]	0.18	6	2	35	0.27
[6]	0.18	6	2	8.8	0.06
[22]	0.18	6	1.6	350	3.4
[23]	0.18	6	1	98	0.76
[24]	0.18	6	0.04	0.54	0.31
[25]	0.18	6	0.6	10	2.05
[26]	0.18	6	0.6	98	11.9
This Work	0.18	6	1.5	4.57	0.047

IV. CONCLUSION

In this work we focused on thermometer to binary decoder in order to improve a new approach by merging of two good works and apply it in a typical 6-bit flash ADC. In this method, both the power consumption and complexity of the proposed decoder are fewer than the features of them in a conventional decoder. Therefore, to eradicate the metastability and bubble errors, the complicated algorithms which increase the power consumption and required chip area and decrease the speed, aren't needed. So, the proposed method may also be applied to high-resolution ADCs to reduce the undesirable effects as well as possible. The proposed structure has been simulated using HSPICE software level 49 parameters in standard 0.18μm CMOS technology which demonstrates 32.50dB SNDR, 5.10 ENOB, and also 4.57mW of power dissipation at the frequency range of 1.5 Giga Sample per Second. The converter operates in 1.8 V supply, yielding 47 fJ/conversion-step of FoM.

REFERENCES

[1] B. Razavi, "The Flash ADC [A Circuit for All Seasons]", *IEEE Solid-State Circuits Magazine*, 9(3), pp. 9–13, 2017.

[2] P. Mroszczyk, J. Goodacre, V.-F. Pavlidis, "Energy Efficient Flash ADC with PVT Variability Compensation through Advanced Body Biasing", *IEEE Trans. Circuits Syst. -II* 66(11), pp. 1775–1779, 2019.

[3] M. Damghanian, S. J. Azhari, "A low-power 6-bit MOS CML flash ADC with a novel multi segment encoder for UWB applications", *Integration, the VLSI journal*, 57, pp. 158–168, 2017.

[4] A. Amini, A. B. Rezaeii, M. Hassanzadazar, "A Novel Online Offset-Cancellation Mechanism in a Low-Power 6-Bit 2GS/s Flash-ADC", *Analog integr. Circuits Signal Process.* 99(2), pp. 219-229, 2018.

[5] T. Aspokeh, A. Amini, A. Baradaranrezaeii, M. Yazdani, "Low-Power 13-Bit DAC with a Novel Architecture in SA-ADC", *25th International Conference on Mixed Design of Integrated Circuits and Systems, (MIXDES)*, 2018.

[6] N. Kalyani, and M. Monica, "Design and Analysis of High Speed and Low Power 6-bit Flash ADC", *2nd International Conference on Inventive Systems and Control (ICISC)*, 2018.

[7] J. Matsuno, M. Hosoya, M. Furuta, T. Itakura, "A 3-GS/s 5-bit flash ADC with wideband input buffer amplifier", *in: International Symposium on VLSI Design, Automation, and Test (VLSI-DAT)*, pp. 1-4, 2013.

[8] Y. Ch. Chen, J. S. Lai, Zh. M. Lin, "A 6Bit 3GS/s two-channel time interleaved interpolating flash ADC", in: *IEEE International Conference of Electron Devices and Solid-State Circuits (EDSSC)*, pp. 1-4, 2013.

[9] J. I. Lee, J. In. Song, "Flash ADC Architecture using Multiplexers to Reduce a Preamplifier and Comparator Count", *in: IEEE Region 10 Conference TENCON*, pp. 1-4, 2013.

[10] Siqiang, F., Tang, H., Zhao, H., Wang, X., Wang, A., Zhao, B., Zhang, G. G. (2011). Enhanced offset averaging technique for flash ADC design. *Tsinghua Science and Technology*. 16 (3), 285–289.

[11] Y. Zhao, Sh. Wang, Y. Qin, Zh. Hong, "A sub-sampling 3-bit 4GS/s flash ADC in 0.13-μm CMOS", in: *10th IEEE International Conference on Solid-State and Integrated Circuit Technology (ICSICT)*, pp. 436–438, 2010.

[12] A. Ismail, M. Elmasry, "A 6-Bit 1.6-GS/s Low-Power Wideband flash ADC Converter in 0.13-μm CMOS Technology", *IEEE Journal of Solid-State Circuits*, 43 (9), 1982-1990, 2008.

[13] Sh. Wang, C. Dehollian, Zh. Hong, "A termination scheme using intended asymmetric spatial filter response for averaging flash A/D converter", *Analog Integrated Circuits and Signal Processing*. 72(1), 251-257, 2012.

[14] Sh. Wang, C. Dehollian, Zh. Hong, "Design of a parallel low power flash A/D converter for the sub-sampling IR-UWB receiver", *Analog Integrated Circuits and Signal Processing*. 74(1), 255-266, 2013.

[15] T. S. Lakshmi, A. Srinivasulu, "A low power encoder for a 5-GS/s 5-bit flash ADC", *in: IEEE Sixth International Conference on Advanced Computing (ICoAC)*, pp. 41-46, 2014.

[16] A. Amini, M. Hassanzadazar, A. B. Rezaeii, "On Improving Accuracy of the Resistor Strings Based on a New Design Technique", *Iranian Journal of Science and Technology, Transactions of Electrical Engineeringin, in press*, 2020.

[17] S. M. Ali, R. Raut, M. Sawan, "A Power Efficient Decoder or 2GHz, 6-bit CMOS Flash-ADC Architecture. *9th International Database Engineering & Applied Symposium (IDEAS'05)*, 2005.

[18] S. Sheikhaei, Sh. Mirabbasi, A. Ivanov, "A 0.18 μm CMOS pipelined encoder for a 5 GS/s 4-bit flash analogue-to-digital converter", *Can. J. Elect. Comput. Eng.*, 30(4), 2005.

[19] V. Hiremath, and S. Ren, "A Novel Ultra High Speed Reconfigurable Switching Encoder For Flash ADC", *IEEE National Aerospace and Electronics Conference (NAECON)*, 2011.

[20] B. V. Hieu, and et al, "A new approach to thermometer-to-binary encoder of flash ADCs- bubble error detection circuit", *54th International Midwest Symposium on Circuits and Systems (MWSCAS)*, 2011.

[21] Ch. N. Yeh, Y. T. Lai, "A novel flash analog-to-digital converter", *in: IEEE International Symposium on Circuits and Systems (ISCAS)*, pp. 2250-2253, 2008.

[22] C. K. Hung, J. F. Shiu, I. C. Chen, & H. S. Chen, "A 6-bit 1.6 GS/s flash ADC in 0.18 μm CMOS with reversed-reference dummy", *IEEE Asian Solid States Conference (ASSCC)*, pp. 335–338, 2006.

[23] Sh. Zhang, Sh. Wang, X. Lin, & G. Ren, "A 6-Bit Low Power Flash ADC with a Novel Bubble Error Correction Used in UWB Communication Systems", *IEEE International Conference on Electron Devices and Solid-State Circuits (EDSSC)*, pp. 1-2, 2014.

[24] G. Huang, & P. Lin, "A 1.0-V 6-b 40 MS/s time-domain flash ADC in 0.18 um CMOS", *Analog Integrated Circuits and Signal Processing*, 77(2), 285-289, 2013.

[25] D. V. Morozov, M. M. Pilipko, and I. M. Piatak, "A 6-bit CMOS inverter based pseudo-flash ADC with low power consumption", *IEEE East-West Design Symposium (EWDTS)*, 2013.

[26] Sh. Zhang, Sh. Wang, and X. Lin, "A 6-Bit Low Power Flash ADC with a Novel Bubble Error Correction Used in UWB Communication Systems", *2015 IEEE International Conference on* (pp. 18-20), 2015.

[27] N. Khiabanmanesh, A. Amini, S. Mihandoost, "Diagnosis of Epilepsy Utilizing Time-Series Distribution of EEG Signals", *25th International Conference on Mixed Design of Integrated Circuits and Systems, (MIXDES)*, 2018.

Proceedings of the 27th International Conference *"Mixed Design of Integrated Circuits and Systems"*
June 25-27, 2020, Łódź, Poland

A Survey on the Application of Parametric Amplification in Next Generation Digital RF Transceivers

Luís Miguel Pires[1,2], João P. Oliveira[1,2]
[1]Centre for Technologies and Systems (CTS) - UNINOVA
[2]Department of Electrical Engineering (DEE)
Universidade Nova de Lisboa (UNL)
Campus FCT/UNL, 2829-516, Caparica, Portugal
l.pires@campus.fct.unl.pt, jpao@fct.unl.pt

Abstract—**The accelerating growth in mobile networks for data communication, covering a diversity of applications, has originated a high demand not only for low-power devices but also for low-cost System on Chip (SoC). An example of this is the next generation 5G, which is under intense development intending to reach an effective ubiquitous connectivity. CMOS technology offers the best trade-off between costs versus performance and facilitates the co-integration of digital-analog. However, integrating analog RF in recent technology nodes remains a challenge, which has pushed the research towards innovative techniques. One of these techniques is the parametric signal conversion, which can offer high speed and low-noise operation. This paper presents a survey on the application of this technique in modern heterodyne receivers.**

Keywords—**Parametric Amplification; Next Generation Digital RF Receivers; 5G; RF Front-End; Heterodyne RF Transceiver; Mixer-First; CMOS; Beamforming Receiver;** *Introduction*

I. INRODUCTION

The accelerating growth in wireless systems (5G) for data communication, for sensing and monitoring applications (ranging from biomedical to environmental targets), has originated a high demand not only for low-power autonomous devices but also for ultra-low cost SoC combining single chip radio with baseband signal and digital data processing [1-2]. Target applications, such as Body Area Networks, Internet of Things (IoT), and high-density environmental sensing networks, will be massively available if the requirement of explicit human action for energy recharge is removed. A second concern for the deployment of large numbers wireless sensor networks is the cost of each individual remote device.

Mobile high digital data rate communications requirements will demand for broadband digital radio channels. This concept underlies the new generation 5G, which will also provide wireless connectivity for an extended portfolio of new applications. Examples are the delivery of high data rate multimedia content in real-time and even IoT applications as wearables, smart homes, and critical infrastructure or industry

applications. 5G can be viewed as a global and massive radio access system that address the future requirements of mobile communication beyond 2020 [3]-[5]. Previous generations will continue to coexist, like the Long-Term Evolution (LTE), that will maintain the development in a legacy position as it will play a significant role in the 5G ecosystem for bands under 6 GHz. However, the available spectrum, Fig.1, in the microwave range, is viewed as promising radio resources for 5G mobile networks to be able to provide multi-gigabit wireless services. Spectrum available at these microwave frequencies can boost several hundred times the actual cellular allocations and capacity [5].

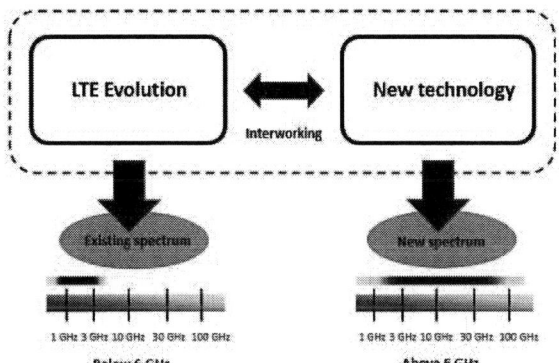

Fig. 1. 5G mobile networks spectrum.

Standard digital CMOS technology offers the best trade-off between cost versus performance and facilitates the integration of the digital and analog circuit blocks [1]-[2]. Moreover, reducing the cost value of the digital radio transceiver implies an increasing need for integration that favors the use of nanoscale CMOS technology. However, this evolution has demanded the use of advanced analog design techniques to allow the co-integration of analog and digital building blocks. A promising approach is the parametric signal conversion and amplification technique. Initially developed in the middle of the 20th century, one of the most important features is his low-noise intrinsic characteristics. Due to the success of the integrated transistor verified over the decades that followed, the technique

---

This work was financially supported by FCT – Fundação para as Ciências e Tecnologia in the scope of the Investigation Unit CTS - Centro de Tecnologias e Sistemas, under the reference UIDB/00066/2020 within the project foRESTER PCIF/SSI/0102/2017.

69

has lost importance. More recently, a growing interest has been observed, especially after the demonstration of its implementation and operation under CMOS technology.

In this paper, we carry out a survey of parametric amplification than can be considered in the design of 5G RF transceivers. We begin by reviewing the basic concepts of the parametric amplification technique in continuous complemented by the recent discrete time domain version. In section 3, we describe one of the possible approaches for the RF receiver using parametric amplification. Section 4 presents the conceptual SoC for RF transceiver and simulations results for the receiver side. Finally, in section 5, we draw the conclusions.

II. PARAMETRIC AMPLIFICATION: CONTINUOUS TIME AND DISCRETE TIME

A. Parametric Amplification: Continuous Time

Firstly, introduced in the middle of the 20th century as a technology that was able to achieve signal amplification with a very low noise footprint, has seen its applications in the radio and microwave range. One example was presented in 1957 in [6], where an equivalent circuit for a cavity type of a parametric amplifier. It consisted of two resonant circuits coupled to each other through a time-varying capacitor or inductor, Fig. 2. Tien [7] and Suhl [6] then worked out the traveling wave version of the parametric amplifier. In their original paper a propagation circuit loaded with time-varying reactor was studied.

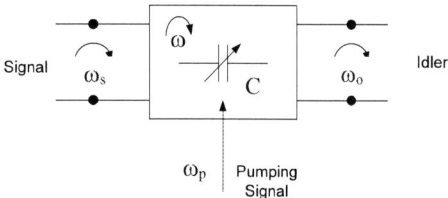

Fig. 2. A parametric amplification circuit model (adapted from [8]).

The parametric amplifier distinguishes from the classical signal amplifiers since transfer power to the output from an AC source rather from a DC power supply. In Fig. 2, the signal frequency, ω_S, and the pump frequency, ω_P, are mixed in a nonlinear capacitor (MOS varactor), a voltage of the fundamental frequencies as well as the sum and difference frequencies $(m.\omega_S \pm n.\omega_P,)$ appears across C. If a resistive load is connected across the terminals of the idler circuit, an output voltage can be generated across the load at the output frequency, ω_O. The output (or idler) frequency in the idler circuit is expressed as the sum and the difference frequencies of the signal frequency ω_S and the pump frequency ω_P.

In 1956, J.M. Manley and H.E. Rowe published a manuscript that analyzed the power flow into and out of a nonlinear reactive element under excitation at its different harmonic frequencies, [8,9]. Manley-Rowe have derived a set of power-conservation relations, eqs. (1) and (2), that are extremely useful in evaluating the performance that can be achieved from a parametric device consisting of a nonlinear reactance. The circuit model considered by Manley-Rowe is shown in Fig. 3. It consists of a nonlinear capacitor driven by two independent signals sources located at frequency ω_S (source signal) and ω_P (the pump signal). The capacitor is also

connected to an array of band-pass filters acting as loads. These filters define the frequencies of the signals that are allowed to drive the nonlinear component, and a diverse of filters with pass frequencies $m.\omega_S + n.\omega_P$ are used [10-13]. The Manley-Rowe relations show the constraints on the power, $P_{m,n}$, absorbed by the nonlinear capacitor at frequencies $m.\omega_S + n.\omega_P$. That is, the power delivered to the capacitor at $m.\omega_S + n.\omega_P$ is denoted by,

$$P_{m,n} = \frac{1}{2} \cdot \mathrm{Re}\left\{V_{m,n} \cdot I_{m,n}^*\right\} \quad , \qquad (7)$$

where the voltage and current represent peak harmonic amplitudes. It is expected that as a result of the nonlinear capacitance, power components at different frequencies will appear. Considering a lossless capacitance, the net power will be zero, [1],

$$\sum_{m,n} P_{m,n} = 0 \ . \qquad (8)$$

The Manley-Rowe relations express the fundamental characteristics as a function of the frequencies allowed to excite the circuit. The band-pass frequencies present in the circuit will determine the characteristics of the parametric based converter [1], [10-13]. The Manley-Rowe relations describe functions as up-conversion (signal at ω_S amplified and mixed to higher frequencies), down-conversion (signal at ω_S amplified and mixed to lower frequencies), or simply direct amplification (signal at ω_S amplified at ω_S), [1].

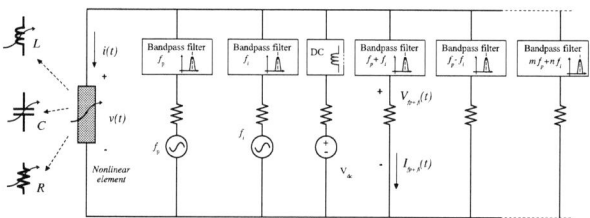

Fig. 3. Circuit model for illustration of the Manley-Rowe relations.

The relations apply to ideal, i.e. lossless, varactors where current and voltage are assumed to exist at frequencies $m \cdot \omega_S + n \cdot \omega_P$, where m and n are integers. The relations are given by, [9],

$$\sum_{m=0}^{\infty} \sum_{n=-\infty}^{\infty} \frac{m \cdot P_{m,n}}{n \cdot \omega_S + m \cdot \omega_P} = 0 \qquad (1)$$

$$\sum_{n=0}^{\infty} \sum_{m=-\infty}^{\infty} \frac{n \cdot P_{m,n}}{n \cdot \omega_S + m \cdot \omega_P} = 0 \qquad (2)$$

where $P_{m,n}$ is the average power flowing into the varactor at the frequency $m \cdot \omega_S + n \cdot \omega_P$. For a frequency multiplier all frequencies are harmonics of only one frequency, ω_S, thus $m=0$, and the Manley-Rowe relations become $P_1+P_n=0$. That is, if the circuit is designed so that only real power can flow at the input frequency ω_S and at the output frequency $m.\omega_P$, the Manley-Rowe relations predict a conversion efficiency of 100%. It should be noted, however, that the relations do not cover the DC power, meaning that the power at any output frequency

$m.\omega_S$ cannot be increase by supplying power at DC frequencies. Interestingly, the above relations also show that besides the application on signal amplifiers, the transfer of energy can also occur between distinct frequencies, thus realizing the operation of low-noise mixing with the possible of having and effective power gain. For the up converter mixing operation, the power gain between the input signal the upconverter version is given by,

$$G_U = \frac{P_U}{P_S} = \frac{\omega_U}{\omega_S} \qquad (3)$$

$$\omega_U = \omega_P + \omega_S \qquad (4)$$

where it shows that the gain is simply the ratio between the up converted frequency, ω_U, and the input signal frequency, ω_S. The upconverted frequency is the sum of the frequency the pump oscillator and the input signal.

A similar relation can be derived for the downconverter case,

$$G_D = \frac{P_D}{P_S} = \frac{\omega_D}{\omega_S} \qquad (5)$$

$$\omega_D = \omega_P - \omega_S \qquad (6)$$

where ω_U corresponds to the downconverted signal frequency.

The gain predicted by the Manley-Rowe relations [9], can be achieved by using a variable reactance, such as variable capacitance (varactor). One of the first device used as varactor in parametric amplifier was the diode. Since this is a two-terminal device, the excitation is applied at the same terminal used for all the other for all signals. In CMOS technology instead of the diode, a MOS varactor can be used in a RF parametric amplification both for RF and other baseband applications, [1]. Moreover, MOS device has a third terminal meaning that there is the possibility to pump the structure using a different terminal than the one for the input signal, thus decoupling them.

In a receiver, a RF signal may be mixed with a signal from the local oscillator in a nonlinear circuit (the mixer) to generate the sum and difference frequencies. In a parametric amplifier the local oscillator is the pumping signal source, as shown in Fig. 4.

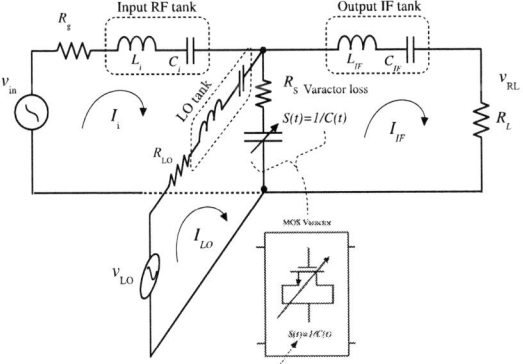

Fig. 4. Equivalent model Manley-Rowe for a parametric amplifier.

B. Parametric Amplification: Discrete Time

The discrete-time operation of a parametric amplifier is obtained by adding switching elements around a nonlinear reactance. The conceptual schematic of this amplifier is depicted in Fig. 5.

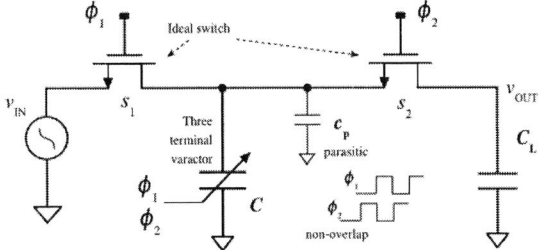

Fig. 5. Conceptual discrete-time parametric amplifier, [14].

The conceptual discrete time parametric amplification diagram is composed of an input voltage source v_{IN}, input and output switches, a capacitive load C_L, and an electrically variable MOS capacitor C which is controlled by a squared voltage pump signal with frequency ω_P, [14]. For this simplified analysis it is considered that the ideal switches are driven by two-phase non-overlapping clocks, ϕ_1 and ϕ_2, [14]. During ϕ_1 a sample of the input signal is acquired by C through switch S_1. During this sampling phase, the output switch S_2 is open and the controlling pump signal is such that the varactor C reaches a high capacitance value, C_{ϕ_1}. During the holding phase, ϕ_2, the voltage amplification is achieved by reducing the value of the capacitance capacitor while conserving approximately the same charge at the top plate of the device. The gain is approximately given by the ratio between the capacitance values observed in each phase.

In contrast with the continues time version, this discrete time parametric cell does not need high quality filters, but it cannot reach the operation at microwave range. Nevertheless, this cell has already be proved to operate in standard CMOS technology in analog-to-digital converters (ADCs), discrete-time filters (FIRs) and mixers, which are fundamental building blocks in modern RF transceivers.

III. USING PARAMETRIC AMPLIFICATION IN 5G RF TRANSCEIVERS

The application requirements for higher data rage in mobile communications is driving the 5G development to support larger capacity. Considering the high occupancy of the radio spectrum up to 6 GHz, one of the most promising evolution is the utilization of the mmWave range, i.e., 28 GHz frequency that can offer larger bandwidth channel, when comparing with the current 4G standard [15-16]. Recently, data rate up to 0.5 Gb/s were demonstrated in outdoor/indoor coverage environments, at 28 GHz band [17, 18].

The potential applications of millimeter wave communications in the 5G network include the small cell access, the cellular access, and the wireless backhaul. In comparison to 4G networks, the millimeter wave 5G network offers significant advantage in data throughput due to the much wider RF bandwidth.

With high carrier frequency and wide bandwidth, there are several technical challenges in the design of circuit components and antennas for millimeter wave communications [5]. In the millimeter bands, high transmit power, i.e., equivalent isotropic radiated power (EIRP), and high bandwidth can increase nonlinear distortion of power amplifiers (PA) [16]. Besides, phase noise and IQ imbalance are also challenging problems faced by radio frequency (RF) integrated circuits [17-18]. Research progress on integrated circuits for millimeter wave communications has been discussed in [22], including in-package antennas, radiofrequency (RF) power amplifiers (PAs), low-noise amplifiers (LNAs), voltage-controlled oscillators (VCOs), mixers, and analog-to-digital converters (ADCs).

The CMOS technology, an attractive option to develop mixed-mode circuits, has been evolving towards smaller devices during the past decades, that is enabling the MOS device to achieve high frequencies of operation. Therefore, the integration of complete transceivers in one single chip (SoC) is now a reality, and ideally without requiring any external components. The signal path model of a wireless communication system (transceiver consist of a transmitter and a receiver) is presented in Fig. 6. There are a number of possibilities of transceivers architectures, namely, heterodyne, and Zero-IF or Low-IF. The later facilitates the full integration in CMOS technology since it does not require external filters, [14].

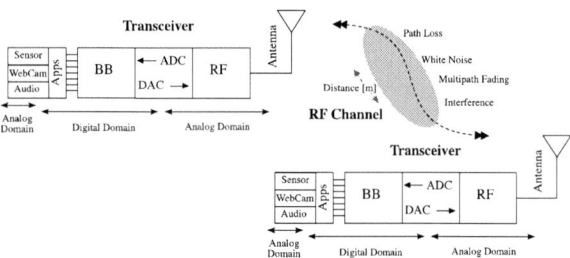

Fig. 6. Transceiver Front-End block diagram.

For higher frequency range, namely for mmWave, the heterodyne architecture is of major interest, that combines mixer-first approach in a software defined radio architecture [19–21]. The main issue, however, with a passive mixer-first receiver front-end is its fairly high noise figure. The challenge is then to minimize the noise figure of the following blocks, which will be imply spending more current, decreasing the on-resistance of the mixer, while keeping the input impedance well matched with the input antenna.

An alternative, for the implementation of a heterodyne architecture for a 5G compatible transceiver, is the use of parametric amplification along the signal processing chain. This can be achieved in both time continuous time and discrete time, as shown in Fig. 7, which can be integrated in a single chip. In a first step, a gain of approximately 15 dB could be achieved by using a 28 GHz RF frequency applied to a parametric based downconverter, in continuous time, to a first intermediate frequency of 1 GHz. A second mixing step, to bring down to baseband or low-IF, can be implemented using a discrete-time parametric based downconverter.

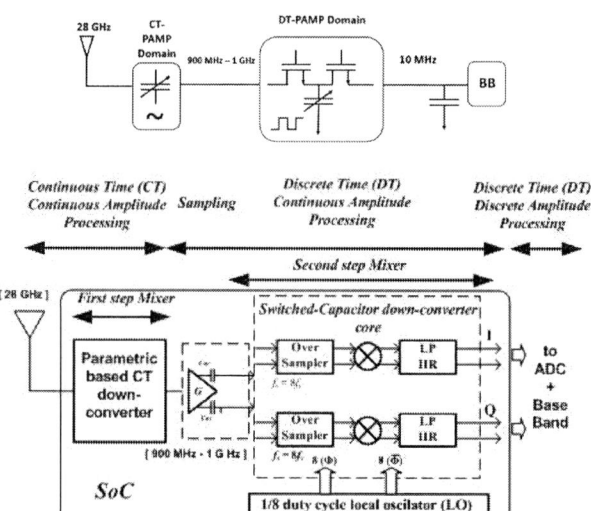

Fig. 7. Conceptual SoC for the RF front-end transceiver, using parametric amplification in both continuous and discrete time domains.

Fig. 8 shows a block diagram for a multiple antenna mixer-first receiver for a front-end transceiver using parametric signal conversion. In fact, it is possible to use a parametric mixer block to adapt a sub 6 GHz direct-converter transceiver architecture to operate at 28 GHz.

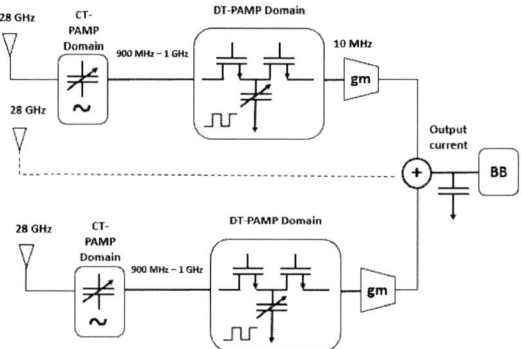

Fig. 8. LNA-less mixer-first RF receiver for a front-end transceiver with MIMO.

In CMOS technology, one can use the variable gate capacitance characteristic of the MOS device (MOSCAP) and put it to operate in parametric mode. Fig. 9 shows the implementation of a downconverter using the MOS device as a two-terminal varactor, [10]. In [10], the circuit implemented in a 130 nm CMOS technology is able act as downconverter from 30 GHz to 10 GHz, with a conversion gain of more than 4 dB, while achieving a noise figure below 1.8 dB, which can be considered low when compared with other type of passive mixers. The circuit achieves these values without the need to use LC thanks with high quality factors, thus favouring its full integration, without external components. The output transconductance buffer at the output can drive the remaining stages of the receiver part of the transceiver, namely the second mixing stage based on the discrete version of the parametric downconverter.

Interesting, one might explore the use of the third terminal of the MOS device to accomplish a similar parametric operation. Fig. 10 shows a possible, simplified, configuration, where the pumping signal is applied to the drain/source terminals of the device. Similar results are achievable using the same level of quality factor of the LC tanks.

IV. CONCLUSION

In this paper, an overview about reactance-based signal amplification and conversion, as well as their system integration targeting modern digital transceivers, is provided. The reactance-based amplifier depends on the capability of a nonlinear capacitor to transfer energy between circuit tanks when conveniently driven by both an input and a pump signal. This technique can be applied in the 5G transceiver design, since it is particularly well adapted to the digital standard CMOS technology. Being compatible with this type of technology, the integration full system on chip for mmWave 5G transceiver is then envisaged. One possibility is combining continuous time and discrete time parametric conversion. The continuous time parametric amplification is possible to implement in modern CMOS technologies with moderate quality factor bandpass filters (LC tank). For lower frequency, these LC type filters tends to occupy large areas thus pushing the implementation of the amplifier into deep GHz band. The alternative presented is the use of MOS varactor in a discrete time parametric mode of operation.

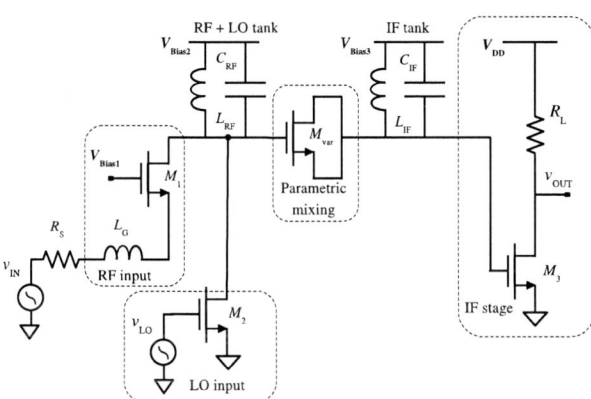

Fig. 9. Parametric downconverter parametric circuit for 28-30 GHz RF frequency, [10].

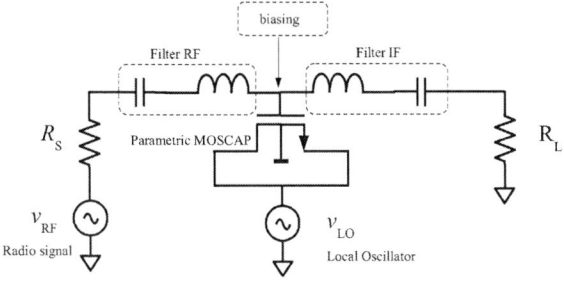

Fig. 10. Parametric downconverter circuit using a MOSCAP.

REFERENCES

[1] K. Iniewski, Ed., Wireless Technologies: Circuits, Systems, and Devices. NY, USA: CRC Press 2007.

[2] I.O. Donnell and R.W. Brodersen, "An ultra-wideband transceiver architecture for low power, low rate, wireless systems", IEEE Transactions on Vehicular Technology, vol.54 no.5, pp. 1623-1631, Sep. 2005.

[3] A. Osseiran, F. Boccardi, V. Braun, K. Kusume, P. Marsch, M.Maternia, O. Queseth, M. Schellmann, H. Schotten, H. Taoka, H. Tullberg, M. Uusitalo, B. Timus, and M. Fallgren, "Scenarios for the 5G Mobile and Wireless Communications: the Vision of the METIS Project", IEEE Communications Magazine, May, 2014.

[4] The 5G Infrastructure Public Private Partnership. (2015) 5G vision. https://5g-ppp.eu/our-vision/. Accessed: 2016-02-10.

[5] "NGMN 5G white paper", Feb 2015, https://www.ngmn.org/home.html. Accessed: 2016-02-10.

[6] H. Suhl, "Proposal for a Ferromagnetic Amplifier in the Microwave Range", Physical Review, Vol.106, No.2, pp-384-385, June 1957.

[7] P.K. Tien, et al.,"Parametric Amplifier, Radio Electronics", Report of US Comission 7, URSI.

[8] P.R. Johannssen, et al., "Theory of Nonlinear Reactance Amplifiers", IEEE Transactions of Magnetics, Vol.3, No.3 September 1967.

[9] J.M. Manley and H.E. Rowe, "Some General Properties of Nonlinear Elements – Part I – General Energy Relations", Proceedings of the Institute of Radio Engineers, Vol.44, pp. 904-913, January 1956.

[10] S. Magierowski, J. Bousquet, Z. Zhao, and T. Zourntos, "RF CMOS Parametric Downconverters," IEEE Transactions on Microwave Theory and Techniques, vol. 58, no. 3, pp. 518–528, Mar. 2010.

[11] S. Magierowski, H. Chan, and T. Zourntos, "Subharmonically pumped RF CMOS paramps," IEEE Transactions on Electron Devices, vol. 55, no. 2, pp. 601–608, Feb. 2008.

[12] S. Magierowski, T. Zourntos, J.-F. Bousquet, and Z. Zhao, "Compact parametric downconversion using MOS varactors," in IEEE MTT-S International Microwave Symposium Digest, 2009, pp. 1377 –1380.

[13] H. Chan, Z. Chen, S. Magierowski, and K. Iniewski, "Parametric conversion using custom MOS varactors," EURASIP J. Wirel. Commun. Netw., vol. 2006, no. 2, pp. 20–20, 2006.

[14] J.P. Oliveira and J. Goes, Parametric Analog Signal Amplification Applied to Nanoscale CMOS Technologies, NY, USA: Springer, 2012.

[15] T. Rappaport, S. Sun, R. Mayzus, H. Zhao, Y. Azar, K. Wang, G. Wong, J. Schulz, M. Samimi, and F. Gutierrez, "Millimeter wave mobile communications for 5G cellular: It will work!", IEEE, vol. 1, pp. 335–349, 2013.

[16] B. Bangerter, S. Talwar, R. Arefi, and K. Stewart, "Networks and devices for the 5G era," Communications Magazine, IEEE, vol. 52, no. 2, pp. 90–96, February 2014.

[17] W. Roh, et al., "Millimeter-Wave Beamforming as an Enabling Technology for 5G Cellular Communications: Theoretical Feasibility and Prototype Results," IEEE Comm. Mag., vol.2, pp.106-113,Feb. 2014.

[18] W. Roh, "Performances and Feasibility of mmWave Beamforming Prototype for 5G Cellular Communications," IEEE International Conf. on Comm. (ICC), Jun. 2013.

[19] C. Andrews and A. Molnar, "A passive mixer-first receiver with digitally controlled and widely tunable RF interface," IEEE Journal of Solid-State Circuits, vol. 45, pp. 2696–2708, Dec 2010.

[20] M. Soer, E. Klumperink, Z. Ru, F. van Vliet, and B. Nauta, "A 0.2-to-2.0GHz 65nm CMOS receiver without LNA achieving >11dbm IIP3 and <6.5 db NF," in Proceedings of IEEE International Solid-State Circuits Conference, 2009, pp. 222–223.

[21] D. Murphy et al., "A blocker-tolerant, noise-cancelling receiver suitable for wideband wireless applications," IEEE Journal of Solid-State Circuits, vol. 47, pp. 2943–2963, Dec 2012.

[22] B. Yang, Z. Yu, J. Lan, R. Zhang, J. Zhou and W. Hong, "Digital Beamforming-Based Massive MIMO Transceiver for 5G Millimeter-Wave Communications," in *IEEE Transactions on Microwave Theory and Techniques*, vol. 66, no. 7, pp. 3403-3418, July 2018, doi: 10.1109/TMTT.2018.2829702.

Proceedings of the 27th International Conference *"Mixed Design of Integrated Circuits and Systems"*
June 25-27, 2020, Łódź, Poland

A W-band SiGe BiCMOS Transmitter
Based on K-band Wideband VCO
for Radar Applications

Maciej Kucharski, Michał Widlok, Radosław Piesiewicz

SIRC Sp. z o.o.

ul. Starowiejska 41-43, 81-363 Gdynia

Emails: m.kucharski@si-research.eu, m.widlok@si-research.eu, r.piesiewicz@si-research.eu

Abstract—**This paper presents an 86-97 GHz transmitter (TX) using a wideband voltage-controlled oscillator (VCO) operating in 21.5-26 GHz range and frequency quadrupler (FQ) fabricated in SiGe BiCMOS technology. The VCO implements a self-buffered common-collector Colpitts topology with binary-weighted varactor ladder for low VCO gain (K_{VCO}) and wide tuning range. Use of high-Q passive components and low-noise heterojunction bipolar transistors (HBT) results in worst-case phase noise of -92.8 dBc/Hz at 1 MHz offset from the carrier. The VCO is loaded by a low-loss transformer that splits the signal between frequency prescaling and multiplying blocks. The prescaler comprise three divide-by-two circuits (DTC) based on D flip-flops (D-FF) providing adequate feedback signal for an external phase-locked loop (PLL). The multiplying section consists of two cascaded Gilbert-cell frequency doublers driving a W-band power amplifier (PA). The TX achieves 0.2 dBm output power at 92 GHz and more than -2.8 dBm in 86-97 GHz range consuming 60 mA from 3.3 V supply. The chip occupies 0.755 mm^2 silicon area.**

Keywords—**multiplier chain, transmitter, radar, wideband VCO.**

I. Introduction

Radar systems have been extensively invstigated over the past years owing to gradual growth of demand for non-military sensor applications. Automotive industry has been undoubtedly a great beneficiary but a rapid development of unmanned aerial vehicles (UAV) such as drones, created an unexplored field for variety of radar components used e.g. cruise control, obstacle mitigation, ground detection etc. Furthermore, enormous increase of drones in aerial spaces put pressure on safety systems designed to counteract unauthorized UAVs entering facilities requiring high security, e.g., airports, military-controlled ranges, sport venues etc. Development of silicon technologies enabled high integration of milimeter-wave (mmW) circuits and realization of cost-effective systems on chip (SoC). Countinuous semiconductor scaling make transistors with f_T/f_{MAX} above 200 GHz easily available, thus allowing a broad set of mmW frequency ranges to be used. W-band (75-110 GHz) seems interesting for wide bandwidth aand weaker attenuation comparing to V-band (50-75 GHz), where strong dioxide molecule absorption limits the maximum radar detection range. Moreover, existing 24 GHz radar systems could be conveniently extended by quadrupling

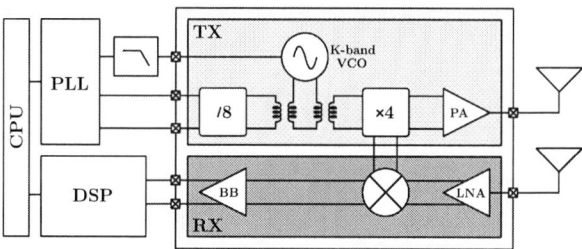

Fig. 1. Radar system block diagram.

the local oscillator (LO) signal up to W-band. This paper presents a 86-97 GHz transmitter (TX) using a K-band voltage-controlled oscillator (VCO) and frequency multiplier chain intended for W-band radar applications.

II. Radar System Concept

Fig. 1 presents W-band frequency-modulated continuous-wave (FMCW) radar system concept using an integrated K-band signal source. Sub-harmonic LO generation offers some advantages over topologies based on fundamental tone sources. First, negative resistance required for oscillation at lower frequencies is reduced due to higher gain of active devices resulting in lower power consumption and noise. Also, LC-tank Q-factor tends to have optimum value at 15-25% of transistor's f_T enhancing VCO phase noise. For that reason we chose K-band as the target VCO frequency range. The LO signal is multiplied by 4 to reach W-band and divided at least by 8 to drive an external commercially available PLL. The up-converted W-band signal is radiated using a power amplifer (PA) and off-chip antenna. The back-scattered signal is sensed by receive antenna, amplified using a low-noise amplifier (LNA) and down-converted in mixer stage. Resulting baseband (BB) signal is fed to a DSP unit, where further processing is performed.

III. Circuit Design

Block diagram of the W-band TX is shown in Fig. 2. A K-band signal is generated using an LC-VCO having three tuning inputs connected to a binary weighted varactor bank. The VCO is loaded with a transformer that splits the signal between the divider an multiplier chains. The former consists

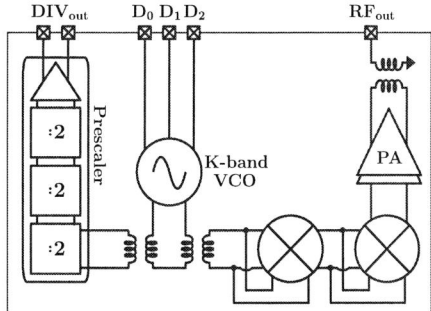

Fig. 2. Block diagram of W-band transmitter.

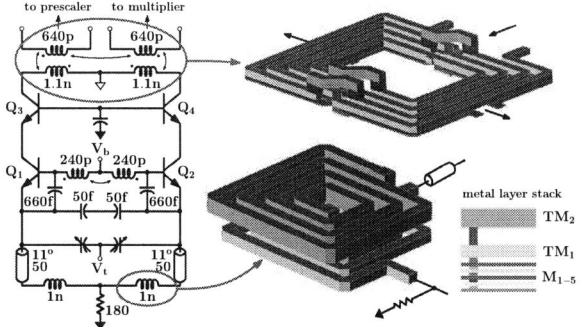

Fig. 3. Schematic of the VCO.

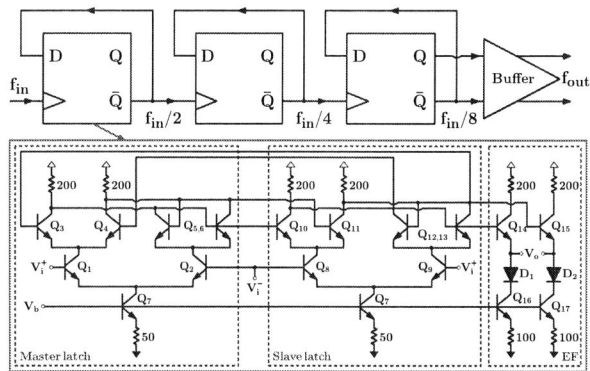

Fig. 4. Prescaler block diagram and schematic of a single divide-by-2 circuit.

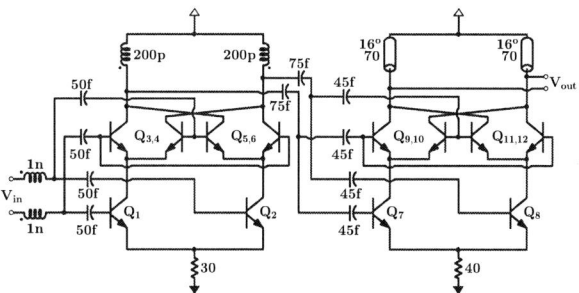

Fig. 5. Schematic of the frequency quadrupler composed of two cascaded Gilbert-cell frequency doublers.

of three cascaded divide-by-2 circuits (DTC) and a buffer to drive an external PLL. The latter is composed of two cascaded frequency doublers and PA.

The VCO (see Fig. 3) implements modified differential common-collector Colpitts topology with a common-base (CB) stage connected to $Q_{1,2}$ [1]. This arrangement improves both isolation of the resonant circuit and the output power. The resonant frequency depends mainly on base inductors and divider formed of base-emitter capacitor and varactor. Transmission line (T-line) and inductor at the emitter must be large enough to present high impedance for the signal. Therefore, the inductor is realized as stacked spiral coil using top metal layers TM1 and TM2. CB stage is terminated by a transformer comprising three closely coupled spiral inductors. The primary coil has four turns and is implemented in TM1 layer. Both secondary coils having two turns each, are realized in M5 layer. The varactor consists of three binary weighted devices resulting in overlapped sub-bands for lower VCO gain [2].

The VCO signal is fed to the prescaling section used to divide the K-band signal down to 3 GHz range and drive an external PLL. A single DTC is basically a D-type flip-flop (D-FF) with feedback as depicted in Fig. 4. Each D-FF has two latches working in master-slave modes and a emitter-follower (EF) buffer. The latches are realized using emitter-coupled logic (ECL) for high-speed operation. The first divider consumes 6 mA whereas the current is halved in each following stage resulting in 10.5 mA total current consumption. The

output buffer is implemented as resistively loaded common-emitter (CE) amplifier capable to deliver -5 dBm to an external PLL.

Part of the VCO output signal drives the multiplier chain composed of two cascaded frequency doublers (FD) and a power amplifier. A single FD is basically a Gilbert-cell mixer where the input signal is connected to the LO switching quad transistors. Self-mixing of the input signal produces second-harmonic tone that drives the next FD.

After frequency multiplication to W-band the signal drives the PA consisting of two stages. Each stage is a differential cascode (CE-CB) amplifier for improved gain and isolation due to Miller effect mitigation with respect to CE topology. Unfortunately, the output capacitance of CE-CB stage increases significantly which is undesired concerning the bandwidth. Therefore, we deliberately terminate the bases of CB stage with fixed capacitance of 100 fF instead of connecting them directly and exploiting virtual ground node. This is beneficial with respect to the bandwidth at cost of smaller gain. The output stage is loaded with a reduced size Marchand balun to convert balanced signal to single-ended for measurement purposes. Size reduction is obtained by increasing characteristic impedance of the T-lines and including shunt 20 fF capacitors. A short-stub is used to compensate for bondpad capacitance and ESD protection.

75

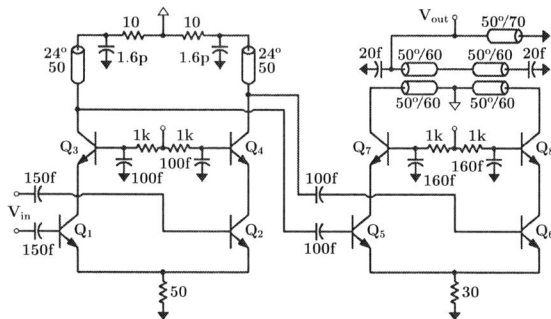

Fig. 6. Schematic of the PA.

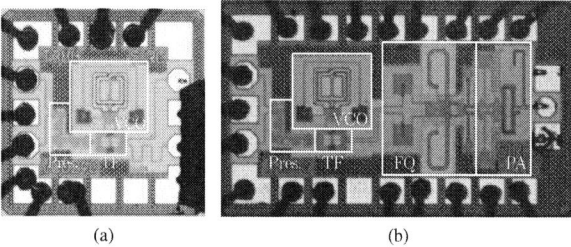

(a)   (b)

Fig. 7. (a) VCO break-out chip photograph. (b) Complete TX die photograph.

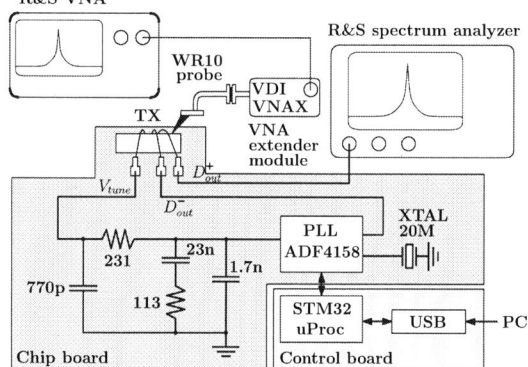

Fig. 8. Setup used to measure TX output power and phase noise.

IV. EXPERIMENTAL RESULTS

The TX was manufactured using SG13S process from IHP [3]. It offers heterojunction bipolar transistors (HBT) with f_T/f_{MAX}=240/330 GHz for RF circuits and CMOS devices for efficient control and digital processing. Die photographs of the VCO break-out circuit and the complete TX are shown in Fig. 7. A simple test board was prepared to characterize the standalone VCO regarding the tuning range and output power. However, in order to reliably measure TX output power and the phase noise a dedicated PCB containing ADF4158 PLL was designed to stabilize LO signal and eliminate drift effects. The chip board contains the TX IC as a bare die mounted on the PCB. The chip generates differential output signals (D_{out}^+ and D_{out}^-) representing the VCO signal divided by 8 using an on-chip prescaler. D_{out}^+ is connected to R&S spectrum analyzer for phase noise measurement whereas D_{out}^- is used to feed the PLL chip. The PLL is programmed using a control board equipped with a STM32 and USB interface for communication with PC. Consequently, the TX output frequency can be conveniently set to the desired value and problem of VCO drift is effectively eliminated. All the TX chip pads except for RF TX output are bonded to the PCB. The TX output is measured using a wafer probe.

Fig. 9 presents standalone VCO tuning range (TR) and single-ended output power measured using spectrum analyzer. There is almost perfect agreement between expected and measured TR due to careful design process based on electromagnetic (EM) simulations. The output power does not deviate by more than 1 dB from expected values. The total VCO TR is almost 5 GHz which corresponds to 21% fractional bandwidth.

Fig. 10 presents VCO phase noise (PN) at 1 MHz offset from the carrier. The measured PN varies from -95.5 to -

92.8 dBc/Hz reaching highest value in the middle of the VCO tuning range due to the largest VCO gain and resulting AM-to-PM noise conversion. A good agreement is observed between measured and simulated results except for 0.5 GHz shift in tuning range comparing to Fig. 9. This is due to fact that the VCO break-out circuit was fabricated in an earlier run and was slightly modified to accomodate the target frequency range.

The TX output power is presented in Fig. 11. The peak output power at 92 GHz is 0.2 dBm and the 3-dB bandwidth is 86-97 GHz. Measurement uncertainty of the RF output power is estimated to ±1 dB including accuracy of the VDI PM4 power detector (±0.2 dB), probe parameters, probe contact repeatability, system noise tracking etc. The chip was powered from a regulated DC power supply with 3.3 V nominal supply voltage. Measured current consumption is 60 mA which corresponds to 198 mW power dissipation. This results in 0.53% TX efficiency (η) calculated as the ratio of TX peak output power to DC power consumption.

V. CONCLUSION

A W-band TX based on K-band wideband VCO for radar applications was designed, manufactured and measured. The TX achieves more than -2.8 dBm output power in 86-97 GHz range and worst-case phase noise of -80.8 dBc/Hz at 1 MHz offset from the 94 GHz carrier. Table I summarizes performance of the fabricated W-band TX circuit.

TABLE I.
SUMMARY OF W-BAND TRANSMITTERS WITH INTEGRATED VCOS.

Ref.	Process	f_{out} [GHz]	# harm.	PN[a] [dBc/Hz]	η [%]	P_{TX} [dBm]	P_{DC} [mW]	Area [mm²]
[4]	SiGe 180 nm	68-94	1	-85.0	0.19	-1	423	3.04[b]
[5]	CMOS 65 nm	76-77	1	-85	1.76	6.4	248[c]	0.49[c]
[6]	CMOS 65 nm	77-79	2	-81.0	10.04	13.2	208[c]	2.09[c]
[7]	SiGe 130 nm	89-95	2	-76.0	1.87	6.4	233[d]	0.72[d]
This Work	**SiGe 130 nm**	**86-104**	**4**	**-80.8**	**0.53**	**0.2**	**198**	**0.76**

[a] worst case at 1 MHz offset, [b] incl. RX, [c] incl. PLL/analog BB, [d] est. for single TX incl. prescaler,

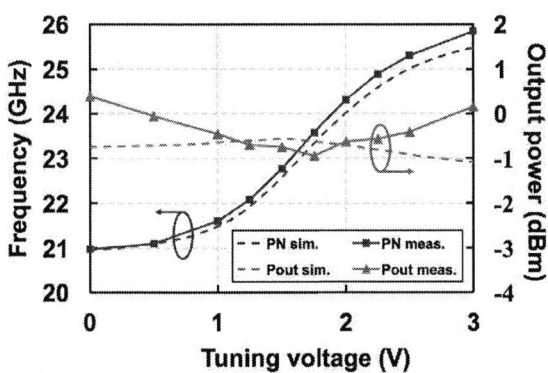

Fig. 9. VCO tuning range and single-ended output power.

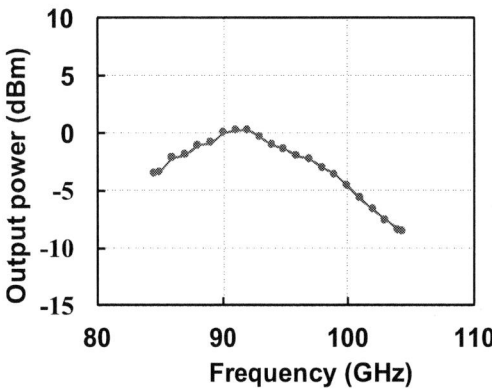

Fig. 11. Measured TX output power versus frequency.

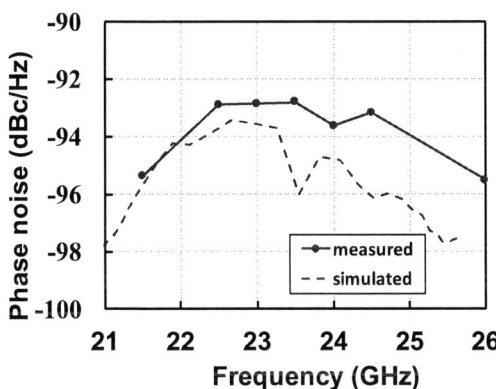

Fig. 10. VCO phase noise at 1 MHz offset from the carrier.

ACKNOWLEDGMENT

We acknowledge European Space Agency funding (ESA Contract No. 4000117820/16/NL/CBi) as well as NCBiR national funding (Contract numbers PBS3/B3/30/2015, PBS3/A3/18/2015, POIR.01.01.01-00-0393/18-00) that led to creation of the results presented in this paper.

REFERENCES

[1] H. Li and H. M. Rein, "Millimeter-wave VCOs with wide tuning range and low phase noise, fully integrated in a SiGe bipolar production technology," *IEEE J. Solid-State Circuits*, vol. 38, no. 2, pp. 184–191, Feb. 2003.

[2] S. A. Osmany, F. Herzel, and J. C. Scheytt, "An integrated 0.6–4.6 GHz, 5–7 GHz, 10–14 GHz, and 20–28 GHz frequency synthesizer for software-defined radio applications," *IEEE Journal of Solid-State Circuits*, vol. 45, no. 9, pp. 1657–1668, Aug. 2010.

[3] H. Rücker *et al.*, "A 0.13 SiGe BiCMOS technology featuring f_T/f_{max} of 240/330 GHz and gate delays below 3 ps," *IEEE J. Solid-State Circuits*, vol. 45, no. 9, pp. 1678–1686, Sep. 2010.

[4] N. Pohl, T. Jaeschke, and K. Aufinger, "An ultra-wideband 80 GHz FMCW radar system using a SiGe bipolar transceiver chip stabilized by a fractional-N PLL synthesizer," *IEEE Trans. Microw. Theory Techn.*, vol. 60, no. 3, pp. 757–765, Mar. 2012.

[5] T.-N. Luo, C.-H. E. Wu, and Y.-J. E. Chen, "A 77-GHz CMOS automotive radar transceiver with anti-interference function," *IEEE Trans. Circuits Syst. I*, vol. 60, no. 12, pp. 3247–3255, Dec. 2013.

[6] H. Jia *et al.*, "A 77 GHz frequency doubling two-path phased-array FMCW transceiver for automotive radar," *IEEE J. Solid-State Circuits*, vol. 51, no. 10, pp. 2299–2311, Oct. 2016.

[7] A. Townley *et al.*, "A 94-GHz 4TX–4RX phased-array FMCW radar transceiver with antenna-in-package," *IEEE Journal of Solid-State Circuits*, vol. 52, no. 5, pp. 1245–1259, Mar. 2017.

Proceedings of the 27th International Conference *"Mixed Design of Integrated Circuits and Systems"*
June 25-27, 2020, Łódź, Poland

Active Feedbacks Comparative Analysis for Charge Sensitive Amplifiers Designed in CMOS 40 nm

Grzegorz Wegrzyn, Rafal Kleczek, Piotr Kmon

Faculty of Electrical Engineering, Automatics, Computer Science and Electronics

AGH University of Science and Technology

Krakow, Poland

grzesweg@student.agh.edu.pl

Abstract—**In this paper we present a comparative analysis of active feedback circuits dedicated to charge sensitive amplifiers (CSA) used in X-ray imaging systems. This work is motivated by the fact there are many papers discussing advantages and disadvantages of using particular CSA feedback but non of them are done in the same process which is very crucial. The presented design, prototype recording channels fabrication employing two the most competing solutions, and their further measurement results may therefore help one in choosing the most suitable feedback for a particular application. The presented circuits are designed in CMOS 40 nm process and are compared in terms of noise contribution, power consumption, area occupation, ability to minimize detectors leakage current, and also CSA stability.**

Keywords—**X-ray imaging, active feedback circuit, CMOS, sensor, energy measurement, charge sensitive amplifier.**

I. INTRODUCTION

X-ray imaging is a widely used solution for biology, medicine, materials analysis, high energy physics, etc. [1], [2]. Noticeably, thanks to the latest technology developments the direct digital X-ray imaging has become a main contestant of, commonly used in medical imaging, indirect medical imaging. This is because contrary to the indirect, the direct imaging allows lowering the patient applied dose and to increase system spatial resolution.

The typical direct X-ray imaging system has a pixelated architecture and is composed of two main parts, i.e. detector and application specific integrated circuit (ASIC), connected together with the use of a flip-chip bonding technique (see Fig. 1a). Thanks to modular architecture it is possible to adapt both parts for different applications requirements. Typically, both the ASIC and the detector are of the same geometry of 100 μm pitch or below.

Direct X-ray imaging is based on converting the incoming X-ray photons into a voltage. This is done with the use of detector (a reversely biased p-n junction) where incoming photons deposit their energy that is then transformed into short current pulses. Current pulses are next converted by the charge sensitive amplifier (CSA) into a voltage step of amplitude equal to Q_{in}/C_F, where C_F is the CSA feedback capacitance and Q_{in} is a charge deposited in the sensor active volume by impinging photon. Regarding the CSA signal processing, there are two main approaches, i.e. a charge integration type (used in CCDs, CMOS imagers, etc.) and a Single Photon Counting (SPC) mode. The SPC mode has become very popular as,

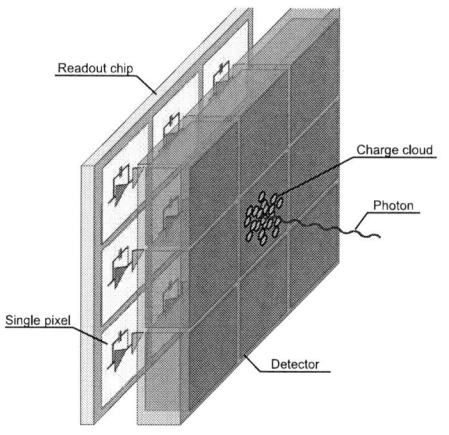

(a)

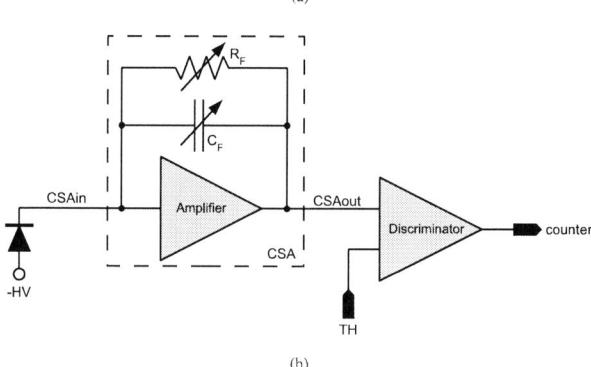

(b)

Fig. 1. a) 2-D hybrid pixel detector system and b) simplified recording channels schematic.

comparing to the charge integration type detectors, it offers very high dynamic range, noiseless imaging, and possibility to count only those pulses which amplitudes are only within a user defined energy window. Though, the SPC mode, comparing to the integration mode, suffers from worse capability of coping with high energy fluxes. Therefore, many research groups worldwide lead intensive research on developing SPC based mode systems overcoming this disadvantage [3]–[7].

The typical readout channel (see Fig. 1b) is composed of CSA, a threshold TH setting block, a discriminator, and a

78

digital counter. Its operation is based on counting only those pulses that are above the user defined threshold value TH. The SPC readout channel is a crucial part of the hybrid detector and it determines the overall SPC main parameters like noise, power consumption, spatial resolution or high count rate performance. Thus, readout channels' particular blocks need to be thoroughly verified not to worsen entire system parameters.

In this paper we present a comparative analysis of two active feedback circuits. The first is a commonly used solution named Krummenacher circuit [8] while the second feedback is an approach presented in [9] and being a very attractive counterpart. The feedback circuits are compared in terms of noise contribution, power consumption, area occupation, detectors leakage current compensation, and CSA stability. Two prototype recording channels were designed in the CMOS 40 nm process and sent for fabrication. The paper is organized as follows. In the Section II we provide CSA feedbacks description while in the Section III a prototype recording channels design is shown. The Section IV provides post-layout simulation results of designed circuits and finally conclusions are provided.

II. CSA FEEDBACKS ARCHITECTURE DESCRIPTION

There are many CSA feedback solutions differing in their complexity, functionality and influence on the overall X-ray imaging systems parameters [6], [9]. One of the most popular solutions is an active feedback named Krummenacher circuit (named in the article as AF-A) which is compared with an active feedback presented in [2], [9] (named in the article as AF-B). Both circuits are presented in the Fig.2.

In general the CSA feedback is responsible for:
- setting the DC operating point of the CSA core,
- discharging the CSA feedback capacitance C_F with a user defined time (often controlled in a wide range to meet different system requirements),
- detector leakage current compensation.

The CSA feedback can be modelled as a parallel connection of two components, i.e. resistor R_F (responsible for capacitor C_F discharging), and inductor L_F (responsible for minimization of detectors leakage current influence).

Regarding the resistance R_F and discussed active feedback circuits small-signal models, it is realized by M1/M2 and M3 transistors in the AF-A and AF-B feedback circuits respectively. The idea of detector leakage current influence compensation is to continuously control the feedback to minimize additional DC current flowing through the resistance R_F. In that way the CSA output DC voltage shift may be minimized which is especially important in low voltage systems. This functionality is realized by C_X capacitor, M2/M3 transistors, and C_X capacitor with M1/M2/M5 transistors in AF-A and AF-B feedbacks respectively. Main active feedbacks parameters are shown in the Tab. I. It can be seen that the AF-B circuit offers better stability with respect to AF-A which is caused by extremely high M5 transistors based resistance (this transistors is biased to work in the cut-off region). Importantly this

directly translates to smaller AF-B area occupation as the C_X capacitor may be meaningfully reduced compared to the AF-A. Also, large inductance of AF-B is formed in a much easier way. Both feedback solutions allow for the $R_f C_f$ constant to be controlled in a broad range and therefore adapted to specific input signal rate conditions. The AF-B shows however DC drift at the output of the CSA as a result of detectors leakage current or feedback settings. This inconvenience, however can be mitigated by using AC coupling or by using an idea proposed in [10].

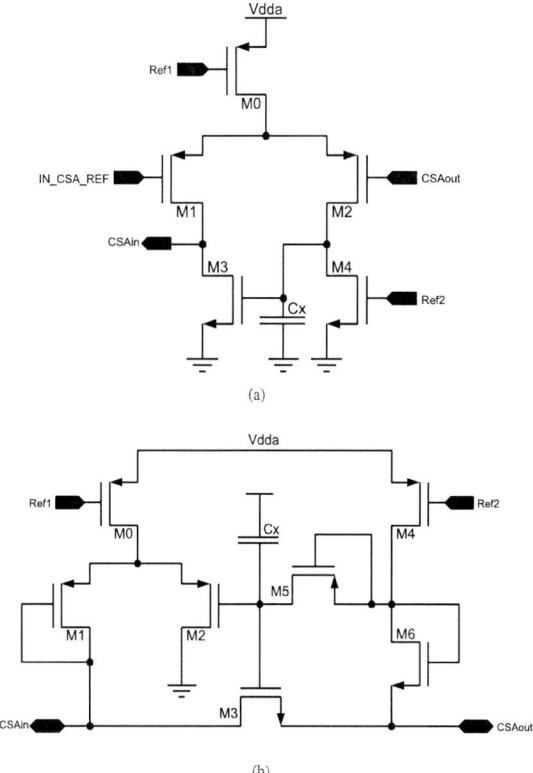

Fig. 2. CSA active feedbacks schematic ideas: a) AF-A circuit and b) AF-B circuit.

TABLE I.
ACTIVE FEEDBACK CIRCUITS PARAMETERS

Parameter	AF-A	AF-B
Small signal equivalent	$R_f \approx \dfrac{2}{g_{m1}}$ $L_f \approx \dfrac{2C_X}{g_{m2}g_{m3eff}}$	$R_f \approx \dfrac{1}{g_{m3}}$ $L_f \approx \left(\dfrac{1}{g_{m1}} + \dfrac{1}{g_{m2}}\right)\dfrac{C_X}{g_{ds5}}$
Stability requirement	$\dfrac{C_X}{g_{m3eff}} \gg \dfrac{2C_f}{g_{m1}}$	$\dfrac{C_X}{g_{ds5}}\dfrac{g_{m1}+g_{m2}}{g_{m1}g_{m2}} > \dfrac{4C_f}{g_{m3}^2}$

III. DESIGN OF THE PROTOTYPE RECORDING CHANNEL

The considered active feedbacks were implemented as a test structures in the one of projects we have lately sent for fabrication (see Fig. 3). We assumed that both feedbacks should be

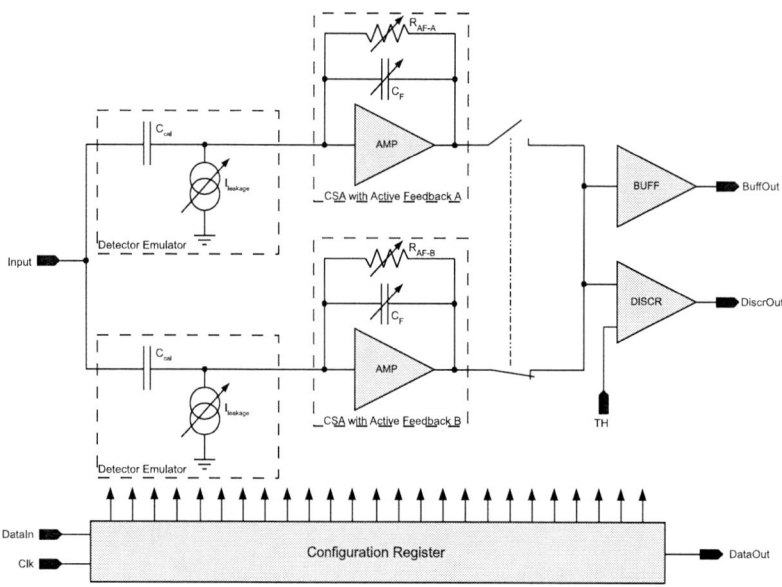

Fig. 3. Active feedbacks prototype recording channels schematic idea.

designed for the same parameters, i.e. power consumption not higher than 30 nW, CSA output pulses time width controlled by the user in the range of 200 ns - 1.5 μs, and phase margin to be about 50 deg. Additionally, the compared feedback circuits should operate with the same CSA core which in our case is a voltage amplifier based on the folded cascode architecture. The main active feedbacks components dimensions are given in the Table II. In order to control the CSAs' output pulses time widths, vary CSA gain, compensate described blocks for PVT influence, provide recording channels input pulses, change discriminators threshold, the configuration register is used (see Fig. 3). Both CSA output signals are provided either in the analog or digital manner by the use of an analog multiplexer followed by discriminator and analog buffer.

TABLE II.
ACTIVE FEEDBACKS KEY TRANSISTORS DIMENSIONING

Block	Transistor	$W[\mu m]$	$L[\mu m]$
AF-A	M_1, M_2	0.5	0.4
	C_X	22	4.7
AF-B	M_0, M_2	0.2	0.2
	M_3	0.4	2
	C_X	3	3

IV. SIMULATION RESULTS

The Fig. 4 shows the designed prototype layout masks view with its main blocks description.

One of the required recording channels parameters is its ability to control CSA output pulse width in a broad range. Thanks to this functionality the overall system may be adapted to different experiments differing in input pulses intensity.

The Fig. 5 shows postlayout simulation results of the CSA output pulses for different active feedback settings. It can be seen that both recording channels allows for setting pulses time widths (defined as 1% amplitude value) in the range of 200 ns - 1.5 μs. As we pointed out in the Section II the AF-B circuit shows slight CSA output DC voltage dependence on the feedback setting.

We also checked the active feedback circuits noise contribution to the CSA overall equivalent noise charge (ENC). It can be seen (see Fig. 6) that both feedback circuits contributes substantial noise to the overall ENC_{TOT}. However, the AF-A circuit noise is almost 50% times higher compared to the AF-B circuit, i.e. the AF-A contributes about 60 e- while AF-B about 40 e- of noise for a given pulse time width range. Fig. 6 shows a ratio of the feedback noise ENC_{AF} to the total noise represented as ENC_{TOT}.

The detectors leakage current influence on the both active feedbacks was verified by supplying the CSA input with a DC current of a controlled current value. It can be seen (see Fig. 7) that the CSA output DC level of AF-B is much more susceptible to the detectors leakage current compared to the AF-A. This is a result of the way the AF-B circuits M2 transistor is controlled, i.e. in order to react on the input detectors leakage current the M2 transistors' gate voltage needs to be changed which is a CSA output DC voltage. However, the AF-B circuit keeps reasonable phase margin even the detectors leakage currents exceed the tranistors M0 DC current. This behaviour however is not a case for AF-A circuit.

Active feedback area occupation is mainly determined by the size of C_X capacitor which is caused by the CSA stability requirement (see Table I). Providing reasonable CSA phase margin directly influences the C_X value and therefore the

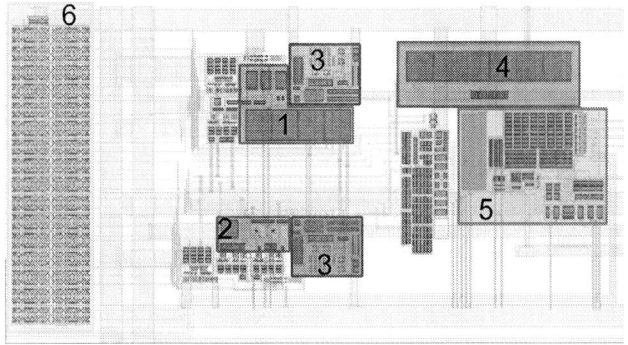

1 – AF-A   3 – CSA   5 – Discriminator
2 – AF-B   4 – Output Buffer 6 – Configuration Register

Fig. 4. Designed circuit layout view.

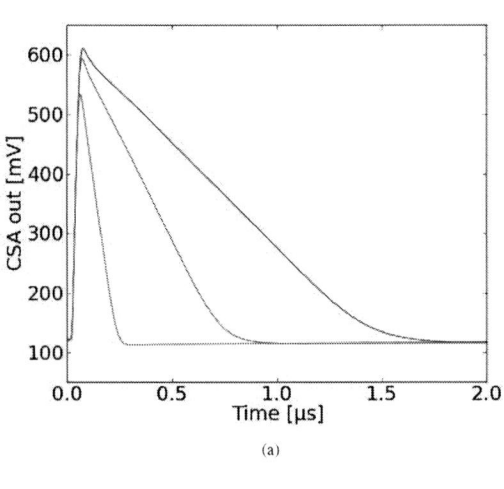

(a)

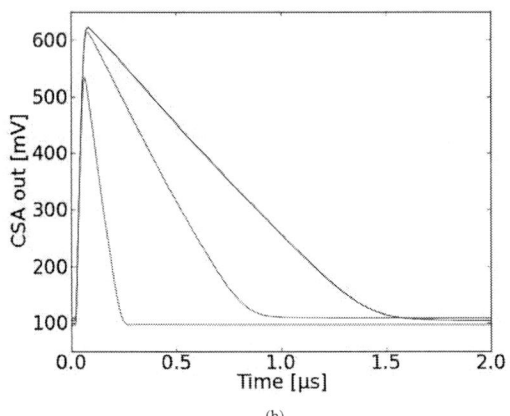

(b)

Fig. 5. Pulse width control of (a) AF-A circuit and (b) AF-B circuit ($Q_{in} \approx$ 15 ke, $C_F = 10fF$).

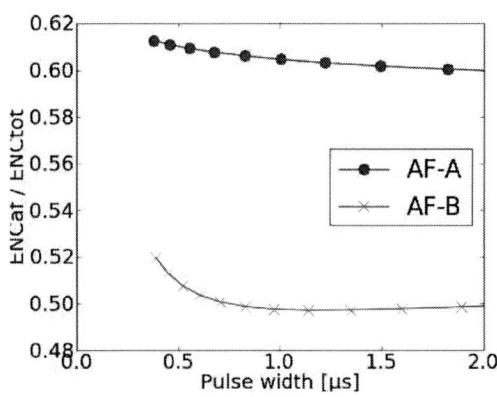

Fig. 6. Feedback noise contribution to the overall CSA noise expressed in ENC units ($C_F = 10fF$, $Q_{in} \approx 2\ ke$).

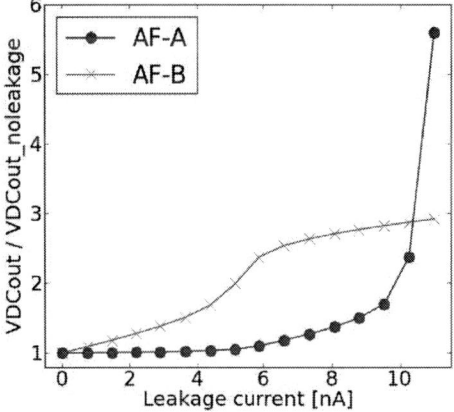

Fig. 7. CSA output DC voltage dependence on the detectors leakage current for AF-A and AF-B circuits (the pulse time width set to 700 ns).

feedback area occupation. In the presented design we decided to use a MOS transistors working as a capacitor C_X and these are PMOS transistors of 22 μm/4.7 μm and 3 μm/3 μm channel area for AF-A and AF-B feedbacks respectively. One should keep in mind that whenever very long pulse time widths are required, i.e. very large feedbacks' resistance needs to be set, the MOS based capacitors' C_X own leakage currents have to be also taken into account.

Active feedback parameters are summarized in the Table III.

TABLE III.
ACTIVE FEEDBACKS PARAMETERS SUMMARY

Parameter	Unit	AF-A		AF-B	
Pulse width	μs	0.2	1.5	0.2	1.5
Power consumption	nW	15	4	22	5
ENC_{AF}/ENC_{TOT}	–	0.61	0.60	0.52	0.5
ENC_{TOT}	e^-	101	100	78	72
Phase margin	deg	54	55	53	55
Occupied area	μm^2	160		50	

CONCLUSIONS

The paper provides comparative analysis one of the most popular active feedbacks used mainly in digital X-ray imaging (Krummenacher circuit) with the active feedback recently proposed in [9]. Both feedbacks were adapted in the recording channel dedicated to medical imaging requiring low noise, detectors leakage current compensation, and small silicon area occupation. Postlayout simulations show that both feedbacks allow for satisfying power consumption and CSAs output pulses time width control range requirements. However, as it was shown the AF-B circuit surpasess the AF-A in terms of lower noise contribution, smaller area occupation, and ability to work with higher leakage currents. Though, the AF-A shows much less detectors leakage current influence on CSA output DC voltage than AF-B. Importantly, this disadvantage can be easily adressed by AC coupling or solution shown in [10].

ACKNOWLEDGMENT

The presented work has been supported by the National Science Center, Poland under Contract No. UMO-2016/23/D/ST7/00488.

REFERENCES

[1] P. Delpierre, "A history of hybrid pixel detectors, from high energy physics to medical imaging," *Journal of Instrumentation*, vol. 9, no. 05, pp. C05 059–C05 059, may 2014. [Online]. Available: https://doi.org/10.1088%2F1748-0221%2F9%2F05%2Fc05059

[2] S. L. Holm, G. W. Deptuch, F. Fahim, P. Grybos, J. R. Hoff, P. Kmon, P. Maj, D. P. Siddons, R. Szczygiel, and T. M., "Minivipic: Pixel readout integrated circuit with on-chip charge cluster reconstruction for x-ray photon science," *Nuclear Science Symposium and Medical Imaging Conference, NSS/MIC*, p. 1, 2015.

[3] H. Kim, S. Han, J. Yang, S. Kim, Y. Kim, S. Kim, D. Yoon, J. Lee, J. Park, Y. Sung, S. Lee, S. Ryu, and G. Cho, "An asynchronous sampling-based 128×128 direct photon-counting x-ray image detector with multi-energy discrimination and high spatial resolution," *IEEE J. Solid-State Circuits*, vol. 48, no. 2, pp. 541–558, Feb. 2013.

[4] R. Ballabriga, J. Alozy. G. Blaj, M. Campbell, M. Fiederle, E. Frojdh, E. H. M. Heijne, X. Llopart, M. Pichotka, S. Procz, L. Tlustos, and W. Wong, "The Medipix3RX: a high resolution, zero dead-time pixel detector readout chip allowing spectroscopic imaging," *J. Instrum.*, vol. 8, no. 02, pp. 02 016–1–02 016–15, feb 2013.

[5] R. Dinapoli, A. Bergamaschi, D. Greiffenberg, B. Henrich, R. Horisberger, I. Johnson, A. Mozzanica, V. Radicci. B. Schmitt, X. Shi, and G. Tinti, "EIGER characterization results," *Nucl. Instrum. Meth.*, vol. A 731, pp. 68–73, 2013.

[6] A. Rivetti, *CMOS: Front-End Electronics for Radiation Sensors*. CRC Press, 2015.

[7] R. Kłeczek, P. Gryboś, R. Szczygieł, and P. Maj, "Single photon-counting pixel readout chip operating up to 1.2 Gcps/mm^2 for digital X-ray imaging systems," *IEEE J. Solid-State Circuits*, vol. 53, no. 9, pp. 2651–2662, Sep. 2018.

[8] F. Krummenacher, "Pixel detectors with local intelligence: an ic designer point of view," *Nuclear Instruments and Methods in Physics Research Section A: Accelerators, Spectrometers, Detectors and Associated Equipment*, vol. 305, no. 3, pp. 527 – 532, 1991.

[9] P. Kmon, G. Deptuch, F. Fahim, P. Gryboś, P. Maj, R. Szczygieł, and T. Zimmerman, "Active feedback with leakage current compensation for charge sensitive amplifier used in hybrid pixel detector," *IEEE Trans. Nucl. Sci.*, p. 1, 2019.

[10] P. Kmon, P. Kaczmarczyk, and Ł. Kadłubowski, "Design of analog pixels front-end active feedback," *Journal of Instrumentation*, vol. 13, no. P01018, pp. 1–8, jan 2018.

Proceedings of the 27th International Conference *"Mixed Design of Integrated Circuits and Systems"*
June 25-27, 2020, Łódź, Poland

ASIC Architecture and Implementation of RED Scheduler for Mixed-Criticality Real-Time Systems

Lukáš Kohútka, Viera Stopjaková

Institute of Electronics and Photonics
Slovak University of Technology in Bratislava
Bratislava, Slovakia
lukas.kohutka@stuba.sk

Abstract—**This paper presents a new ASIC design of a coprocessor that performs process scheduling for embedded mixed-criticality real-time systems consisting of processes of various criticality and various real-time attributes. The proposed solution is implementing Robust Earliest Deadline (RED) algorithm and previously developed hardware architectures used for scheduling of real-time processes. Thanks to the on-chip implementation of the scheduler in a form of a coprocessor, the scheduler operations can be completed in two clock cycles regardless of the process amount within the system contains. The proposed scheduler was verified by simulations that applied millions of random inputs. Chip area costs are evaluated by synthesis into an ASIC using 28 nm process by TSMC. Two versions of real-time process schedulers were compared: EDF scheduler designed for hard real-time processes only and the proposed RED scheduler. The RED algorithm handles variations of process execution times better, achieves higher CPU utilization and can be used for scheduling of hard real-time, soft real-time and non-real-time processes combined within one system that is not possible using the other scheduling algorithms.**

Keywords—**scheduling, mixed-criticality, ASIC, RED.**

I. INTRODUCTION

Real-time (RT) embedded systems are systems that execute RT processes, also known as RT tasks. Successful execution of a real-time process depends not only on the correct results of the computation inside the process but on the time when the process is finished as well. Even if we use a processor that disposes the highest performance, there is no guarantee that all RT processes are finished before their deadlines. Therefore, a process scheduler that is taking into consideration process deadlines must be used [1].

Ideally, RT process schedulers always create an optimum schedule with the best possible order of processes so that all processes will be finished before their deadlines. In addition to that, an ideal process scheduler introduces a minimum CPU overhead, which means that the CPU time is consumed for the scheduled processes, not their scheduling. An ideal scheduling should not only consume minimum CPU time, this time should be a constant time too! In fact, this requirement is the most important because only this way, it can is guaranteed that the whole RT system remains deterministic and well-predictable [1-3].

The major motivation to implement a process scheduler in ASIC is that the software (SW) implemented process schedulers are usually based on user-defined priorities, not the actual deadlines of processes. Therefore, these SW schedulers are not enough robust nor efficient enough for deterministic support of processes of various criticality levels that furthermore are combined in the same system (without hypervisor). Luckily, process scheduling algorithms can be accelerated by hardware (HW) in order to meet the performance and determinism requirements [1].

Another problem is that the existing schedulers implemented in HW are generally applied in only simple systems that consist of hard RT processes only. These solutions are not suitable enough for RT systems with higher complexity, robustness, size and with variety of process types that are mixed within the same system. Thus, a more robust scheduler supporting various types of processes is required. Such a scheduler can be achieved by implementing and existing algorithm called Robust Earliest Deadline (RED), which is designed for handling of system overheads [1-10].

Mixed-criticality is an active research field, as reflected in the review by Burns and Davis [11]. Mixed-criticality is resulting from the global trend in microelectronics that lies in increasing integration of growing number of components on chip belonging to safety-critical domain. Until not so long time, process isolation was required within safety-critical certification to imply performing of critical processes in a separated hardware. Due to the process isolation, RT systems were under-utilized and based on the pessimistic analysis of worst-case execution time for each process. As average execution time is generally significantly smaller than the worst-case time, CPU resources are usually reserved much more than it is really needed for execution of RT processes [12-15].

Mixed-criticality RT systems can contain any combination of critical and/or non-critical processes in the same platform without complete isolation of the processes as it is considered to be too inefficient (e.g. using hypervisors). Non-isolation is recommended as a demanding and efficient way to improve CPU utilization, which is achieved by executing low-priority and non-critical processes within slack time that becomes available whenever high-priority and high-critical processes are finished sooner than it was predicted according to worst-case execution time. This is happening relatively frequently in real-world applications. The problem is that safety-critical processes and best-effort processes have usually conflicting requirements, being an important research challenge [16-18].

The topic of mixed-criticality is considered as a very active research field, which is mentioned in the review from Burns and Davis. The mixed-criticality research is trying to solve the problem of scheduling policies that are supposed to maximize CPU utilization and simultaneously to maintain the schedulability of every safety-critical process [11].

II. RELATED WORK

One of the most popular scheduling algorithms for RT systems is Earliest Deadline First (EDF), which is proven to always create an optimum schedule for systems containing hard RT processes only [1, 2]. The main idea of this algorithm is to order all processes according to their deadlines in such a way so that a process with the earliest deadline is the process that should be executed first. The processes that are ready for execution are stored in a structure called "process queue". Process queues can be implemented by various sorting architectures, such as shift registers architecture [19] or systolic array architecture [20].

In our previous work published in [21-24], we already presented novel RT process schedulers implemented in a form of coprocessors. A slightly modified EDF algorithm was used to support killing of processes according to their ID, which is important in recent operating systems. The possibility to kill any process in the system according to the process ID increases flexibility and extensibility of the proposed solution, such as process synchronization and support of non-RT processes.

Beside of our research, there have been also other solutions presented. One solution employs EDF algorithm with the maximum number of processes being 64 [25], and the other approach uses priorities instead of deadlines, which is less efficient for RT systems in terms of CPU utilization [26]. There are other solutions based on priorities or static scheduling [8-10, 27-29] as well. These solutions are suitable for systems consisting of hard RT processes only.

Since various types of processes may co-exist together in complex RT mixed-criticality system, a requirement for a robust process scheduler that supports all types of processes arises. For this reason, we decided to create an ASIC-based process scheduler that implements Robust Earliest Deadline (RED) algorithm, which is suitable for all types of processes [1].

The RED algorithm is an extension of the EDF algorithm. Both algorithms are performing sorting of processes based on process deadlines, which is shown in Fig. 1. EDF accepts all requests for adding and scheduling a freshly created process, but RED is predicting if any of the processes is able to miss a deadline and if so, then it chooses the most suitable process to temporarily reject it and move it to Reject queue. By this, we can guarantee that all hard RT processes always meet deadlines and additionally, as many soft RT processes as possible meet deadlines too. In case when any of the ready processes is finished sooner than the worst-case timing analysis predicts, and there is sufficient free time to reclaim and finish on time any of the rejected processes, then RED uses its reclaiming policy for moving one of the rejected processes back from Reject Queue to the Ready queue [1, 2].

The method for monitoring of execution time of RT processes and prediction of their deadline misses was also presented in [30-32].

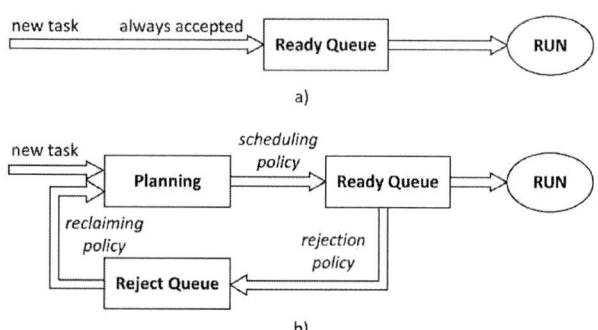

Fig. 1. EDF (a) and RED (b) scheduling algorithms [1].

III. PROPOSED RED SCHEDULER

A. Top module

Fig. 2 shows a top module of the proposed scheduler represents a coprocessor that contains 3 components: Ready Queue, Control Unit and Reject Queue.

The top module has only one input (called INSTR) – the instruction provided by CPU. There are two types of instructions: *schedule_process* and *kill_process*.

The output of the top module is PROCESS_TO_RUN, which provides information back to CPU about which process is selected by the scheduler for execution at the moment.

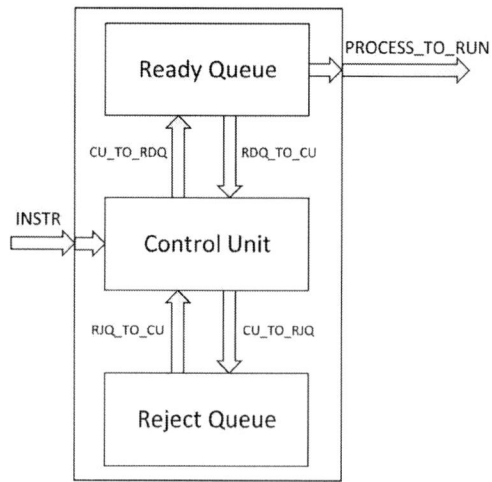

Fig. 2. Top module of RED scheduler.

B. Ready Queue

This component is based on an existing sorting architecture called Shift Registers, which consists of process cells. Each process cell contains one process comparator, control logic and a register to store one process. The process cells can exchange processes with adjacent process cells. Each process cell receives the same instruction in parallel from the shared bus. An example of the Shift Registers with four cells is given in Fig. 3 [19].

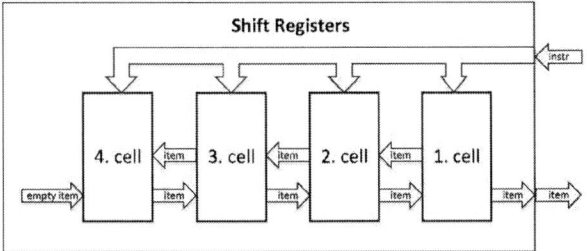

Fig. 3. Shift Registers [19].

The Ready Queue component was extended by a feature called Overload Analysis by applying the following changes:

1. Extend each process cell with register for execution time

2. Add overload detection into each process cell

3. Add combination of overload bits for all process cells

4. Add victim selection

Change 1 is to insert an execution time register into every process cell within the Ready queue component. It is used as a remaining worst-case execution time for the actual process cell + the sum of the remaining worst-case execution times of all processes that are scheduled before the actual process.

If a new process is inserted to the process cell, then the new value of the execution time register is calculated by adding the worst-case execution time of the new process + the value of the execution time register stored in the process cell located before the actual process cell, where the new process is supposed to be inserted. In addition to this, the worst-case execution time value of the new process is added to the execution time registers of all processes scheduled behind the new process.

In an existing process is removed from Ready queue, then all processes that are scheduled behind the removed process must decrease the values of execution time registers by the current value of execution time register of the removed process.

Change 2 is to implement a logic that calculates the overload bit of every process cell. The logic checks whether the worst-case execution time of the new process + the value stored in the execution time register of the actual process cell is greater than the deadline value of the actual process. If yes, then scheduling the new process before the process in the

actual process cell can cause the actual process to miss its deadline. This situation is called overload and the overload bit of the corresponding process cell is set to high in such a situation.

Change 3 is to add an OR gate within the Ready queue component but outside of the process cells. The OR gate takes as input all overload bits provided from the process cells. If at least one overload bit is '1', then the OR gate output is '1' too, meaning that the system is overloaded. This result is then given to the Control Unit component via RDQ_TO_CU.

Change 4 is to add a logic that selects one process within the ready processes stored in the Ready Queue component. The selected process is removed from Ready Queue and inserted to Reject Queue. This process is called "victim process" as well because it is sacrificed by the scheduler because the system would be overloaded otherwise.

C. Control Unit

This component is responsible for controlling Reject Queue and Ready Queue components. Whenever a new valid instruction is provided from CPU, Control Unit performs this instruction by forwarding it to Ready Queue. If no instruction is provided from CPU, Control Unit can automatically move processes between Reject Queue and Ready to maximize the number of processes to be executed on time but to keep the system to be not overloaded.

Whenever the system is overloaded (i.e. in overload state), the Control Unit gives an order to remove one process within the processes stored in Ready Queue (called a victim) and to store this process to Reject Queue. The victim is chosen according to priority values and position of the process within the Ready Queue in such a way that the process with the lowest priority within those processes that are located before the first process reporting the overload is chosen as the victim.

Whenever an existing process is killed (i.e. removed from the Ready Queue) and the system is not overloaded, the Control Unit tries to reclaim one process from Reject Queue. Thus, one process is selected among the processes in Reject Queue and moved back to Ready Queue. Then, the overload state is checked again. Control Unit moves back the reclaimed process from Ready Queue to Reject Queue if the system is overloaded again. However, the reclaiming is successful and the reclaimed process remains in the Ready Queue if the system is not overloaded.

There are 4 criticality levels supported by the proposed scheduler. They are used for the rejection and reclaiming policy described above. These 4 criticality levels are defined by 2 bits the following way:

- "00" – non-RT processes and soft RT processes with low priority
- "01" – soft RT processes with medium priority
- "10" – soft RT processes with high priority
- "11" – safety-critical hard RT processes

D. Reject Queue

The Reject Queue component is a queue that consists of rejected processes. These processes are ordered in such a way that the process with the highest priority and the latest deadline is selected and sent to output of the queue. The selection of such a process is performed in 1 clock cycle. The Reject Queue is an implementation of Shift Registers architecture, just like the Ready Queue but without the new changes performed only for Ready Queue. Also, the Shift Registers architecture is configured as a MAX queue.

Every Reject Queue item is mapped 1:1 to one rejected process. The items of the Shift Registers are ordered according to a value that is a combination of two values that are concatenated: process priority and process deadline. The process priority is used for the higher bits and the process deadline is used for the lower bits. Thus, process priority has higher importance than process deadline when choosing which process is supposed to be reclaimed as the next one.

Shift Registers can remove any process depending on the given process ID. However, this feature is not needed for Reject Queue component because only the process that is at the output of the queue is always removed. Thus, the Reject Queue component was simplified to always extract the 1st process only.

IV. VERIFICATION

The proposed RED scheduler was described using SystemVerilog language. This language was used to verify the proposed solution in simulations as well. These simulations are based on Universal Verification Methodology in principle, but only using the SystemVerilog language without any UVM libraries. One UVM transaction is mapped to 1 instruction, which takes 2 clock cycles to execute. No UVM agents were needed for interfacing of DUT (Design Under Test).

We created one test procedure can randomly generate instructions for DUT input and one predictor that is predicting expected outputs of DUT. The expected outputs provided by predictor ale compared to the real output of DUT and the comparison results are written to scoreboard.

Our test procedure generates millions of instructions consisting of fixed instruction opcodes and process IDs but random process deadlines, worst-case execution times and real execution times.

The predictor generates expected outputs of the DUT according to the instructions sent to DUT input. The predictor logic is described in a pure sequential manner (same as a software) using much higher level of abstraction as well. For example, our predictor adopts existing SystemVerilog keyword called queue and a built-in function for processes sorting called *sort()*.

Correct behavior of the proposed solution was verified by several millions of test cycles, each consisting of 1000 instructions. One half is *schedule_process* and the second half is *kill_process*. Full capacities of the Ready Queue and Reject Queue were tested. The DUT output and expected output from predictor were the same for 100% of used instructions.

V. SYNTHESIS RESULTS

The proposed RED scheduler and original EDF scheduler were both synthesized into TSMC 28nm high performance mobile (HPM) technology using Cadence Genus. A clock frequency of 500 MHz and the power supply of 0.9 V were used for the synthesis. The static timing analysis report contained no negative setup slack after the synthesis.

The chip area costs of the EDF scheduler and RED scheduler are presented in Table I. The Maximum Number of Processes column defines the maximum number of processes supported by the synthesized scheduler, which is equal to the Ready Queue size. The chip area cost of EDF scheduler and RED scheduler are presented in column EDF Chip Area and RED Chip Area, respectively. Table I also shows a relative overhead (in column Overhead), which represents a relative increase of chip area cost caused by extending the EDF scheduler into RED scheduler.

Both schedulers use the minimum required process ID bit width, e.g. 3 for 8 processes or 5 for 32 processes. The obtained results show that the RED scheduler has notably higher chip area cost, which is caused by the fact that the RED scheduler implements significantly more complex algorithm.

TABLE I.
CHIP AREA COST OF EDF AND RED
WITH VARIOUS MAXIMUM NUMBER OF PROCESSES

Maximum Number of Processes	EDF Chip Area [μm²]	RED Chip Area [μm²]	Overhead
8	1702	13214	+676,38%
16	3637	28531	+684,47%
24	5979	50218	+739,91%
32	9085	69389	+663,78%
40	13307	92389	+594,29%
48	19387	108522	+459,77%
56	26787	130421	+386,88%
64	33340	146146	+338,35%

The power consumption results are presented in Table II. These results represent the total power consumption that consists of the leakage and dynamic power. The dynamic power consumption was calculated using an average toggle rate of every bit in the chip, which is not very precise metric because the real power consumption would depend a lot on the frequency and order of coprocessor instruction calls. However, the same test case with the same toggle rate of bits was applied for both, EDF and RED schedulers. Therefore, the comparison is still valid and relevant. As one can see from the table, these power consumption results are proportionally similar to the increase of chip area cost that was already presented in Table I. The dependency between the chip area results and power consumption results is apparently linear.

TABLE II.
POWER CONSUMPTION OF EDF AND RED
WITH VARIOUS MAXIMUM NUMBER OF PROCESSES

Maximum Number of Processes	EDF Power Consumption [µW]	RED Power Consumption [µW]	Overhead
8	2044	12121	+493,00%
16	3809	25913	+580,31%
24	6356	43373	+582,39%
32	9881	66559	+573,61%
40	14711	77434	+426,37%
48	20693	93948	+354,01%
56	28263	115748	+309,54%
64	36535	124162	+239,84%

The overall comparison of the EDF and the novel RED schedulers is presented in Table III. The support of all types of processes and their combinations in one scheduler is provided at a cost of higher chip area and power consumption.

TABLE III.
COMPARISON OF EDF AND RED SCHEDULERS

Criterion	Selected Scheduler	
	EDF	RED
Chip area	A	(4.38 – 8.40) x A
Power	P	(3.40 – 5.82) x P
Execution time	2 clock cycles	2 clock cycles
Hard RT processes	yes	yes
Soft RT processes	no	yes
Non-RT processes	yes	yes
Mixed-criticality supported	no	yes

The presented results showed that the proposed RED scheduler requires from three to five times higher chip area than the EDF scheduler. On the other hand, the RED scheduler is much more robust since supporting of hard RT processes and soft RT processes is provided simultaneously.

Both schedulers perform their operations in the same constant time, which is only 2 clock cycles. Thus, the latency and throughput remain unchanged. The presented schedulers can be applied for periodic and sporadic processes the same way as for the aperiodic processes.

VI. FUTURE IMPROVEMENTS

The RED scheduler was optimized for RT systems that are using single-core CPUs only. However, the scheduler can be extended to support two or four CPU cores simultaneously by applying the same approaches that were already used for EDF based schedulers [9-12].

Further improvements can be achieved by adding hardware-accelerated management of periodic processes (i.e. automatic rescheduling of periodic processes after each period), inter-process communication and process synchronization.

Additionally, an optimization of the scheduler in order to make it more scalable with respect to the maximum number of processes would be beneficial too, especially with focus on increasing the maximum clock frequency of the design.

VII. CONCLUSION

We proposed a new process scheduler that performs RED scheduling algorithm and is implemented on chip. The proposed scheduler is suitable for robust mixed-criticality RT systems due to the ability to distinguish between various types of processes, including their priority and deadlines. The proposed solution used an approach of reusing and extending of existing EDF schedulers. The Ready Queue as well as the Reject Queue (i.e. both queues) is implemented using an existing sorting architecture called Shift Registers. Both queues perform an operation of process insertion or process removal in constant time (1 clock cycle) regardless of number of processes used in the system. The RED algorithm extended the EDF algorithm with two features: rejection of process and reclaiming of process reclaiming. These features are combined with a possibility to use various types of processes and various criticality levels.

The RED scheduler correctly and efficiently handles any combination of safety-critical hard RT processes, high-priority soft RT processes, medium-priority soft RT processes, low-priority soft RT processes and non-RT processes. This means that all safety-critical hard RT processes meet their deadlines, and as many as possible soft RT processes meet their deadlines, considering the priority within the soft RT processes as well. Therefore, the proposed RED scheduler is very suitable for mixed-criticality RT systems.

The RED scheduler is providing 1024 priority levels for the non-RT processes and 4 criticality levels for the RT processes. The 4 criticality levels are dividing RT processes to the abovementioned hard RT, high-priority soft RT, medium-priority soft RT and low-priority soft RT processes. Thus, there are 1028 levels of criticality in total (1024 + 4). The priority level, type of task, deadline value and worst-case execution time are provided from system (CPU + operating system + user) to the proposed scheduler via coprocessor interface.

The proposed solution is consuming higher, but still acceptable resource costs, (i.e. chip area and power consumption) which are scaling still linearly with growing maximum number of processes. If we compare the proposed solution to existing solutions implemented in software, the ASIC implementation is able to execute instructions (*schedule_process* and *kill_process*) in constant time (2 clock cycles) and with dramatically higher throughput. Thanks to that, the system using the proposed RED scheduler would be much more efficient and deterministic, resulting in lower amount of processes missing their deadlines, especially if they are high-criticality processes. The ASIC-implemented scheduler causes CPUs to waste minimum possible time for scheduling and to focus only on execution of the scheduled processes instead.

The proposed process scheduler is not intended to fully replace existing software implementations of kernels. Instead of that, it is supposed to be combined with existing software-implemented operating systems in a reasonable way.

ACKNOWLEDGMENT

This work was supported in part by the Ministry of Education, Science, Research and Sport of the Slovak Republic under grant VEGA 1/0905/17, and ECSEL JU under project PROGRESSUS (876868).

REFERENCES

[1] G. Buttazzo, "Hard Real-Time Computing Systems: Predictable Scheduling Algorithms and Applications," 2011.

[2] G. Buttazzo, J. Stankovic, "Adding robustness in dynamic preemptive scheduling," In D. S. Fussell and M. Malek (eds.), Responsive Computer Systems: Steps Toward Fault-Tolerant Real-Time Systems. Boston: Kluwer Academic Publishers, 1995.

[3] M. Caccamo and G. Buttazzo, "Optimal scheduling for fault-tolerant and firm real-time systems," Proceedings Fifth International Conference on Real-Time Computing Systems and Applications (Cat. No.98EX236), Hiroshima, Japan, 1998, pp. 223-231.

[4] G. C. Buttazzo and F. Sensini, "Optimal deadline assignment for scheduling soft aperiodic tasks in hard real-time environments," in IEEE Transactions on Computers, vol. 48, no. 10, pp. 1035-1052, Oct. 1999.

[5] G. Buttazzo, F. Conticelli, G. Lamastra and G. Lipari, "Robot control in hard real-time environment," Proceedings Fourth International Workshop on Real-Time Computing Systems and Applications, Taipei, Taiwan, 1997, pp. 152-159.

[6] M. Spuri, G. Buttazzo, F. Sensini, "Robust aperiodic scheduling under dynamic priority systems," Proceedings 16th IEEE Real-Time Systems Symposium, Pisa, 1995, pp. 210-219.

[7] M. Gergeleit, L. B. Becker, E. Nett, "Robust scheduling in team-robotics," Proceedings International Parallel and Distributed Processing Symposium, Nice, France, 2003, pp. 8.

[8] A. Norollah, D. Derafshi, H. Beitollahi and M. Fazeli, "RTHS: A Low-Cost High-Performance Real-Time Hardware Sorter, Using a Multidimensional Sorting Algorithm," in IEEE Transactions on Very Large Scale Integration (VLSI) Systems, vol. 27, no. 7, pp. 1601-1613, July 2019.

[9] D. Derafshi, A. Norollah, M. Khosroanjam and H. Beitollahi, "HRHS: A High-Performance Real-Time Hardware Scheduler" in IEEE Transactions on Parallel & Distributed Systems, vol. 31, no. 04, pp. 897-908, 2020. doi: 10.1109/TPDS.2019.2952136.

[10] C. Suman and G. Kumar, "Performance Enhancement of Real Time System using Dynamic Scheduling Algorithms," 2019 IEEE 5th International Conference for Convergence in Technology (I2CT), Bombay, India, 2019, pp. 1-6.

[11] A. Burns and R. Davis, "Mixed-Criticality Systems - A Review", 10th edition, University of York, UK, Jan. 2018.

[12] R. Arbaud, D. Juhász, A. Jantsch, "Resource Management for Mixed-Criticality Systems on Multi-Core Platforms with Focus on Communication," 21st Euromicro Conference on Digital System Design, 2018 IEEE. doi: 10.1109/DSD.2018.00108.

[13] R. Ernst, M. di Natale, "Mixed-Criticality Systems - A History of Misconceptions ?", in IEEE Design and Test, vol. 33 issue 5, Oct. 2016. doi: 10.1109/MDAT.2016.2594790.

[14] O. Kotaba, J. Nowotsch, M. Paulitsch, S. M. Petters, H. Theiling, "Multicore in Real-Time Systems - Temporal Isolation Challenges due to Shared Resources", CISTER technical report, 2014.

[15] E. Wandeler, A. Maxiaguine, L. Thiele, "Quantitative Characterization of Event Streams in Analysis of Hard Real-Time Applications", in Real-Time Systems, vol. 29, pp. 205-225, 2005 Springer. doi: 10.1007/s11241-005-6885-x.

[16] M. Neukirchner, P. Axer, T. Michaels, R. Ernst, "Monitoring of Workload Arrival Functions for Mixed-Criticality systems", in 34th Real-Time Systems Symp., 2013 IEEE. doi: 10.1109/RTSS.2013.17.

[17] M. Neukirchner, S. Quinton, R. Ernst, K. Lampka, "Multimode Monitoring for Mixed-Criticality Real-Time Systems", in 2013 Int. Conf. on Hardware/Software Co-Design and System Synthesis, 2013 IEEE. doi: 10.1109/CODESISSS.2013.6659021.

[18] G. Giannopoulou, N. Stoimenov, P. Huang, L. Thiele, B. Dupont de Dinechin, "Mixed-criticality scheduling on clusterbased manycores with shared communication and storage resources", in Real-Time Systems, vol. 52, 2016 Springer. doi: 10.1007/s11241-015-9227-y.

[19] G. Bloom, G. Parmer, B. Narahari, and R. Simha, "Shared Hardware Data Structures for Hard Real-Time Systems," Proceedings of the tenth ACM international conference on Embedded software, 2012.

[20] S.W. Moon, "Scalable Hardware Priority Queue Architectures for High-Speed Packet Switches," IEEE Transactions on Computers, 2000.

[21] L. Kohutka, M. Vojtko, and T. Krajcovic, "Hardware Accelerated Scheduling in Real-Time Systems," Engineering of Computer Based Systems Eastern European Regional Conference, 2015.

[22] L. Kohutka, V. Stopjakova, "Hardware-Accelerated Task Scheduling in Real-Time Systems: Deadline Based Coprocessor for Dual-Core CPUs," International Symposium on Design and Diagnostics of Electronic Circuits and Systems, 2016.

[23] L. Kohutka, V. Stopjakova, "Task Scheduler for Dual-Core Real-Time Systems," International Conference on Mixed Design of Integrated Circuits and Systems, 2016.

[24] L. Kohutka, V. Stopjakova, "Improved Task Scheduler for Dual-Core Real-Time Systems," Euromicro Conference on Digital System Design, 2016.

[25] Y. Tang, and N.W. Bergmann, "A Hardware Scheduler Based on Task Queues for FPGA-Based Embedded Real-Time Systems," IEEE Transactions on Computers, 2015.

[26] J. Starner, J. Adomat, J. Furunas, and L. Lindh, "Real-Time Scheduling Co-Processor in Hardware for Single and Multiprocessor Systems," Proceedings of the EUROMICRO Conference, 1996.

[27] C. Ferreira, and A.S.R. Oliveira, "Hardware Co-Processor for the OReK Real-Time Executive," 2010.

[28] S.E. Ong, and S.C. Lee, "SEOS: Hardware Implementation of Real-Time Operating System for Adaptability," Computing and Networking (CANDAR), 2013 First International Symposium, 2013.

[29] K. Kim, D. Kim, and Ch. Park, "Real-Time Scheduling in Heterogeneous Dual-core Architectures," Proceedings of the 12th International Conference on Parallel and Distributed Systems, 2006.

[30] A. Kritikakou, C. Pagetti, O. Baldellon, M. Roy, C. Rochange, "Run-time Control to Increase Task Parallelism in MixedCritical Systems," in 26th Euromicro Conf. on Real-Time Systems, 2014 IEEE. doi: 10.1109/ECRTS.2014.14.

[31] A. Kritikakou C. Pagetti, C. Rochange, M. Faugere, S. Girbal, D. Gracia Perez, "Distributed Run-time WCET Controller for concurrent critical tasks in Mixed-Critical Systems" in Proc. of the 22nd Int. Conf. on Real-Time Networks and Systems, 2014 ACM. doi: 10.1145/2659787.2659799.

[32] Angeliki Kritikakou, Thibaut Marty, and Matthieu Roy. 2017. DYNASCORE: DYNAmic Software COntroller to Increase REsource Utilization in Mixed-Critical Systems. ACM Trans. Des. Autom. Electron. Syst. 23, 2, Article 13 (October 2017), 26 pages. DOI: https://doi.org/10.1145/3110222

Proceedings of the 27th International Conference *"Mixed Design of Integrated Circuits and Systems"*
June 25-27, 2020, Łódź, Poland

CMOS Interface for Capacitive Sensors
with Custom Fully-Differential Amplifiers

Mariusz Jankowski, Piotr Zając, Piotr Amrozik, Michał Szermer
Department of Microelectronics and Computer Science
Lodz University of Technology
Lodz, Poland
jankowsk@dmcs.pl

Abstract—**In many applications it is crucial to design reliable and efficient analog readout circuits for micro-electromechanical (MEMS) capacitive sensors. In this paper, we describe the switched-capacitor, open-loop, capacitive-sensing readout circuit, which was designed and manufactured in 0.18 μm technology. Non-standard application of a fully differential amplifier structure is also presented. The post-layout simulation results are described to show the proper operation of the circuit. They show that with the proper symmetrical design of the differential signal path the output offset voltage can be kept at acceptable level.**

Keywords—**switched-capacitor circuit, fully-differential amplifier, MEMS accelerometer, capacitive sensing**

I. INTRODUCTION

Intelligent systems for detection and monitoring of balance disorders have recently gained a lot of interest [1]. The most popular solutions are based on multiple sensing devices that are put on patient's body and allow continuous tracking of the patient's movement [2]. The data from sensors is wirelessly transmitted to the central database and can be subsequently used for diagnosis by a human doctor or artificial intelligence.

Such a system is currently being developed within the frame of "InnoReh" (Innovative Rehabilitation) project [3]. One of the crucial parts of the sensing device is the capacitive MEMS accelerometer with analog sensing interface. Several parts of the system have already been described in literature [4][5]. In this paper, we concentrate on custom solutions used for the design of the analog readout circuit (ARC), which is a switched-capacitor, open-loop interface circuit. Its main goal is to convert the MEMS capacitance to a digital signal. The block diagram of the ARC is shown in Fig. 1 and the corresponding layout is presented in Fig. 5.

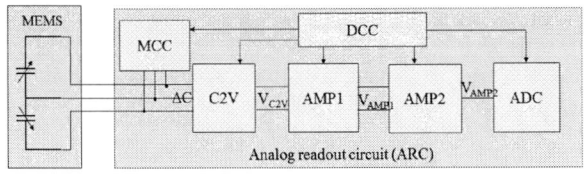

Fig. 1. Block diagram of the designed readout circuit

The MCC block is responsible for mismatch compensation, i.e., the difference in two MEMS capacitances which may occur after MEMS production. The MCC uses a well-known solution of digitally-programmable capacitance ladder [6] with 8 bits. It allows compensating for as much as 10 % of the total MEMS capacitance.

The design of the amplifiers is also based on common designs [ref]. The first amplifier has constant gain and is fully differential while the second has programmable gain one converts the differential signal to single-ended signal. The amplified signal is then sampled by the 10-bit successive-approximation analog-to-digital converter (ADC), operating at 800 kHz. The digital control circuit (DCC) enables dynamic configuration of circuit parameters (clock frequencies, gain settings, etc.) and controlling the ADC.

II. CAPACITANCE-TO-VOLTAGE CONVERTER

In this section, we will describe the most important part of the ARC, namely the capacitance-to-voltage converter (C2V, see Fig. 2). The role of the C2V circuit is to produce the output voltage proportional to the difference in MEMS capacitances ΔC. In the first phase Φ_1, the switches S1, S3 and S4 are closed, which enables charging the MEMS capacitors C_{s1} and C_{s2}. In the second phase Φ_2, the charge from top and bottom MEMS capacitors flows to capacitors C_{i1} and C_{i2}, respectively. The I_1 block is a fully differential op-amp, which works as an integrator. The integrator's output is then equal to the difference of voltages on the C_{i1} and C_{i2} capacitors, which is basically equivalent to being proportional to the difference between C_{s1} and C_{s2}. The circuit is based on pseudo-differential operation: the voltage is applied to the middle node of the MEMS accelerometer and the signal between the other two nodes is measured. Therefore, the problem of the common-mode shift had to be prevented by implementing a common-mode feedback amplifier I_{cmfb} and capacitors C_{fb1} and C_{fb2}. Another potential issue is that the offset voltage of the I_1 amplifier can potentially influence its output. It was mitigated by using a correlated double sampling technique [7] using the switches S6, S7, S8 and S9 with capacitors C_{cds1} and C_{cds2}. In the Fig. 2 the parasitic capacitances between the MEMS and ARC were also shown. The impact of these parasitics is discussed in section IV.

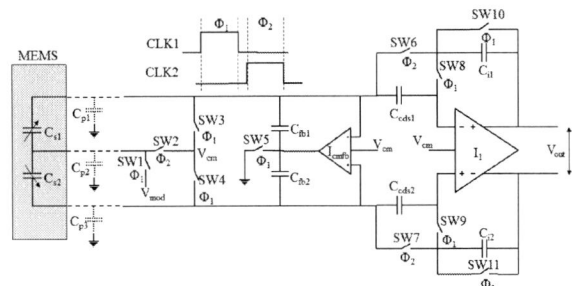

Fig. 2. Design of the capacitance-to-voltage (C2V) converter

89

III. CUSTOM FULLY DIFFERENTIAL OP-AMP DESIGN

The readout circuit uses a fully differential signal flow for possibly highest processing quality. The most complex parts of the designed differential signal path are fully differential (FD) operational amplifiers (op-amps). They all are based on the same topology, however the particular instances are individually tuned according to their place in the signal path and the parameters of processed signals. The CMOS design process selected for implementation of the readout circuitry was XFAB's xp018 technology, as it provides an analog component library. However, this library comprises only single-ended op-amps. Fully-custom design of a complex block like a FD op-amp is a risky task, consequently in the first design, library-based single-ended op-amps simulations have shown that such an approach results are not satisfying so the second design attempt involved a fully-custom FD op-amp.

Several various FD op-amps have been analyzed and simulated in search for the one (or more) topology properly serving its task for expected amplitude and speed parameters of its input signals. One part of typical FD op-amps not present in single ended op-amps is CMFB (common mode feedback) circuitry used for assuring proper level of common voltage at the FD op-amp outputs. Even though such property of FD op-amps makes it possible to set this common voltage to desired value, it also requires additional circuitry for sensing output voltages and re-biasing of the FD op-amp. Thus, internal feedback is present in FD op-amps, making their design even more complicated. At first, continuous time (CT) CMFB circuits were taken into account but as tests progressed and the signal-path vision matured, discrete time (DT) CMFB circuits were also taken into account. One common solution is to use a structure that uses resistor divider to produce average (common) output voltage for FD op-amps, which in turn is used by a control op-amp. Presence of DC current path between outputs is one of main drawback, here. Various resistorless CMFB structures provide better operation quality, as presented in [8]. A few selected resistorless CMFB structures tested [9][10]. One stage differential pair based CMFB circuits like one used in [9] were found to offer inadequate signal processing quality. Two stage op-amp based structures like used in [10] fulfill requirements of the designed FD op-amp, though more complex solutions offering higher signal processing quality have also been proposed [11].

As mentioned before, when works on the FD op-amp substantially progressed, it was found that DT CMFB circuits could be used, which further improves the FD op-amp speed parameters. Various types of such structures have already been proposed. While some of them use mixed approach that uses both continuous time and discrete-time subcircuits [12], most promising ones are the simplest switched-capacitor structures consisting only of switches (NMOS or CMOS transmission gates) and capacitors, as reported in [13]. DT CMFB structure finally used in the design is presented in Fig. 2. Some papers proposed one stage folded-cascode op-amps suitable for our design [14]. The analysis of such op-amp architectures it was unfortunately found that one stage op-amps are not applicable in our case. The reason is that, in general, one stage op-amps can be treated as OTAs (Operational Transimpedance Amplifiers). In such structures, any current path connected to their outputs disrupts the OTA operation. In our case outputs of the required op-amp work with capacitive loads, which just makes settling times too long.

Therefore, various topologies of two stage op-amps have been tested. First, two-stage folded-cascode FD op-amps were extensively tested and it was found that CMFB control loops that drives first stage of such op-amp s is too slow and unstable. CMFB control circuits used for driving the second stage of the designed op-amps show promising behavior worth deeper insight. Further tests of two stage folded-cascode FD op-amps, involving use of Monte Carlo and corner simulations, have shown that overall operation quality strongly depends on process-based mismatches between devices used in the first stage of such op-amps. This effect occurs because the first cascode-based stage has high voltage gain and the CMFB loop is unable to control this stage of op-amp due to speed limitations. FD op-amps with classic structure of their first stage have been found more promising for our design. It usually has lower voltage gain, which is compensated by higher voltage gain of the second stage. Because the second stage of such FD op-amps can be controlled by the CMFB circuit loop, operation of such op-amps in presence of device mismatches in their first stage can be retained. One remaining problem with discussed classic FD op-amp topologies has been caused by the supply voltage equal 5 V. Such high value was selected for the ADC maximum resolution and precision sake. It was found that such high supply voltage causes high current flows output stages of simulated DT op-amps. Current flowing through output stage transistors in absence of any loads were reaching up to several milliamperes, which was unacceptable. A sufficient reduction of these cross-currents by transistor resizing was impossible. Therefore, the problem was solved by the implementation of cross-current limiting structures in the first stage of the designed amplifier, like proposed in [15]. Final schematic of the FD op-amp obtained this way is presented in Fig. 3.

Studies of topologies used in readout systems of MEMS generated signals showed application of a two stage FD op-amp utilizing a specialized first stage, as shown in Fig. 4 [13]. Internal positive feedback is used in this stage for lowering common mode voltage gain, which potentially simplifies common mode output voltage control and topology of CMFB circuitry used for providing such control means. This specific FD op-amp is optimized for low and ultra-low supply voltage applications. There are reported applications of such or similar op-amps for 1 V supply applications [13][16], even for 0.8 V or 0.6 V [17] supply voltages. Even though there are different applications of internal positive feedback into FD op-amps, utilizing both NMOS and PMOS differential pairs, there seems no analysis or remark on operation of this kind of FD op-amp input stages in much higher supply voltages. Adaptations of internal positive feedback for FD op-amps designed for operation with higher supply voltage are focused rather on topology improvement, like cascode FD op-amps [18].

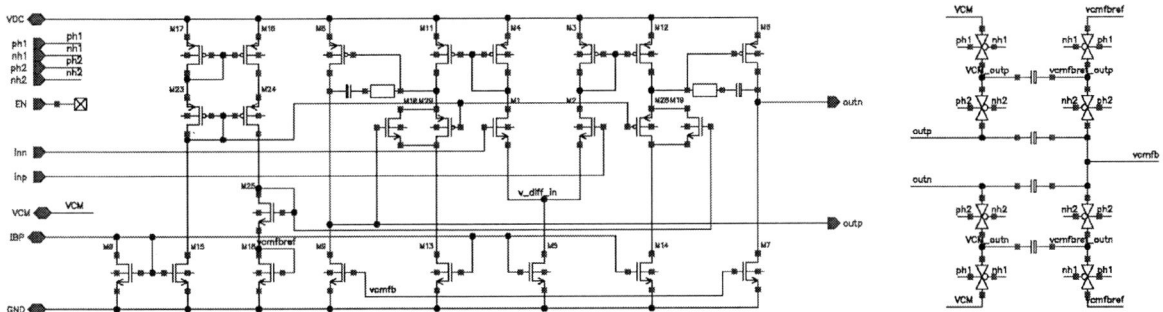

Fig. 3. Two-stage op-amp with fully implemented crosscurrent limitation circuitry in first stage output (left), discreet-time CMFB circuit (right).

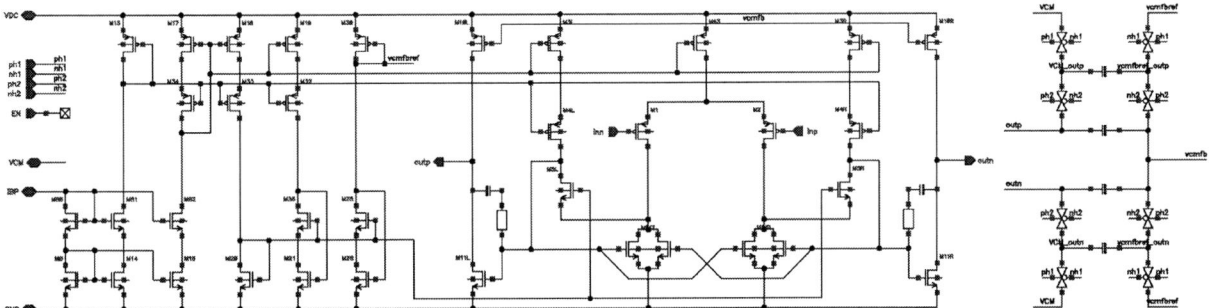

Fig. 4. Two-stage low-voltage op-amp with internal positive feedback.

The insight into topology of the FD op-amp presented in Fig. 4 and its subsequent simulations have shown that such FD op-amps with properly resized transistors can successfully operate with substantially higher supply voltages. Apart from retaining its operation, FD op-amp of Fig. 4 provides additional asset: internal positive feedback structure built into the first stage of this op-amp efficiently limits crosscurrents in its output stages. This effect is achieved by keeping gate-source voltages of relevant MOS transistors only weakly dependent on supply voltage values. Thus, the control of output stage crosscurrents is obtained with no additional circuitry as opposed to FD op-amp shown in Fig. 3.

Most probably, the design and application of FD op-amps with internal positive feedback for systems with very low supply voltages made the crosscurrent limiting effect to be missed in its typical applications because low crosscurrents are expected in such systems.

IV. POST-LAYOUT SIMULATIONS

The final designed chip consists of three separate readout blocks (one for each axis of the accelerometer). The layout of the readout circuit is presented in Fig. 5 while its properties are summarized in Table I. It is worth mentioning that the layout of the differential part of the circuit had to be almost perfectly symmetrical to reduce the impact of parasitic capacitances. Fig. 6 illustrates the operation of one readout block. The topmost signal is the voltage at ADC input as shown in Fig. 1.

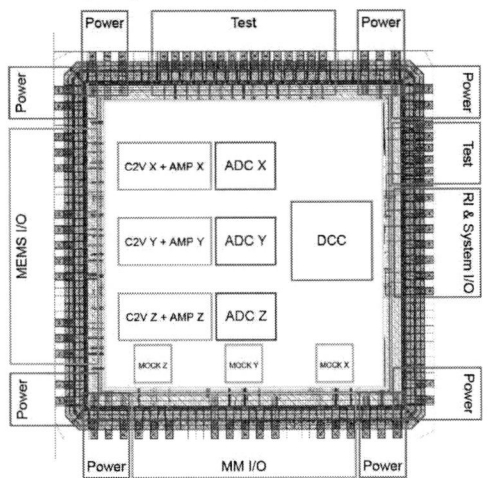

Fig. 5. Designed chip layout manufactured in 0.18 μm technology. The inside of the chip is not shown because of pending patent.

TABLE I.
SUMMARY OF LAYOUT PARAMETERS

Chip layout parameters			
Process	CMOS 180 nm	No. of I/O pads	25
Core area	9 mm²	No. of power pads	24
Supply voltage (digital part)	1.8 V	No. of test pads	21
Supply voltage (analog part)	5 V	Pad pitch	120 μm/ 90 μm
Power dissipation	< 10 mW	Pad size	57 μm / 70 μm
Package	FOQ 80		

91

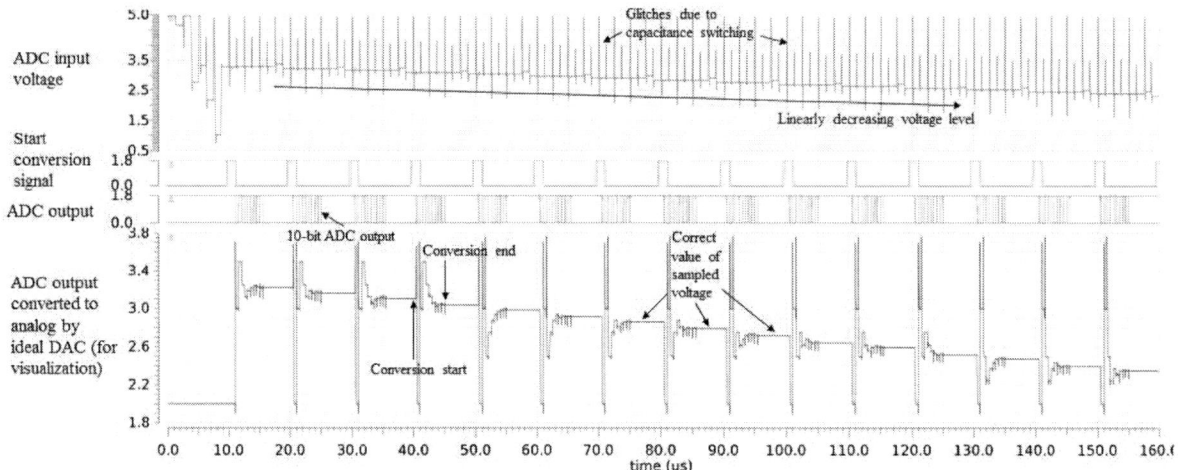

Fig. 6. Simulation results of the entire chip layout. Linearly decreasing acceleration is provided as input.

This signal represents the capacitance change of the MEMS sensor, converted to voltage by C2V circuit and amplified by AMP1 and AMP2 circuits. In the simulations, linearly decreasing acceleration was provided as input and, therefore, the ADC input signal is also linearly decreasing. Note that the signal contains many glitches, however it is perfectly normal in switched-capacitor circuits, as long as the voltage levels is correct during sampling by the ADC. The second signal shows the moments when conversion is initiated and the third signal depicts the ADC output. For better visualization, it was decided to plot the ouput converted by the ideal DAC (the bottommost signal). The conversion process can be then clearly seen. Moreover, the subsequent voltage levels decrease linearly, as expected for the provided input. Therefore, the obtained results show that the circuit functions correctly. One potential issue is the offset voltage: ideally it should be zero, however in the observed output it is equal to about 0.3 V (for a ADC range of 0-4 V). Although the offset does not cause the saturation of the output, it suggests that the chip should be calibrated before use or the offset should be trimmed using the designed MCC block.

It also proves that even when the chip layout is very symetrically designed, the parasitic capacitances of the pads and metal interconnections can still cause significant offset in the ADC output.

V. MANUFACTURING AND MEASUREMENT PREPARATION

The MEMS and ASIC have been manufactured and provided to the design team. Both these structures are available as silicon dies for self-made packaging procedures (Fig. 7) or bonded in QFN100 packages. A portion of MEMS and ASIC structures is placed in separate packages, while another bunch of MEMS and ASIC sets is placed and bonded together inside same enclosures (Fig. 8).

Due to input configuration of the ASIC, it is possible to analyze operation of both the ASIC alone and the ASIC driven by the MEMS. The designed ASIC is equipped with a test output that provides input voltage of the internal ADC for external examination, and it has been found fully operational. Owing to this feature it is possible to trace and distinguish operation quality issues of the MEMS structure, analog part of the ASIC and digital part of the ASIC. To exploit this possibility, set of PCB boards has been designed and manufactured, making it possible to measure all manufactured versions of packaged MEMS and ASIC structures, as well as unpackaged MEMS and ASIC dies (Fig. 9).

Fig. 7. MEMS and ASIC structures provided as dies.

Fig. 8. MEMS and ASIC structures provided in QFN100 packages.

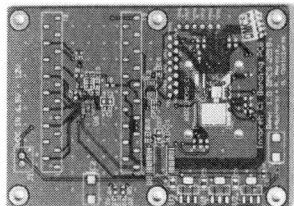 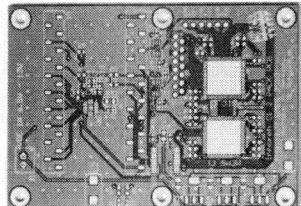
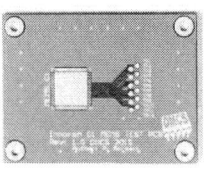 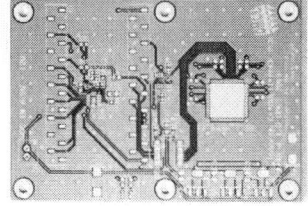

Fig. 9. Designed and manufactured versions of test PCB boards.

Details of the MEMS specimen bonding to one of PCB boards of Fig. 9 is presented in Fig. 10. Complex set of measurements is planned for the manufactured structures and accompanying PCB boards. Tests of the analog part of the ASIC are among the most important ones to be conducted.

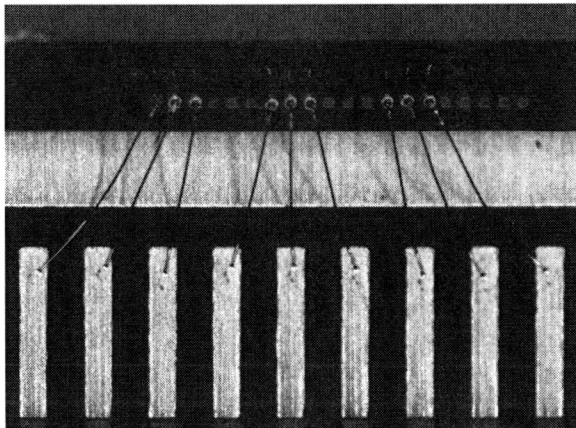

Fig. 10. Details of the MEMS specimen to PCB board bonding.

VI. CONCLUSIONS

The readout circuit for capacitive sensors has been designed and manufactured. Although the analog blocks are based on designs found in literature, the important novelty presented in this paper is their combination and application of a low-voltage positive feedback op-amp topology as cross-current control. The post-layout simulations prove that the circuit operates correctly and emphasize the importance of symmetry in the design of the differential parts of the circuit to reduce the impact of parasitics. With a carefully realized symmetrical design we were able to reduce the output offset voltage to an acceptable value of 0.3 V.

Tests of the manufactured and already received MEMS and ASIC structures will prove correctness of presented design principles and solutions.

ACKNOWLEDGMENT

The presented research is supported by the project STRATEGMED 2/266299/19NCBR/2016 funded by The National Centre for Research and Development in Poland.

REFERENCES

[1] D. Basta, M. Rossi-Izquierdo, A. Soto-Varela, et al. "Mobile posturography: posturographic analysis of daily-life mobility", Otol Neurotol, 34 (2013), pp. 288-297.

[2] L. Chiari., "Wearable systems with minimal set-up for monitoring and training of balance and mobility," 2011 Annual International Conference of the IEEE Engineering in Medicine and Biology Society, Boston, MA, 2011, pp. 5828-5832.

[3] M. Jankowski, M. Szermer and P. Amrozik, „Concept of an Electronic Device Dedicated to Imbalance Disorders Monitoring," 15th International Conference The Experience of Designing and Application of CAD Systems in Microelectronics (CADSM), Lviv, 2019, pp. 1-4.

[4] M. Szermer, P. Amrozik, P. Zając, C. Maj and A. Napieralski, „Capacitive MEMS accelerometer with open-loop switched-capacitor readout circuit," International Journal of Microelectronics and Computer Science, 2017, vol. 8, no 4, p. 139-145

[5] P. Zając, M. Szermer, P. Amrozik, C. Maj, G. Jabłoński, „Coupled Electro-mechanical Simulation of Capacitive MEMS Accelerometer for Determining Optimal Parameters of Readout Circuit," 20th International Conference on Thermal, Mechanical and Multi-Physics Simulation and Experiments in Microelectronics and Microsystems (EUROSIME), Hannover, 2019, pp. 399-403.

[6] C. Tse, "Design of a Power Scalable Capacitive MEMS Accelerometer Front End", PhD dissertation, 2013

[7] C. Wang, C. Chen and K. Wen, "A monolithic CMOS MEMS accelerometer with chopper correlated double sampling readout circuit," 2011 IEEE International Symposium of Circuits and Systems (ISCAS), Rio de Janeiro, 2011, pp. 2023-2026.

[8] J. F. Duque-Carisso, "Control of the common-Mode Component in CMOS Continuous-Time Fully Differential Signal Processing", Analog Int. Circuits and Signal Processing 4, pp. 131-140, 1993.

[9] V. Rao P. P. and Kishore K. L., "A 3.3V 11mW 200 Ms/sec switched capacitor sample and hold amplifier for pipelined A/D converters", JHPCSN 4, 2012, pp. 65-71.

[10] S. Koziel, S. Szczepanski, "Design of highly linear tunable CMOS OTA for continuous-time filters", IEEE Transactions on Circuits and Systems II: Analog and Digital Signal Processing, Volume 49, Issue 2, Feb 2002, pp. 110-122.

[11] H. Alzaher, M. Ismail, "A CMOS fully balanced differential difference amplifier and its applications", IEEE Transactions on Circuits and Systems II: Analog and Digital Signal Processing, Volume 48 , Issue 6, Jun 2001. pp. 614 - 620.

[12] E. Ghodsevali, S. Morneau-Gamache, J. Mathault, H. Landari, É. Boisselier, "Miniaturized FDDA and CMOS Based Potentiostat for Bio-Applications", Sensors 17 (4), 810, 2017.

[13] Tsung-Sum Lee, Hua-Yuan Chung, Sheng-Min Cai, "Design techniques for low-voltage fully differential CMOS switched-capacitor amplifiers", Proceedings of 2006 IEEE International Symposium on Circuits and Systems, 21-24 May 2006, pp. 2825-2828.

[14] F. Lacerda, et al., "A Differential Switched-Capacitor Amplifier with Programmable Gain and Output Offset Voltage", SBCCI Proceedings of the 19th annual symposium on Integrated circuits and systems design, Aug. 28-Sep. 1, 2006.

[15] C. F. Lee, P. K. T. Mok, "A monolithic current-mode CMOS DC-DC converter with on-chip current-sensing technique", IEEE journal of solid-state circuits 39 (1) 2004, pp. 3-14.

[16] T. S. Lee and C. C. Lu, "Two 1-V fully differential CMOS switched-capacitor amplifiers", Circuits, Systems and Signal Processing, Vol. 29, No. 2, pp. 195-207, April 2010.

[17] Tsung-Sum Lee and Chi-Chang Lu, "A 0.6-V subthreshold-leakage suppressed fully differential CMOS switched-capacitor amplifier", Analog Integrated Circuits and Signal Processing, February 2013, Volume 74, Issue 2, pp. 409-416.

[18] S. Farahmand, H. Shamsi, "Positive feedback technique for DC-gain enhancement of folded cascode Op-Amps", Proceedings of 10th IEEE International NEWCAS Conference, 17-20 June 2012, pp. 261-264.

Proceedings of the 27th International Conference *"Mixed Design of Integrated Circuits and Systems"*
June 25-27, 2020, Łódź, Poland

Comparative Analysis of Power Consumption of Parallel Prefix Adders

Ireneusz Brzozowski

AGH University of Science and Technology
Department of Electronics
Krakow, Poland
ireneusz.brzozowski@agh.edu.pl

Abstract—**This paper presents results and conclusions derived from simulations of tens structures of Parallel Prefix Adders considering over a dozen activity scenarios of input vector changes. Based on extended power model of static CMOS gates accurate analysis is done, thanks to the fact, that the model take into consideration changes of input vectors, not only switching activity of signals. Various structures of PG tree have been examined: regular, non-regular, with grey cells only, with both grey and black and with higher valency cells. Obtained results shows that some structures are better for some kind of summed data, but general remarks for adders design can be derived.**

Keywords—**power consumption, low-power design, parallel prefix adders, integrated circuits, CMOS technology**

I. INTRODUCTION

Power consumption is still important issue in design of integrated circuits. But achievement of reduction needs more effort, detailed analysis and specialist tools. It is not enough to take into consideration general conditions of a circuit work. Efficient reduction of power consumption needs to individual consideration of work conditions of designed circuit.

Arithmetic circuits (for example adders, multipliers etc.) are very important because they are essential part of microcontrollers and other data processing systems. In many cases, they work with specific and known data. So some properties of the data can be well defined. And there is a temptation to design circuits especially for specific kind of data.

One of the fastest adders is Parallel Prefix Adders (PPA). They are well described in many papers [1], [2], [3]. Numerous works concern these adders in aspect of power or area reduction and speed improvement. In papers [4] and [5] authors deal with parallel prefix adders in the aspect of power dissipation reduction, but they use traditional power model with switching activity factor for description of a circuit activity. In [6] and [7] authors analyzed power dissipation of adders using extended power model of CMOS circuits. The model taking into account probability of input vector changes, so it is independent of spatial and temporal correlation between inputs signals.

In the next section parallel, prefix adders are shortly described. Section III describes extended power model of static CMOS gates used in this study. In the next section blocks

needed to build various adders are described. Section V contains description of input data for adders simulations. Section VI presents description of tested adders and results of simulations. Finally conclusions and remarks are given.

II. PARALLEL PREFIX ADDERS

Multi-bit binary addition, besides cascading of 1-bit full adders, can be realized as a parallel prefix operation. At the beginning, for a full adder, we remind definition of signals, used to describe this operation: *generate* (G), *propagate* (P), and *kill* (K) signals. The adder generates a carry when C_{out} is 1 independent of C_{in}: $G = A \cdot B$, where A and B are inputs. The adder kills a carry when C_{out} is 0 independent of C_{in}: $K = \bar{A} \cdot \bar{B} = \overline{A + B}$. The adder propagates a carry: it produces a carry-out if and only if it receives a carry-in. It occurs when exactly one input is 1: $P = A \oplus B$. The same meaning of these signals are for a group spanning bits $i...j$, inclusive. So now the signals can be defined recursively for $i \geq k > j$ as follows:

$$\begin{aligned} G_{i:j} &= G_{i:k} + P_{i:k} \cdot G_{k\text{-}1:j} \\ P_{i:j} &= P_{i:k} \cdot P_{k\text{-}1:j} \end{aligned}, \quad (1)$$

with the base case:

$$\begin{aligned} G_{i:i} &= G_i = A_i \cdot B_i \\ P_{i:i} &= P_i = A_i \oplus B_i \end{aligned}, \quad (2)$$

and G_i, P_i are bitwise generate and propagate signals. Observe, that carry-out at i-th position G_{OUTi} is equal to group generation $G_{i:0}$. Hence, group generate signals and carries can be used synonymously. Finally sum for bit i can be calculated:

$$S_i = P_i \oplus G_{i\text{-}1:0} . \quad (3)$$

Thus, addition can be described as three-step operation: precomputation (calculation of bitwise G_i, P_i), **prefix calculation** (PG block, PG tree), and postcomputation (summation S_i) [8]. This second step (prefix calculation) is the main point of our interests in the context of power consumption.

When considering parallel prefix adders the PG block is usually drawn as a graph using black and grey cells. Their meaning is presented in Fig. 1 and they realize operations described by (1). This is a case when the cells are valency-2. It combines pairs of smaller groups.

This work has been supported by AGH University of Science and Technology under subvention funds (no. 16.16.230.434).

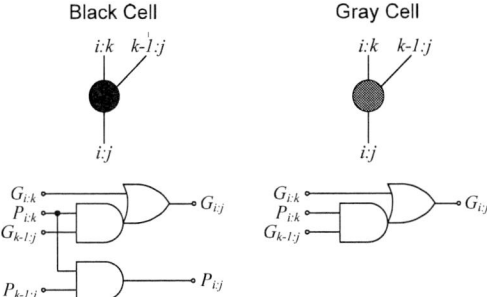

Fig. 1.   The valency-2 black and grey cells of PG block.

There is possible to define higher valency PG block, which demands more complex gates, but can results in reduction of stages. and for example PG block of valency-4 is described as follows:

$$G_{i:j} = G_{i:k} + P_{i:k} \cdot G_{k-1:l} + P_{i:k} \cdot P_{k-1:l} \cdot G_{l-1:m} + $$
$$+ P_{i:k} \cdot P_{k-1:l} \cdot P_{l-1:m} \cdot G_{m-1:j} \qquad , \qquad (4)$$
$$P_{i:j} = P_{i:k} \cdot P_{k-1:j} \cdot P_{l-1:m} \cdot P_{m-1:j}$$

for $i \geq k > l > m > j$. The generate signal can be realized by AND-OR gate, what is described by:

$$G_{i:j} = G_{i:k} + P_{i:k} \cdot (G_{k-1:l} + P_{k-1:l} \cdot (G_{l-1:m} + P_{l-1:m} \cdot G_{m-1:j})). \qquad (5)$$

The symbol of the cell and schematic corresponding to (4) is presented in Fig. 2.

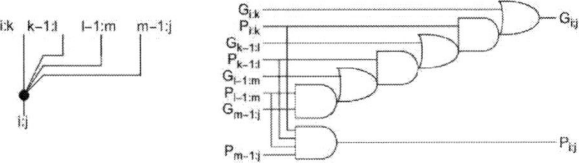

Fig. 2.   The valency-4 black cell of PG block [8].

Using black and grey cells diagram of PG block, which represent almost final schematic (except buffers used for delay balancing), can be easy drawn for different adders. For instance Fig. 3 shows PG diagram of Sklansky 8-bit adder. Axes presents number of: levels – vertical, bits – horizontal.

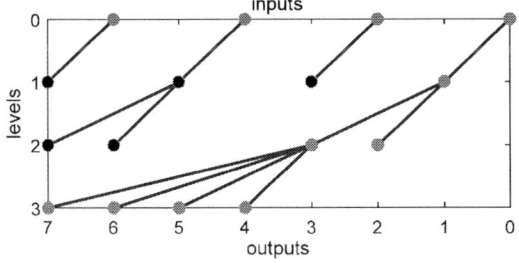

Fig. 3.   The PG diagram of Sklansky 8-bit adder

Some authors do not distinguish color of cells, because it is easy to determine, which cells should gray and which black. If a cell is preceded by others, not including the first input (LSB),

then propagation signal must be calculated in such cell, so it is black [1]. But others use more compound marks describing propagation block, giving more information [9].

III.   EXTENDED POWER MODEL OF STATIC CMOS GATES

Power dissipation model of gates used in this work, take advantage of information about changes of vector of a circuit primary input. So it is extended against to traditional one, usually found in literature, which uses switching activity and constant nodal capacitance. However detailed analysis of CMOS gates behavior shows that power dissipation depends on reason of the gate switching. It means that changes of whole input vectors should be considered as activity measure, not independent input signals as in traditional model. So, new model of energy consumed by static CMOS gates was introduced [10]. The model consists of capacitors representing amount of consumed energy as a function of a gate switching reason called *gate driving way* (Fig. 4). Capacitors represent *equivalent capacitance*, which are calculated from current flowing directly from supply source or through previous gates for each possible change of input vectors. So the values depend on the gate driving way.

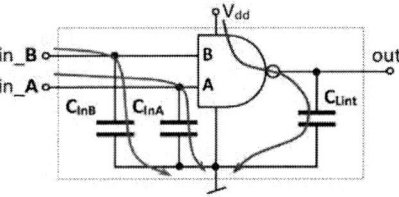

Fig. 4.   Power dissipation model of NAND with *equivalent capacitance*.

In traditional model is supposed that in CMOS gates energy is consumed only during switching, but in fact power consumption occurs even when gate output state is stable but input vector changes. So, all possible input vectors changes are essential. The table in Fig. 5 shows all driving ways for 2-input gate. Arrows denote change of input signals – rising or falling edges. Thus it is the second component of the model – probability of a gate driving way.

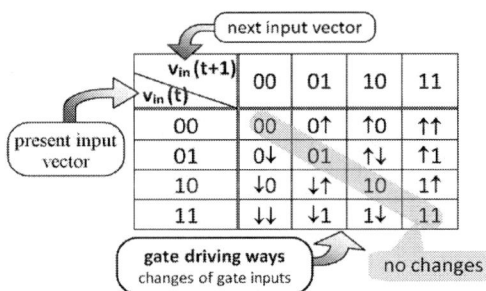

Fig. 5.   All possible changes of vectors for 2-input gate – *driving ways*.

Parameters of the model – equivalent capacitance – are calculated based on simulation results of current flowing through the gate terminals. Thus, each gate is characterized by set of tables (for each input and supply). But considering dependencies in a circuit all values can be summed giving one table with total equivalent capacitance for the gate.

A probability of a gate driving way is treated as contribution factor of particular equivalent capacitance in the total power dissipation of whole circuit. Therefore, following equation characterizes the model:

$$C_{T\_equ}(g) = \sum_{dw_g} c_{t\_equ}(dw_g) \cdot p(dw_g), \qquad (6)$$

where: C_{T_equ} is total equivalent capacitance of gate g, $c_{t\_equ}(dw_g)$ is all equivalent capacitance value regarding terminal X for driving way dw_g, and $p(dw_g)$ is driving way probability.

Obviously probability of driving way for circuit primary inputs has to be known, as activity measure of a circuit. Next, basing on circuit simulations or calculation methods driving ways for all gates in the circuit can be calculated.

IV. ADDERS BUILDING BLOCKS

A. Cells

Directly from previously presented equations (1), (4), and (5) flows the use of complex gate AND-OR and gate AND to realize PG block of an adder. But from other hand, static CMOS technology naturally utilizes gates with negation, e.g.: NOT, NAND, etc. Hence we have complex gates AND-OR-INVERT and OR-AND-INVERT, which be used alternately to avoid additional inverters. It can be done, because a rule similar to de'Morgan's laws works for the complex gates:

$$\begin{aligned}\overline{AOI(x_i)} &= OAI(\overline{x_i}) \\ \overline{OAI(x_i)} &= AOI(\overline{x_i})\end{aligned} \qquad (7)$$

It means that negation of function AOI is function OAI performed for negated variables and vice versa, negation of OAI is AOI of negated variables.

In this work we are focused on the PG block and its power consumption. But the first stage of the parallel prefix adder computing bitwise generate and propagate signals (2) needs AND and EX-OR gates. And the third stage, calculating final sum, needs EX-OR gate. In static CMOS we can use NAND gate and complex gates according to (7), but design of EX-OR gate is quite problematic. So it can be realized as serial connection of NOR and AOI gates ($y = \overline{\overline{a + b} + a \cdot b}$). In this work the first and the last stage of all adders were the same, and we consider various structures of the prefix computation block (PG), and assess their power consumption.

In order to design various PG blocks a set of gates was designed in such way that allows easy and fast creation of the final adders layouts. All designed cells have the same high, so they can be used as standard cells. Supply lines (vdd, gnd) are matched to each other. Inputs and an output are made with metal 3 for easy further connections. Internal connections are made with metal 1 and occasionally metal 2. Generally 15 gates were designed: NOT, 2-, 3-, and 4-input NAND and NOR, and complex gates described in Table I. The high of cells is 5.24 µm; their width is in fourth column of the table (in µm). Layouts of cells were designed using Cadence Virtuoso in UMC 180 nm CMOS technology. Minimum dimensions of transistors were used, except width of PMOS transistors, which was increased by factor 4.74. Two exemplary layouts of cells are presented in Fig. 6.

TABLE I.
COMPLEX GATES DESIGNED FOR ADDERS CREATION

Cell Name	Function	Num. of Transis.	Layout Width
AOI21	$y = \overline{a + b \cdot c}$	6	3.48
AOI2111	$y = \overline{a + b \cdot (c + d \cdot e)}$	10	5.20
AOI211111	$y = \overline{a + b \cdot (c + d \cdot (e + f \cdot g))}$	14	6.92
AOI21111111	$y = \overline{a + b \cdot (c + d \cdot (e + f \cdot (g + h \cdot i)))}$	18	8.64
OAI21	$y = \overline{a \cdot (b + c)}$	6	3.88
OAI2111	$y = \overline{a \cdot (b + c \cdot (d + e))}$	10	5.60
OAI211111	$y = \overline{a \cdot (b + c \cdot (d + e \cdot (f + g)))}$	14	7.33
OAI21111111	$y = \overline{a \cdot (b + c \cdot (d + e \cdot (f + g \cdot (h + i))))}$	18	9.04

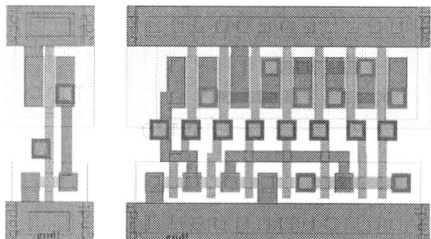

Fig. 6. Layouts of NOT and AOI211111 gates

After functional verification of designed cells, netlists with parasitic elements (capacitors and resistors) were generated for further simulations and calculation of equivalent capacitance for extended power model of gates.

B. Assessment of Power Model Parameters

Extended power model mentioned in Section III needs to calculate values of equivalent capacitance for all the cells versus their possible driving ways (6). In order to consider all possible changes of input vectors suitable input sources were designed. For N-input gate there are $2^{2 \cdot N}$ changes of the input vectors – driving ways. So for gates with more inputs simulations can take some time. Time of input signals edge (rising and falling) were chosen in such way that quasi-short power was not occurred. Moreover, we can assume that for 180 nm technology static losses can be neglected. Thus, values of equivalent capacitance corresponding with all terminals of a gate were calculated and next summed appropriately, because we are interested in total power dissipation of a circuit. Obviously, it is possible to consider all component of power dissipation separately e.g. input, internal etc. Obtained results are collected in tables. But numbers collected in tables are difficult to analysis, but are useful for simulation programs. Therefore, assessment results of bigger gates can be presented as color map in graphs. All equivalent capacitances (for inputs, output and the total, refer to Fig. 4) for AOI21 gate is shown in Fig. 7. And Fig. 8 shows the total equivalent capacitance for AOI2111 gate.

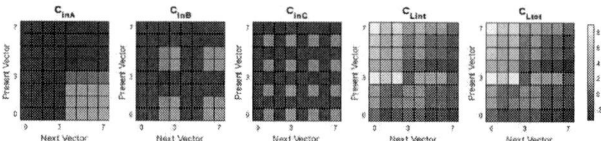

Fig. 7. Equivalent capacitances of AOI21 gate shown as colour maps.

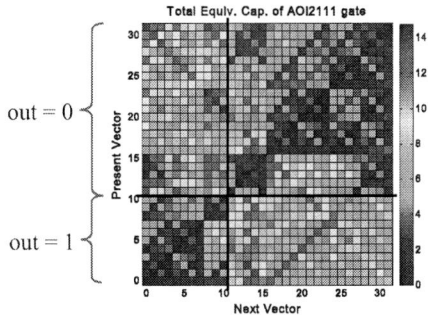

Fig. 8. Total equivalent capacitance of AOI2111 gate [fF].

Analysing above graph areas with lower (blue) or higher (brown/red) values can be noticed. They are placed accordingly to the output logic state of the gate. In case of AOI2111, for vectors from 0 to 10 the output state is logic "1" and "0" for the rest. For vectors switching the output from zero to one the equivalent capacitance takes larger values (top left). Opposite changes gives lower values. Moreover, it is visible, that such input vectors changes, which not affect the gate output cause power consumption too. And values of equivalent capacitance are lower than previously, but not zero (bottom left and top right).

V. TEST VECTORS PREPARATION

Power consumption in CMOS static circuits strongly depends on activity profile of input signals. Extended model of power consumption presented in Section III utilizes probability of input vector changes (*gate driving way*) as a circuit activity measure. Thus for adders examination some distributions of input vector changes are needed. Obviously, the easiest case is uniform distribution and it can be used as general or reference one. On the other hand random distribution gives similar results. But adders in many cases work with specific kind of data. In this work 27 distributions of input vector changes were prepared based on assumed values of added numbers. The following scenarios of dependencies between summed data were considered. At the beginning two series (S_1, S_2) of disjoint numbers were randomly generated (20000 vectors for 4-bit adder and it was doubled increasing the number of bits by 1). The first series included numbers from 0 to half of the range ($0 \div 0.5 \cdot 2^N - 1$) and the second included values from half of the range to the end ($0.5 \cdot 2^N - 1 \div 2^N - 1$), assuming N-bit adder (Fig. 9). Histograms for these series (4-bit) are shown in Fig. 10. Next, consecutive numbers were merged into one 2N-number, using all possible combinations: $A=S_1$, $B=S_2$; $A=S_2$, $B=S_1$; $A=S_1$, $B=S_1$ and $A=S_2$, $B=S_2$, giving four scenarios of input vectors distributions. Finally, for the first case the distribution of 8-bit (merged) vector changes is presented in Fig. 11.

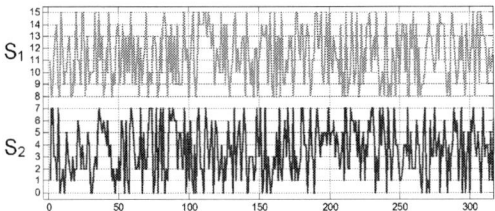

Fig. 9. Exemplary series of disjoint numbers for 4-bit adders (frag.).

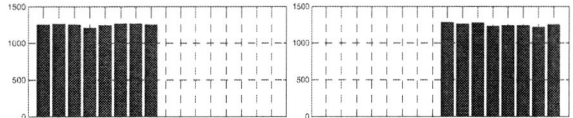

Fig. 10. Histograms of disjoint series S_1 (left) and S_2 (right) for 4-bit adder.

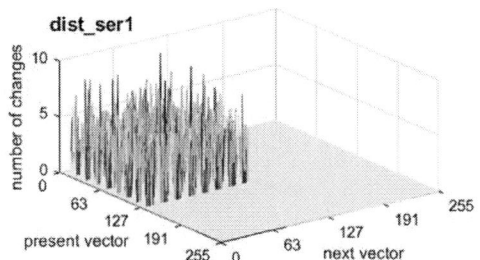

Fig. 11. The distribution of merged (series S_1 and S_2) vector changes.

Next four cases were obtained in similar way but series of disjoint vectors were generated each time. Another four distributions were created assuming small overlap of numbers, e.g. $S_1 = [0 \div 60\%\ max]$, $S_2 = [40\%\ max \div max]$ and series were generated each time. Finally 12 distributions were obtained.

Moreover another one scenario was considered – addition of sinusoidal waveforms. Fifteen cases were considered. The final distribution of the first one, merged sin(t) and sin(2·t) with maximum amplitude is presented in Fig. 12. All considered cases of sinus waveforms addition are described in Table II using simplified equation.

When specified data is added some vectors are absent at an adder input. Such situation take pale is cases described above and can be seen in figures presented distribution of vector changes (Fig. 11, Fig. 12).

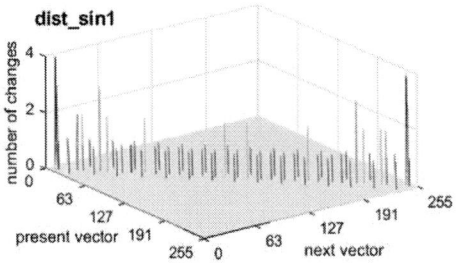

Fig. 12. Distribution of vector changes for two sinus waveforms (1st case).

TABLE II.
DESCRIPTION OF VETORS CHANGES DISTRIBUTIONS

Name	Simplified Equation	Name	Simplified Equation
dist_sin1	sin(t) + sin(2·t)	dist_sin9	sin(t) + 0.5·sin(t-0.12)
dist_sin2	sin(t) + sin(3·t)	dist_sin10	sin(t) + sin(t+0.12)
dist_sin3	sin(t) + 0.5·sin(2·t)	dist_sin11	sin(t) + cos(t)
dist_sin4	sin(t) + 0.5·sin(3·t)	dist_sin12	sin(t) + cos(2·t)
dist_sin5	0.5·sin(t) + sin(2·t)	dist_sin13	sin(t) + cos(3·t)
dist_sin6	sin(t) + 0.2·sin(t)	dist_sin14	sin(t) + 0.5·cos(t)
dist_sin7	sin(t) + sin(t-0.12)	dist_sin15	sin(t) + 0.5·cos(2·t)
dist_sin8	0.5·sin(t) + 0.5·sin(t-0.12)		

VI. TESTS

A. Circuits Under Test

Using building blocks described in Section IV as much as possible adders were created. At the beginning only 4-, 5- and 6-bit circuits were considered. In parallel prefix adders the first stage (bitwise propagation and generation) and the third (final summation) can be realized in a few ways but realization of the second part (carry propagation) gives many possibilities. Thus in this paper main focus is put on the PG block and its power consumption. According to theory presented in Section II for simplification dashed diagram represents structure of PG block (see Fig. 3). So, for tests as much as possible such structures were created. Appropriate diagrams are shown in Fig. 13 (4-bit adders), Fig. 14 (5-bit adders), Fig. 15 (6-bit adders with gray cells only), and Fig. 16 (6-bit adders with gray and black cells). They are consequently named, and "xp" means x-level block and letter "c" means, that in the structure black cells were used.

Considering these diagrams we can see realizations with various number of levels, gray and black cells with valency from 2 up to 5 (so, no 1-level 6-bit adder). Some structures are regular and others no. And we can find structures described in literature (e.g. in [1]).

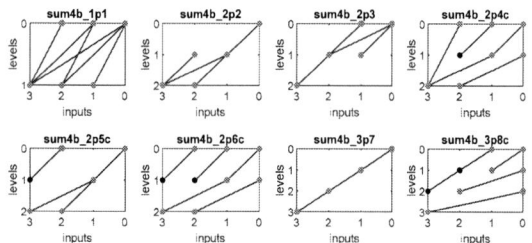

Fig. 13. Diagrams of PG blocks for 4-bit parallel prefix adder.

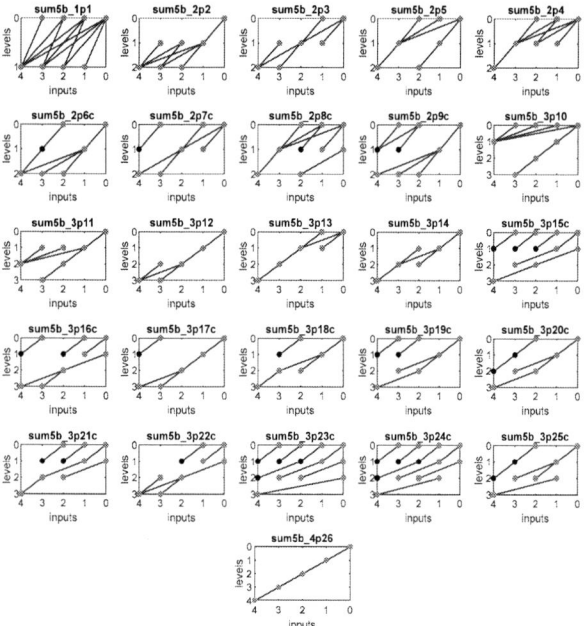

Fig. 14. Diagrams of PG blocks for 5-bit parallel prefix adder.

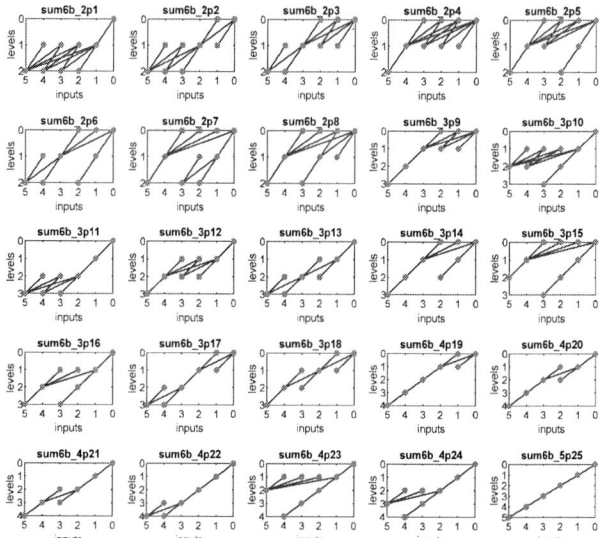

Fig. 15. Diagrams of PG blocks with only grey cells for 6-bit PPA.

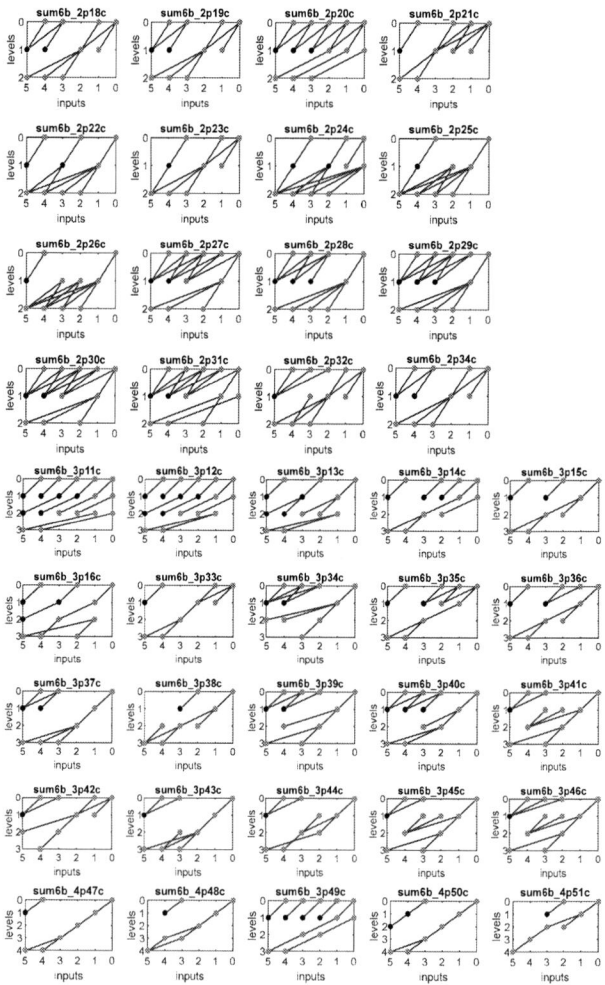

Fig. 16. Diagrams of PG blocks with grey and black cells for 6-bit PPA.

B. Results of Power Consumption Estimation

All circuits described in the previous point were simulated using Matlab environment for distributions of input vectors changes described in Section V. In total, there were 28 cases: uniform, 12 with number series and 15 with sinus signals. Results, the total equivalent capacitance of PG tree for all 100 circuits are collected in Table III, Table IV and Table V for 4-, 5- and 6-bit adders respectively. Adders included black cells are marked in gray. For easier comparison values in tables are marked with blue stripes. Additionally, minimum values are written in green italic and maximum values in red bold. Moreover, the minimum values except Ripple Carry Adder (sum4b_3p7, sum5b_4p25 and sum6b_5p25) are marked in yellow. It is seen that this adder is the best realization in many cases. It has the smallest number of transistors and the largest number of levels, thus it usually has the largest delay. The second realization (marked in yellow) is usually only a few percent worse than RCA, usually about 1 or 2 %. The 4-bit adders are worse than others in this ranking, and differences go up to 7.9% (dist_sin13).

Considering obtained data, especially for 5- and 6-bit adders, it can be seen, that in general, larger power consumption is for structures utilized black cells, but some exception exist: e.g. sum5b_2p7, sum6b_2p21, _3p33, _3p42. They have similar structure – only one black cell small redundancies. Another observation is that bigger number of levels usually gives smaller power consumption. In adders with small number of levels cells with higher valency have to be used. It takes more number of transistors causing bigger power consumption but some exceptions can be found. In structures with more levels good idea is to avoid redundancies. Generally structure of PG logic in PPA depends on kind of summed data.

VII. CONCLUSIONS

In the paper comparison of power consumption of PPA under advanced analysis were presented. Only propagation-generation tree was considered. Analyzed structures are shown as diagrams and results are collected in tables. Structures of adders in aspect of kind of summed data should be taken into consideration during design for low power.

TABLE III.
RESULTS OF 4-BIT ADDERS ASSESMENT – TOTAL EQUIVALENT CAPACITANCE OF PG TREE [FF]

Adder Name	unif.	ser1	ser2	ser3	ser4	ser5	ser6	ser7	ser8	ser9	ser10	ser11	ser12	sin1	sin2	sin3	sin4	sin5	sin6	sin7	sin8	sin9	sin10	sin11	sin12	sin13	sin14	sin15
sum4b_1p1	19,0	18,7	18,7	10,9	10,9	19,0	19,1	14,4	14,5	19,1	18,9	14,8	18,7	8,21	10,60	4,80	6,90	7,38	3,82	4,82	2,72	4,53	4,82	6,66	10,21	11,04	5,26	7,22
sum4b_2p2	16,8	16,1	16,1	10,9	11,0	16,4	16,4	12,7	12,8	16,8	16,6	12,9	16,4	6,74	8,59	3,80	5,51	5,62	3,00	3,81	2,02	3,68	3,81	5,18	8,71	8,60	4,13	6,03
sum4b_2p3c	16,7	15,0	15,0	10,9	10,9	15,3	15,2	13,3	13,3	16,4	16,2	13,4	16,3	6,66	8,84	4,08	6,09	6,11	3,27	3,84	2,15	3,80	3,84	5,32	8,84	9,00	4,69	6,11
sum4b_2p4c	24,2	23,5	23,5	15,3	15,4	23,9	23,9	20,2	20,4	24,4	24,1	19,8	24,0	9,28	12,33	5,53	8,17	8,28	4,25	5,70	2,84	5,40	5,70	7,38	12,21	13,06	6,21	8,18
sum4b_2p5c	22,4	19,9	19,9	10,9	10,9	20,2	20,3	12,7	12,7	22,1	22,0	15,4	20,0	7,34	9,11	4,05	5,69	6,16	3,26	4,18	2,40	4,05	4,18	5,58	9,24	9,45	4,48	6,39
sum4b_2p6c	28,3	25,7	25,7	13,6	13,7	26,1	26,2	18,7	18,7	28,3	27,9	20,9	27,9	9,02	11,80	5,39	7,63	8,23	4,17	5,61	3,02	5,31	5,61	7,18	11,56	12,92	6,10	7,75
sum4b_3p7	16,4	14,7	14,7	11,7	11,8	15,0	14,9	13,0	13,0	16,0	15,9	13,0	15,9	6,23	8,23	3,73	5,61	5,46	2,96	3,57	1,91	3,53	3,57	4,80	8,45	8,15	4,24	5,83
sum4b_3p8c	28,2	23,7	23,7	13,7	13,7	24,0	24,0	18,4	18,4	27,8	27,5	21,2	26,4	9,04	11,97	5,65	8,16	8,55	4,33	5,67	3,23	5,50	5,67	7,40	11,85	13,49	6,45	8,22

TABLE IV.
RESULTS OF 5-BIT ADDERS ASSESMENT – TOTAL EQUIVALENT CAPACITANCE OF PG TREE [FF]

Adder Name	unif.	ser1	ser2	ser3	ser4	ser5	ser6	ser7	ser8	ser9	ser10	ser11	ser12	sin1	sin2	sin3	sin4	sin5	sin6	sin7	sin8	sin9	sin10	sin11	sin12	sin13	sin14	sin15
sum5b_1p1	29,1	29,4	29,4	16,5	16,5	29,5	29,6	24,1	24,0	29,5	29,6	23,6	28,6	17,0	20,1	11,7	13,6	15,3	10,6	14,1	6,2	11,0	14,8	16,2	19,9	20,3	12,4	16,4
sum5b_2p2	25,1	25,1	25,1	15,7	15,6	25,1	25,2	20,6	20,4	25,3	25,4	20,3	24,4	13,7	15,6	8,8	10,8	11,8	8,2	13,1	4,5	8,8	13,7	12,8	16,0	16,1	9,2	12,2
sum5b_2p3	23,0	22,5	22,5	14,6	14,8	22,6	22,6	19,0	18,9	22,9	23,0	18,7	22,2	12,4	14,3	8,3	10,1	10,6	7,8	10,8	4,1	7,9	11,4	11,3	14,3	14,2	8,7	11,8
sum5b_2p4	24,8	23,2	23,2	15,7	15,6	23,3	23,4	21,5	21,3	24,3	24,4	20,8	24,3	13,8	16,6	9,8	11,8	12,6	9,0	11,9	4,9	9,2	12,5	13,2	16,6	16,6	10,4	13,7
sum5b_2p5	22,6	21,0	21,0	14,9	14,8	21,1	21,1	19,3	19,1	22,0	22,2	18,7	22,1	11,5	14,4	8,3	9,9	10,7	7,4	9,8	4,2	7,8	10,3	11,1	14,4	14,3	8,6	11,5
sum5b_2p6c	33,1	32,6	32,6	21,3	21,7	32,6	32,7	29,1	28,9	33,1	33,2	27,9	32,5	15,3	17,9	10,1	12,2	13,7	9,1	14,4	4,9	10,3	15,1	13,8	18,7	19,5	10,9	13,9
sum5b_2p7c	28,6	26,4	26,4	14,9	14,8	26,4	26,5	18,9	18,8	28,2	28,3	20,7	25,0	13,0	14,8	8,6	10,3	11,1	8,0	11,2	4,5	8,3	11,8	11,7	14,8	15,0	9,0	12,1
sum5b_2p8c	30,4	28,9	28,9	19,3	19,3	28,9	29,0	27,1	26,9	29,8	30,0	26,1	30,0	16,9	20,3	11,4	14,4	15,4	10,6	14,5	6,1	10,9	15,2	15,3	20,6	20,3	12,6	16,3
sum5b_2p9c	38,0	37,5	37,5	20,9	20,9	37,5	37,6	28,1	27,9	38,3	38,4	28,5	34,1	16,5	18,9	10,4	12,6	14,3	9,6	15,3	5,5	10,7	16,0	14,9	19,4	20,7	11,4	14,6
sum5b_2p10	26,5	26,8	26,8	17,3	17,3	26,9	27,0	21,5	21,4	26,9	27,0	21,3	25,6	14,7	16,8	9,6	11,6	12,6	8,9	13,3	5,0	9,3	13,9	13,5	16,9	16,9	10,0	13,4
sum5b_3p11	24,7	24,6	24,6	16,5	16,4	24,7	24,8	20,2	20,0	24,9	24,9	19,9	23,8	13,2	15,0	8,4	10,5	11,1	8,0	12,5	4,3	8,4	13,2	12,0	15,3	15,2	8,8	11,8
sum5b_3p12	22,7	22,2	22,2	15,7	15,6	22,2	22,3	18,7	18,6	22,6	22,7	15,4	21,8	12,0	13,6	7,7	9,7	10,9	7,4	11,4	3,8	7,6	12,0	10,9	13,9	13,5	8,0	10,8
sum5b_3p13	22,5	20,9	20,9	15,7	15,6	21,0	21,1	19,2	19,1	22,0	22,1	18,8	21,7	11,8	14,0	8,2	10,2	10,4	7,7	10,5	4,0	7,7	11,2	10,9	14,1	13,8	8,6	11,6
sum5b_3p14	22,6	21,1	21,1	15,7	15,6	21,1	21,2	19,4	19,2	22,1	22,2	18,8	22,0	11,9	13,9	8,0	10,1	10,6	7,6	11,6	4,0	8,0	12,3	11,2	14,3	14,1	8,4	11,0
sum5b_3p15c	40,2	38,0	38,0	21,3	21,2	38,0	38,1	30,6	30,4	39,8	40,0	31,5	36,9	17,9	20,9	11,5	14,3	16,1	10,6	15,5	6,2	11,5	16,1	15,7	21,4	22,4	13,0	16,4
sum5b_3p16c	32,6	30,3	30,3	16,8	16,7	30,4	30,5	22,9	22,8	32,1	32,3	24,5	29,1	15,3	17,5	9,8	12,3	13,3	9,2	13,1	5,4	9,6	13,6	13,3	17,8	18,1	10,8	14,1
sum5b_3p17c	28,3	26,0	26,0	15,7	15,6	26,1	26,2	18,7	18,5	27,9	27,9	20,5	24,7	12,6	14,1	8,0	9,9	10,5	7,7	11,7	4,2	8,0	12,3	11,3	14,3	14,4	8,4	11,2
sum5b_3p18c	26,6	25,0	25,0	17,7	17,6	25,1	25,1	23,1	23,1	26,1	26,2	22,6	26,0	13,0	15,1	8,8	10,7	11,6	8,0	12,5	4,3	8,8	13,1	12,1	15,7	16,0	9,3	11,9
sum5b_3p19c	34,3	32,0	32,0	18,5	18,4	32,1	32,2	24,7	24,5	33,9	34,0	26,0	30,9	14,4	16,5	9,3	11,2	12,7	8,6	13,4	4,8	9,5	14,0	13,1	16,9	17,7	10,1	12,9
sum5b_3p20c	35,9	31,7	31,7	20,1	20,0	31,7	31,9	26,1	25,9	35,0	35,1	27,8	33,4	15,1	17,5	9,9	12,0	13,3	9,1	14,1	5,1	10,1	14,8	13,7	18,2	19,1	10,7	13,8
sum5b_3p21c	32,5	30,9	30,9	20,5	20,4	31,0	31,1	29,2	29,0	32,0	32,2	28,1	32,0	16,5	19,5	10,9	13,7	15,0	10,0	14,6	5,7	10,8	15,2	14,7	20,2	20,6	12,2	15,5
sum5b_3p22c	27,0	26,5	26,5	16,8	16,8	26,5	26,6	22,9	22,8	26,9	27,0	22,4	26,2	14,7	17,0	9,5	12,1	12,8	9,0	12,7	5,0	9,2	13,3	12,9	17,4	17,3	10,4	13,7
sum5b_3p23c	42,7	40,9	40,9	22,1	22,0	41,0	41,2	31,6	31,4	42,7	42,8	33,1	38,7	19,2	22,1	12,8	15,1	17,3	11,5	16,6	6,8	12,5	17,3	17,0	22,9	24,1	14,0	17,6
sum5b_3p24c	44,4	42,6	42,6	23,8	23,6	42,7	42,9	33,3	33,1	44,4	44,5	34,7	40,5	20,8	23,5	13,4	16,2	18,6	12,2	18,3	7,3	13,3	19,0	18,2	24,5	25,4	14,7	18,7
sum5b_3p25c	34,3	30,0	30,0	18,5	18,4	30,1	30,2	24,5	24,3	33,4	33,5	26,3	31,7	14,6	16,6	9,7	11,6	13,1	8,8	13,8	5,0	9,9	14,5	13,5	17,5	18,4	10,5	13,3
sum5b_4p26	22,2	20,6	20,6	16,5	16,4	20,7	20,7	18,9	18,7	21,7	21,8	18,5	21,3	11,4	13,3	7,6	9,8	9,8	7,3	11,0	3,7	7,5	11,7	10,5	13,6	13,2	7,9	10,7

TABLE V.
RESULTS OF 6-BIT ADDERS ASSESMENT – TOTAL EQUIVALENT CAPACITANCE OF PG TREE [FF]

Adder Name	unif.	ser1	ser2	ser3	ser4	ser5	ser6	ser7	ser8	ser9	ser10	ser11	ser12	sin1	sin2	sin3	sin4	sin5	sin6	sin7	sin8	sin9	sin10	sin11	sin12	sin13	sin14	sin15
sum6b_2p1	35,6	36,0	36,0	21,1	21,2	35,9	35,9	30,5	30,3	36,1	36,1	30,4	34,6	24,4	24,6	15,7	20,2	21,7	15,7	20,8	13,6	18,7	21,6	22,2	25,1	25,8	19,8	20,5
sum6b_2p2	31,4	31,5	31,5	19,4	19,5	31,5	31,4	26,8	26,6	31,7	31,8	26,9	30,4	19,6	20,3	12,9	16,9	18,2	13,5	19,4	12,0	16,3	20,2	19,2	21,8	21,3	16,7	16,8
sum6b_2p3	31,2	30,8	30,8	19,5	19,5	30,7	30,7	27,2	27,0	31,3	31,3	27,2	30,1	20,7	20,8	14,1	18,0	18,7	14,5	18,0	12,6	16,5	18,9	19,6	22,1	21,8	18,1	17,8
sum6b_2p4	35,0	33,6	33,6	21,2	21,2	33,5	33,5	31,8	31,5	34,7	34,7	31,2	34,4	24,6	24,6	17,1	21,7	22,8	17,3	20,3	14,7	19,4	21,3	23,3	26,5	26,0	22,0	20,9
sum6b_2p5	34,7	33,3	33,3	22,0	22,0	33,2	33,2	31,4	31,2	34,4	34,4	30,9	34,0	24,3	24,3	16,8	21,3	22,2	17,0	20,1	15,3	19,4	21,1	22,8	26,0	25,6	21,3	20,8

TABLE V. (CONT.)

Adder Name	Distribution Name																											
	unif.	*ser1*	*ser2*	*ser3*	*ser4*	*ser5*	*ser6*	*ser7*	*ser8*	*ser9*	*ser10*	*ser11*	*ser12*	*sin1*	*sin2*	*sin3*	*sin4*	*sin5*	*sin6*	*sin7*	*sin8*	*sin9*	*sin10*	*sin11*	*sin12*	*sin13*	*sin14*	*sin15*
sum6b_2p6	30.9	30.5	30.5	20.3	20.3	30.4	30.4	26.9	26.7	31.0	31.0	26.9	29.8	20.3	20.5	13.8	17.6	18.1	14.2	17.8	13.2	16.4	18.7	19.0	21.6	21.4	17.4	17.8
sum6b_2p7	32.9	31.4	31.4	21.2	21.2	31.3	31.3	29.6	29.3	32.5	32.5	29.1	32.1	22.4	22.6	15.2	19.7	20.4	15.5	19.2	14.5	18.0	20.2	20.8	24.0	23.8	19.4	19.3
sum6b_2p8	32.7	31.3	31.3	21.2	21.2	31.2	31.2	29.5	29.2	32.4	32.4	29.0	32.0	21.6	22.1	14.9	19.4	20.2	15.4	19.8	13.3	17.9	20.8	20.8	24.1	23.4	19.3	18.6
sum6b_3p9	30.7	29.3	29.3	20.3	20.3	29.2	29.1	27.4	27.1	30.4	30.4	27.3	29.6	20.2	20.4	13.9	18.1	18.6	14.4	17.7	12.5	16.4	18.6	19.1	21.9	21.5	18.0	17.7
sum6b_3p10	35.1	35.6	35.6	21.9	22.0	35.5	35.5	30.0	29.9	35.7	35.7	30.0	34.1	23.3	23.9	15.0	19.5	20.9	15.4	21.2	15.0	18.6	22.0	21.7	24.7	25.0	19.1	19.7
sum6b_3p11	31.1	31.2	31.2	20.2	20.4	31.2	31.1	26.5	26.3	31.4	31.4	26.6	30.0	19.3	20.1	12.6	16.5	17.6	13.2	19.2	12.6	16.3	20.0	18.6	21.3	21.1	16.0	16.7
sum6b_3p12	31.0	29.5	29.5	20.3	20.3	29.5	29.5	27.7	27.5	30.7	30.7	27.4	30.2	20.7	21.1	13.6	18.0	18.7	14.0	18.3	13.8	16.7	19.1	19.0	22.0	22.1	17.5	18.0
sum6b_3p13	29.0	28.6	28.6	19.4	19.5	28.5	28.5	25.0	24.8	29.1	29.1	25.1	27.9	18.4	18.8	12.2	15.9	16.3	12.7	16.9	12.4	15.1	17.7	17.1	19.6	19.6	15.5	16.2
sum6b_3p14	30.4	28.9	28.9	21.1	21.1	28.9	28.8	27.1	26.8	30.1	30.0	27.0	29.3	19.8	20.2	13.6	17.7	18.0	14.1	17.5	13.1	16.3	18.4	18.6	21.4	21.1	17.3	17.6
sum6b_3p15	32.4	31.0	31.0	22.0	22.0	30.9	30.9	29.1	28.9	32.1	32.1	28.7	31.7	21.3	21.9	14.6	19.0	19.6	15.1	19.7	13.9	17.9	20.6	20.3	23.6	23.0	18.6	18.5
sum6b_3p16	30.6	29.1	29.1	21.1	21.1	29.1	29.0	27.3	27.1	30.2	30.2	27.0	29.7	19.6	20.3	13.0	17.4	17.9	13.7	18.7	13.2	16.5	19.6	18.5	21.6	21.3	16.7	17.2
sum6b_3p17	28.9	28.5	28.5	19.5	19.5	28.5	28.4	24.9	24.7	29.0	29.0	25.0	27.7	17.7	18.3	11.8	15.7	16.1	12.6	17.5	11.2	15.0	18.4	17.1	19.7	19.1	15.4	15.5
sum6b_3p18	28.9	27.4	27.4	19.5	19.5	27.4	27.3	25.6	25.3	28.5	28.5	25.4	27.9	17.9	18.6	12.1	16.2	16.9	12.8	17.9	11.4	15.4	18.7	17.5	20.3	19.7	15.8	15.7
sum6b_4p19	28.4	27.0	27.0	20.3	20.3	26.9	26.9	25.1	24.9	28.1	28.1	25.1	27.3	17.2	18.0	11.7	15.8	16.0	12.5	17.2	*11.1*	14.9	18.1	16.7	19.6	18.8	15.3	15.4
sum6b_4p20	28.5	27.1	27.1	20.3	20.3	27.0	27.0	25.2	25.0	28.2	28.2	25.2	27.4	17.9	18.5	12.0	16.1	16.2	12.6	*16.6*	12.3	15.0	*17.5*	16.6	19.5	19.2	15.4	16.1
sum6b_4p21	28.6	27.1	27.1	20.3	20.3	27.0	27.0	25.3	25.0	28.2	28.2	25.1	27.6	17.5	18.3	11.8	15.8	16.2	12.5	17.7	12.0	15.3	18.5	16.9	19.8	19.3	15.1	15.7
sum6b_4p22	28.6	28.2	28.2	20.3	20.3	28.1	28.1	*24.6*	*24.4*	28.7	28.6	*24.7*	27.4	17.3	18.1	11.5	*15.3*	15.5	12.3	17.4	11.8	14.9	18.2	16.5	19.2	18.8	14.7	15.4
sum6b_4p23	32.7	33.1	33.1	21.9	22.0	33.0	33.0	27.5	27.4	33.2	33.2	27.8	31.3	20.6	21.3	13.4	17.5	18.3	13.9	19.5	13.5	16.9	20.4	19.3	22.2	22.1	17.0	17.8
sum6b_4p24	30.7	30.8	30.8	21.1	21.2	30.7	30.6	26.0	25.9	31.0	31.0	26.3	29.4	18.5	19.5	12.2	16.1	16.7	12.9	18.6	12.3	15.8	19.4	17.8	20.5	20.3	15.5	16.3
sum6b_2p18c	45.5	45.2	45.2	24.5	24.6	45.1	45.1	35.6	35.4	46.0	46.1	36.7	41.6	22.7	24.6	14.5	18.6	21.2	14.8	21.4	13.0	18.2	22.2	21.3	25.0	27.0	18.9	19.2
sum6b_2p19c	45.5	45.2	45.2	24.5	24.6	45.1	45.1	35.6	35.4	46.0	46.1	36.7	41.6	22.7	24.6	14.5	18.6	21.2	14.8	21.4	13.0	18.2	22.2	21.3	25.0	27.0	18.9	19.2
sum6b_2p20c	53.9	53.6	53.6	29.2	29.3	53.5	53.5	44.0	43.7	54.5	54.6	44.9	49.8	31.1	32.7	20.7	27.4	29.2	21.2	26.2	18.2	24.2	27.3	29.0	33.7	35.1	27.4	26.0
sum6b_2p21c	39.6	37.6	37.6	20.9	20.9	37.5	37.5	30.1	29.8	39.4	39.5	31.8	36.0	21.6	21.9	14.5	18.5	19.7	14.8	18.5	13.0	17.0	19.4	20.2	22.7	23.1	18.7	18.3
sum6b_2p22c	50.7	49.1	49.1	27.8	27.9	49.0	48.9	40.4	40.1	50.7	50.7	41.7	47.2	28.4	29.7	17.6	24.2	26.2	17.9	24.3	16.5	21.2	25.2	25.3	29.5	31.0	22.8	23.1
sum6b_2p23c	42.8	39.2	39.2	23.7	23.8	39.2	39.2	33.8	33.7	42.1	42.1	35.0	40.6	21.4	23.2	14.1	18.3	20.2	14.3	20.5	12.6	17.7	21.4	20.3	24.3	25.5	18.4	18.7
sum6b_2p24c	56.2	53.3	53.3	31.9	32.0	53.2	53.2	46.4	46.1	55.9	55.9	47.3	54.1	32.6	34.4	21.6	28.6	31.2	22.1	29.2	20.4	26.8	30.4	30.8	36.3	36.8	28.0	27.9
sum6b_2p25c	52.3	49.4	49.4	28.8	29.0	49.3	49.3	42.5	42.3	51.9	52.0	43.4	50.3	30.2	32.0	19.0	24.9	28.1	18.6	25.0	17.6	22.7	25.9	27.1	32.2	34.1	24.3	25.2
sum6b_2p26c	53.9	52.6	52.6	29.1	29.2	52.5	52.5	43.4	43.1	53.9	54.0	44.4	50.9	31.5	33.9	19.0	25.1	28.6	19.0	25.7	18.0	22.9	26.7	28.0	32.4	34.7	24.3	25.3
sum6b_2p27c	53.9	53.9	53.9	27.7	27.9	53.9	53.9	43.7	43.5	54.6	54.7	44.8	49.6	33.4	35.0	22.3	28.3	31.0	21.5	26.4	18.8	24.7	27.3	30.7	34.9	36.6	27.7	26.7
sum6b_2p28c	55.4	55.5	55.5	29.7	29.9	55.5	55.5	45.3	45.1	56.2	56.3	46.2	51.3	32.4	34.6	20.7	27.3	30.4	19.9	26.5	18.2	23.7	27.3	28.8	34.1	36.3	26.3	25.8
sum6b_2p29c	59.7	59.8	59.8	31.4	31.5	59.7	59.7	49.5	49.3	60.5	60.5	50.4	55.5	36.6	39.4	24.4	31.9	33.8	24.0	29.7	20.1	27.0	30.8	33.8	38.5	41.1	30.5	29.5
sum6b_2p30c	53.9	53.9	53.9	27.7	27.9	53.9	53.9	43.7	43.5	54.6	54.7	44.8	49.6	33.4	35.0	22.3	28.3	31.0	21.5	26.4	18.8	24.7	27.3	30.7	34.9	36.6	27.7	26.7
sum6b_2p31c	54.9	55.0	55.0	29.4	29.5	55.0	54.9	44.8	44.6	55.7	55.7	45.9	50.7	34.4	35.8	22.8	29.0	32.5	22.4	26.7	20.3	25.7	27.8	31.7	36.1	37.4	28.7	27.6
sum6b_2p32c	47.0	47.1	47.1	25.0	25.1	47.0	47.0	36.8	36.6	47.6	47.7	37.9	43.0	24.2	26.8	15.4	19.8	22.4	15.5	22.2	13.7	18.9	23.1	22.5	26.2	28.2	19.6	19.9
sum6b_2p34c	45.5	45.2	45.2	24.5	24.6	45.1	45.1	35.6	35.4	46.0	46.1	36.7	41.6	22.7	24.6	14.5	18.6	21.2	14.8	21.4	13.0	18.2	22.2	21.3	25.0	27.0	18.9	19.2
sum6b_3p11c	61.6	60.1	60.1	35.8	35.8	60.1	60.0	50.6	50.2	61.9	61.9	52.2	57.5	37.4	37.9	25.5	33.1	35.0	25.4	33.4	23.6	29.8	34.6	34.8	40.0	40.8	32.2	31.2
sum6b_3p12c	64.9	63.4	63.4	39.1	39.1	63.3	63.3	53.9	53.4	65.2	65.2	55.4	60.8	41.1	40.7	28.6	36.7	37.6	28.3	38.1	27.0	33.3	39.4	38.5	43.6	44.3	35.7	34.9
sum6b_3p13c	58.0	56.5	56.5	34.8	34.8	56.4	56.3	47.0	46.6	58.1	58.2	48.6	54.0	35.8	35.3	24.1	31.3	32.2	23.8	32.8	23.3	28.0	33.8	32.6	37.0	37.8	30.3	30.1
sum6b_3p14c	46.9	45.0	45.0	25.9	25.9	44.9	44.8	36.1	35.8	46.8	46.9	38.6	43.2	28.0	28.9	18.4	24.5	26.4	18.9	24.3	17.2	21.9	25.2	26.0	29.8	29.8	23.6	23.3
sum6b_3p15c	42.7	40.7	40.7	22.9	22.9	40.6	40.6	31.9	31.6	42.6	42.6	34.4	39.0	24.3	25.4	15.0	20.3	22.3	15.3	20.6	14.2	18.1	21.4	21.6	25.2	26.2	19.4	19.8
sum6b_3p16c	45.8	44.3	44.3	27.2	27.2	44.2	44.2	34.7	34.4	45.9	46.0	37.1	41.5	28.2	27.2	19.3	24.5	24.9	19.4	26.8	19.7	22.9	27.7	25.9	29.1	29.9	24.1	24.5
sum6b_3p33c	36.9	35.0	35.0	*19.3*	*19.2*	34.9	34.8	26.1	25.8	36.8	36.8	28.8	33.3	18.9	19.9	12.2	15.8	17.2	12.8	17.8	11.5	15.2	18.5	17.7	20.3	21.1	15.9	16.1
sum6b_3p34c	47.8	47.8	47.8	25.3	25.4	47.8	47.8	37.7	37.5	48.5	48.6	39.0	43.5	26.9	28.8	17.6	22.6	25.2	16.9	22.6	15.7	20.3	23.3	24.5	28.7	30.1	22.2	21.6
sum6b_3p35c	42.2	40.2	40.2	21.5	21.4	40.1	40.1	31.4	31.0	42.1	42.2	33.9	38.4	24.0	24.8	16.2	21.2	21.8	16.8	21.1	14.1	18.7	22.0	22.6	25.3	26.2	20.9	20.4
sum6b_3p36c	44.8	42.8	42.8	23.8	23.7	42.7	42.7	34.0	33.7	44.7	44.8	36.5	41.1	26.3	27.7	17.0	22.7	24.1	17.6	22.2	15.2	19.9	23.1	24.3	27.5	28.5	21.8	21.8
sum6b_3p37c	45.7	45.7	45.7	23.9	23.9	45.6	45.6	35.0	34.7	46.4	46.5	36.5	41.5	23.4	24.8	14.6	18.5	22.1	14.5	20.8	13.4	18.1	21.5	21.5	25.0	27.1	18.7	19.1
sum6b_3p38c	35.9	35.5	35.5	22.9	23.0	35.4	35.4	31.9	31.7	36.0	36.0	31.7	34.9	23.5	24.7	14.6	20.1	21.8	15.0	20.1	13.7	17.6	20.9	21.1	24.8	25.2	18.9	19.3
sum6b_3p39c	46.5	46.5	46.5	24.5	24.6	46.5	46.5	35.8	35.5	47.3	47.4	37.4	42.2	25.1	26.8	16.0	21.1	23.6	15.9	21.6	14.6	19.3	22.3	23.0	27.0	28.4	20.8	20.4
sum6b_3p40c	55.0	55.1	55.1	29.2	29.2	55.0	55.0	44.3	44.0	55.9	56.0	45.4	50.9	30.8	33.1	19.4	26.1	29.2	19.0	25.7	16.7	22.7	26.4	27.5	32.9	35.0	25.3	24.5
sum6b_3p41c	48.8	48.9	48.9	25.8	25.8	48.8	48.8	38.1	37.9	49.5	49.6	39.3	45.1	26.4	28.8	15.9	21.2	24.4	16.2	22.7	15.2	19.8	23.5	23.9	27.7	30.0	20.5	21.3
sum6b_3p42c	37.8	37.5	37.5	20.1	20.1	37.5	37.4	27.9	27.7	38.2	38.3	29.6	33.7	20.0	21.2	12.8	16.6	18.4	13.4	18.9	12.0	16.0	19.7	18.9	21.7	22.8	16.7	17.0
sum6b_3p43c	38.9	38.9	38.9	20.1	20.1	38.8	38.8	28.1	27.9	39.6	39.6	30.2	34.5	21.6	22.7	13.3	17.4	20.1	13.7	19.3	13.0	16.7	20.0	20.0	22.8	24.0	17.1	17.7
sum6b_3p44c	43.7	43.8	43.8	23.4	23.4	43.7	43.7	33.0	32.7	44.4	44.5	34.8	39.5	25.0	26.5	15.2	20.0	23.0	15.6	21.5	14.7	18.7	22.3	22.6	26.0	27.8	19.5	20.6
sum6b_3p45c	48.8	48.9	48.9	25.8	25.8	48.8	48.8	38.1	37.9	49.5	49.6	39.3	45.1	26.4	28.8	15.9	21.2	24.4	16.2	22.7	15.2	19.8	23.5	23.9	27.7	30.0	20.5	21.3
sum6b_3p46c	50.7	51.1	51.1	26.6	26.6	51.0	51.0	40.1	39.8	51.5	51.6	41.2	46.7	28.3	31.0	17.5	22.4	26.2	17.0	23.7	16.0	20.8	24.5	25.4	29.5	31.9	21.8	22.3
sum6b_3p49c	53.8	53.8	53.8	30.6	30.6	53.7	53.7	45.1	44.7	55.8	55.9	46.8	52.5	31.2	32.7	20.4	26.6	29.7	20.4	27.1	18.3	24.4	28.1	28.8	33.6	34.8	26.2	25.7
sum6b_4p47c	38.4	36.3	36.3	21.5	21.5	36.2	36.2	28.8	28.5	38.1	38.1	30.6	34.7	19.7	21.1	12.3	16.5	18.1	12.8	17.4	12.0	15.4	18.2	18.0	20.9	21.4	15.8	16.2
sum6b_4p48c	43.8	40.2	40.2	23.0	23.2	40.1	40.1	35.0	34.8	43.1	43.1	36.4	41.5	21.9	23.8	13.9	18.2	20.7	13.9	19.7	12.8	17.4	20.5	20.3	24.1	25.3	17.9	18.6
sum6b_4p50c	45.4	41.9	41.9	22.8	22.9	41.8	41.8	31.7	31.5	45.0	45.1	34.7	40.1	21.4	23.1	13.5	17.6	20.2	13.6	18.8	12.7	16.9	19.6	19.5	23.4	25.2	17.5	18.1
sum6b_4p51c	38.3	34.2	34.2	22.0	22.1	34.1	34.1	30.4	30.2	37.2	37.3	31.9	36.0	23.1	24.2	14.5	20.1	21.7	14.9	19.8	13.8	17.5	20.6	20.8	24.7	25.0	18.8	19.2
sum6b_5p25	*28.1*	*26.7*	*26.7*	21.1	21.1	*26.6*	*26.5*	24.8	24.6	*27.7*	*27.7*	24.8	*26.9*	*16.8*	*17.7*	*11.4*	15.4	*15.4*	*12.2*	17.1	11.7	*14.8*	17.9	*16.1*	*19.0*	*18.4*	*14.6*	*15.3*

REFERENCES

[1] S. Knowles, "A family of adders," Proc. 14th IEEE Symp. on Computer Arithmetic, Adelaide, Australia, pp. 30–34, April 1999.

[2] D. Harris, "A taxonomy of prefix networks," Proc. 37th Asilomar Conf. Signals, Systems, and Computers, pp. 2213–2217, Nov. 2003.

[3] S. Rakesh and K.S. Vijula Grace, "A comprehensive review on the VLSI design performance of different Parallel Prefix Adders", Materials Today: Proceedings, vol. 11, part 3, pp. 1001–1009, 2019.

[4] B.R. Zeydel, D. Baran, and V.G. Oklobdzija, "Energy-efficient design methodologies: high-performance VLSI adders," IEEE Journal of Solid-State Circuits, vol 45, no. 6, pp. 1220–1233, June 2010.

[5] M. Aktan. D. Baran, and V.G. Oklobdzija, "Minimizing energy by achieving optimal sparseness in parallel adders," IEEE 22nd Symp. on Computer Arithmetic, Lyon, France, pp. 10–17, June 2015.

[6] I. Brzozowski and A. Kos, "Designing of low-power data oriented adders," Microelectronics Journal, vol. 45 no. 9, pp. 1177–1186, Sept. 2014.

[7] I. Brzozowski, "An analysis of Parallel Prefix Adders Regarding the design of low-power data oriented adders," Proc. of Int. Conf. on Signals and Electronic Syst. (ICSES'18), Kraków, Poland, pp. 7-12, Sept. 2018.

[8] N. Weste and D. Harris, CMOS VLSI Design, 4th ed., Boston, MA: Pearson, 2011, pp. 387–416.

[9] D. Harris, "A taxonomy of parallel prefix networks," Record of the Thirty-Seventh Asilomar Conf. on Signals, Systems and Computers, pp. 2213-2217, Nov. 2003.

[10] I. Brzozowski, A. Kos, "A new approach to power estimation and reduction in CMOS digital circuits," Integration, the VLSI Journal, vol. 41, pp. 219–237, Febr. 2008.

Proceedings of the 27th International Conference *"Mixed Design of Integrated Circuits and Systems"*
June 25-27, 2020, Łódź, Poland

Low Hardware Complexity Filters for On-Chip Algorithm Used in Air Pollution Sensors for Dense Urban Areas in Smart Cities

Zofia Długosz[1], Michał Rajewski[1], Marzena Banach[2], Tomasz Talaśka [1], Rafał Długosz[1]

[1] UTP University of Science and Technology
Faculty of Telecommunication, Computer Science and Electrical Engineering
ul. Kaliskiego 7, 85-796, Bydgoszcz, Poland
[2] Poznan University of Technology
Institute of Architecture and Spatial Planning
Nieszawska 13C, 61-021 Poznań, Poland
E-mail: zosia.dlugosz@gmail.com, michalrajon@gmail.com, marzena.banach@put.poznan.pl,
tomasz.talaska@gmail.com, rafal.dlugosz@gmail.com

Abstract—**The paper presents a method of transistor level implementation of a reconfigurable filter for the application in the algorithm responsible for processing air pollution data. The assumption of the proposed solutions is the realization of the algorithm that uses such filters directly in the wireless sensor, along with other components of such devices. Thanks to this, the amount of data exchanged between the sensors and the base station can be reduced. In the proposed filter structure, a special emphasis was placed on reducing the hardware complexity of the filter. The objective is to reduce the chip area of the overall device. The filter features a modular reconfigurable structure, which allows to achieve different filter orders, with almost linear increase in the hardware complexity. Target application of the proposed solution is in wireless sensors networks (WSN) that consist of large numbers of devices distributed, e.g. in dense urban areas in cities.**

Keywords—**Air pollution sensors; Air pollution data processing; Filters; Edge computing**

I. INTRODUCTION

Air pollution belongs to one of the key problems in cities today. According to the United States Environmental Protection Agency, various types of air pollutants, especially those designated as particulate matter $PM_{2.5}$ and PM_{10} (smog), are hazardous to human health [1]. This applies especially to the particles with diameters below 2.5 μm, as their size allows them to enter the human blood directly. Prevention of the described problems is one of the basic assumptions and ecological goals of theoreticians of the, so-called, smart cities.

This problem may be observed through more and more popular networks of air pollution sensors. Such networks allow to create air pollution maps showing the levels of pollution in cities, with a relatively frequent update of current situation. Selected systems of this type are briefly described and compared in following Section. The maps are available on the Internet for free.

One of main sources of air pollution is transportation in cities, in particular its combustion part. Traffic intensity on roads significantly affects daily pollution levels. In practice, however, there is no simple (linear) relationship between these two factors. Pollution levels depend heavily also on weather conditions, insolation level, wind strength and direction, season of the year, etc.

A factor that substantially impacts the course of the air pollution over time in cities is the structure of urban development. Urban density and height of buildings have an impact on ventilation abilities. In some parts of the city there may be observed an increase in the wind speed, due to the so-called tunnel effect. In larger cities one can also observe an effect of the, so-called, urban heat islands. It results in higher temperatures in some areas of the city. It causes differences in atmospheric pressure between particular areas of the city, which in turn is the source of air suction in directions toward the heat islands, where the pressure is usually smaller. The size and spatial structure of the urban development is a crucial parameter here. The specificity of tightly built-up urban interiors may also cause a phenomenon of air stagnation or air circulation around without replacing it [2]. The impact of the described factors on pollution levels may vary daily and be dependent on the season. In winter, for example, apartments heating becomes one of noticeable factors.

Pedestrians and cyclists belong to the group of people, moving in an urban space, that is particularly exposed to the factors described above. Usually they move around the city without any cover. Considering this, we believe that maps with higher spatial resolution are needed than those currently offered. These maps should be also updated more frequently. This would allow the mentioned users of the cities to plan their routes in such a way to avoid the most polluted areas.

The described problem may be addressed through the development of air pollution monitoring systems based on wireless sensor network (WSN) composed of miniaturized, low power devices densely distributed in the urban environment. The idea is to enable the monitoring at the level even of particular streets. More dense maps require a large number of sensors, which should be substantially cheaper than devices available today. They should also allow for an easy installation in the urban infrastructure of the cities. This may be achieved by eliminating the need for an external power supply.

Low power operation and the ability to achieve compact sizes requires the optimization of particular circuit components of the sensors. The purpose of this work is to contribute to the development of such solutions. It is a continuation of our earlier work presented in [3]. Here, we focus specifically on filters, which are one of the main components of the previously proposed algorithm responsible for analysis of recorded data.

The paper is structured as follows. In the next section we present briefly selected air monitoring systems. Then we provide details of the proposed algorithm structure and the reconfigurable filter. The filter has been designed so that its parameters can be easily changed, depending on the noise levels in the measured signal. Conclusions are drawn at the end of the paper.

II. STATE-OF-THE ART AIR POLLUTION MONITORING SYSTEMS

Networks of sensors used for monitoring the air quality have become very popular in recent years. Several such networks already operate in Europe. Particular networks offer measuring devices with different abilities and thus with different prices. They differ in the scope of measurements and the accuracy of the measurements. Some of the companies offering the devices charge an additional monthly subscription fee for each device installed and active. Below we briefly present the most popular networks of this type.

A. Airly network

Airly is a company created by graduates of the AGH University of Science and Technology in Kraków, Poland. The device that is offered by this company, using a laser sensor measures the following quantities:

- dust levels $PM_{2.5}$ and PM_{10},
- weather conditions: humidity, temperature, air pressure.

The measurement results are transferred to a base station, where after calibration, they are published through the Airly company's web page [4]. The price of a single device is approximately 400 Euro. In addition, a monthly subscription fee of 10 Euro per a single device is charged.

B. Looko2 network

ZAS company, which is the owner of the LookO2 network, offers several external sensors at prices that vary in-between about 140 and 430 Euro. Depending on the type of the sensor, they measure:

- dust levels PM_1, $PM_{2.5}$ and PM_{10},
- weather conditions: humidity and temperature,
- in more advanced devices also HCHO (formaldehyd).

The customers receive daily reports on selected measured quantities. The measured data are also available through the company's web page [5]. An advantage here is the lack of the subscription fee.

C. Luftdaten

Luftdaten is a German network that originates from the 2015 citizens' project from the city of Stuttgard. Over time, the network covered more and more cities in Germany. Then it began to be spread out in other countries. Currently it exists in over 70 countries. A characteristic feature here is the necessity of own construction of the device from a provided assembly kit. The SDS 011 laser sensor enables measuring:

- dust levels $PM_{2.5}$ and PM_{10},
- weather conditions: humidity and temperature.

The promotional price of the device for the network members is about 35 Euro, while the standard price is 44 Euro. The project's assumption was to create an alternative to more expensive measuring devices. The measurement accuracy is worse than from the more expensive devices. Reported test results show that for smaller levels of the $PM_{2.5}$ quantity, below 20 mg / m^3, the sensor lowers the measured values. As the pollutant level increases, the differences become smaller. The results are offered via a dedicated web page [6].

D. SmogTok

SmogTok is a project initiated in February 2018 in the Wawer district of Warsaw, which is an area with majority of family houses. Currently, the network includes about 200 external sensors. They are located in Warsaw and in some other Polish cities. SmogTok sells its sensors at the price of about 70-75 Euro. This is the only necessary expense, as the company does not charge the subscription fee. The sensor provided by SmogTok measures:

- dust levels $PM_{2.5}$ and PM_{10},
- weather conditions: humidity and temperature.

It uses a laser sensor. It requires protection against direct contact with solar radiation. If the installation location does not meet this condition, SmogTok suggests adding a Drumik passive radiation shield to the smog sensor, which costs about 30 Euro. The recorded data may be verified in this case throughout an analysis of former recordings. The measurement results are also presented via a web page [7] and a mobile application.

E. Syngeos

The Syngeos company, based in Katowice, offers certified measuring stations that allow to monitor:

- dust levels $PM_{2.5}$ and PM_{10},
- weather conditions: humidity and temperature,
- optionally other pollutants: carbon monoxide, nitrogen dioxide, sulfur dioxide and benzene.

The measurement results are available via the "Syngeos – Nasze Powietrze" (Our Air) application and on a web page [8]. The communication is performed via LoRaWAN network. The price of a single Syngeos device is about 400 Euro. As with the Airly devices, a monthly subscription fee of 16 Euro is also charged here. The user gains access to the platform containing reports and visualization of collected and analyzed data.

F. Summary

Table I presents a comparison of particular pollution monitoring systems discussed above. Attention was paid to selected technical parameters such as the access method to the base

TABLE I
COMPARISON BETWEEN SELECTED AIR POLLUTION MONITORING SYSTEMS

Model	Price €	Subscription fee €	Communication	Power supply	Measurement time interval [minute]
Airly	395	10	Wi-Fi	power grid	$1-5^{(*)}$
Looko2	140-430	none	Wi-Fi	power grid	1
Luftdaten	44	none	Wi-Fi	power grid	2.5
SmogTok	73	none	Wi-Fi	min USB	10
Syngeos	400	16	LoRaWAN network	no data	no data

$(*)$ Airly on-line maps are updated every hour

station and the power supply. All the described devices require access to external power supply. In all the described cases the communication is performed through existing Wi-Fi networks. The use of low power and low-cost WiFi modules becomes a standard today. Such networks are now widely available in urban areas [4], [9], [10].

The observed tendency is to increase the measurement frequency, however, it can be concluded that more frequent measurements than once per minute are not necessary. For this reason, we assumed a minute time interval in our investigations.

III. PROPOSED CONTRIBUTION TO AIR POLLUTION SENSOR DEVELOPMENT

To enable a high efficiency in functioning of the most important systems on which the concept of smart cities is based, a large amount of data is needed, often delivered in real time from different locations in the city [11]. This requires the use of a large number of different types of sensors. Air pollution sensors belong to this group. To simplify the installation process of such devices and their maintenance, a paramount feature is the one, in which the devices will not require an access to external power source. To make it possible, the sensors should offer simple structure, small sizes, as well as low energy consumption. Such trend may be already observed in the market [12]. The latter feature is crucial from the point of view of energy self-sufficiency. Low energy consumption may allow to operate with energy scavenged from the environment [13], [14], [15]. Energy sources that are potentially useful here include heat and solar.

One of the possibilities is the development of the sensor as a system-on-chip (SoC) or system-in-package (SiP). In the first case, in a single integrated circuit are placed the blocks responsible for measuring a given physical quantity, anti-aliasing filtering, conversion of measured analog signal to digital domain, and analysis of the collected data and their transmission to the base station.

Examples of devices that meet these criteria can be already found in the literature. An example of the sensor used to measure air pollution is the device provided by Texas Instruments [16]. This device uses an optical measurement method of the $PM_{2.5}$ and the PM_{10} particles. This is a fairly popular method, frequently used in devices described in the previous Section. In this case, the light stream passes through the chamber that is exposed to air and thus contains the air sample. The stream scatters on the pollutant particles. As a result, the current from the photodiode located on the other side of the chamber is correlated with the concentration of pollutants suspended in the air.

Assuming the measurement frequency even at the level of once a minute, most of the time the device may remain in the standby mode, in which it consumes only a minimal amount of energy. Another possible way of saving energy, is the use of an edge computing concept, popular in recent time, in which even advanced signal processing algorithms are implemented directly in the sensor devices [17], [18], [19]. Thanks to this, only part of the measured data needs to be transmitted to the base station. It is important, as the RF data transmission block may consume up to 80-90 % of total energy consumed by the device [20], [21], [22].

In one of our earlier works we proposed an algorithm of this type [3]. Its input signal are samples of the measured pollutant. These samples within a specified time horizon are saved in the internal memory of the algorithm. Its role is to directly control the block responsible for data transmission. Its operation principle is based on detecting periods, in which the level of a given pollutant does not change significantly. During these periods the data transmission is performed less frequently.

Below we briefly present an overview of the proposed algorithm [3]. This description is necessary to properly present our recent research results. In this work, we focus specifically on digital filters, which are one of its main components of the algorithm.

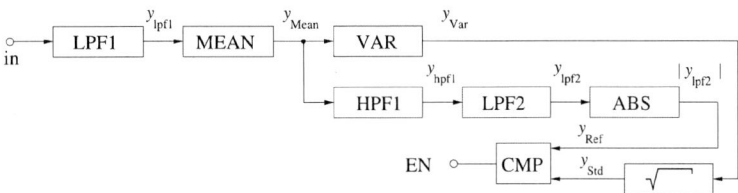

Fig. 1. A general structure of the proposed algorithm used in the analysis of the recorded air pollution data. The block provides a direct control to the radio-frequency (RF) communication.

A. Proposed algorithm for analysis of air pollution data

A general block diagram of the algorithm is shown in Figure 1. It is based on several operations, which include filtering, computing of variance and standard deviation, comparison of signals, etc. Initially, the input signal is processed in a single channel. At this stage, the signal is enhanced with the use of low order low-pass filter (LPF1). Then calculated is an average (Mean) value from the input samples over a window with a given length. To reduce the complexity of internal memory block, we applied a low pass infinite impulse response (IIR) filter of low order. As a result, the Mean value is computed recursively. This requires keeping in memory a previous mean value and the new sample.

Then the signal processing takes place in two parallel channels. In one of them, the variance is calculated, also in an iterative way [23]. In the second channel, the signal is first filtered using the 1 st order high-pass filter (HPF), with the coefficients [0.5, -0.5]. The output signal from this filter illustrates how much the input signal changes in-between two adjacent measurements. If the input signal is noisy, the output signal from the HPF filter is also noisy. For this reason we use another low-pass filter (LPF2), whose role is to extract a trend of the changes, even when the input signal is very noisy. The absolute value of the signal is calculated after the LPF2 filter. This operation is performed, as the algorithm should react in the same way to signal changes in both directions.

The signal from the first channel is used as the adaptive threshold, which is compared with the signal from the second channel. As a threshold, one can directly use the computed variance, or the standard deviation, which may be obtained by calculating the square root of the variance. In our previous work, both these signals were used for testing purposes. They were compared either directly with the output signal from the ABS block or with its square.

The resultant enable signal (EN) from the comparator indicates when data should be transmitted to the base station. Either the directly measured signal or the signal after its filtration may be sent do the base station. The second one facilitates its further processing on the base station side.

B. Proposed reconfigurable filter

The algorithm has been designed so that it can be reconfigured depending on the input signal parameters, mostly noise parameters. In practice, this largely means the possibility of changing the parameters of the applied filters. For larger noise levels, higher filter orders are used, however at the price of larger group delays (a trade-off).

To keep the filter structure compact, finite impulse response (FIR) filters with flat frequency responses have been applied. Comprehensive software level investigations, presented in [3], show that such filters are sufficient in the described application. In filters of this type coefficients have simple integer values and thus may be easily implemented in hardware in fixed point arithmetic.

The proposed filter has been implemented in a modular way. It consists of serially connected 1 st order sections with equal absolute values of the coefficients. As a result, each subsequent section increases the steepness of the transient band of the frequency response, thus improving the selectivity of the filter. At the same time, an approximately linear relationship between the filter length and the complexity of its hardware structure is maintained.

The structure of the programmable filter is shown in Figure 2. It is a modified version of the structure that was only briefly presented in [3]. In the current version, we introduced the ability to achieve both the low-pass and the high-pass filtering mode in a single circuit.

To make it possible, each of the 1st order sections should enable switching their coefficients between $[1, 1]$ and $[1, -1]$. An advantage of this solution is that the filter does not need multiplying blocks. A single 1st order section requires only a single multi-bit asynchronous summing or subtracting operation, depending on the filtering mode. The block that allows the realization of these two operations is marked as SUMB in the diagram, shown in Fig. 2. It is switched over with the use of a single 1-bit SLH signal.

Depending on the required filter order, N, a corresponding signal m_k is set to 1, while only one of the m_k signals may be set to '1'. Each of these signals controls its corresponding switch, thus providing output from a given 1st order section to the input of the block, which is marked as DIV. The role of the DIV block is to normalize the signal at the filter output. This is carried out throughout the division (thus DIV) of the output from a given 1st order section by the sum of the absolute values of the used filter coefficients. In case of the flat frequency response, this sum is always one of the powers of the number 2 (2^k). In practice, the division operation boils down in this case to a simple shift of the bits by k positions to the right. The DIV block may be realized either as a field of switches, controlled by the m_k signals, or as a set of AND logic gates.

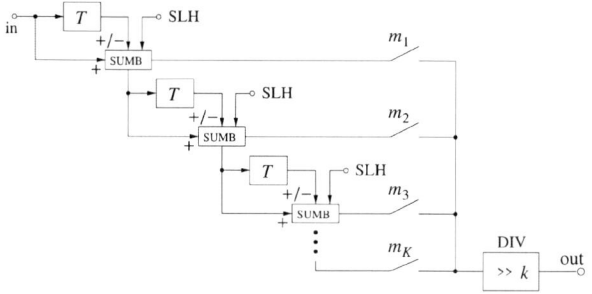

Fig. 2. Structure of the reconfigurable filter with flat frequency response (version 1). The 1-bit SLH signal switches the filter between low-pass and high-pass modes.

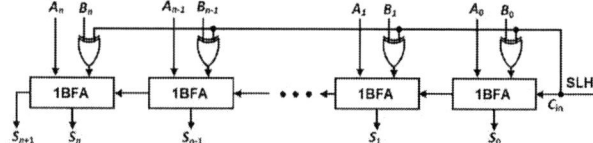

Fig. 3. An n-bit, parallel, asynchronous summing / subtracting circuit composed of a series of 1-bit full adders.

Two approaches can be used here. In the first one, presented directly in Fig. 2, all 1st order sections are active, which means that their delay lines and the SUMB blocks are working independently on which m_k signal is '1'. A disadvantage of this approach is that all 1st order sections consume energy, even though part of them do not participate in computing the filter output signal. However, this approach has one important advantage. When the noise level changes and it is necessary to switch the filter order to a higher value, it can be done immediately, without introducing any distortion into the output signal. It can be said that the filter internally calculates the output signals (non-normalized) for all possible filter orders, resulting from the number of the used sections. Since the normalization is an asynchronous operation, it can be performed quickly. In this case, for a given input signal sequence, one can receive, practically in parallel, output signals for all or several selected filter orders, N. To achieve this, an appropriate m_k sequence should be applied for each new input sample. For each following m_k signal, it is necessary only to read the resultant sample from the filter output.

In the second approach, shown in Fig. 4, to save energy, one can turn off unused 1st order sections. In this case, additional switches are placed in-between particular sections. These switches may be realized as logic AND gates. They are controlled with the use of additional M_k signals, determined asynchronously on the basis of the m_k signals. The block responsible for computing the M_k signals is shown in Fig. 5. A chain of OR gates causes that always all M_l signals become '1', where $l \leq k$.

The delay element (T) in each of the 1st order sections was implemented as a block of D flip-flops (one per each bit).

These memories were selected in this case because of the stability of the stored information. The filter structure is activated relatively rarely (once per minute). For this reason, memories with a long storage time are mandatory here. However, this block itself can also be implemented in a different way. Delay elements in all sections are switched at the same time, so they can be controlled by a simple 1-phase clock.

C. Implementation details and verification

The filter was initially tested by means of the software model along with the overall proposed algorithm. The results of these investigations are presented and commented, in detail, in [3], for different levels of the noise in the input signal. In [3] we also proposed a general structure of the filter, without implementation details.

The filter has been recently realized at the transistor level in the Hspice environment. Selected simulation results are provided in Fig. 6, for a single 1st order filter section. To facilitate the presentation, the results are shown for 4-bit input signals (adjacent signal samples). However, in general, the overall circuit was designed to process 8-bit input signals, varying in the range from 0 to 255. The resolution can be easily extended, by increasing the sizes of the SUMB and the T (delay) blocks. In Fig. 6 two modes are tested (low-pass and high-pass filter), witched by the V_{contr} signal. In the subtraction mode, at the filter output we obtain directly the absolute value of the difference between samples. This means that an additional ABS block is not required.

Without a signal normalization, each section doubles the gain, adding one bit. As a result, in following sections the signal resolutions were 9, 10, etc. bits. The normalization could be performed after each section, by dividing its output

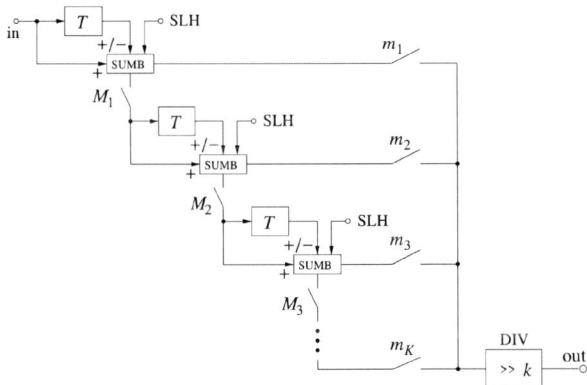

Fig. 4. Structure of the reconfigurable filter with flat frequency response (version 2).

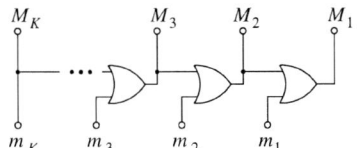

Fig. 5. An asynchronous circuit used to compute M_k signals used in the filter on the basis of m_k signals.

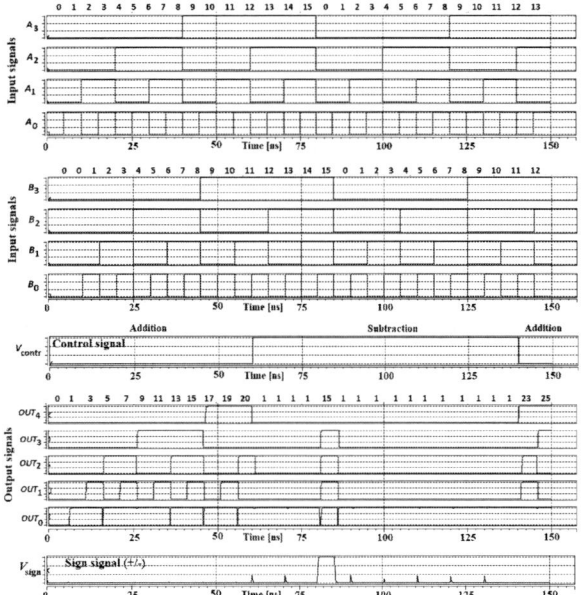

Fig. 6. Selected simulation result of a single 1st order section, illustrating the performance abilities for large data rate.

signal by 2. Such division would not require any additional circuit. It would be sufficient to simply ignore the least significant bit (LSB). In this approach each section would have equal structure, thus simplifying the implementation of the overall filter. It is suitable for larger resolutions of the input signal (e.g. 16-bit). In the tested case i.e. for small signal resolutions, the signal was normalized only at the final stage of data processing (DIV block), to avoid too large distortion of the signal at early processing stages.

Since the output signal from the HPF filter may have negative values, therefore at this stage the signal is coded in the two's complement code. For this reason, LPF1 and LPF2 filters work in a slightly different way. However, the differences are not significant.

Each section requires only a single summing and subtracting block, as well as a given number of 1-bit memory cells. It can be shown that the number of transistors per a single bit equals about 70 (26 in DFF and 44 in 1BFA). For the filter order of 16, assuming signal normalization at the end of the filter in DIV block, for an initial 8-bit input signal, the average number of bits in a single section is 16. For a filter order $N = 15$, 16 described sections are required, so the number of transistors in the overall circuit equals about $16 \cdot 16 \cdot 70 = 17920$, not counting the DIV block.

IV. CONCLUSIONS

The paper presents the implementation of a flexible FIR filter for the application in systems requiring relatively low hardware complexity (in this case air pollution sensor). The flexibility here means the ability to adjust the frequency response of the filter to current properties of the input signal, which may vary in a wide range.

The design of the filter is part of a larger project, which at this stage aims to implement the overall described algorithm. The filter is one of its main components.

The filter can be easily switched between low-pass and high-pass modes. One can also introduce an independent control of selected filter sections (separate SLH signals). For example, the first section could be implemented as a high-pass filter, while the next as low-pass filters. In this case, a single presented structure would be used as a substitute for both the HPF and the LPF2 filters.

REFERENCES

[1] United States Environmental Protection Agency, „Particulate Matter (PM) Basics", https://www.epa.gov/pm-pollution/particulate-matter-pm-basics. (access 2020.02.17)

[2] Perera N., Emmanuel R., Mahanama P.K.S., „Projected urban development, changing 'Local Climate Zones' and relative warming effects in Colombo, Sri Lanka", *International conference on Cities, People and Places*, Colombo, 2013.

[3] Banach, M., Długosz, R., Pauk, J., Talaśka, T. "Hardware Efficient Solutions for Wireless Air Pollution Sensors Dedicated to Dense Urban Areas". *Remote Sensing*, Vol. 12, Issue 776, 2020.

[4] Airly net of air pollution sensors: https://airly.eu/en/

[5] Looko2 net of air pollution sensors: https://www.looko2.com/heatmap.php

[6] Luftdaten net of air pollution sensors: https://luftdaten.info/en/home-en/

[7] https://smogtok.com/

[8] https://panel.syngeos.pl/sensor/pm2_5

[9] Migos, T.; Christakis, I.; Moutzouris, K.; Stavrakas, I., „On the Evaluation of Low-Cost PM Sensors for Air Quality Estimation", *In Proceedings of the 2019 8th International Conference on Modern Circuits and Systems Technologies (MOCAST)*, Thessaloniki, Greece, 13-15 May 2019, pp. 1-4.

[10] Postolache, O.A., Pereira, J.M.M., Girao, P.M.B.S., „Smart Sensor Network for Air Quality Monitoring Applications", *IEEE Transactions on Instrumentation and Measurement*, **2009**, *58*, 9, pp. 3253-3262.

[11] Johnsen, F.T., „Using Publish Subscribe for Short-lived IoT Data", *In Proceedings of the 2018 Federated Conference on Computer Science and Information Systems, Annals of Computer Science and Information Systems*, Poznań, Poland, 9–12 Sept. 2018, 15, pp. 645-649.

[12] Cypress Semiconductor, „Lowest-Power Energy Harvesting Power Management ICs for Battery-Free Wireless Sensor Nodes", https://www.cypress.com, (access 2020.02.17).

[13] Bosse S., „Smart Micro-scale Energy Management and Energy Distribution in Decentralized Self-Powered Networks Using Multi-Agent Systems", *In Proceedings of the 2018 Federated Conference on Computer Science and Information Systems, Annals of Computer Science and Information Systems*, Poznań, Poland, 9–12 Sept. 2018, 15, pp. 203-213.

[14] Loubet, G.; Takacs, A.: Dragomirescu, D., „Implementation of a Battery-Free Wireless Sensor for Cyber-Physical Systems Dedicated to Structural Health Monitoring Applications", *IEEE Access*, **2019**, 7, pp. 24679-24690.

[15] La Rosa, R.; Zoppi, G.; Di Donato, L.; Sorbello, G.; Di Carlo, C. A.,; Livreri, P., „A battery-free smart sensor powered with RF Energy", *In Proceedings of the 2018 IEEE 4th International Forum on Research and Technology for Society and Industry (RTSI)*, Palermo, Italy, 10-13 Sept. 2018, pp. 1-4.

[16] Texas Instruments, „PM2.5 and PM10 Particle Sensor Analog Front-End for Air Quality Monitoring Reference Design", *Texas Instruments Incorporated*, http://www.ti.com/tool/TIDA-00378. (access 2020.02.17).

[17] Zhirong Xu; Ming Cai; Xiaoyan Li; Tianlei Hu; Qianshu Song, „Edge-Aided Reliable Data Transmission for Heterogeneous Edge-IoT Sensor Networks", *Sensors*, MDPI, **2019**, *19*, 2078.

[18] Sitton-Candanedo, I.; Alonso, R.S.;. García, O.: Munoz, L.: Rodríguez-Gonzalez, S., „Edge Computing, IoT and Social Computing in Smart Energy Scenarios", *Sensors*, MDPI, **2019**, *19*, 3353. pp. 1-20.

[19] Talaśka, T., „Components of Artificial Neural Networks Realized in CMOS Technology to be Used in Intelligent Sensors in Wireless Sensor Networks", *Sensors*, MDPI, **2018**, *18*, 4499, pp. 1-19.

[20] Yan, J.; Zhou, M.; Ding, Z., „Recent Advances in Energy-Efficient Routing Protocols for Wireless Sensor Networks: A Review", *IEEE Access*, **2016**,*4*, pp. 5673-5686.

[21] Srbinovska, M.; Dimcev, V.; Gavrovski, C., „Energy consumption estimation of wireless sensor networks in greenhouse crop production", *In Proceedings of the IEEE EUROCON 2017 -17th International Conference on Smart Technologies*, Ohrid, Macedonia, 6-8 July 2017, pp. 870-875.

[22] Mogi, R.; Nakayama, T. ; Asaka, T., „Load balancing method for IoT sensor system using multi-access edge computing", *In Proceedings of the 2018 Sixth International Symposium on Computing and Networking Workshops (CANDARW)*, Takayama, Japan, 27-30 Nov. 2018, pp. 75-78.

[23] West D.H.D., „Updating Mean and Variance Estimates: An Improved Method", *Communications of the ACM*. **1979**, *22*, 9, pp. 532-535.

Proceedings of the 27th International Conference *"Mixed Design of Integrated Circuits and Systems"*
June 25-27, 2020, Łódź, Poland

Low Power Preamplifier for Biomedical Signal Digitization

Nawaf ALjehani[1], Mohamed Abbas[1,2]

[1] EE Department, College of Engineering, King Saud University, Kingdom of Saudi Arabia
[2] EE Department, Faculty of Engineering, Assiut University, Egypt
mohabbas@ksu.edu.sa
437105870@student.ksu.edu.sa

Abstract—This paper presents a low power, low noise preamplifier stage with simple common mode desensitization circuit for dynamic comparators. The target application of the proposed circuit is analog to digital converter for biomedical applications. Adopting TSMC 0.18μm technology, the proposed circuit is designed to work in weak inversion using g_m/I_D design methodology. The simulation results show that the preamplifier stage consumes less than 32nW using power supply of 0.75V. The input referred noise is 17μV, DC gain of 43.15dB and unity gain frequency of 300 kHz.

Keywords—component; formatting; style; styling; insert (key words)

I. INTRODUCTION

Ultra low power consumption is an essential specification for biomedical devices especially portable and wearable ones where they designed to operate with a small non-rechargeable battery [1]. A typical interface to biomedical sensor consists of analog-to-digital converter (ADC) and digital signal processing block. The ADC can be considered the most critical component in such an interface, because the system behavior depends on ADC output. Typical specifications of ADC for biomedical applications are moderate resolution (8-12 bits) and sampling rate of about (1-100kS/s). Figure 1 shows typical biomedical signal frequency and amplitude. [2], [3].

At the heart of the ADC is the comparator circuit or the decision making circuit, which must be designed carefully to satisfy low power consumption, low input referred noise, high resolution to detect the minimum change, and a minimum delay time for a relative small input (small overdrive recovery) [4-6].

The comparator function is to convert small signal, which is millivolt or microvolt higher or lower than a reference level, to digital signal swing from rail to rail. Dynamic comparator is an excellent choice for low power consumption since no static current is consumed [7].

Usually the Dynamic comparator consists of regerentive latch circuit preceded by a preamplifier stage. With a high gain of the preamplifier stage, the influence of noise and mismatch can be reduced [8]. The power consumption of portable biomedical devices is a crucial specification. Targeting a low power comparator for biomedical applications, this paper presents a small area, low power comparator with simple common mode desensitization circuit. The consumed power is utilized efficiently by designing the circuit such that the

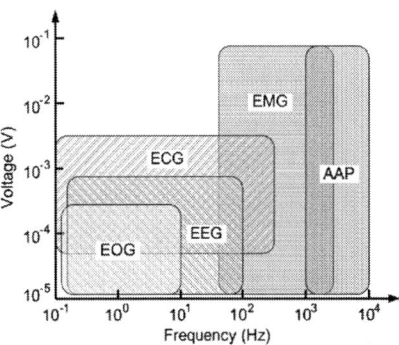

Figure 1. Biomedical signal frequency range [3]

transistors work in weak inversion region. The design is done using g_m/I_D design methodology. Where the g_m/I_D design methodology has been proven to be a robust design method for modern MOS transistor that deviates from the classical square low behavior [9-11].

The paper is organized as following; section II describes MOS transistor model in the subthreshold region. The gm/Id design methodology is described in section III. In sections IV and V, the circuit design, simulation results and discussions are presented followed by the conclusion.

II. MOS TRANSISTOR IN SUBTHRESHOLD REGION

By definition, for MOS transistor when V_{GS} is less than threshold voltage V_{th}, the MOS transistor works in subthreshold or weak inversion region where the current can be described by [12]

$$I_D = \mu C_{ox} \frac{W}{L} V_T^2\, e^{\frac{V_{GS}-V_{th}}{nV_T}}\left(1 - e^{\frac{-V_{DS}}{V_T}}\right) \qquad (1)$$

where n is sub-threshold slope factor (~0.5 to1.5), V_T is the thermal voltage. W, L. are the transistor width and channel length. V_{DS}, V_{GS} and V_{th} are the drain-source voltage, gate-source voltage and threshold voltage respectively. Weak inversion operation results in highest transcendence efficiency which make the most energy efficient implementations for biomedical applications taking into account that in these applications the processing bandwidth is less than 1 MHz, therefore, there is no need for high speed circuit [13].

A small signal model for MOS transistor in weak inversion region was developed in [14].

107

$$g_m = \frac{I_D}{nV_T} \qquad (2)$$

$$g_{ds} = \frac{I_D}{V_T} \, e^{\frac{-V_{DS}}{V_T}}, \quad V_{DS} \leq 4V_T \qquad (3)$$

$$g_{ds} = \frac{I_D}{V_A}, \quad V_{DS} > 4V_T \qquad (4)$$

The power spectral noise of the MOS transistor in subthreshold region is given by [14] and shown in Equation (5).

$$PSD_{\overline{i_{nd}^2}} = 2KTn g_m + \frac{K_f g_m{}^2}{fWLC_{ox}} \qquad (5)$$

The first term in Equation (5) represents thermal noise while the second term represents the flicker noise. As it can be inferred, the flicker noise term can be reduced or make it negligible if the input pair transistor size is made sufficiently large.

III. G_M/I_D Design Methodology

As technology advance, i.e. the channel length of MOS transistor decreased, the classical square law fails to accurately describe the modern MOS transistor drain current, wherein, the short channel affect must be taken into account [9]. Also, as it is indicated in Equation (1), the current-voltage relation is an exponential dependence and the MOS works much like a BJT [10].

g_m/I_D design method based on the inversion level of the MOS transistor has been proposed to overcome these issues and others [9-11]. Where the ratio of g_m/I_D can be used as a proxy to design analog circuit in all operating region as shown in Figure 2.

The design procedure for MOS transistor based on g_m/I_D in the weak inversion introduced by [11] has been used to design the preamplifier stage. The proposed design procedure is based on pre-computed lookup tables, where sweeping all transistor node voltages and channel length L, then compute small signals parameters (g_m, f_T, gds, C_{gd}, etc...). After that using a numerical optimization Software like MATLAB, the designer can optimize the circuit performance based on these data and the transistor width and channel length can be decided accordingly.

IV. Comparator Sub-circuit Design and Simulation

A. Preamplifier Circuit

The preamplifier circuit is the preceding stage of the latch stage. The purpose of the preamplifier stage is to increase the comparator resolution, overcome the kickback noise and offset voltage of the latch [16]. The preamplifier circuit has been chosen with a minimum number of transistor, to minimize the output noise.

The proposed preamplifier circuit is shown in Figure 3. The common mode output is quite well defined and equal to the difference between V_{DD} and V_{SG} of PMOS transistor. While for difference mode output the gate of PMOS transistors is virtually ground. The branch composed of (M7 and M7) or (M8 and M9) represents a very high resistance since all transistors work in subthreshold region. The resistance of each transistor (R_{eq}) can be given by Equation (6) [17]

$$R_{eq} = \frac{V_0}{\mu C_{ox} \frac{W}{L} V_T^2} \, e^{\frac{-(V_{GS}-V_{th})}{nV_T}} \qquad (6)$$

where V_0 is constant positive voltage

Since the resistance of the branch (M6-M7) or (M8-M9) is much higher than the output resistance of M1 or M3, the DC gain of the circuit can be given by Equation 7.

$$A_v = -\frac{\dfrac{g_{m1}}{I_{D_1}}}{\dfrac{g_{ds1}}{I_{D_1}} + \dfrac{g_{ds3}}{I_{D_1}}} \qquad (7)$$

The design procedure for the preamplifier stage in weak inversion using the g_m/I_D design procedure is as follow:

- Compute $g_m = 2 * \pi * f_u * C_L$; where C_L is assumed to be 50fF and unity gain frequency f_u is 1MHz
- Sweep J_D to determine g_m/I_D using Lookup tables
- Determine $I_D = \frac{g_m}{\frac{g_m}{I_D}}$
- Finds $W = \frac{I_D}{J_D}$

TSMC 0.18µm technology was adopted. Following the mentioned design procedure, the sizes of transistors are as shown in Table I.

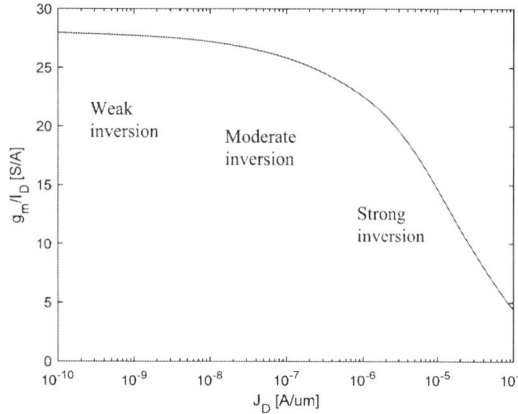

Figure 2. Transconductance efficiency versus current density of the selected technology

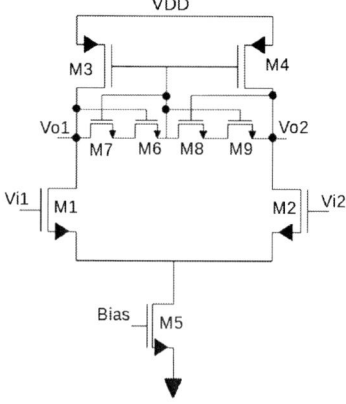

Figure 3. Proposed preamplifier circuit diagram

TABLE I
PREAMPLIFIER TRANSISTOR SIZES WHERE DIMENSION IN μM

Transistor	$M_{1,2}$	$M_{3,4}$	M_5	$M_{6,7,8,9}$
W/L	$25/2$	$5/2$	$5/2$	$5/0.18$

B. Latch Design

The StrongARM latch circuit, shown in Figure 4, finds a wide range of applications such as sense amplifier, and comparator. It is popular because of it can produce rail to rail output and consumes zero static power [18]. The circuit design procedure and design operation can be found in [18 - 20]. The following lines briefly describe the operation of the latch.

- When the clock is low, transistors M_{14}, M_{15}, M_{16}, and M_{17} are used to pre-charge the nodes, which are connected to their drains, to V_{DD}. This is called pre-charging phase.
- Once the clock switches to high state, the evaluation phase starts. The nodes connected to the drains of input transistors M_8 and M_9 start discharging at different rates depending on the inputs V_{in1} and V_{in2}.
- The cross-coupled inverters composed of (M_{10} and M_{12}) and (M_{11} and M_{13}) act to decide the final outputs of the latch.

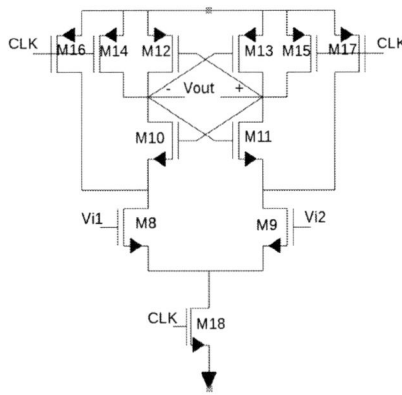

Figure 4. StrongARM circuit diagram

V. SIMULATION RESULTS AND DISSCUSSION

The simulation results of the preamplifier circuit show that the DC gain of the preamplifier stage is 43.15dB, -3dB frequency is 2.13 kHz, power consumption is 32nW and total equivalent input referred noise is 17.13uV in the range of [1Hz-10KHz]. The frequency response of the circuit is shown in Figure 5.

Table II shows a comparison with pervious works. The obtained simulation results show that the presented design is competitive with the previous works in terms of power consumption and input referred noise due to the compact design. The power consumption and gain has been tested against the process variation of the adopted technology using Monte Carlo simulation for 500 runs. The results are shown in Figure 6, 7.

The complete comparator circuit is shown below in Figure 8. The reference voltage $V_{ref} = \frac{V_{DD}}{2} = 0.375V$ is

chosen as a threshold voltage of the comparator. The test signal is shown in Figure 9. Delta is set to be 50μV. Where Delta is the difference between threshold voltage and input signal. V_{Out} is the difference between outp and outn, Vin is connected to V_{i1} while V_{th} is connected to V_{i2}). The delay of the comparator is function of differential input signal, i.e. if the input signal get smaller the delay time will increase, so to detect a small differential signal the frequency of clock must be reduced. This is shown on Figure 10 where a relation between the minimum resolvable differential input and the clock frequency.

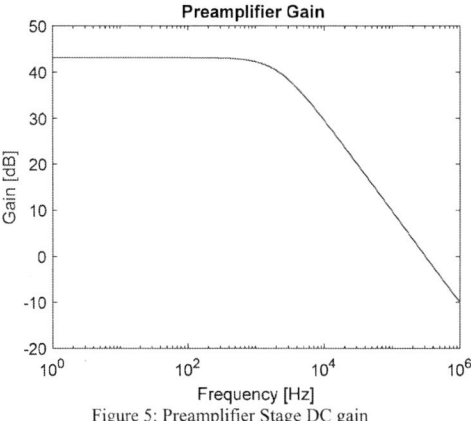

Figure 5: Preamplifier Stage DC gain

TABLE II
PREAMPLIFIER COMPARISON TABLE

Reference	[23]	[4]	[24]	This work (Simulation)
Technology [nm]	130	180	500	180
Power supply [V]	1.2	0.75	3.3	0.75
DC gain [dB]	40	45	47	43.15
Transistors number	8	13	12	9
Input referred noise [μV]	1.4 (10-100K)	63 (1Hz-5KHz)	1.8 (1Hz-10KHz)	17.13 (1Hz-10KHz)
Power consumption [W]	35.64u	NA	53.13u	31.94n

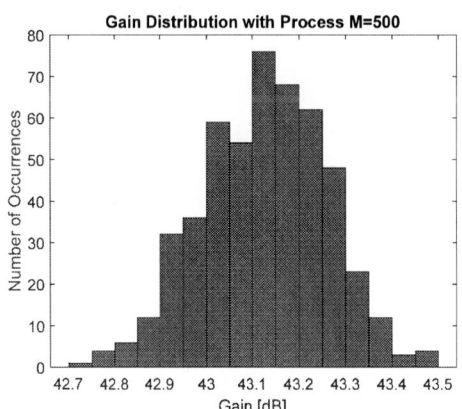

Figure 6. Monte Carlo gain distribution

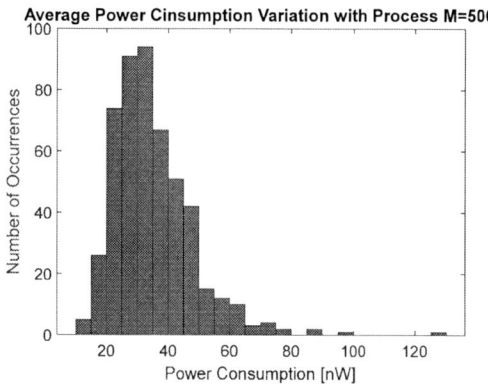

Figure 7. Power consumption distribution

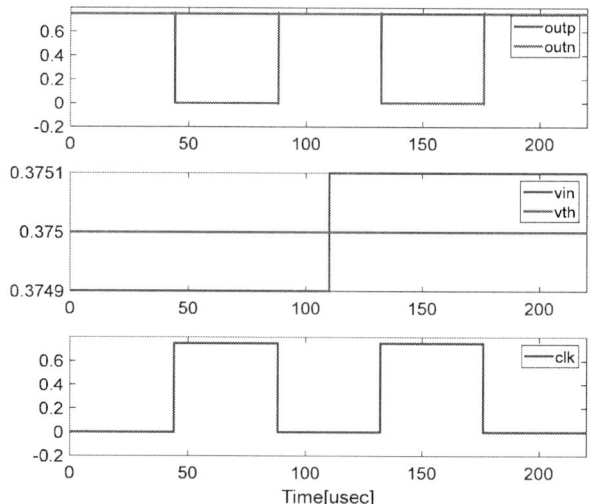

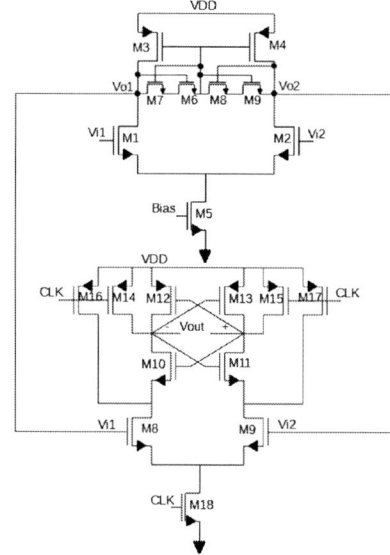

Figure 9. Sample simulation of outputs, inputs and control clock of the comparator

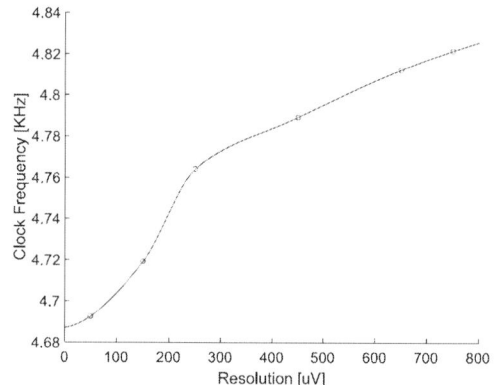

Figure 10. Minimum resolvable input versus sampling clock frequency

Figure 8. Comparator circuit

VI. CONCLUSION

A low power preamplifier stage for biomedical ADC application has been introduced. The proposed preamp is combined with StrongARM latch and the whole comparator is designed to work in weak inversion using g_m/I_D design methodology adopting TSMC 0.18μ technology. The results show that the preamplifier stage gain is 43.15dB and an input referred noise of 17.13 μV. The proposed circuit is competitive with the previous works in terms of power consumption, active area and input referred noise.

ACKNOWLEDGEMENT

This project was funded by the Basic Research Program-The National Transformation Program in King Abdulaziz City for Science and Technology- Kingdom of Saudi Arabia – Grant No. 0005-001-03-18-5.

REFERENCES

[1] D. Zhang, A. Bhide, and A. Alvandpour; " A 53-nW 9.12-ENOB 1-kS/s SAR ADC in 0.13μm CMOS for Medical Implant Devices, " IEEE J. Solid-State Circuit, vol.47 no.7,pp. 1585-1593, Jul. 2012

[2] Z. Liu, W. Wang, P. Wan, J. Geng and Z. Chen, "An Offset Calibration Technique in a SAR ADC for Biomedical Applications," 2018 12th IEEE International Conference on Anti-counterfeiting, Security, and Identification (ASID), Xiamen, China, 2018, pp. 217-220.

[3] Medical instrumentation: Application and Design, J. G Webster, Ed., 3rd edition. New York: Wiley , 1998, p259.

[4] M. Sadollahi, K. Hamashita, K. Sobue and G. C. Temes, "An 11-Bit 250-nW 10-kS/s SAR ADC With Doubled Input Range for Biomedical Applications," in IEEE Transactions on Circuits and Systems I: Regular Papers, vol. 65, no. 1, pp. 61-73, Jan. 2018.

[5] Y. Hwang and D. Jeong, "Ultra-low-voltage low-power dynamic comparator with forward body bias scheme for SAR ADC," in Electronics Letters, vol. 54, no. 24, pp. 1370-1372, 29 11 2018.

[6] J. Guerber, H. Venkatram, T. Oh and U. Moon, "Enhanced SAR ADC energy efficiency from the early reset merged capacitor switching algorithm," 2012 IEEE International Symposium on Circuits and Systems (ISCAS), Seoul, 2012, pp. 2361-2364.

[7] Y. Hirai et al., "A Biomedical Sensor System with Stochastic A/D Conversion and Error Correction by Machine Learning," in IEEE Access, vol. 7, pp. 21990-22001, 2019.

[8] M. Pelgrom, Analog to Digital Conversion, 3rd ed., Springer Publishing Company, Incorporated, 2015

[9] P. Jespers, the gm/ID Methodology, a Sizing tool for low-voltage analog CMOS Circuits: the Semi-empirical and Compact model approaches, Springer Publishing Company, Incorporated, 2010

[10] F. Silveira, D. Flandre and P. G. A. Jespers, "A g_m/I_D based methodology for the design of CMOS analog circuits and its application to the synthesis of a silicon-on-insulator micro-power OTA," in IEEE Journal of Solid-State Circuits, vol. 31, no. 9, pp. 1314-1319, Sept. 1996

[11] P. Jespers, B. Murmann, Systematic Design of Analog CMOS Circuits: Using Pre-Computed Lookup Tables. Cambridge: Cambridge University Press. 2017

[12] B. Razavi, Design of Analog CMOS Integrated Circuits. Boston, MA: McGraw-Hill, 2001

[13] A.P. Chandrakasan, N. Verma, D.C. Daly, "Ultralow-power electronics for biomedical applications". Annual Review of Biomedical Engineering. Vol. 10, pp. 247-274, Aug.2008.

[14] R. Sarpeshkar, Ultra low Power Bioelectronics. Cambridge, U.K: Cambridge University Press, 2010

[15] N. Reynders, W. Dehaene, Ultra-Low-Voltage Design of Energy-Efficient Digital Circuits, Analog Circuits and Signal Processing, Springer Publishing Company, Incorporated, 2015

[16] T. C. Carusone, D. Johns, K. W. Martin, & D. Johns, Analog integrated circuit design. Hoboken, NJ: John Wiley & Sons, 2012

[17] C. Enz, E. Vittoz, "Static drain current," in Charge-Based MOS Transistor Modeling, UK: Wiley, 2006, pp.33-54.

[18] A. Abidi and H. Xu, "Understanding the regenerative comparator circuit," Proceedings of the IEEE 2014 Custom Integrated Circuits Conference, San Jose, CA, 2014, pp. 1-8.

[19] B. Razavi, "The Strong ARM Latch". IEEE Solid-State Circuits Magazine. Vol. 7, no. 2, pp. 12-17, spring .2015.

[20] H. Xu, and A. A. Abidi, "Analysis and Design of Regenerative Comparators for Low Offset and Noise," in IEEE Transactions on Circuits and Systems I: Regular Papers, vol. 66, no. 8, pp. 2817-2830, Aug. 2019.

[21] Y. Tao and Y. Lian, "A 0.8-V, 1-MS/s, 10-bit SAR ADC for Multi-Channel Neural Recording," in IEEE Transactions on Circuits and Systems I: Regular Papers, vol. 62, no. 2, pp. 366-375, Feb. 2015

[22] H. Tang, Z. C. Sun, K. W. R. Chew and L. Siek, "A 1.33 µW 8.02-ENOB 100kS/s Successive Approximation ADC With Supply Reduction Technique for Implantable Retinal Prosthesis," in IEEE Transactions on Biomedical Circuits and Systems, vol. 8, no. 6, pp. 844-856, Dec. 2014.

[23] R. Caballero, G. Carozo, M. C. Costa-Rauschert, P. Aguirre, C. Rossi-Aicardi and J. Oreggioni, "Bio-potential Integrated Preamplifier," 2020 IEEE 11th Latin American Symposium on Circuits & Systems (LASCAS), San Jose, Costa Rica, 2020, pp. 1-4, doi: 10.1109/LASCAS45839.2020.9069006.

[24] J. Oreggioni, P. Castro-Lisboa and F. Silveira, "Enhanced ICMR amplifier for high CMRR bio-potential recordings," 2019 41st Annual International Conference of the IEEE Engineering in Medicine and Biology Society (EMBC), Berlin, Germany, 2019, pp. 3746-3749, doi: 10.1109/EMBC.2019.8856656.

Proceedings of the 27th International Conference *"Mixed Design of Integrated Circuits and Systems"*
June 25-27, 2020, Łódź, Poland

Multichannel Programmable Readout IC for Photodiodes Array

Paweł Pieńczuk, Cezary Kołaciński, Andrzej Szymański, Paweł Janus,
Krzysztof Kucharski, Dariusz Obrębski, Michał Zbieć, Mariusz Jakubowski
Sieć Badawcza Łukasiewicz — Instytut Technologii Elektronowej
Warsaw, Poland
E-mails: pawel.pienczuk@ite.waw.pl, ckolacin@ite.waw.pl, aszym@ite.waw.pl, janus@ite.waw.pl,
kucharsk@ite.waw.pl, zbiec@ite.waw.pl, obrebski@ite.waw.pl, mariusz.jakubowski@ite.waw.pl

Abstract—**This paper describes the design of the 17-channel readout integrated circuit targeted to front-end operation for photodiodes array. Proposed system contains an analog front-end electronics digitally configured using built-in Serial Peripheral Interface. Each signal channel has been designed for one particular set of photodiodes. Chip was designed and fabricated utilizing CMOS 0.18 μm technology and occupies area of 4960 μm x 1525 μm. Total current consumption is expected to be less than 33 mW (with 3.3 V supply).**

Keywords—**CMOS, ParCour, photodiodes, readout.**

I. INTRODUCTION

A. ParCour Project

Actual particle counters are bulky and costly devices. Their sizes, weights and power consumption make them stationary devices without an option of mobility.

Under the Particle Counter (ParCour — 2/POLBER-3/2018) project, the established consortium develops new solutions, applicable in mobile devices. A measurement apparatus, which is based on a new and unique particle counting method, allows to minimize the size, cost and power consumption with the comparable level of accuracy. Readout circuit described in this paper is meant to be an important part of the final device.

The basics of the measurement system are presented in Fig. 1. Implemented particle detection mechanism is based on a Fraunhofer diffraction. Generated LED light is transmitted via an input optical system (2) to form a parallel light beam. Obtained flux is fed to the test cell (3) containing particles to be counted. Due to the scattering, the diffraction image is created on a detecting photodiodes (PDs) (6) layer, formed as an array shown in Fig. 3. As a result, the PDs create a current signal observed by the measurement tools [1].

The measurement system is depicted in Fig. 2. Developed PDs array generates the current response, proportional to the intensity of received light. Then it is converted into a voltage signal and amplified using dedicated integrated circuit (AMP). In order to control and calibrate the main parameters of AMP, a standard microcontroller (MCU) will be used. System output signal can be observed with oscilloscope (OSC). In target application, MCU and AMP units are going to be assembled on the same printed circuit board (PCB). Dedicated algorithm, also developed under ParCour project, allows simultaneous counting of different sizes particles.

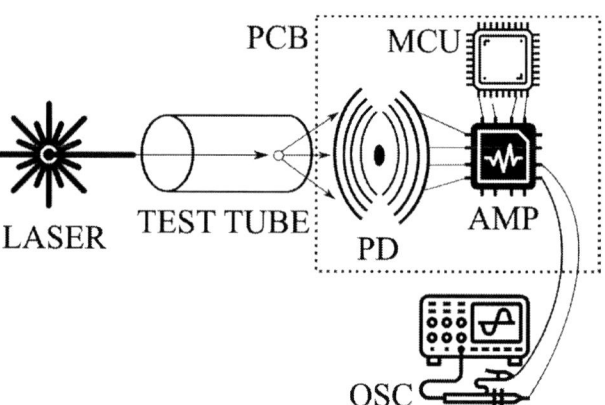

Fig. 2. Overall measurement system (icons from www.flaticon.com).

B. Photodiodes Array

Photodetector array was designed in order to detect the distribution of the light diffraction pattern, generated by the scattering on a single particle. The PDs must be able to detect the maxima and minima of the diffraction images. Therefore PDs array is organized as a ring structure (Fig. 3a) and consists of 17 sections with different sizes and positions corresponding to Bessel function distribution. The sizes (r, R – internal and external radius, A – PD area) and basic parameters (I_{max} – max. current, C_d – PD capacitance) are presented in Table I.

The smallest, central PD section (20um diameter circle) is used for calibration procedure and the remaining 16 sections form 120 deg arms. Such a layout fulfills the design rules

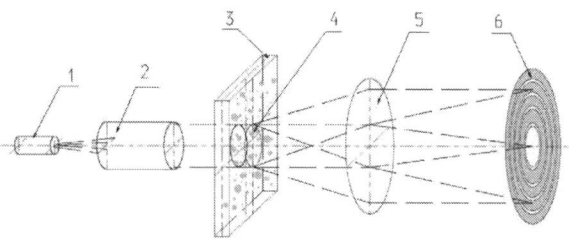

Fig. 1. Particle counting technique idea developed under ParCour project.

TABLE I
PHOTODIODES PARAMETERS

no	r	R	A	I_{max}	C_d
unit	$[\mu m]$	$[\mu m]$	$[mm^2]$	$[\mu A]$	$[pF]$
1	34.14	54.14	0.002	0.046	0.046
2	74.14	111.5	0.007	0.180	0.180
3	131.5	192.3	0.020	0.512	0.512
4	212.3	304.3	0.049	1.237	1.237
5	324.3	457.7	0.108	2.712	2.712
6	477.7	664.6	0.222	5.550	5.550
7	684.6	938.4	0.428	10.71	10.71
8	958.4	1286	0.765	19.12	19.12
9	43.25	67.37	0.003	0.070	0.070
10	87.37	131.1	0.010	0.249	0.249
11	151.1	257.6	0.027	0.683	0.683
12	251.6	348.1	0.093	2.318	2.318
13	368.1	523.0	0.144	3.588	3.588
14	543.0	161.9	0.297	7.426	7.426
15	781.9	1083	0.584	14.601	14.60
16	1103	1500	1.074	26.868	26.87
0	0	14.14	0.0006	80	0.016

regarding the distances between each PD section, as well as metal paths.

Manufacturing technology is based on a well established die PD CMOS process, developed at Łukasiewicz-ITE [2]. This technology allows the relatively easy tuning of the sensitivity function maximum by the modification of the passivation dielectric layers thickness.

In order to facilitate the PDs integration with designed readout electronics, PDs array has been assembled in a standard PLCC32 package (Fig. 3b).

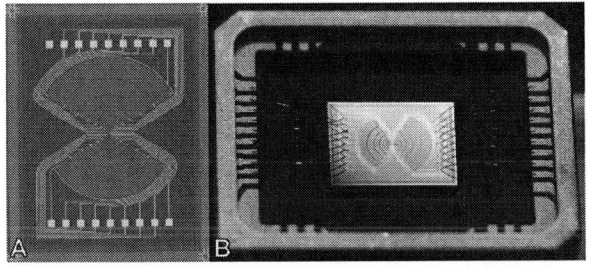

Fig. 3. Developed photodiodes: a) array structure, b) packaged sensor.

II. OVERALL ARCHITECTURE

The overall structure of the PDs readout IC is presented in Fig. 4. Designed chip is equipped with 17 pure analog signal inputs, 17 pure analog signal outputs, Serial Peripheral Interface (SPI) and several diagnostic and auxiliary ports. THe analog path of the system can be reconfigured using implemented digital block, controlled via already mentioned SPI. Both domains (analog and digital) are separated from each other, in order to minimize the noise and distortion effects.

Designed analog block consists of 17 separate channels, one for each photodiode forming the detection array. Parameters of particular channel has been suited to the expected characteristic of corresponding PD. In general, developed front-end

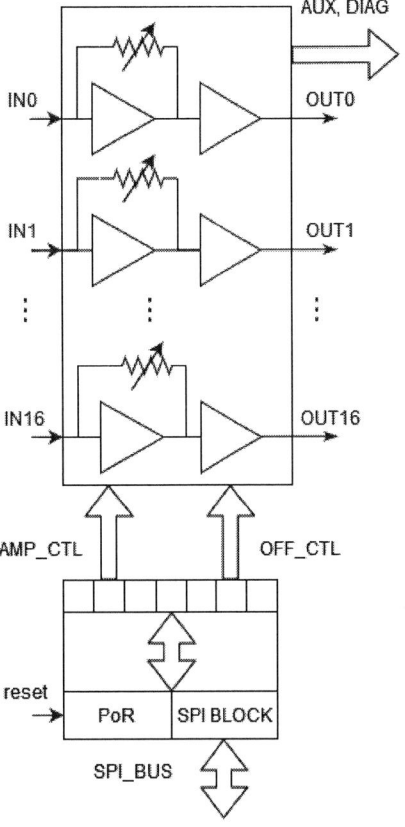

Fig. 4. Overall structure of the PDs readout IC.

circuit provides a conversion from input current signals to the output voltage domain.

Digital interface provides configuration capabilities of the mentioned analog block parameters. It is composed of internal calibration registers for each channel and a SPI slave receiver. It also ensures the reset process at every power supply event, with the use of implemented Power-on Reset (PoR) block. Due to the utilization of SPI bus, chip can be easily controlled using the wide range of programming devices (microcontrollers, processors, etc.).

Presented readout IC also consists of built-in bandgap reference (BGR) source, providing internal reference voltage. Several internal signals can be observed via dedicated ports, enabling diagnostic and more detailed measurement process.

III. ANALOG PART

Each of the analog channels is based on the same architecture, presented in Fig. 5). Single signal path is composed of two amplification blocks: a transimpedance amplifier (TIA) and a voltage amplifier (buffer). Gain and output offset values of these two stages can be controlled (within a limited range) by digital block (see Section IV). The only exception from the scheme described above is channel number 0, dedicated to the smallest, central photodiode. Due to the smallest internal

113

capacitance and the largest expected output current, a separate signal path structure had to be used.

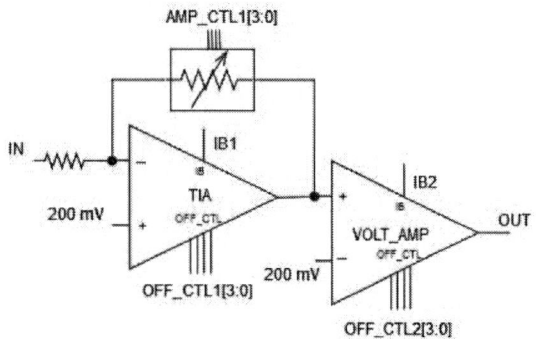

Fig. 5. Single channel architecture.

A. Block Scheme

The TIA block is based on a standard two-stage operational amplifier (opamp). Several auxiliary ports were added for diagnostic and calibration purposes. A configurable resistor is used in the feedback loop for purpose of controlling the current-to-voltage gain of this stage.

Input stage of an opamp itself is based on PMOS differential pair. Reference 10 μA bias current must be fed to the amplifier via current input IB1. An active load of the device is realized by a current mirror. A class-A output buffer with NMOS transistor is used in order to minimize input-reffered noise [3].

OFF_CTL1 4-bit port provides configuration of an opamp output offset voltage. First bit controls the sign of this offset (+/-), the remaining 3 switch on/off the additional PMOS transistors connected in parallel to the input differential pair. Effectively, the width of each transistor is controlled and the amplifier output offset value can be adjusted within the (-10 mV; +10 mV) range.

The current-to-voltage gain of TIA can be controlled by the circuit depicted in Fig. 6. 4-bit control signal A[3:0] is used for on/off switching of NMOS transistors, resulting in connection changes for particular resistors.

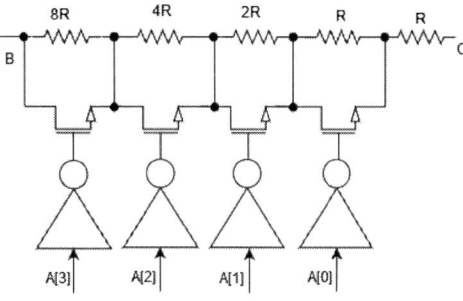

Fig. 6. Implemented digitally-controlled resistor. Terminals B and C are connected to the TIA feedback loop.

At the end of each channel signal path, an voltage amplifier is placed, providing desired gain value and proper output load. The amplifier is based on a standard non-inverting architecture [4], with several additional components needed for proper stability. Similarly to TIA block, the output offset of voltage amplifier also can be adjust utilizing 4-bit digital port (OFF_CTL2).

B. Channel Characteristics

Designed IC has been comprehensively simulated using software from Cadence Design Systems and adequate PDK (Process Design Kit). Assumed photodiodes parameters were in accordance with values specified in Table I.

The Table II shows simulated channels parameters. Wide range of input currents results in different characteristics of each channel.

TABLE II
ANALOG CHANNELS MAIN PARAMETERS

channel	max. input current $[\mu A]$	max. gain $[V/\mu A]$	output settling time $[\mu s]$
1	0.055	30.813	12
2	0.225	7.972	6
3	0.615	2.842	6
4	1.506	2.349	5
5	3.289	1.047	4.5
6	6.897	0.510	4
7	12.90	0.270	3.5
8	23.26	0.150	2.5
9	0.087	19.71	12
10	0.309	5.600	7.5
11	0.833	2.084	5
12	2.703	1.238	5
13	4.397	0.778	4
14	9.091	0.374	4
15	17.70	0.204	2.5
16	32.26	0.110	2.5

Exemplary simulation results for channel number 3 are presented in Fig. 7.

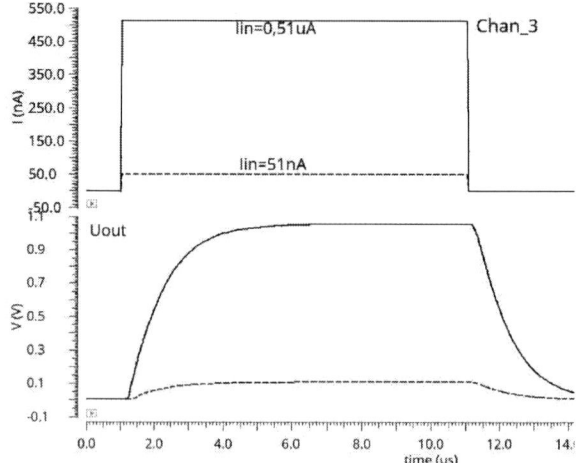

Fig. 7. Simulation results for channel number 3: current input pulses (top) and corresponding voltage output signals (bottom).

114

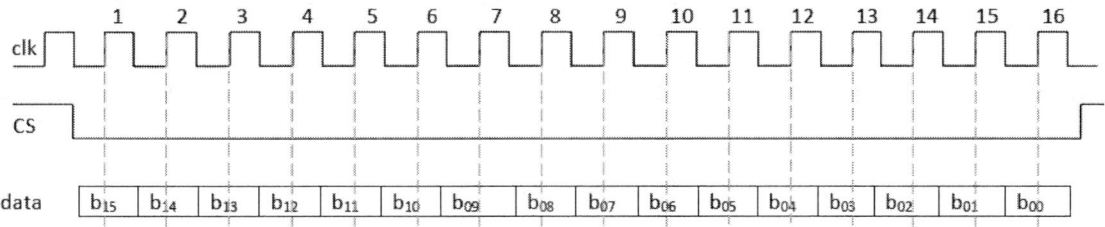

Fig. 8. Single SPI transmission sequence.

Similar analyses have been performed for all remaining channels. It has to be mentioned that values presented in Table I are preliminary and will be verified during the further measurements of fabricated PDs structures. Any disparities between predictions and final parameters values can affect readout IC performance in the target application.

IV. DIGITAL PART

As it has been already mentioned, the main role of the digital part is to provide proper calibration settings for the whole readout IC. Each of the 17 analog channels is equipped with 3 4-bit configuration registers, which results in 51 registers in total.

A. SPI Interface

Every single configuration register can be accessed using SPI protocol [5], implemented in designed IC. Particular register configuration requires transmission of a single data frame. Such a frame contains 16 bits, but only first 11 of them (starting from the LSB) are used for configuration - last 5 bits of each data word are redundant and their values can be ignored. A single SPI transmission sequence takes 16 clock cycles, so proper configuration of all internal registers lasts $17 \cdot 16 \cdot 3 = 816$ cycles. In Fig. 8 a transmission sequence of a template 16-bit data frame is shown.

Fig. 9 presents the overall structure of a single data frame. First 5 bits describes channel number, the next 2 distinguish register type, and then a 4-bit configuration word is given. Every one of 51 implemented registers must be set individually - independently from others.

b00 b01 b02 b03 b04	b05 b06	b07 b08 b09 b10	b11 - b15
channel number	register number		
00000 – channel 0			
00001 – channel 1	00 – TIA gain		
.	01 – TIA offset	configuration word	not used
.	10 – amp offset		
.	11 – not used		
10000 – channel 16			

Fig. 9. 16-bit data frame for SPI.

It has to be noticed that data bits should be fed to the serial data input in the reversed order (compared to the data structure presented in Fig. 9): MSB first, LSB last. For example, frame XXXXX11111010000 stands for configuration word 1111 that is stored in TIA offset register (01) for channel 1 (00001).

B. Calibration Registers

For each channel, the first calibration register provides TIA gain control, while the second and third ones allow voltage offset adjustment of — respectively — the TIA and voltage amplifier blocks. The calibration mechanism has been described in the previous section.

Fig. 10 presents the summary of the registers data structure for a single channel (names correspond with ports in Fig. 5).

REG0	REG1	REG2
AMP_CTL	OFF_CTL1	OFF_CTL2
(TIA gain register)	(TIA offset register)	(AMP offset register)
r_{11} r_{10} r_9 r_8	r_7 r_6 r_5 r_4	r_3 r_2 r_1 r_0

Fig. 10. Single channel registers set.

V. CHIP OVERVIEW

Designed IC was fabricated in UMC CMOS 180nm technology, occupies area of 4960 μm x 1525 μm and is equipped with 57 I/O pads. Total current consumption has been estimated at 10mA for the complete chip (at 3.3 V power supply).

Fig. 11 shows the ASIC layout (left) and single structure mounted in 68-lead PLCC package (right).

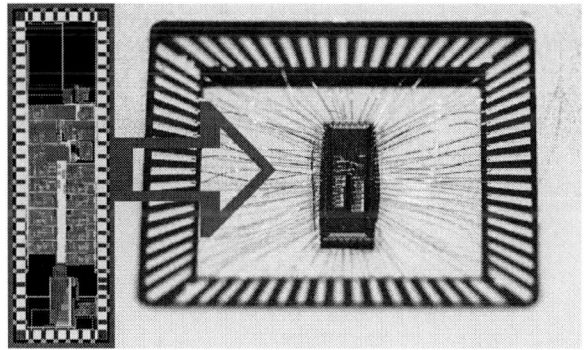

Fig. 11. Designed readout ASIC: layout (left) and mounted structure (right).

VI. CHIPS MEASUREMENTS

A dedicated test setup (Fig. 12) was developed to handle functional tests and parametric characterization of the manufactured ICs. It is built in a form of a complete, modular device, placed in the shielded enclosure. Its main board contains the USB to UART converter, featuring galvanic isolation and the local MCU responsible for translation of the UART data frame to the SPI bus messages, to be sent to the device under test (DUT).

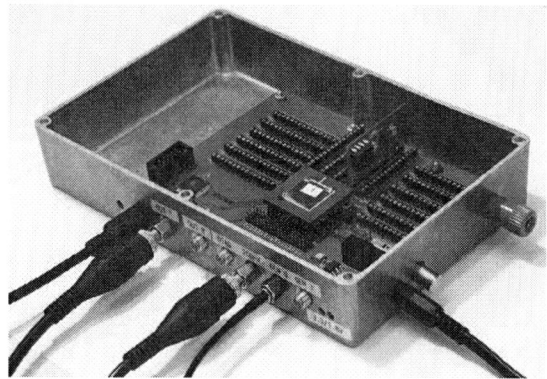

Fig. 12. Test setup for ICs measurements.

The most important task carried out by the test setup is generation of very small, current-type, controllable AC stimuli signals, tailored for each channel. This is performed by the analog circuitry located at the add-on cards, individual for each channel. The first variant of the add-on card deploys the HCNR201 — high-linearity optocoupler from Avago [6]. This device incorporates in a single package the LED and two closely-matched photodiodes. The input photodiode is placed in the feedback loop controlling the LED operation. In this manner, the non-linear light emitter behavior and its thermal drift is virtually eliminated. The output photodiode is connected to the input of the DUT, while the LED driver can be controlled by the external DC or AC signal source. To examine the channels of the DUT intended to handle the smallest photodiodes within the sensor array, the second type of the add-on card was developed. It deploys the fast-photodiode-LED couple, separated from the external light, however, the linearizing feedback loop based on the second photodetector was not possible in this case.

The dedicated application for a PC was developed to provide the convenient, graphical, test setup control.

At the time of writing this article, characterization of the fabricated chips is still not finished. The authors ensure that the results of chips measurements will be included in the presentation during the conference.

VII. CONCLUSION

A. Design Results

The multichannel, programmable integrated circuit for photodiodes array readout has been designed. It handles different sizes of photodiode with considerable range of calibration. Integration on silicon reduces area occupied by the full developed particle counting system. Its low power consumption allows to use battery power supply, if needed.

B. Next Steps

At this moment (end of February) fabricated chips are still under characterization. First results are promising but full measurement result will be available within the next few months.

Finally, the integration of entire particle counting system will be executed. The project can be extended by adding analog-to-digital (ADC) converter and processing auxiliaries for purpose of elimination of other measurement tools (like oscilloscope).

ACKNOWLEDGMENT

The presented experiments were supported by the Poland-Berlin-Brandenburg Project no. 2/POLBER-3/2018 Parcour — "Particle counter".

REFERENCES

[1] Shi, G. et al., "Laser diffraction application on detection technology of online tool setting," in *2015 International Conference on Optoelectronics and Microelectronics (ICOM)*, 2015.
[2] Węgrzecki, M. et al., "New silicon photodiodes for detection of the 1064 nm wavelength radiation," in *Proc. of SPIE Vol. 10175*, 2016.
[3] R. Baker, *CMOS Circuit Design, Layout, and Simulation, 3rd Edition*. Hoboken, New Jersey, USA: Wiley-IEEE Press, 2010.
[4] B. Razavi, *Design of Analog CMOS Integrated Circuit*. McGraw-Hill Higher Education, 2001.
[5] A. Oudjida, "SPI-Slave Specification Document," 11 2008.
[6] Avago (Broadcom). (2014) HCNR200 and HCNR201 High-Linearity Analog Optocouplers — Datasheet. [Online]. Available: https://docs.broadcom.com/docs/AV02-0886EN

Relocatable Partial Bitstreams For Virtual Overlay Architectures atop Field-Programmable Gate Arrays

Zbigniew Mudza
Department of Microelectronics and Computer Science
Lodz University of Technology
Lodz, Poland

Abstract—Intermediate virtual architecture overlays atop physical FPGA chips provide convenient abstraction level, which can increase productivity in FPGA-targeted application development. Individual reconfigurable modules of the overlay can be reprogrammed independently using partial reconfiguration. Homogeneous reconfigurable modules can be programmed using common configuration data, on condition that appropriate implementation constraints and proper floorplanning of the virtual architecture are provided. This paper presents methodology that can be used to generate relocatable bitstreams for Xilinx 7 series FPGA devices. The methodology is based on using constraints to force Xilinx Vivado Design Suite tools to implement multiple reconfigurable partition in the same way. Partial Reconfiguration Flow is used to implement multiple variants of individually reconfigurable partitions and Isolation Design Flow is used for feed-through prevention.

Keywords—FPGA, Overlay, Virtual Architectures, Partial Reconfiguration, Design Methodology, CGRA, Productivity

I. INTRODUCTION

Field-Programmable Gate Arrays (FPGAs) are seldomly used in systems that require frequent changes of functionality. There are rational reasons responsible for this state.

A single development cycle (a develop-implement-test iteration) of an FPGA-based application is much longer than in the case of CPU-based software. Performing full implementation of an FPGA project (logical synthesis, optimization, mapping, placement and routing) takes more time and effort than software compilation. Moreover, the conservative FPGA development approach (manual, low-level, hardware-aware description of complex systems in HDL languages) is focused on quality-of-result rather than productivity. Consequently, it is regarded as difficult, inconvenient and time-consuming. Considering that time-to-market parameter is of great importance nowadays, addressing the productivity issues is crucial. To make things worse, even with ready-to-run configurations, swapping between different applications takes time, as it requires reprogramming FPGA memory. Fortunately, if functionality change can be restricted to particular subsections of the FPGA fabric, they could be reconfigured individually using partial reconfiguration.

Productivity gain can be obtained by exploiting the concepts of high-level of abstraction, modularity and reusability. Operating on higher level of abstraction is not limited to behavioral description (High-Level Synthesis,

MyHDL) but can also be applied to the architecture model. Instead of instantiating virtual application-specific circuits directly in the FPGA fabric, an intermediate abstraction layer in a form of virtual overlay architecture can be used [1,9]. In such a scenario, the developers operate on the simplified virtual platform regardless of the actual hardware underneath. As overlays can consist of any components that can be implemented in the underlaying logic - from simple single-purpose blocks to fully functional microprocessors; developers can be provided with any "virtual hardware" they need. What is more, the same virtual architecture can theoretically be implemented on various physical platforms [1] including multi-chip systems. Consequently, functionality of a module could potentially be transferred to different chip without additional development, just by using another implementation of the same overlay.

An intermediate virtual architecture can be used to exploit reconfigurability of FPGA. In particular, an array of coarse-grained reconfigurable modules can be implemented as an overlay architecture. Functionality of each module can be changed by partially reconfiguring it with its individual module-specific partial bitstream for the particular function. If a set of homogeneous blocks could be configured from a single common configuration data (rather than individual bitstreams), storage space for configuration memory could be greatly reduced.

This paper presents a methodology for designing reconfigurable overlays targeting Xilinx 7 series FPGAs that can be used to obtain configuration data that can be relocated and applied to any equivalent blocks. Theoretically, the general concept could be adapted to other FPGAs with partial reconfiguration support. However, due to differences between individual architectures and tool flows implementation details presented in this paper may be inapplicable.

II. PARTIAL BITSTREAMS IN XILINX 7 SERIES DEVICES

Xilinx 7 series FPGAs comprise four 28nm FPGA families – Spartan-7, Artix-7, Kintex-7 and Virtex-7, targeting different ranges of system requirements – from low cost to high performance [7]. All the 7 series devices offer advanced high-performance reconfigurable logic based on 6-input look-up tables (LUT), enhanced with: 36kB dual-port RAM blocks, DSP slices with 25x18 multipliers and 48-bit accumulators and additional integrated blocks (like XADC and PCIe controller).

Resources in all of the 7 series devices are organized on a non-regular two-dimensional grid – defined by X,Y coordinates (referred to as horizontal (X) and vertical (Y) for the scope of this paper). Internal structure of all 7 series devices is divided into clock domain regions (with far more horizontal than vertical splits). Resources of each type (excluding clock resources) are organized in form of vertical columns (Fig.1). A single column contains only one type of elements (LUTs, BRAMs, switching blocks etc.). Each column is divided into smaller sections associated with each clock domain region the column is crossing. Entire section is either populated with appropriate resource type or left empty (Fig. 1).

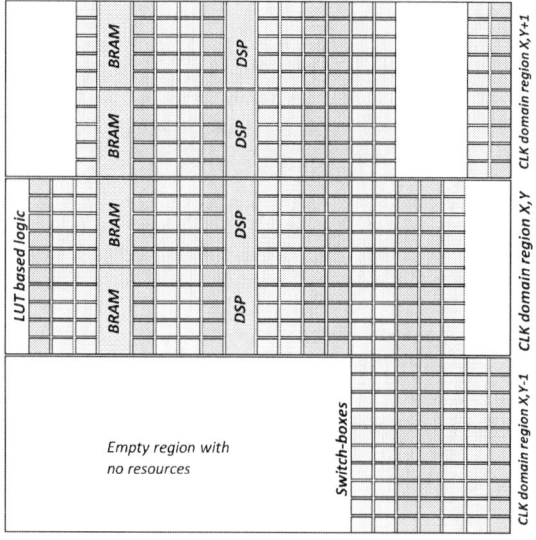

Fig. 1. General model of spatial distribution of resources in Xilinx 7 series devices

FPGA reconfiguration is performed by overwriting obsolete configuration data with a new content bitstream. Xilinx 7 series FPGAs (excluding 7A12T, 7A25T and Spartan-7 family) support partial reconfiguration – reprogramming a single section of an FPGA with no effect on the rest of the fabric. In general, any design using partial reconfiguration can be divided into a set of partially reconfigurable modules (RM) and the static part (that remain unchanged during partial reconfiguration). As the configuration is written in chunks corresponding to physical distribution of the programmed memory, each reconfiguration module needs to be individually assigned to a specific section of a chip called partition [5].

The smallest addressable segment of the configuration memory space is called a configuration frame [5, 6]. As the configuration frame are indivisible, a single frame cannot be assigned to more than one reconfigurable module. In the case of Xilinx 7 series, configuration frame corresponds to individual (clock domain region high) section of a resource columns. The location of memory that is going to be programmed with the configuration data is indicated by Frame Address Register (FAR) value in bitstream file [6]. Theoretically, partial bitstreams can be relocated (i.e. written

to a different location), by applying appropriate FAR offsets and updating CRC value accordingly [2,3]. If the new FAR value points to a region identical to the original bitstream location region, configuration data would be valid, thus can be accepted by reconfiguration controller.

If the system design could provide that a partial bitstream (for a certain reconfigurable module) is relocatable (i.e. operates properly and does not affect the rest of the system when written to another location), a single configuration file could be used for a set of equivalent blocks. Needless to say, the more identical reconfigurable modules in the system, the greater the storage space gain. Particularly, in the case of homogeneous array of reconfigurable coarse blocks assigned to identical resource sets, a single partial bitstream, applicable to all the modules, is required for each functionality.

The methodology described in this paper is based on a research conducted using Xilinx Artix-7 (XC7A200T). However, as fundamentals of operations and software flow are the same in other Xilinx 7 series devices and resource organization follows same patterns, this methodology could be applied to all Xilinx 7 series devices that support partial reconfiguration.

III. ANALYSIS

Although many works on relocatable partial bitstreams have been published [10-15], very few target Xilinx 7 series devices [2, 3]. Unlike software for some older Xilinx devices, Vivado toolflow for 7 series FPGAs does not support relocation of reconfigurable partitions [5,11]. Relocatability can still be obtained by constraining the design properly [2,3]. Oomen et. al. [2] proposed a methodology for bitstream relocation in Virtex-7, while Rettkowski et. al. [3] presented automatic tool flow for Zynq-7000, based on very similar approach. Both focus on different aspects and different applications and do not address some of the challenges highlighted in this paper. In particular, the methodology proposed in this paper targets overlay design and focuses on instantiating larger numbers of relocatable modules and increasing productivity. It assumes that a virtual architecture is fully established prior to reconfigurable module functionality development, hence the relocatability must be addressed at the overlay architecture development stage.

Partial bitstreams (pb) of two reconfigurable modules (RM) A and B can be interchangeably relocatable (pb A relocatable to RM B and pb B relocatable to RM A) only if the modules: are assigned to identical partitions, have identical interfaces with all ports placed and connected identically in relation to the partition and no other logic is placed within the span of partitions' frames (Fig. 2). For relocatability in one direction (i.e. pb A need to be relocatable to RM B, but rb B does not have to be relocatable to RM A) partition does not need to be identical, provided that a subset of pb B is identical to pb A and that other conditions are met (Fig. 2). For the purpose of this study only relocation of bitstreams among identically structured partitions is considered. The proposed methodology assumes targeting an array of virtual coarse reconfigurable blocks, with relocatability applied to functionally equivalent modules, so the restriction seems reasonable.

All reconfigurable modules need to be assigned to individual partition so they can be reprogrammed individually. In order to assure equal mappability, all partitions blocks assigned to the same type of RM (or to all RMs for homogeneous overlays) need to comprise of the same set of resources. All such blocks can implement the same functionality. A group of functionally equivalent partitions, with identical spatial organization of resources can be treated as interchangeable, provided that they can be connected to the static design identically. Each group requires its own bitstream for each functionality, but multiple partitions in the same relocatable group can use the same files (with added offset). If all RMs are functionally equivalent and use identical partition blocks – architecture is homogenous logically and physically – a single bitstream could be used for all blocks.

In order to ensure identical connection to static for a group of partition blocks, identical (relatively to partition location) constraints values for partition pins, placement and connection routing can be applied to all such blocks. Theoretically, the constraint values could be set manually, but it is not advisable – as it require additional work effort and usually lead to finding worse quality of results (specifically – poorer routability) than results obtained by automated tools. Instead of that, constraints can be extracted from automated tools results for a single block (reference) and then applied to other blocks once implementation is re-run.

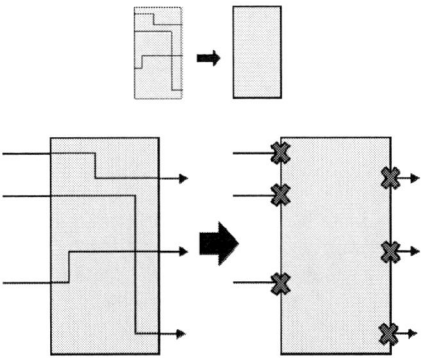

Fig. 3. Replacing a feed-through connected black-box with another black-box that does not contain the feed-through result in net disconnection.

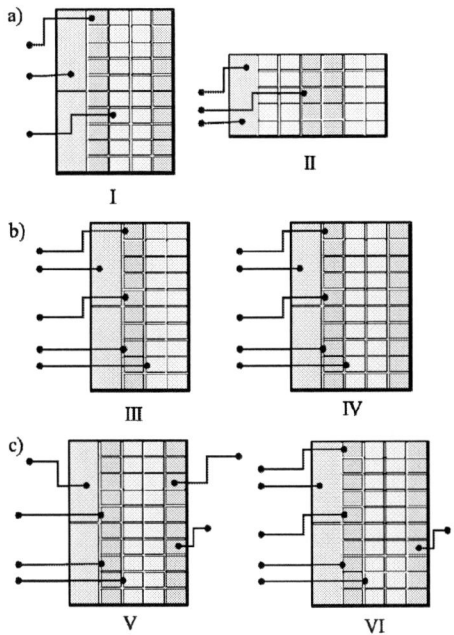

Fig. 2. Bitstream relocatability between partitions: a) Blocks I and II contain different resources arranged in different way, this block pair does not support bitstream relocation. b) All resources of block III are contained within block IV, both blocks have identical port connections. Bitstream of block III can be relocated to block IV but not vice versa. c) Blocks V and VI comprise identical set of resources arranged in the same way. However, they have different port connections. Bitstream relocatability between these blocks is not supported.

Although Vivado Partial Reconfiguration Flow ensures that internal placement and routing of each reconfigurable module is contained within the partition block, and that no other logic is placed within partition boundaries, it allows global routing feed-through. A block containing a feed-through connection cannot be programmed with configuration data that does not contain such a connection (Fig. 3). Consequently, such blocks do not support not relocatability. In small designs with limited global routing connections, feed-through might not occur ([2] do not address this problem at all), but in the course of this research only feed-though did not occur only if it was prevented by using feed-through avoidance techniques. It is necessary to take such steps.

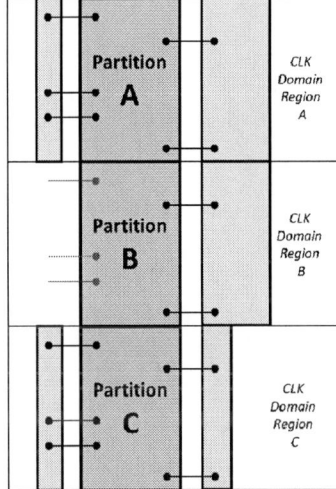

Fig. 4. Vertically adjacent partitions use connection across the horizontal borders. Partition A cannot be relocated to partition B as no resources for port connection are present on the left side of partition B.

Knowing that only logic and routing of the reconfiguration module can reside within the span of its configuration frames, the partition blocks should theoretically be aligned to utilize the entire frames (vertical span of clock domain regions) for best design density. As resource distribution in Xilinx 7 Series is aligned with columns and, as mentioned before, the more identical blocks can be designated the better – partitions can be in close vertical proximity, or even adjacent. Consequently, all connections to static design should be conducted across the horizontal borders (Fig. 4). In order to make identical connections possible, there must be resources left next to partition for static design anchor points - to which reconfigurable modules are connected (Fig. 4).

Specificity of routing arrangement in Xilinx 7 series causes that some relocatable modules aligned to clock regions might be unrelocatable. Vertical routing channels connected to switching blocks in the proximity of clock domain region borders are normally used for connections with the adjacent clock domain region. However, if no switching blocks are present beyond the clock domain region border (e.g. in case of terminal clock domain region or empty space in chip), the routing channels use a loop-back – outgoing and incoming routes that would normally connect to a switching block in the adjacent clock domain region are hardwired to each other (Fig.5). Relocating configuration data for switching blocks using loop-back to a location with no loopback would result in net failure. If all modules need to be interchangeably relocatable, partition blocks should be constrained in a way so that the loopback is not contained in the partition. However, if it is not required any block without the routing loop-back can be used as reference block. Its content can be safely relocated to any other location as the internal routing cannot use switch-blocks outside of the partitions – hence the problematic channels would never be used.

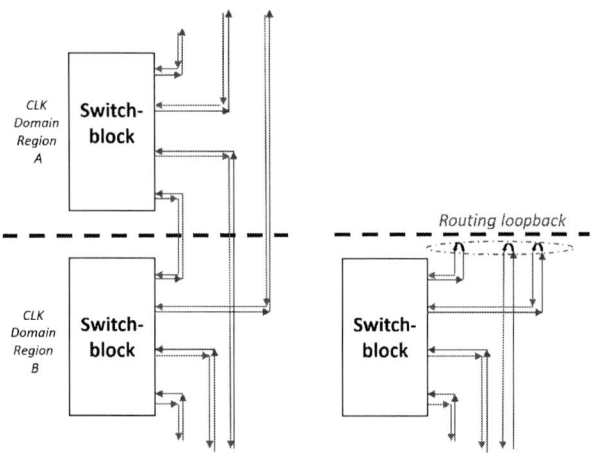

Fig. 5. Simplified visualization of routing loopback. Routing channels used to connect to adjacent clock domain region are looped if no switch-block is present in the adjacent clock domain region.

In order to reduce the complexity of connections between reconfigurable blocks and static, the nets crossing partition borders should be of point-to-point type (with one driver and a single fun-out). It can be easily obtained for all nets by inserting buffers on each net. The buffers might be realized in a form of single input LUT [3] or as a flip-flop register. The latter solution can helps preserving timings which may be crucial in a large network with uneven net lengths that might have to be dealt with in virtual architecture with different logical and physical structure. Thus, it is strongly advised, even if it leads to increased latency.

In Xilinx Vivado design flow routing is relative and depends on previously executed placement stage. For the proper constraining of input and output ports connections,

placement of the anchor point components should also be constrained. Considering that same relative constraints are applied to blocks distributed spatially in the entire chip, the anchor points must be positioned in a way that both placement and routing can be applied to different locations. For the purpose of easier constraining, relocatable partitions should be designated so that they are surrounded by additional identical resources used to place static anchor points. Last but not least, the routing connecting RM ports to static cannot exceed clock domain region boundaries, otherwise expected resources might be unavailable in different location (Fig.6).

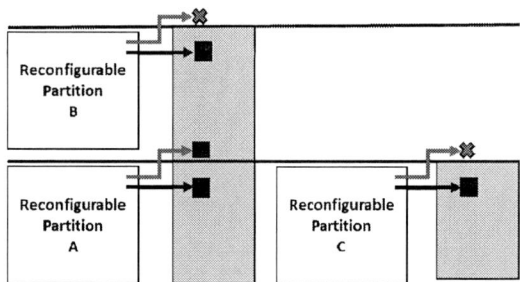

Fig. 6. The routing for Reconfigurable Partition A crosses clock domain regions, it cannot be exported to Reconfigurable Partition B and Reconfigurable Partition C as the resources expected by routing are not present there.

IV. SOLUTION

The process of implementing an application with an intermediate virtual architecture can be divided into the following stages:

- Logical design of the overlay with black-box components

- Physical implementation of the virtual architecture

- Mapping application to virtual architecture

- Physical implementation of application in the overlay

- Generating and uploading configuration data

The first stage defines logical structure of the overlay and specifies entire static functionality but provide no reconfigurable module specification other than interfaces. There is a single design restriction. Due to specifics of further described feed-through avoidance technique, top-level logic should include only clock logic – all other static logic should be placed in a static logic module on same hierarchy level with reconfigurable modules. The final product is a platform-independent, logically synthesized netlist or design checkpoint of the overlay architecture with all reconfigurable blocks instantiated in form of undefined black-box components.

Further specification of virtual architecture requires platform dependent manipulations. Multiple implementations of the synthesized overlay can be obtained by repeating the physical implementation stage. Each implementation can target different chip.

Once a target FPGA platform is selected, floorplanning of the virtual architecture takes place. First of all, partition blocks for all reconfigurable modules needs to be designated. Resource pool sizes for partitions are be determined by number of reconfigurable modules, their potential requirements and device capacity. Partitions for functionally equivalent modules should be assigned to chip sections with identical floorplan (comprising same resources organized in the same way), as bitstreams can be relocated only between such partitions. In order to allow identical placement of anchor points and prevent feed-through connections (as described later), the actual identical section should be extended with two additional columns at either end of its horizontal span. It is advisable to distribute partitions vertically to cover entire span of clock domain region, as otherwise the resources not included in partitions would remain unused (as explained in chapter 3). If multiple reconfigurable partitions are placed in a single column – at least one clock domain region with routing resources in that column should be used as interspace for global routing. Otherwise feed-through connections cannot be avoided. If the number of partitions that can be designated following these rules is less than number of logically equivalent RMs, another partition floorplan should be proposed for the remaining number of blocks.

Once reconfigurable design partitions are designated, the remaining resources can be assigned to the static part of overlay, but firstly feed-through avoidance needs to be addressed. As all the top-level logic resides in the static module (as mention earlier in this chapter) – Vivado Isolation Design Flow (IDF) can be used to isolate reconfigurable modules from the static logic [3, 4, 10]. From the perspective of the IDF tools global routing lines connects logic assigned to one isolated module (static), hence must be prevented from crossing modules it is isolated from (reconfigurable modules). In order for the module isolation to work an interspace (commonly referred to as fence) of at least a resource column (vertical fence) or row of basic elements (horizontal fence) needs to be separate the modules. The resources in fence regions must remain unassigned [4].

Finally, identical locations and connections of ports must be forced, and implementation of the overlay can be run. For this purpose, empty reconfigurable modules (containing only LUT or flip-flop-based port buffers) with forced optimization prevention option can be used in place of the black-boxes. Theoretically, partition pins, port buffers and anchor points placement could be fixed to certain locations. However, not only does it require more work effort but leaving more freedom to placer and router can actually result in better results. Alternatively, an additional implementation can be run. Then tool generated placement and routing for a reference block and its anchor points can be applied as constraints to other blocks. Nevertheless, anchor points placement must be restricted to the area adjacent to isolation fence, as it allows identical placement in other locations. Although LOC constraints allow forcing logic element placement only to a particular localization, placement assignment to a range can be performed by a non-reconfigurable nested partition block assignment (Fig. 7). This method can be combined with HD.PARTPIN_RANGE (range-based pin placement) for better control on how reconfigurable connections are placed, without forcing any specific location [7, 8].

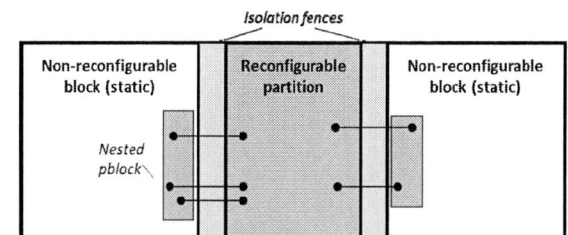

Fig. 7. Constraining locations of anchor points by assigning anchor buffers to non-reconfigurable nested partition blocks.

Even with properly placed anchors the tool generated routing might cross clock domain region borders, which would make it impossible to apply to some locations as there is no isolation between anchor point and the rest of static module. However, as the initial implementation step is conducted only to determine reference block ports connections – it can be conducted using different set of constraints unacceptable in the case of final design, with additional isolation (Fig. 8). Once the initial implementation is completed PARTPIN_LOC, BEL and LOC properties for ports and anchor point buffers should be extracted and applied to other blocks (by offsetting values with XY position difference between each partition and the reference block) [2, 3, 7, 8]. Similarly, routing can be forced by setting FIXED_ROUTE property based on values of ROUTE property extracted from the reference block. However, as routing is described by relative values no offset is applied [7, 8].

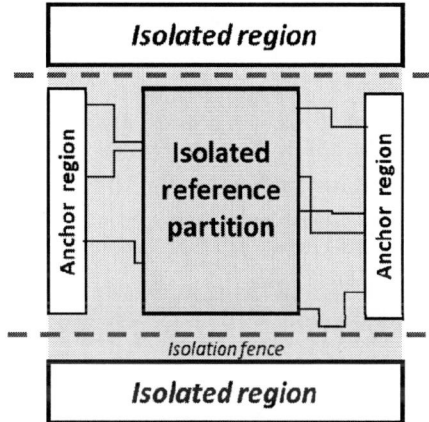

Fig. 8. Isolating anchor region from other static components prevents port routing from going out-of-scope.

Once all the required constraints are established virtual architecture with empty reconfigurable modules can be implemented using Vivado Isolation Design Flow. Resultantly a complete placed-and-routed design is obtained. Finally, locked placement and routing for the virtual architecture only can be exported to a design checkpoint by setting all reconfigurable modules to be treated as black-boxes. This design checkpoint file is a ready-to-use placed-and-routed virtual architecture with bitstream relocation support.

Application development for a platform with intermediate overlay layer starts with virtual architecture selection. A proper overlay is chosen based on interfaces, static functionality and reconfigurable blocks resources pool. Simple tasks can be assigned directly to reconfigurable modules, complex task can be decomposed to a structure of overlay compatible modules manually or mapped by automated tools. All reconfigurable modules need to instantiate buffers in form of either single-input LUT or flip-flop registers for port-placement compatibility. Any variant of reconfigurable module functionality can be synthesized out-of-context (treating RM as a top-level unit rather than an element of complex design). Result of such synthesis in a form of netlist or design checkpoint could then be used in later stages.

---

1. *Conduct logical synthesis of the overlay with black-box reconfigurable modules*

2. *Assign reconfigurable modules to identical partitions.*

3. *Provide interspaces for isolation fences between static and reconfigurable partitions.*

4. *Insert reconfigurable-to-static connection buffers assigned to subregions adjacent to isolation fences.*

5. *Select a reference partition and isolate its proximity to prevent routing from going out-of-scope.*

6. *Run a Isolation Design Flow implementation for the reference block.*

7. *Extract tool-generated placement and routing properties and apply to other partitions as constraints.*

8. *Run a complete Isolation Design Flow implementation of the overlay.*

9. *Export placed and routed overlay with black-box partitions.*

10. *Implement any variants of reconfigurable blocks using Partial Reconfiguration Flow with the overlay.*

11. *Generate partial bitstreams.*

12. *Relocate partial bitstream by applying FAR offset, update CRC value.*

Fig. 9. Simplified algorithm for generating relocatable bitstreams for reconfigurable partitions of an overlay architecture.

Vivado Partial Reconfiguration Flow (PRF) [5] can be used to implement synthesized reconfigurable modules in previously established virtual architecture. Placed-and-routed design checkpoint of the selected overlay targeting chosen physical platform can be linked with synthesized reconfigurable modules. In addition, partition pin and port buffer placement constraints previously applied to reconfigurable modules during overlay implementation should still be used (for the purpose of PRF compatibility). Multiple versions of reconfigurable modules might be added and combined in form of different implementation configuration. As the design checkpoint of placed and routed static components is initially provided, PRF implementation is limited to reconfigurable modules implementation only [5]. Once implementation finishes successfully bitstream for entire chip and partial bitstreams for each reconfigurable module can be generated.

Theoretically, any partial bitstreams, for a partition with no loop-back connections, obtained using this methodology should be relocatable. However, stable operation of reconfigurable module is not guaranteed – as different combinational path lengths may lead to occurrence of timing issues. The problem can be eliminated by using only flip-flop register buffers on nets connecting relocatable modules with static overlay. In order to verify that correct system operation is maintained, placement and routing constraints for the entire reference module can be extracted and applied to the same reconfigurable module variant instantiated in another partition (similarly to what was performed on ports during in overlay implementation process). Then system operation for relocated bitstream and newly generated bitstream could be compared.

The proposed methodology was tested on a proof-of-concept virtual arrays of reconfigurable modules targeting Xilinx Artix-7 XC7A200T device using Xilinx Vivado Suite 2019.2. Identical modules were assigned to configuration frames span – i.e. boundaries of clock domain region. A block with no routing loopback was selected as a reference block. Placement and routing constraints were extracted from implemented reference module and applied to other identical modules. Later, the Partial Reconfiguration Flow implementation was re-run with the constraints and partial bitstreams were generated. Files were than compared using binary files comparison tool. For a group of identical reconfigurable blocks bitstream files differing only in FAR and CRC values were obtained [6]. Considering that partial bitstreams for the constrained blocks generated by Vivado tools are exact equivalents of FAR offsetted reference block partial bitstreams – no further validation of those configuration files is required to prove correct operation. Nevertheless, partial bitstreams were identical for a group but not all reconfiugurable modules. Analysis of results for multiple test configurations indicate that bitstream discrepancy occurrences were consistent, independent from module configuration and limited to the same partitions. Placement and routing results for reconfigurable blocks corresponding to different bitstreams were compared and uniformity of the partitions was verified. As of now, no reasonable cause of bitstream differences has been found, but a research aimed at resolving this issue is undergoing.

Moreover, in the course of conducted tests some of the designs failed due to routability problems. It needs to be stressed, that forcing fixed placement and routing restrains automated tools and makes their task more difficult. Some of the problems were presumably caused by the design of the overlays used in tests. Nevertheless, further work on the methodology should address possible improvements that could help with this issue.

V. CONCLUSIONS

Virtual reconfigurable architecture overlays implemented atop off-the-shelf FPGAs offers more convenient, productivity-oriented approach to reconfigurable computing. Partial reconfiguration can greatly decrease application switching time as only modules that are subject to configuration change are programmed individually. Reprogramming multiple modules using common configuration data reduces storage space required for FPGA configuration data files. It is possible to obtain equivalent partial bitstreams for identical partitions located in different regions of FPGA chip.

Bitstream relocatability support can be provided at the level of intermediate virtual architecture. Identical placement and routing of a reconfigurable block can be obtained by constraining the implementation process. Feed-through avoidance is necessary for partitions to support relocatability. Isolation Design Flow can be used to prevent global routing from crossing reconfigurable partitions.

The methodology presented in this paper was tested in proof-of-concept designs targeting Xilinx Artix-7 XC7A200T FPGA. The results were partially successful – as for a group of partitions relocatable bitstreams with offset indicating partition location were identical to partial bitstreams generated individually for that partition. However, it relocatability did not apply to all of the apparently identical partitions. What is more, not all tested designs were successfully implemented due to routing congestion problems.

The presented methodology exploits modularity and reusability. Individual tasks can be performed independently by different working teams. Reconfigurable modules' functionality changes do not require interfering with static virtual architecture and retargeting application to a different overlay or different physical platform is limited to reimplementing reconfigurable blocks set in different placed-and-routed static design. Nevertheless, addressing bitstream inconsistency and routability issues is required to make it work using.

REFERENCES

[1] Jain A. K,. Maskell D. L., Fahmy S. A., *"Are Coarse-Grained Overlays Ready for GeneralPurpose Application Acceleration on FPGAs?"*, IEEE International Symposium on Dependable Autonomic and Secure Computing (DASC), 2016

[2] Oomen R., Nguyen T., Kumar A., Corporaal H., *"An automated technique to generate relocatable partial bitstreams for Xilinx FPGAs"*, 25th International Conference on Field Programable Logic and Applications, 2015

[3] Rettkowski J., Friesen K., Gohringer D., *"RePaBit: Automated Generation of Relocatable Partial Bitstreams for Xilinx Zynq FPGAs"*, International Conference on Reconfigurable Computing and FPGAs (ReConFig), 2016

[4] Xilinx, Hallet E., "Isolation Desing Flow For Xilinx 7 Series FPGAs or Zynq-700 AP SoCs (Vivado Tools)", 2016

[5] Xilinx, "Vivado Design Suite User Guide Partial Reconfiguration", UG909 (v2018.3), 2018

[6] Xilinx, "7 Series FPGAs Configuration User Guide", UG470 (v1.13.1), 2018

[7] Xilinx, "Vivado Design Suite User Guide Implementation", UG904 (v2019.2), 2019

[8] Xilinx, "Vivado Design Suite User Guide Using Constraints", UG903 (v2018.1), 2018

[9] Khanzadi H., Savaria Y., David J. P., *"A Data Driven CGRA Overlay Architecture with Embedded Processors"*, IEEE International New Circuits and Systems Conference (NEWCAS), 2017

[10] Gantel L., Benkhelifa M. E. A., Lemonnier F., Verdier F., *"Module relocation in Heterogeneous Reconfigurable Systems-on-Chip using the Xilinx Isolation Design Flow"*, International Conference on Reconfigurable Computing and FPGAs (ReConFig), 2012

[11] Drahonovsky T., Rozkovec K., Novak O., *"Relocation of reconfigurable modules on Xilinx FPGA"*, IEEE International Symposium on Design and Diagnostics of Electronic Circuits & Systems (DDECS), 2013

[12] Hannachi M., Rabah H., Jovanovic S., Abdelali A. B., Mtibaa A., *"Efficient relocation of variable-sized hardware tasks for FPGA-based adaptive systems"*, International Conference on Microelectronics (ICM), Doha, , 2014

[13] Lavin C. et al., *"RapidSmith: Do-It-Yourself CAD Tools for Xilinx FPGAs"*, International Conference on Field Programmable Logic and Applications (FPL), 2011

[14] Beckhoff C., Koch D., Torresen J., *"Go Ahead: A Partial Reconfiguration Framework"*, IEEE Annual International Symposium on Field-Programmable Custom Computing Machines (FCCM), 2012

[15] Koch D., Beckhoff C., Teich J., *"ReCoBus-Builder — A novel tool and technique to build statically and dynamically reconfigurable systems for FPGAS"*, International Conference on Field Programmable Logic and Applications, 2008

Thermal Issues
in Microelectronics

Proceedings of the 27th International Conference *"Mixed Design of Integrated Circuits and Systems"*
June 25-27, 2020, Łódź, Poland

Comparison of Set-ups Dedicated to Measure Thermal Parameters of Power LEDs

Krzysztof Górecki, Przemysław Ptak

Department of Marine Electronics
Gdynia Maritime University
Gdynia, Poland
k.gorecki@we.umg.edu.pl, p.ptak@we.umg.edu.pl

Marcin Janicki

Department of Microelectronics and Computer Science
Łódź University of Technology
Łódź, Poland
janicki@dmcs.pl

Abstract—The paper is devoted to measurements of thermal parameters of power LEDs. Two methods of measurements of transient thermal impedance and thermal resistance for the considered semiconductor devices as well as the measurement set-ups implementing these methods are described. The results of measurements of the parameters obtained using both set-ups for selected power LEDs are compared and discussed. Moreover, selected properties of software operating with both the considered measurement set-ups are also analysed.

Keywords—power LEDs; thermal parameters; measurement set-ups; compact thermal models.

I. INTRODUCTION

Power Light Emitting Diodes (LEDs) are basic components of modern lighting systems [1, 2]. Their operating parameters, e.g. the emitted luminous flux, strongly depend on temperature [3]-[5]. the internal temperature of all semiconductor devices, including LEDs, is the sum of the ambient temperature and the increase of this temperature caused by self-heating phenomena [5, 6]. The value of this increase depends on power dissipated in a considered device and on the efficiency of generated heat removal. This efficiency is characterized by thermal parameters such as the transient thermal impedance $Z_{th}(t)$ and the thermal resistance R_{th} [7]-[9].

Values of thermal parameters of semiconductor devices depend on cooling conditions of these devices, which change depending on the applied cooling system [10]-[12]. Therefore, in order to characterize thermal properties of semiconductor devices operating under certain cooling conditions it is indispensable to measure thermal parameters of such devices.

In the case of power LEDs the standard thermal resistance measurements method is described in the norm by JEDEC [13]. This norm contains also a description of measurement set-ups rendering possible the realization of this method. The T3Ster equipment [14], which was designed in compliance with this norm is used also in the Department of Microelectronics and Computer Science (DMCS) in Łódź University of Technology, Poland. However, such equipment is relatively expensive and it is used mostly in industry. Therefore, in Gdynia Maritime University (GMU, Poland), a custom system was developed for measurements of power LEDs thermal and optical parameters. The measurement results obtained using this system were already described, e.g. in [12, 14, 15].

The main disadvantage of the system developed at GMU is that it renders possible measurements of examined devices only with free convection cooling. On the other hand, the standard version of the T3Ster does not allow measurements of radiant power of the emitted light, what is indispensable to determine thermal resistance of power LEDs.

For both measurement set-ups a dedicated software for the estimation of thermal parameters as well as compact models of investigated electronic devices was developed. Such models were previously discussed, e.g. in [6, 10, 16].

This paper provides the description of both measurement set-ups and presents the detailed results of thermal parameter measurements obtained for selected power LEDs. These results are compared for both systems and their usefulness to perform measurements of thermal parameters of the considered class of semiconductor devices is evaluated.

II. MEASUREMENT SET-UPS

According to the standard described in [5, 13] the thermal and optical parameters of power LEDs should be measured simultaneously using the measurement set-up shown in Fig. 1.

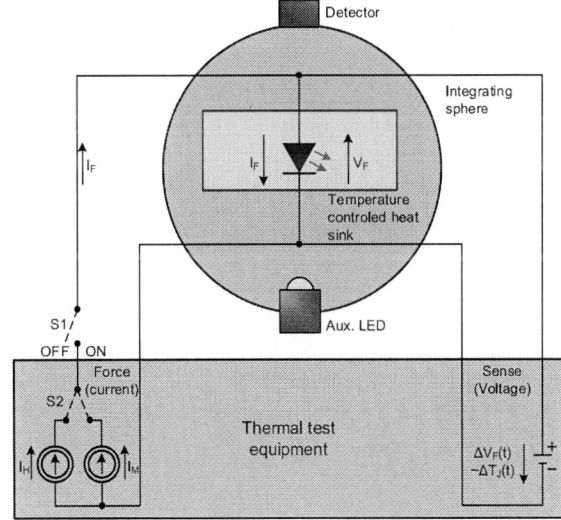

Fig. 1. Design of a combined thermal and radiometric LED testing station [13].

127

During measurements the device under test must be placed in an integrating photometric sphere on the heat-sink with the forced liquid cooling. The main advantage of this method is the possibility of simultaneous measurements of the LED junction temperature T_j, the luminous flux Φ_V, the forward current I_F and voltage drop V_F. An obvious disadvantage of this method, especially for the academics, is the necessity to use expensive instrumentation, such as the sphere and the necessity to use the forced liquid cooling system to stabilize the diode temperature.

A. T3Ster set-up

The transient thermal tester T3Ster available at the DMCS renders possible registration of electronic system dynamic temperature responses with sub-microsecond time resolution. Usually, the temperature sensitive parameter is the voltage drop across a p-n junction or a thermistor measured for a constant current value forced by the tester. According to the principles of the Network Identification by Deconvolution (NID) method, measurements are taken at the time instants equidistant on the logarithmic time scale, e.g. with 20 samples per decade. The software implementing the NID method and provided together with the tester offers the entire range of thermal analysis tools, such as the thermal time constant spectra, cumulative structure functions or thermal impedance Nyquist plots. A view of this tester is shown in Fig. 2 on the left side.

The NID method is intended to measure thermal parameters of typical semiconductor devices, e.g. p-n diodes or transistors, but this system, at least in its basic version, does not allow measurements of power LED optical parameters, hence with this system it is possible to measure only the electric thermal impedance defined in the JEDEC standard [14].

When only the electric thermal impedance is measured, the influence of the LED radiant power on measurement results is neglected. Therefore, in order to measure the actual thermal impedance of the considered class of semiconductor devices, additionally a dedicated light-tight chamber and a radiometer should be used. This thermal impedance can be measured using the method for determining radiant power described in [17].

The results of measurements performed with the use of the described thermal tester are processed with the commercially available MASTER software, which renders possible thermal analyses and the estimation of thermal model parameter values for investigated devices.

B. Custom set-up

Thermal and optical parameters of LEDs can be measured also using the measurement set-up developed at the GMU. The block diagram of this set-up is presented in Fig. 3. This set-up renders possible measurements of the LED thermal impedance as well as the illuminance and power density of emitted light. During measurements the tested diodes are placed in the light-tight chamber. A luxmeter is used to measure the illuminance of emitted light and a radiometer is used to measure the power density of this light.

The thermal impedance is measured here using the indirect electrical method described in [18]. In this method, a voltage drop V_D across the DUT for a constant value I_M of its forward current is used as a temperature sensitive parameter. The fixed value of current I_M is produced by the voltage source E_M and the resistor R_M, whereas the voltage source E_H and the resistor R_H set the heating current I_H flowing through a tested device. The switch S is closed while heating a DUT and it is opened while cooling it. The waveforms of the diode forward voltage are recorded with instrumentation amplifiers, an A/D converter module and a PC, whereas the heating current is measured using an ammeter.

The measurements of thermal impedance are taken in four steps. Initially, the thermometric characteristic, describing the dependence of the device forward voltage V_D on temperature is measured. During this measurement step the switch S is open and the current I_M flows through the diode. In the second step, the switch S is closed and the heating current I_H flows through the investigated diode increasing its junction temperature as the result of the self-heating phenomenon.

Due to the temperature increase, the diode forward voltage changes its value, which is measured by a computer containing an A/D converter and measurement-amplifiers. In the thermal steady-state, the value of current I_H, the diode forward voltage V_H and the emitted light power density I_e are measured. At the third step the switch S is open at the time $t = 0$ and the cooling of the tested diode begins. During this step the A/D converter measures the waveform of the diode forward voltage V_L at the current I_M until the initial ambient temperature is reached. Finally, the transient thermal impedance is calculated using the following formula:

$$Z_{th}(t) = \frac{V_L(t=0) - V_L(t)}{\alpha_T \cdot P_{th}} \tag{1}$$

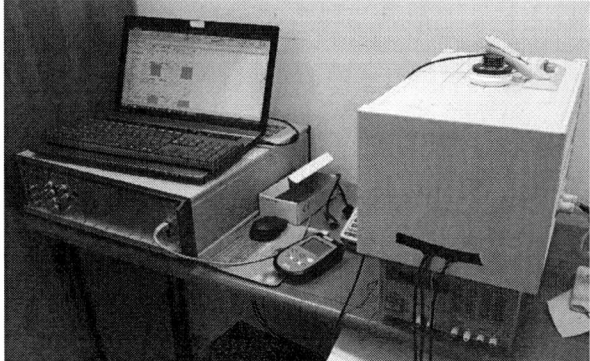

Fig. 2. View of the measurement set-up available at the DMCS.

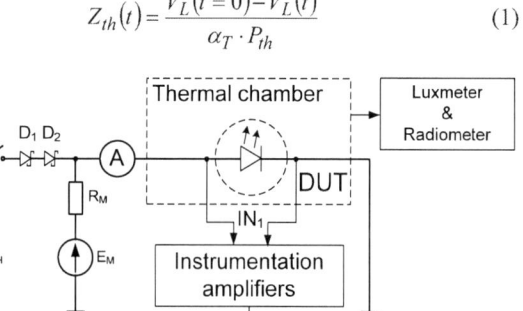

Fig. 3. Block diagram of the measurement set-up developed at GMU.

where α_T denotes the slope of a thermometric characteristic and P_{th} denotes the real heating power. The thermal resistance R_{th} is equal to the value of transient thermal impedance $Z_{th}(t)$ at the steady state.

For power LEDs the real heating power is the difference between the electrical power P_{el} taken from the electrical grid and the radiant power P_{opt} emitted by a tested diode in the form of light. The way of determining the radiant power is described in the following section. When the radiant power is not taken into account, using in (1) the electric power P_{el} instead of the heating power P_{th}, one obtains the electric thermal impedance $Z_{th\ el}(t)$ or the electric thermal resistance $R_{th\ el}$.

C. Determination of radiant power

The authors previously proposed a relatively simple method of determining radiant power emitted by tested LEDs in [17]. According to this method radiant power density I_e is measured by a radiometer situated over the tested LED at the distance r. Then, the characteristics showing the dependence of relative luminous intensity on the observation angle α, which is usually provided by device manufacturers, is approximated within the entire range of light emission angles from $-\alpha$ to α by a constant value equal to the average value of luminous intensity I_{avg} and calculated based on the original characteristics. Finally, the radiant power is determined using simple geometrical relations provided in (2). The correctness of this method and its practical usefulness was demonstrated for selected types of power LEDs in [17]-[18].

$$P_{opt} = I_e \cdot 2\pi \cdot r^2 \cdot (1 - \cos\alpha) \cdot I_{avg} \qquad (2)$$

III. TESTED POWER LEDS

The properties of both considered measurement set-ups were compared for different types of power LEDs. This paper will present and discuss chosen results of investigations carried out for three different types of diodes produced by Cree, Inc., i.e. XPL, XPE and MCE. The values of selected parameters of the tested diodes are provided in Table I [19]-[21], whereas their views are shown in Fig. 3. As can be seen, the admissible electrical power of the tested LEDs is in the range from 2.8 W to 10 W, the emitted luminous flux at forward current I_F equal to 0.35 A varies from 100 lm to 341 lm. Such a big difference in the values of these parameters results mainly from different surface areas of these devices.

During the measurements all the tested LEDs are mounted on small Metal Core Printed Circuit Boards (MCPCBs). Two different types of cooling conditions were taken into account; once the measurements were carried out for the diodes attached to a heat sink having dimensions 175 mm x 118 mm x 10 mm and then without any heat-sink.

TABLE I
VALUES OF SELECTED PARAMETERS OF TESTED POWER LEDS [19]-[21].

diode	P_{tot} [W]	I_{Dmax} [A]	Φ_V [lm]	Viewing Angle [°]
XML	10	3	341 @ I_F=0.35A	125
XPE	3	1	122 @ I_F=0.35A	115
MCE	2.8	0.7	100 @ I_F=0.35A	110

Fig. 3. Views of the investigated XML, XPE and MCE LED packages.

IV. EXPERIMENTAL RESULTS

The thermal parameters of tested diodes were determined using both measurement set-ups considered here. The selected results of these measurements are presented in Figs. 4-7. The solid lines, denoted by DMCS, are used for the results obtained with the T3Ster, whereas the dashed ones, denoted by GMU, the results from the custom system designed by the authors. The top chart in the figures always presents the results obtained for devices with the heat sink and the bottom one without any heat sink.

Additionally, Tables II-III present values of thermal power, radiant power and thermal resistance of tested diodes measured in both types of cooling conditions considered here. The value of radiant power, as can be seen, is comparable with the real thermal heating power, hence neglecting this quantity while determining the transient thermal impedance could be a source of large errors of even up to 30%.

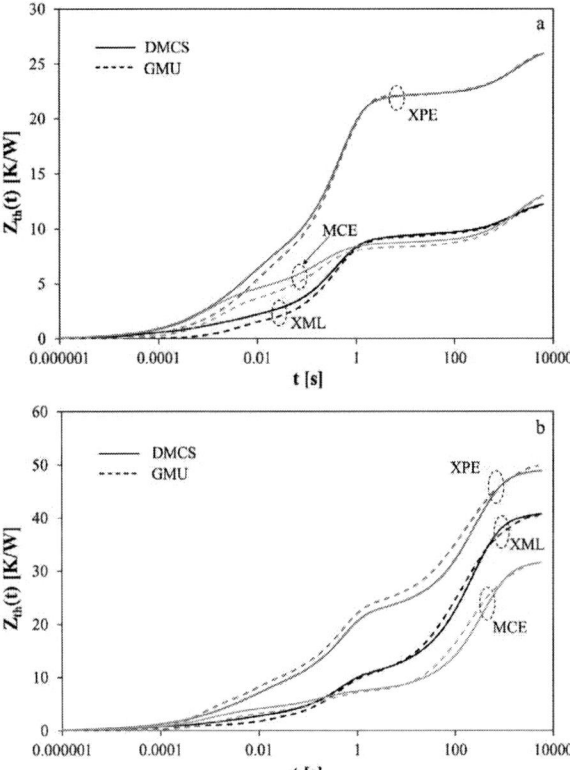

Fig. 4. Waveforms of transient thermal impedances of the tested LEDs measured with the use of both measurement set-ups situated on the heat-sink (a) and operating without any heat-sink (b).

TABLE II
VALUES OF THERMAL POWER, RADIANT POWER AND THERMAL RESISTANCE
OF THE TESTED POWER LEDs SITUATED ON THE HEAT-SINK

diode	P_{th} [W]	P_{opt} [W]	R_{th} [K/W]
XML	3.406	2.106	12.23
XPE	2.339	0.882	25.94
MCE	1.688	0.623	13.07

TABLE III
VALUES OF THERMAL POWER, RADIANT POWER AND THERMAL RESISTANCE
OF THE TESTED POWER LEDs OPERATING WITHOUT ANY HEAT-SINK

diode	P_{th} [W]	P_{opt} [W]	R_{th} [K/W]
XML	3.392	1.836	40.77
XPE	2.266	0.796	48.91
MCE	1.623	0.611	31.56

As can be observed, in all the considered cases, the results of measurements obtained with both set-ups are very similar. At the steady state, the values of $Z_{th}(t)$ measured with both set-ups are almost indistinguishable. Some small differences are observed only in the range of short time instants, but they never exceeded 1 K/W. Slightly higher values of thermal impedance are measured with the T3Ster for the diodes with the heat sink, whereas for the diodes operating without any heat sink higher impedance values are obtained for the custom system designed by the authors.

Comparing the measurements results obtained in different cooling conditions, it is visible that owing to the use of a heat sink the value of thermal resistance is reduced for all the diodes at least by 50% (for the XPE diode) and even by 66% (for the XML diode). On the other hand, the presence of the heat sink increases the time necessary to reach the steady state.

The cumulative thermal structure functions shown in Fig. 5 were obtained using the MASTER software. These functions represent the relation between the thermal resistance R_{th} and the cumulated thermal capacitance C_{th} accumulated along the heat flow path. As can be observed, for all the diodes situated on the heat-sink the cumulative thermal resistance is smaller and the cumulative thermal capacitance is bigger than for the diodes operating without any heat-sink. For both measurement set-ups the curves are very similar and they are interweaving.

The MASTER software allows also the computation of the thermal time constants spectra, which for the tested LEDs are shown in Fig. 6. It can be easily observed that all the longest thermal time constants obtained with the T3Ster equipment are nearly the same as with the authors' set-up. The differences are visible only in the range of short thermal time constants below 1 s. It is also worth noticing that spectra presented in Fig. 6 are continuous functions. However, in the classic definition of the transient thermal impedance of semiconductor devices there exist discrete values of thermal time constants $\tau_{th\,i}$. This classic definition has the following form [10, 22:]:

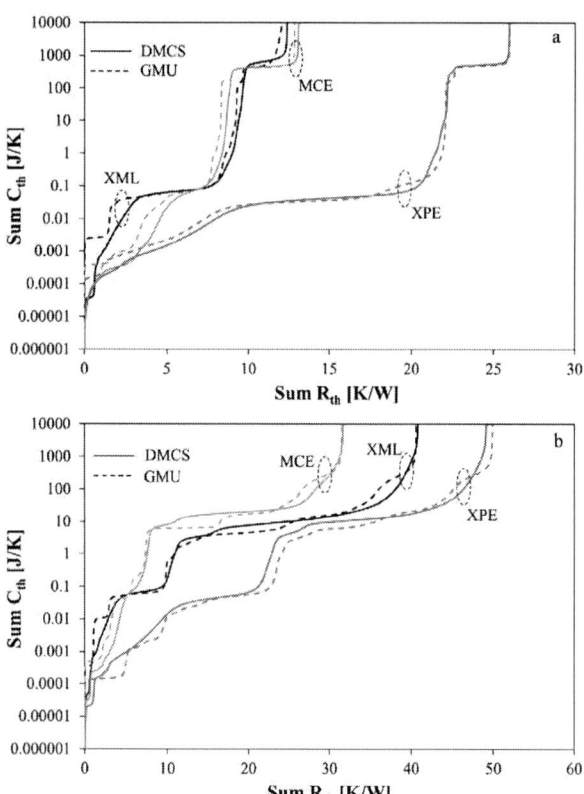

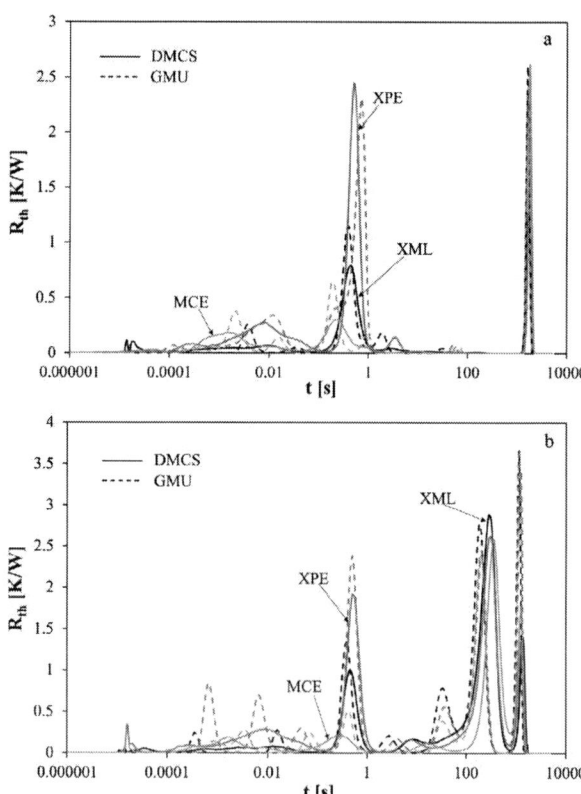

Fig. 5. Structure functions of the tested LEDs situated on the heat sink (a) and operating without any heat sink (b).

Fig. 6. Spectra of thermal time constants of the tested LEDs situated on the heat-sink (a) and operating without any heat-sink (b).

$$Z_{th}(t) = R_{th} \cdot \left(1 - \sum_{i=1}^{N} a_i \cdot \exp\left(-\frac{t}{\tau_{thi}}\right)\right) \qquad (3)$$

where N is the total number of thermal time constants and a_i are the coefficients corresponding to a particular constant τ_{thi}.

The values of parameter appearing in (3) can be estimated using the software ESTYM developed at GMU and described e.g. in [10, 23]. The input data for this software are the results of measurements performed with the use of the authors' set-up. The spectra of thermal time constants obtained for the tested diodes using the ESTYM software are presented in Fig. 7. The estimation results demonstrate that in order to describe thermal properties of tested diodes and to obtain accurate simulation results it is required to use at least 5 thermal time constants for LEDs situated on a heat sink and even 8 thermal time constants for LEDs operating without any heat sink.

V. CONCLUSIONS

This paper described two experimental set-ups dedicated to the measurement of semiconductor device thermal parameters. One of them is a commercially available equipment, whereas the other one is a custom authors' design. Both set-ups render possible the measurements of the radiant power of light emitted by the tested LEDs, hence they can be used to measure the transient thermal impedance of these devices.

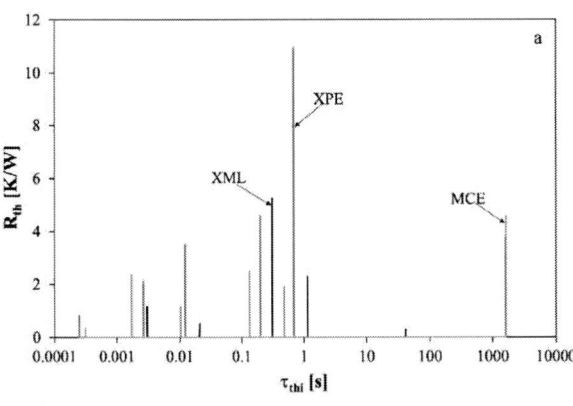

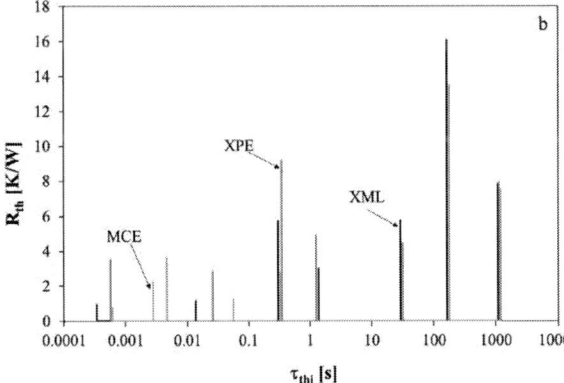

Fig. 7. Spectra of thermal time constants of the tested LEDs situated on heat-sink (a) and operating without any heat-sink (b) obtained using ESTYM software and the custom set-up.

The paper presented measurement results obtained for three types of power LEDs operating at different cooling conditions. It was proved that nearly the same measurements results could obtained using both the considered set-ups. Therefore, it could be stated that the simple measurement set-up realized by the authors operates properly. Additionally, the authors' software ESTYM renders possible the estimation of thermal parameters describing the waveforms of thermal impedance $Z_{th}(t)$.

The present investigations illustrate the thermal properties of power LEDs operating only with the free convection cooling conditions. Therefore, the authors' measurement set-up is now being extended with the system of forced cooling, which will make it possible to measure thermal parameters of investigated devices at constant temperature values of the mounting base.

ACKNOWLEDGMENT

The project financed within the program of the Ministry of Science and Higher Education called "Regionalna Inicjatywa Doskonałości" in the years 2019-2022, the project number 006/RID/2018/19, the sum of financing 11 870 000 PLN.

REFERENCES

[1] B. Weir: Driving the 21st Century's Lights. IEEE Spectrum, Vol. 49, No. 3, 2012, pp. 42-47.

[2] P.S. Martin: High power white LED Technology for Solid State Lighting. Lumileds, 2005.

[3] K. Górecki, P. Ptak: Modelling LED lamps in SPICE with thermal phenomena taken into account. Microelectronics Reliability, Vol. 79, 2017, pp. 440-447.

[4] E.F. Schubert: *Light emitting diodes. Second edition.* Cambridge University Press, New York, 2008.

[5] C.J.M. Lasance, A. Poppe: *Thermal Management for LED Applications.* Springer Science+Business Media, New York, 2014.

[6] K. Górecki: Modelling mutual thermal interactions between power LEDs in SPICE. Microelectronics Reliability, Vol. 55, No. 2, 2015, pp. 389-395.

[7] V. D'Alessandro, N. Rinaldi: A critical review of thermal models for electro-thermal simulation. Solid-State Electronics, Vol. 46, 2002, No. 4, pp. 487-496.

[8] D.L. Blackburn, "Temperature Measurements of Semiconductor Devices – A Review", 20th IEEE Semicon. Thermal Measur. and Menagement Symp. SEMI-THERM, San Jose, 2004, pp. 70-80.

[9] M. Janicki, T. Torzewicz, P. Ptak, T. Raszkowski, A. Samson, K. Górecki: Parametric Compact Thermal Models of Power LEDs. Energies, 2019, Vol. 12, No. 9, 1724, doi: 10.3390/en12091724

[10] K. Górecki, J. Zarębski, P. Górecki, P. Ptak: Compact thermal models of semiconductor devices – a review. International Journal of Electronics and Telecommunications, Vol. 65, No. 2, 2019, pp. 151-158.

[11] K. Górecki, P. Górecki: A new form of the non-linear compact thermal model of the IGBT. 12th IEEE International Conference on Compatibility, Power Electronics and Power Engineering CPE-POWERENG 2018, Doha, 2018, doi: 10.1109/CPE.2018.8372563.

[12] K. Górecki, P. Ptak, M. Janicki, T. Torzewicz: Influence of cooling conditions of power LEDs on their electrical, thermal and optical parameters. Proceedings of 25th International Conference Mixed Design of Integrated Circuits and Systems MIXDES 2018, Gdynia, 2018, pp. 237-242.

[13] JEDEC Standard JESD51-51: Implementation of the electrical test method for the measurement of real thermal resistance and impedance of light-emitting diodes with exposed cooling, April 2012.

[14] JEDEC Standard JESD51-52: Guidelines for Combining CIE 127-2007 total Flux Measurements with Thermal Measurements of LEDs with Exposed Cooling Surface, April 2012.

[15] M. Janicki, T. Torzewicz, A. Samson, T. Raszkowski, A. Napieralski, "Experimental identification of LED compact thermal model element values", *Microelectronics Reliability,* Vol. 86, 2018, pp. 20-26.

[16] B. Dziurdzia, K. Górecki, P. Ptak, "Influence of a soldering process on thermal parameters of large power LED modules", *IEEE Transactions on Components, Packaging and Manufacturing Technology,* Vol. 9, No. 11, 2019, pp. 2160-2167.

[17] K. Górecki, P. Ptak, "New dynamic electro-thermo-optical model of power LEDs", *Microelectronics Reliability*, Vol. 91, 2018, pp. 1-7.

[18] K. Górecki, P. Ptak, "New method of measurements transient thermal impedance and radiant power of power LEDs", *IEEE Transactions on Instrumentation and Measurement*, Vol. 69, No. 1, 2020, pp. 212-220.

[19] Datasheet XLamp XM-L2, www.farnell.com/datasheets/1755855.pdf.

[20] Datasheet XLamp XP-E, http://www.cree.com/led-components/media/documents/XLampXPE.pdf.

[21] Datasheet XLamp MC-E, www.led-tech.de/produkt-pdf/cree/xlamp_mce.pdf.

[22] V. Szekely: A New Evaluation Method of Thermal Transient Measurement Results. Microelectronic Journal, Vol. 28, No. 3, 1997, pp. 277-292.

[23] K. Górecki, M. Rogalska, J. Zarębski, "Parameter estimation of the electrothermal model of the ferromagnetic core", *Microelectronics Reliability*, Vol. 54, No. 5, 2014, pp. 978-984.

Proceedings of the 27th International Conference *"Mixed Design of Integrated Circuits and Systems"*
June 25-27, 2020, Łódź, Poland

Investigations Properties of Selected Methods of Measurements of Thermal Parameters of the IGBT

Krzysztof Górecki, Paweł Górecki

Department of Marine Electronics
Gdynia Maritime University
Gdynia, Poland
k.gorecki@we.umg.edu.pl, p.gorecki@we.umg.edu.pl

Abstract—**In this paper usefulness of electric and optical measurement methods to determine reliably values of thermal parameters of the IGBT is analysed. Factors influencing a measuring error of the considered methods are discussed. The results of measurements of the considered parameters obtained with the use of the considered methods for different cooling conditions of the tested transistor are presented and discussed. It is shown, in what operating conditions each measuring method makes it possible to obtain reliable results.**

Keywords—**thermal parameters; measurements; IGBT; thermal phenomena; power semiconductor devices**

I. INTRODUCTION

IGBTs are used in many electronic and power electronics circuits [1, 2, 3]. Their properties strongly depend on temperature [4, 5]. Internal temperature T_j of the considered transistor exceeds ambient temperature T_a as a result of self-heating phenomena [4, 6, 7]. These phenomena cause, among other things, shortening of life-time of semiconductor devices and they influence changes in values of their exploitive parameters [8, 9].

In order to estimate an increase of internal temperature of the semiconductor device over ambient temperature at dissipation of power of the fixed value in this device, thermal parameters are used. These parameters are transient thermal impedance $Z_{th}(t)$ and thermal resistance R_{th}. These parameters characterise efficiency of removal of heat generated in the examined semiconductor device to the surrounding and their values depend on cooling conditions of this device [10, 11]. Reliable determination of these parameters demands performance of suitable measurements.

The considered thermal parameters are measured with the use of suitable definitional formulas. Particularly, thermal resistance is described by the dependence of the form [12, 13]

$$R_{th} = \frac{T_j - T_a}{P} \qquad (1)$$

where P is power dissipated in the tested device.

The scientific work was financed with the Polish science budget resources in the years 2017–2021, as the investigation project within the framework of the program "Diamentowy Grant".

The project financed in the framework of the program by Ministry of Science and Higher Education called "Regionalna Inicjatywa Doskonałości" in the years 2019 – 2022, the project number 006/RID/2018/19, the sum of financing 11,870,000 PLN.

Values of temperature T_a and power P can be measured in an easy way, whereas the value of temperature T_j cannot be measured directly. The value of this temperature is measured indirectly with the use of optical or electric methods [13, 14]. Optical methods make it possible to measure temperature of the semiconductor structure only in the case of laboratory semiconductor devices without the case, whereas for commercially made devices with cases optical methods make it possible only to measure the case temperature T_C [15].

In turn, indirect electric methods allow determining internal temperature of the device basing on measurements of the value of the selected electric parameter univocally dependent on temperature - a thermo-sensitive parameter [10, 11, 16, 17, 18, 19]. In the literature [16, 17, 18, 19] possibilities of the use of different thermo-sensitive parameters to measure internal temperature of different semiconductor devices are considered. In the case of IGBTs voltage between the gate and the emitter of the transistor operating in the active range, voltage between the collector and the emitter on the switched-on transistor or voltage on the forward biased anti-parallel diode are used [18].

In the paper [19], there is an analysis of differences between values of internal temperature of the power MOS transistor obtained with the use of indirect electric methods, in which voltage on the forward biased p-n junction or threshold voltage of the transistor are thermo-sensitive parameters. The problem of nonlinearity of thermometric characteristics while using threshold voltage as a thermo-sensitive parameter is pointed out. On the basis of the results of simulations in the ANSYS software it is shown that differences in the temperature value on the surface of the semiconductor structure of the square shape of the length of 7 mm could reach even 50 K. In turn, from measurements with the electric method and both thermo-sensitive parameters differences in the value of internal temperature of the transistor within the range from 4 to 11 K were observed.

In the paper [16] usefulness of selected thermo-sensitive parameters to measure internal temperature of power semiconductor devices is analysed. It is shown that the most universal among the considered thermo-sensitive parameters is voltage on the forward biased p-n junction, through which current of a small value flows.

In the paper [18] the results of measurements of internal temperature of IGBT modules obtained with the use of optical and indirect electric methods realised by means of three

133

different thermo-sensitive parameters are compared. It is shown that while using collector-emitter voltage or voltage of the IGBT operating in the saturation range as a thermo-sensitive parameter is the best choice to estimate internal temperature of parallely connected IGBTs in the module.

In the paper [17] new thermo-sensitive parameters of the IGBT, which can be used to monitor its internal temperature during its operation in a switch-mode power converter are proposed. The results of measurements obtained with the use of these parameters imperceptibly differ from maximum values of temperature of the device obtained by means of optical measurements.

In the papers [18, 19] properties of measuring set-ups using the mentioned thermo-sensitive parameters are analysed. However, one does not take into account the fact that the diode and the transistor are situated in the common case, yet they are separate semiconductor structures [20].

In this paper influence of the choice of a thermo-sensitive parameter on the results of measurements of thermal parameters of the IGBT is examined. Measuring set-ups making possible realisation of measurements of thermal parameters with the use of two thermo-sensitive parameters are analysed. Influence of the choice of these parameters on the measuring error is discussed. The results of measurements of thermal parameters performed with the use of the considered thermo-sensitive parameters for the selected IGBT operating at different cooling conditions are compared. The obtained results are compared also to the results of measurements realised with the use of the optical method. The obtained findings are discussed.

II. OPTICAL AND ELECTRICAL MEASUREMENT METHODS

Thermal parameters considered in this paper characterise the ability of the semiconductor device to remove heat dissipated in it both at dynamic conditions ($Z_{th}(t)$) and dc conditions (R_{th}). In order to simplify considerations, in the further part of the paper only dc properties are taken into account and attention is paid to measurements of thermal resistance.

In the case of optical methods of measuring thermal resistance of the commercially accessible transistor the value of its case temperature T_C is measured with the use of the pyrometer. Measurements are performed at the thermally steady state after heating the examined device as a result of dissipation of power of the well-known value in it. Of course, in this state the case temperature is lower than internal temperature of the device, because in the heat flow path material of the case is found.

While measuring thermal resistance with the optical method additionally two problems appear. The first is inequality of temperature distribution on the surface of the case [19] and the other is emissivity of the case smaller than 1. Measurements realized with the optical method can be performed in the classical set-up to measure dc characteristics of the IGBT.

In the case of indirect electrical methods special measuring set-ups are indispensable. In Fig. 1 the diagram of a

measurement set-up of thermal resistance using as a thermo-sensitive parameter of the gate-emitter voltage V_{GE} is shown, whereas in Fig. 2 - a measurement set-up using voltage on the anti-parallel diode V_D is presented.

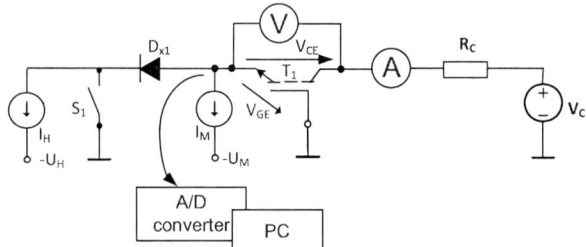

Fig. 1. Measurement set-up to determine thermal resistance of the IGBT with the use of the gate-emitter voltage as a thermally sensitive parameter

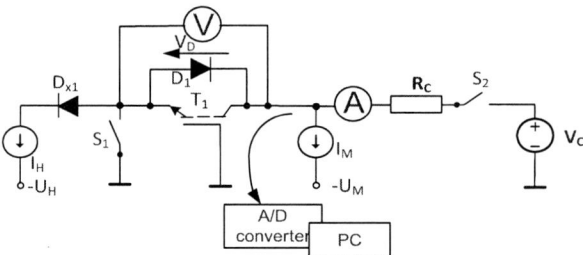

Fig. 2. Measurement set-up to determine thermal resistance of the IGBT with the use of forward voltage of the anti-parallel diode as a thermally sensitive parameter

In both the considered measuring set-ups thermal resistance is determined in three steps. The first is calibration of the thermometric characteristic realised at closed switches S_1 and opened switch S_2. In this case through the examined device flows current of a small value fixed by current source I_M. During calibration temperature of the transistor is regulated by means of the thermostat, wherein this transistor is situated. From this characteristic the slope of this characteristic α_{TSEP} is determined.

In the second step of measurements the examined transistor operates within the active range, and its internal temperature increases as a result of self-heating phenomena. In this stage switches S_1 are opened, and switch S_2 is closed. Through the transistor flows current I_H. This step ends after obtaining the thermally steady state.

The third step begins after a change of the state of switches and consists in measuring the value of the thermo-sensitive parameter immediately after this switch. Then, through the examined transistor flows current of the same value as during calibration. The value of the thermo-sensitive parameter $TSEP_H$ is measured by means of the analogue-to-digital converter and is recorded in the computer.

With the use of the thermometric characteristic measured during calibrations and the value of the thermo-sensitive parameter measured in the third step the value of thermal resistance is calculated using the following formula

$$R_{th} = \frac{TSEP_H - TSEP_K}{\alpha_{TSEP} \cdot I_H \cdot V_{CE}} \qquad (2)$$

where $TSEP_K$ means the value of the thermo-sensitive parameter measured during calibration for ambient temperature, while V_{CE} - voltage between the collector and the emitter measured at the end of the second step of measurements.

III. ANALYSIS OF MEASUREMENT ERROR

In order to estimate usefulness of the considered measuring methods in practice, the error of measurements of thermal resistance of the IGBT by means of these methods are analysed. This error can be estimated using the complete differential method in relation to equation (1). As a result, the following formula describing the relative error of R_{th} measurement is obtained [21]

$$\delta_{Rth} = \frac{\Delta T_j}{T_j - T_a} + \frac{\Delta T_a}{T_j - T_a} + \frac{\Delta P}{P} \qquad (3)$$

where ΔT_j, ΔT_a, ΔP denote the absolute errors of measurements of temperatures T_j, T_a and power P.

Equation (3) shows that at the fixed accuracy of measurements of the mentioned quantities, the measuring error of R_{th} is a decreasing function of both power and temperature difference $T_j - T_a$. Typically, the relative error of measurements of power $\Delta P/P$ with the use of typical laboratory multimeters and correct selection of measuring-ranges does not exceed 0.1 % [21]. The absolute error of ambient temperature ΔT_a is very small and it does not exceed 0.5 K [21].

Using the idea of calculating the error of measurements realised by the indirect method with the complete differential method in relation to equation (2), the following dependence describing the measuring error of $Z_{thT1}(t)$ is obtained

$$\delta_{Rth} = 2 \cdot \frac{\Delta TSEP}{TSEP_H - TSEP_K} + \frac{\Delta \alpha_{TSEP}}{\alpha_{TSEP}} + \frac{\Delta I_C}{I_C} + \frac{\Delta V_{CE}}{V_{CE}} \qquad (4)$$

where $\Delta TSEP$, $\Delta \alpha_{TSEP}$, ΔI_C and ΔV_{CE} denote the absolute errors of measurements of TSEP, of the slope of the thermometric characteristic, of the collector current and of the collector-emitter voltage.

Next, the value of TSEP is measured with the use of the fast A/D converter with the absolute measurement error equal to 2.5 mV. In turn, the error of estimating the slope of the thermometric characteristic is $\Delta \alpha_{TSEP} \leq 20$ µV/K [21]. The difference of the value of $TSEP_H$-$TSEP_K$ depends on an increase of internal temperature of the considered transistor while heating it. For example, assuming that the considered increase is equal to 100 K, the slope of the thermometric characteristic α_{TSEP} amounts to -2 mV/K, and the remaining errors accept values given above, the minimum value of the error of measurements R_{th} amounts to 3.7%. If an increase of temperature is equal to only 20 K, this error increases up to 13.7%. The obtained values of the error of measurements are close to the values of the error of measurements of thermal

parameters of IGBTs realised with the optical method and calculated in the paper [21].

In turn, in the optical method, the measuring error of temperature T_C obtained by means of the thermo-hunter, given by the producer is typically $\Delta T_j = 2$ K. The error ΔT_j results from the fact that in the optical methods the case temperature is measured instead of internal temperature, which can be described with the dependence

$$\Delta T_j = R_{thj-c} \cdot P \qquad (5)$$

where R_{thj-c} is the value of thermal resistance between the junction and the case, given by the producer.

IV. INVESTIGATIONS RESULTS

Investigations were performed for the IGBT of the type IRG4PC40UD by International Rectifier operating at different cooling conditions. This transistor is mounted in TO-247 case and its parameters are described in [20].

For this transistor thermometric characteristics were measured using forward voltage of the anti-parallel diode and gate-emitter voltage as thermally sensitive parameters. With the use of these thermally sensitive parameters dependences of thermal resistance of this transistor operating at both the mentioned cooling conditions were performed. The obtained results of measurements were compared to the results of measurements obtained with the use of optical parameters. In all the figures presented in this section points represent the results of measurements, whereas lines denote functions approximating the results of measurements.

Fig. 3 illustrates thermometric characteristics of the considered transistor while using forward voltage of the diode as TSEP in a wide range of forward current.

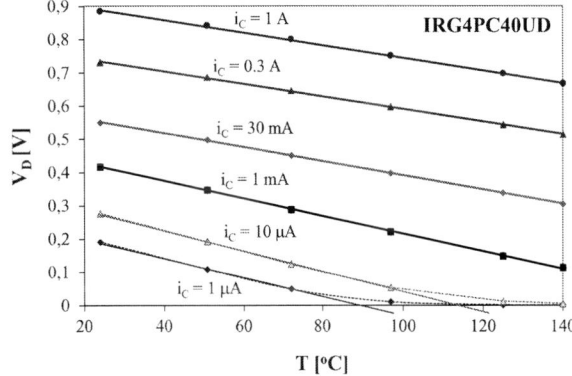

Fig. 3. Thermometric characteristics of the anti-parallel diode for selected values of the collector current

As it is visible, the considered characteristics are linear in a wide range of temperature for forward current higher than 1 mA. For smaller values of this current the range of linearity of the considered characteristics is narrower than for higher current. The slope of these characteristics decreases with an increase of forward current.

In Fig. 4 thermometric characteristics of the anti-parallel diode mounted in the common case with the tested transistor are shown. As can be seen, characteristics are practically linear and their slope changes from -2.6 mV/K to -2.4 mV/K.

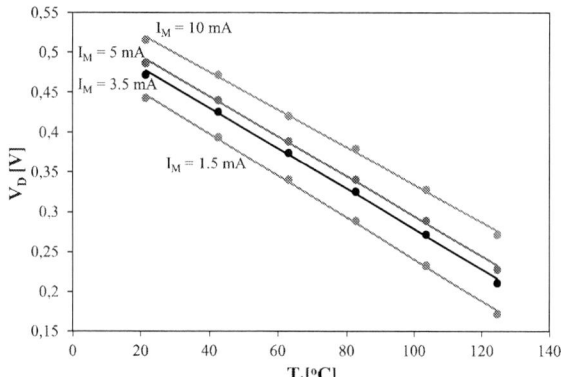

Fig. 4. Thermometric characteristics of the anti-parallel diode for selected values of the collector current

Fig. 5 shows thermometric characteristics $V_{GE}(T)$ measured at selected values of collector current i_C. As it is visible, these characteristics are not linear. They can be approximated with the use of the square function. For example, for the collector current equal to 1 mA this formula has the following form

$$V_{GE} = 5.4038 V - 7.8 \frac{mV}{^oC} \cdot T + 10^{-5} \frac{V}{\left(^oC\right)^2} \cdot T^2 \qquad (6)$$

As can be noticed, characteristics $V_{GE}(T)$ show considerably greater (almost four times) sensitivity of the thermo-sensitive parameter at changes in temperature than characteristics $V_D(T)$. This means that at the same accuracy of measurements of both voltages, four times higher resolution of measurements of internal temperature using V_{GE} voltage as a thermo-sensitive parameter can be obtained.

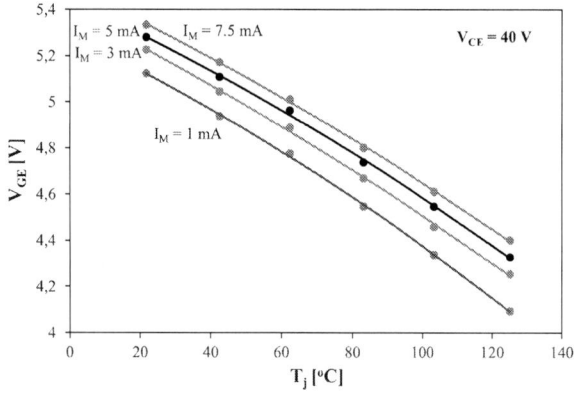

Fig. 5. Thermometric characteristics $V_{GE}(T)$ for a selected values of the collector current

Values of internal temperature of the examined transistor operating at both the considered cooling conditions using the measuring set-ups described in Section 2 and the case temperature of this transistor with the use of the pyrometer Optex PT3S are measured. Using the obtained values of internal temperature and the case temperature of this transistor

dependences of its thermal resistance on power are marked. The results of these measurements, performed at ambient temperature $T_a = 21^oC$ are shown in Figs. 6-9. In these figures with symbols T_{jGE} and R_{thGE} the results of measurements obtained with the use of V_{GE} voltage as a thermo-sensitive parameter are described. In turn, symbols T_{jD} and R_{thD} correspond to measurements, in which a thermo-sensitive parameter was voltage on the anti-parallel diode. The results of optical measurements are marked with symbols T_C and R_{thC}. In the case of optical measurements the highest values of temperature on the surface of the case of the examined transistor are registered.

Fig. 6 illustrates the measured dependences of internal temperature and the case temperature of the examined transistor situated on the heat-sink on power dissipated in this transistor.

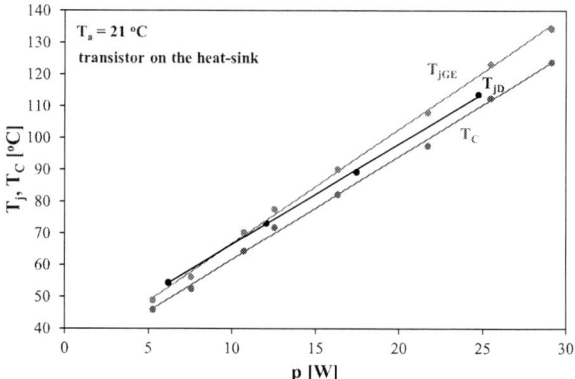

Fig. 6. Measured dependences of internal and case temperature of the tested transistor situated on the heat-sink on power dissipated in this transistor

As can be observed, for all the considered measuring methods an increasing function describing dependence of temperature of the device on power is obtained. Differences between values of temperature measured by means of the considered methods are bigger together with an increase in power attaining maximally 10°C between internal temperature T_{jGE} obtained with the use of voltage V_{GE} as a thermo-sensitive parameter and with the case temperature T_C measured with the pyrometer. Internal temperature T_{jD} measured with the use of voltage on the diode as a thermo-sensitive parameter is lower than temperature T_{jGE} by no more than 7 °C.

In Fig. 7 measured dependences of thermal resistance of the considered transistor situated on the heat-sink on power dissipated in this transistor are shown.

Dependences $R_{th}(p)$ measured with the use of each considered thermo-sensitive parameter and using the optical method are monotonically decreasing functions. The biggest changes of the value R_{th} are visible for measurements performed with the electrical method making use of voltage on the diode as a thermo-sensitive parameter. They exceed even 30%. In other cases these changes do not exceed 25%. It is proper to notice that the difference in values of thermal resistance measured with the electrical method using voltage V_{GE} as a thermo-sensitive parameter and by means of the optical method is practically constant and amounts 0.5 K/W.

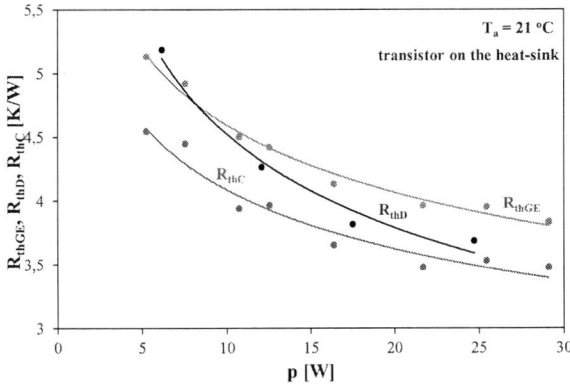

Fig. 7. Measured dependences of thermal resistance of the tested transistor situated on the heat-sink on power dissipated in this transistor

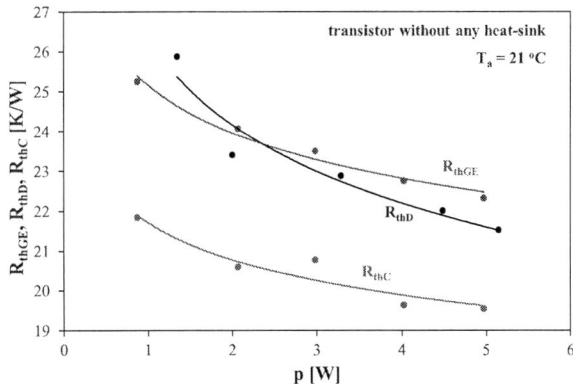

Fig. 9. Measured dependences of thermal resistance of the tested transistor operating without any heat-sink on power dissipated in this transistor

In Fig. 8 dependences of internal temperature and the case temperature of the examined transistor operating without any heat-sink on power dissipated in this transistor are illustrated.

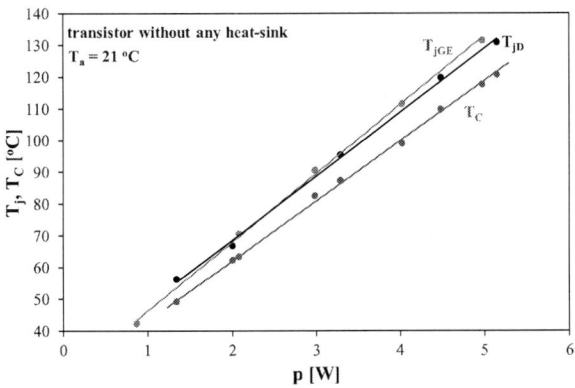

Fig. 8. Measured dependences of internal and case temperature of the tested transistor operating without any heat-sink on power dissipated in this transistor

The obtained dependences of temperatures T_{jGE}, T_{jD} and T_C on power are monotonically increasing functions. Differences between the measured values of temperatures T_{jGE} and T_{jD} are not big and they do not exceed 3 K. In turn, the measured values of the case temperature T_C are smaller than temperature T_{jGE} even by 10 °C.

Figs. 9 illustrates the measured dependences of thermal resistance of the examined transistor operating without the heat-sink on power dissipated in it.

By means of all the used measuring methods decreasing dependences of thermal resistance on power are obtained. Values of thermal resistance measured by means of both electric methods do not differ more than by about 1 K/W. Changes in the value of thermal resistance in the considered range of changes of power do not exceed 15%. Values of thermal resistance determined by means of the optical method are by about 3 K/W smaller than values of this parameter obtained with the use of the electrical method, in which a thermo-sensitive parameter is voltage V_{GE}.

V. CONCLUSIONS

In the paper influence of the method of measurements on thermal resistance of the IGBT is analysed. The indirect electrical and optical methods are considered. In the electrical method, in turn two thermo-sensitive parameters: gate-source voltage V_{GE} and voltage V_D on the forward biased anti-parallel diode are used.

On the basis of the obtained results it is stated that thermometric characteristics $V_{GE}(T)$ are non-linear and can be approximated with big accuracy by a square function. In turn, thermometric characteristics $V_D(T)$ are practically linear. The slope of the characteristic $V_{GE}(T)$ is even twice higher than the slope of the characteristic $V_D(T)$. This means that with the use of voltage V_{GE} it is possible to obtain a smaller value of the measurement error of thermal resistance than with the use of voltage V_D.

It is also shown that the measuring error of thermal resistance determined with the use of the method of the complete differential is a decreasing function of an excess of internal temperature over ambient temperature. If this increase attains 100 K, the measuring error of thermal resistance with the electric method does not exceed 4%. This error is higher for the optical method due to occurrence of thermal resistance between the semiconductor structure and the case.

The obtained results of measurements of thermal resistance of the considered transistor show that with the use of the indirect electrical method the difference between values of the considered parameter obtained with the use of voltages V_{GE} and V_D as thermo-sensitive parameters is observed. Due to the fact that the diode is the separate semiconductor structure situated in the common case with the transistor and power is dissipated in the transistor its temperature is lower than for the IGBT. Values of thermal resistance obtained with the use of these thermo-sensitive parameters do not differ between each other more than about 5%. This difference is only imperceptibly higher than the estimated error of measurements. In turn, the value of this parameter obtained by means of the optical method is underestimated by about 10%. It is proper also to notice that the measured dependences of thermal resistance on power are decreasing functions and a drop in the value of this parameter in the considered range of changes of power reaches even 25%.

The obtained results show that the use of the optical method of measurements can cause indeed understating of the value of thermal resistance. The results of measurements of this parameter with the use of the electrical method and one of the considered thermo-sensitive parameters do not differ from each other.

REFERENCES

[1] M.H. Rashid, Power Electronic Handbook, Academic Press, Elsevier, New York, 2007.

[2] R. Perret, Power electronics semiconductor devices. John Wiley & Sons, Hoboken, 2009.

[3] M.K. Kazimierczuk, Pulse-width modulated DC-DC power converters. John Wiley & Sons, Chichester, 2008.

[4] A.R. Hefner and D.L. Blackburn, An analytical model for the steady state and transient characteristics of the power Insulated Gate Bipolar Transistor, Solid-State Electronics, Vol. 31, No. 10, 1988, pp. 1513-1532.

[5] P. Górecki, K. Górecki, Influence of thermal phenomena on dc characteristics of the IGBT. International Journal of Electronics and Telecommunications, Vol. 64, No. 1, 2018, pp. 71-76.

[6] P. Górecki, Investigation of the influence of thermal phenomena on characteristics of IGBTs contained in power modules. 24th International Conference Mixed Design of Integrated Circuits and Systems Mixdes 2017, Bydgoszcz 2017, pp. 355-359.

[7] K. Górecki and P. Górecki, Modelling dynamic characteristics of the IGBT with thermal phenomena taken into account, Microelectronics International, Vol. 34, No. 3., 2017, pp. 160-164.

[8] A. Castellazzi, R. Kraus, N. Seliger, D. Schmitt-Landsiedel, Reliability analysis of power MOSFET`s with the help of compact models and circuit simulation. Microelectronics Reliability, Vol. 42, 2002, pp.1605-1610.

[9] A. Castellazzi, Y.C. Gerstenmaier, R. Kraus and G.K.M. Wachutka, Reliability analysis and modeling of power MOSFETs in the 42-V-PowerNet, IEEE Transactions on Power Electronics, Vol. 21, 2006, No. 3, pp. 603-612.

[10] P.E. Bagnoli, C. Casarosa, M. Ciampi, E. Dallago, Thermal resistance analysis by induced transient (TRAIT) method for power electronic devices thermal characterization. IEEE Transactions on Power Electronics, I. Fundamentals and Theory Vol. 13, No. 6, 1998, pp. 1208-19.

[11] K. Górecki, P. Górecki, J. Zarębski: Measurements of parameters of the thermal model of the IGBT module. IEEE Transactions on Instrumentation and Measurement, Vol. 68, No. 12, 2019, pp. 4864-4875.

[12] D.L. Blackburn, Temperature measurements of semiconductor devices – A review", 20th IEEE Semicon. Thermal Measur. and Menagement Symp. SEMI-THERM, San Jose, 2004, pp. 70-80.

[13] F. F. Oettinger and D. L. Blackburn, Semiconductor measurement technology: Thermal resistance measurements, U. S. Department of Commerce, NIST/SP-400/86, 1990.

[14] J. Zarębski, K. Górecki, A method of measuring the transient thermal impedance of monolithic bipolar switched regulators. IEEE Transactions on Components and Packaging Technologies, Vol. 30, No. 4, 2007 pp:627 – 631.

[15] K. Górecki, P. Górecki: Non-linear compact thermal model of the IGBT dedicated to SPICE. IEEE Transactions on Power Electronics, in press, doi: 10.1109/TPEL.2020.2995414.

[16] Y. Avenas, L. Dupont, Z. Khatir: Temperature measurement of power semiconductor devices by thermo-sensitive electrical parameters – a review. IEEE Transactions on Power Electronics, Vol. 27, No. 6, pp. 3081-3092, 2012.

[17] L. Dupont, Y. Avenas: Preliminary evaluation of thermo-sensitive electrical parameters based on the forward voltage for online chip temperature measurements of IGBT devices. IEEE Transactions on Industry Applications, vol. 51, No. 6, pp. 4688-4698, 2015.

[18] L. Dupont, Y. Avenas, P.-O. Jeannin: Comparison of junction temperature evaluations in a power IGBT module using an IR camera and three thermosensitive electrical parameters. IEEE Transactions on Industry Applications, Vol. 49, No. 4, pp. 1599-1608, 2013.

[19] G. Zeng, H. Cao, W. Chen, J. Lutz: Deffirence in device temperature determination using p-n junction forward voltage and gate threshold voltage. IEEE Transactions on Power Electronics, Vol. 34, No. 3, 2019, pp. 2781-2793

[20] IRG4PC40UD, Insulated gate bipolar transistor with ultrafast soft recovery diode, Data sheet, International Rectifier, http://www.irf.com/product-info/datasheets/data/irg4pc40ud.pdf.

[21] K. Górecki, P. Górecki, The analysis of accuracy of selected methods of measuring the thermal resistance of IGBTs. Metrology and Measurement Systems, Vol. 22, No. 3, 2015, pp. 455-464.

Proceedings of the 27th International Conference *"Mixed Design of Integrated Circuits and Systems"*
June 25-27, 2020, Łódź, Poland

Thermal Characterization of Electronic Components Using Single-detector IR Measurement and 3D Heat Transfer Modelling

Michał Kopeć, Bogusław Więcek

Institute of Electronics
Lodz University of Technology
Lodz, Poland
michal.kopec@edu.p.lodz.pl, boguslaw.wiecek@p.lodz.pl

Abstract—A novel methodology of thermal impedance measurement by temperature monitoring out of the heat source in a power transistor is presented. A low-cost Infra-Red (IR) head is used to register evolution of temperature after step-function powering. A dedicated power generator has been developed to synchronize temperature recording with power dissipation in a device. Estimation of temperature in the heat source is performed by 3D FEM modelling of multilayer transistor structure. It allows fitting the measurement and simulation results to achieve the classically-defined thermal impedance in the heat source.

Keywords—Thermal impedance; IR temperature measurement; 3D thermal modelling; Laplace transform; Forster thermal network

I. Introduction

Thermal impedance concept is still used in electronics for thermal testing and management both in production and exploitation. There are standard procedures and methodology of measuring thermal impedance in practice [1,2]. This methodology has the fundamental limits. It can be effectively applied only for devices with dominant one dimensional heat transfer. In addition, in practice, a Device Under Test (DUT) can only be powered by the approximated Heaviside step function excitation, what can lead to the erroneous results, especially in the high frequency range of the thermal impedance [3-5,6-7]. Another main engineering problems in such measurements is to evaluate temperature in the heat source, what is impossible in the general case [1,2,3,4,8,9]. Moreover, the overall methodology is based on the inverse thermal problem solution, which is by definition, ill-conditioned. In the existing commercially available apparatus, temperature is measured using a contact method [2]. It has an advantage of the high speed temperature monitoring at the beginning of a thermal process [2, 8-9].

There have already been attempts undertaken to measure temperature and thermal impedance using IR thermography, especially in medical applications, where the cooling provocation is applied to the surface of the skin [10-15]. The similar experiments were performed for electrical cables and integrated-circuit inductors, where the IR camera had the optical access to the heat source [5,16,17].

There are different mathematical approaches of thermal impedance calculation and finally modelling a DUT using Foster and Cauer approximations [3,4,6,7,12,18-24]. They are based on either time-series analysis [3,4,18,19,21,23] or integral transforms, such as Gardner, Fourier and Laplace ones, and in consequence, the analysis in frequency domain [12,21].

II. Methodology

In the research presented this paper, the fast-response, low-cost, Peltier-cooled single-detector IR system was used to measure temperature on the outer surface of the case of a power transistor [24]. The DUT was powered using self-developed power generator which allows synchronizing the starting point of powering and temperature registration. The signal from the IR head is noisy, and therefore a kind smoothing of temperature curves is required. In order to smooth the temperature signal, the Savitzky-Golay filter was implemented [25].

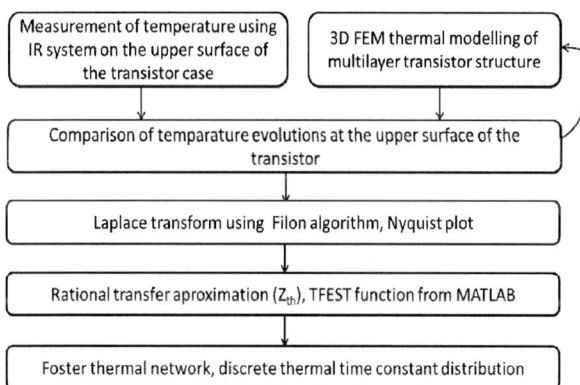

Fig. 1. The algorithm of the proposed method thermal characterization of power devices using IR measurement

Simultaneously, the 3D FEM thermal model of a DUT was elaborated to calculate the step-function temperature response vs. time on the outer surface of the transistor's case. Modelling was realized using the Comsol environment [26]. Fitting of temperature rise by changing thermal parameters of a DUT, obtained from measurement and simulation was then made using optimization procedure. Having the trimmed 3D thermal model, it was possible to appoint the temperature rise in the heat source.

Next, the thermal impedance, in the form of the Nyquist plot was numerically calculated using the Filon integration. In addition, converting the Nyquist plot given as the frequency-series, into the analytical rational transfer function (Zth) was performed. It is the crucial step of the proposed methodology, and for it, the TFEST function from Matlab System Identification Toolbox was used [27]. Finally, in order to estimate the discrete thermal time constant distribution, the Foster network approximation was calculated. The proposed methodology of thermal analysis of power devices is presented in Fig. 1.

III. MEASUREMENT SETUP

The measurement setup enables precise placement of a tested object under the IR system. It consists of single-detector IR head [28,29], developed electronic controller, z-axis moveable stand and x-y micrometer table. The important function of the developed controller is the precise synchronization of temperature measurement and power dissipation in a DUT. Software of the developed controller enables setting of sampling period Ts, acquisition time t_M and dissipated power P_D in a DUT. The single-detector IR system for thermal characterization of electronic device is shown in Fig. 2.

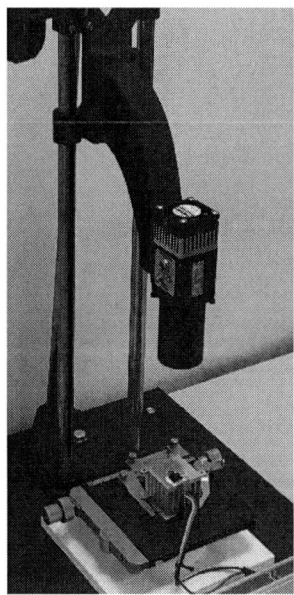

Fig. 2. The measurement single-detector IR system for thermal characterization of electronic devices

IV. SIMULATION MODEL

The simulation model was developed using COMSOL Multiphysics software [26]. The Heat Transfer in Solids interface of the Heat Transfer Module was used to carry out time dependent analysis. The modeled system consisted of a

transistor mounted to the heat sink, the screw and the substrate layer. The dimensions of the model were consistent with transistor and heat sink manufacturers' datasheets. The developed model of the transistor in TO-220 case is shown in Fig. 3.

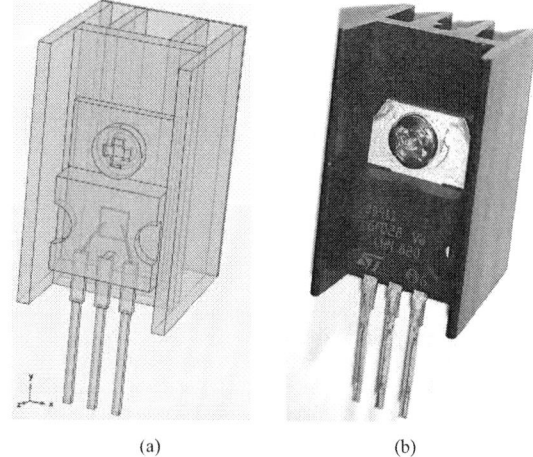

(a)   (b)

Fig. 3.   Geometry of the model (a) and the tested object (b)

The convective heat transfer coefficient h described the boundary conditions at the outer surface of the transistor and radiator. The same power $P_0 = 2.68$ W was dissipated in the heat source while modelling and measurement. Examples of the temperature distributions during heating-up obtained from simulations are shown in Fig. 4.

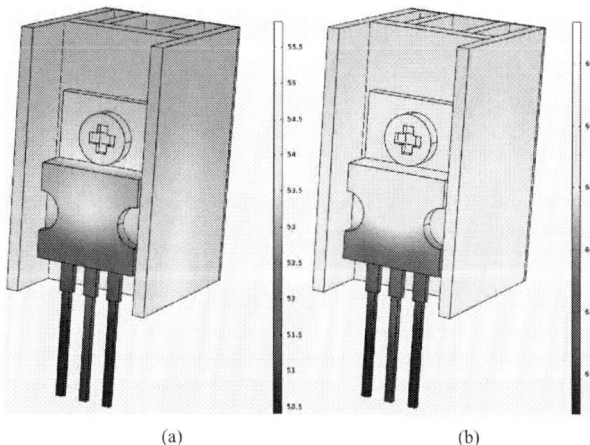

(a)   (b)

Fig. 4.   Simulated temperature distribution for t=110 s (a) and t=600 s (b)

The first step of the proposed method was to fit simulated and measured temperature curves for the heating-up process. The black line in Fig. 5 represents the simulation result, while the red line corresponds to the IR measurement. The sampling rate of temperature recording fs = 2 kHz was identical both in simulation and measurement. The goal of the optimization was to select appropriate values of thermal parameters and boundary conditions in order to obtain the best fitting of simulated and measured temperature curves. The chosen thermophysical parameters after fitting are shown in Tab. I.

140

TABLE I.
THERMOPHYSICAL PARAMETERS AFTER FITTING THE SIMULATION
AND MEASUREMENT RESULTS

Layer	Thermal conductivity (Wm^{-1}K^{-1})	Density (kg/m^3)	Heat capacity (J·kg^{-1}K^{-1})
case	1.2	950	1000
die	148	2330	720
bond wire	200	2700	900
lead	237	2700	950
tab	237	2700	950
heat sink	237	2800	950
screw	58	7900	500
substrate	58	7900	500

Figure 5 shows the simulated and measurement temperature curves in the middle point on the transistor case obtained from simulation and measurement. After 500 seconds, the transistor reaches the quasi steady state, and temperature of the case is converging approximately to 28.8 °C above ambient. It should be noted that the temperature doesn't rise immediately inside the case. The rise of temperature at the outer surface of the transistor's case begins after 0.165 s. This happens because heat flows from the die to the outer side of the case along the distance of a few millimeters and it takes time.

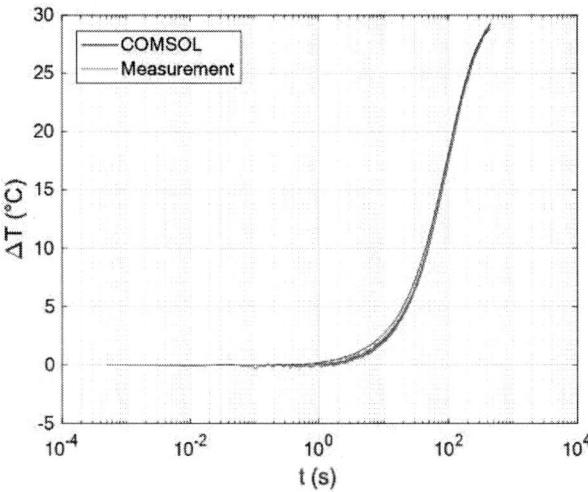

Fig. 5. Temperature on the outer side of the transistor's case: red line – measured by the single-detector IR head, blue line – single-detector IR measurement and simulation result from COMSOL

Measuring temperature vs. time evolution on the outer surface of the case allows calculating so-called transfer thermal impedance presented in Fig. 6. It can be noticed that in case of the transfer thermal impedance, the Nyquist plot crosses the imaginary axis and has the negative real part in high-frequency range. The transfer impedance can be approximated by the Foster RC network, but in such a case the amplitudes of the thermal time constants can be negative.

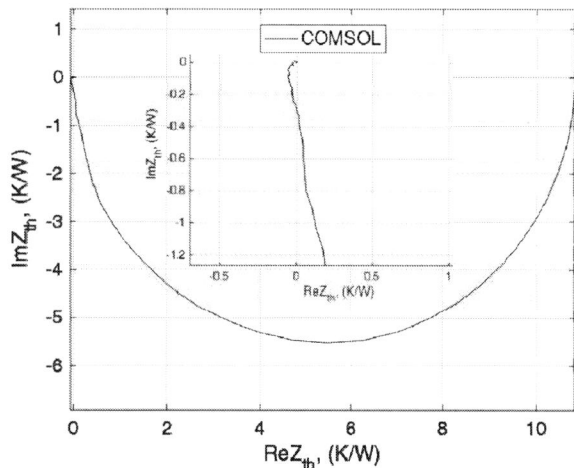

Fig. 6. The Nyquist plot on the outer side in the transistor's case – from the single-detector IR measurement and simulations in COMSOL

V. RESULTS AND ANALYSIS

The next step of the analysis leads to obtain the simulated temperature in the die, as shown in Fig. 7 – black line. For this purpose, the observation point in the simulation was placed in the die layer, and the simulation was performed for previously identified thermophysical parameters and boundary conditions of the DUT. Comparison of the results obtained from simulation and measurement using the contact method is almost perfect – Fig. 7.

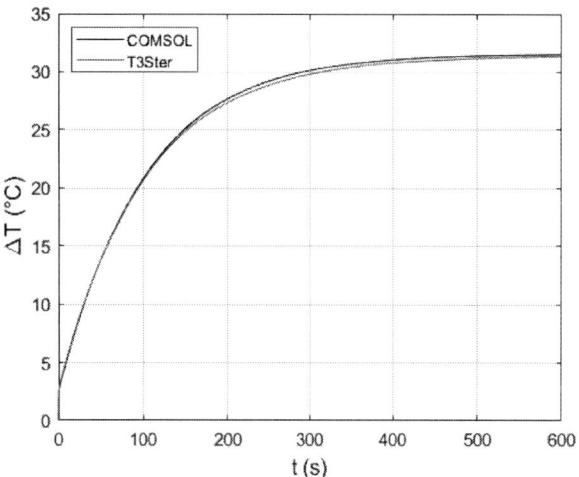

Fig. 7. Temperature rise after step-function powering in the transistor's junction: red line – T3Ster measurement, blue line – single-detector IR measurement and COMSOL simulation

The simulated temperature curve (black line) agrees with the measured temperature curve (red line) as shown in Fig. 7. The contact temperature measurement performed by T3Ster apparatus uses voltage over the emitter die of BJT, and it needs calibration [2].

The temperature curves in the logarithmic scale for better understanding of heat transfer in the multilayer structure of the transistor are shown in Fig. 8. It is visible, that first increase of temperate starts for $t \approx 10^{-3}$ s. It corresponds to the shortest thermal time constants of the internal structure of the transistor.

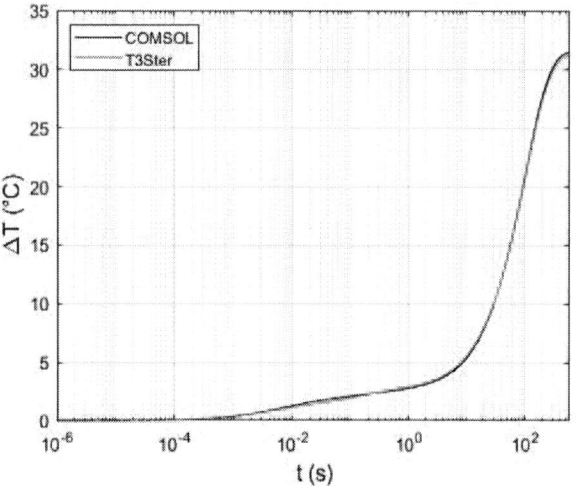

Fig. 8. Temperature rise in the junction: red line – measured by T3Ster system, black line – single-detector IR measurement and simulation in COMSOL case

The simulation results indicate an impact of bond wires on the rapid increase of temperature in the initial period of the thermal heating process.

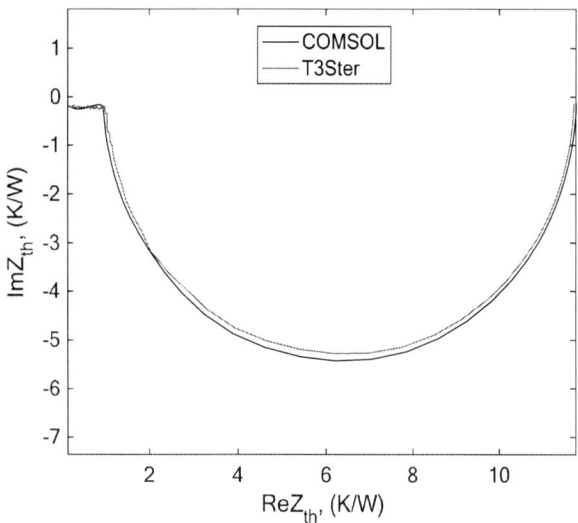

Fig. 9. Nyquist plot in the die: red line – T3Ster measurement, black line – single-detector IR measurement and COMSOL simulation

The next step of the proposed analysis allows calculating the Nyquist plots of the thermal impedance for simulation and measurement results using the Laplace transform [12]. The thermal impedance plots for the heat source obtained using two different methods are presented in Fig. 9, and they confirm the acceptable accordance. Each curve in Fig. 9 can be roughly split up in two parts. Both parts have the shape of circular arcs. The internal structure of the transistor is attributed to the part of the Nyquist plot with high value of angular frequency ω – small arc. In turn, the heat sink is represented by the large arc in Fig. 9, and it corresponds to the part with low values of angular frequencies ω.

The last step of the proposed methodology allows representing the thermal impedance as the ratio of low-order polynomials. Then, calculation the poles of Z_{th} leads to the Foster RC network approximation and the thermal time constant distribution, as it is shown in Fig. 10 and Tab. II.

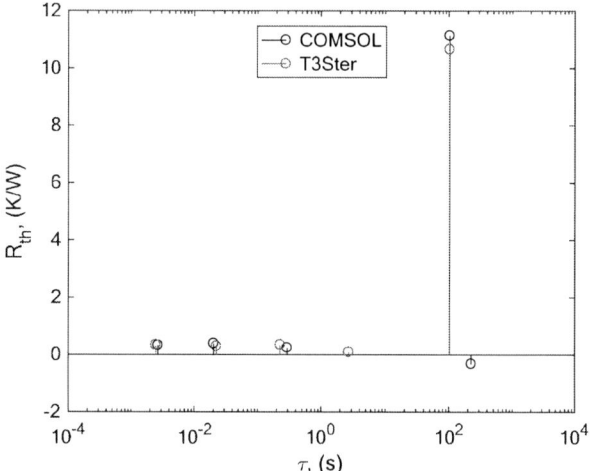

Fig. 10. Thermal time constants: red line – T3Ster, black line – single-detector IR measurement

TABLE II.
VALUES OF THERMAL TIME CONSTANTS IN THE DIE

T3Ster		Single-detector IR system	
R_i, K/W	τ_i, s	R_i, K/W	τ_i, s
0.338	0.0025	0.327	0.0027
0.282	0.0222	0.383	0.0201
0.340	0.2269	0.235	0.2925
0.083	2.7124	11.366	102.3120
10.663	101.5537	-0.324	224.8329

There is a slight disagreement between the results obtained by calculating thermal time constants from temperature curves measured by T3Ster inside the structure and the developed single-detector IR head on the outer side of the transistor. The proposed method of thermal analysis allowed getting the negative amplitude of the longest time constant. It denotes that the thermal process is still depending on a heat source located away from the semiconductor die. This phenomenon can be attributed to additional heating by the electrical contacts at the end of collector and emitter pins delivering the current to the transistor. It's noticeable that measurements made by T3Ster system never give the negative amplitudes of thermal time constants. Nevertheless, the agreement of the results obtained by both systems is acceptable.

VI. CONCLUSIONS

In this paper, the application of IR temperature measurement on the surface of a power device and 3D thermal simulation are presented to estimate the thermal impedance. The low-cost IR single-detector system was developed to measure the temperature synchronously with the power dissipated in a DUT using step-function excitation. The example of measurement and comparison of results obtained from the new low-cost and commercially available systems are presented showing the acceptable accordance.

REFERENCES

[1] EIA/JEDEC standard: Integrated Circuits Thermal Measurement Method – Electrical Test Method (Single Semiconductor Device). Electronic Industries Alliance Engineering Department, no. 51-1, pp. 1-27, 1995.

[2] T3Ster-Master Thermal Evaluation Tool – User's Manual Version 2.2, Mentor Graphics Corporation.

[3] V. Szekely, "On the representation of infinite-length distributed RC oneports", IEEE Trans. Circuits Syst., vol. 38, pp. 711–719, 1991.

[4] V. Szekely V, "Identification of RC networks by deconvolution: Chances and limits", IEEE Trans. Circuits Syst., vol. 45, no. 3, pp. 244–258, 1998.

[5] M. Kałuża, B. Więcek, G. De Mey, A. Hatzopoulos, V. Chatziathanasiou, "Thermal impedance measurement of integrated inductors on bulk silicon substrate", Microelectronics Reliability, vol. 73, pp. 54-59, 2017.

[6] K. Górecki, J. Zarebski, "The influence of the selected factors on transient thermal impedance of semiconductor devices", Proceedings of the 21st International Conference Mixed Design of Integrated Circuits and Systems MIXDES, Lublin, pp. 309-314, 2014.

[7] K. Górecki, M. Rogalska, J. Zarębski, "Parameter estimation of the electrothermal model of the ferromagnetic core", Microelectronics Reliability, vol. 54, no. 5, pp. 978-984, 2014.

[8] M. Janicki, T. Torzewicz, P. Ptak et al., "Parametric Compact Thermal Models of Power LEDs", ENERGIES vol. 12, issue: 9, 2019.

[9] M. Janicki, T. Torzewicz, A. Samson and et al., "Experimental identification of LED compact thermal model element values", MICROELECTRONICS RELIABILITY, vol. 86, pp. 20-26, 2018.

[10] P. Chatzipanagiotou, M. Strąkowska, G. De Mey, V. Chatziathanasiou, B. Więcek, M. Kopeć, "A new software tool for transient thermal analysis based on fast IR camera temperature measurement", Measurement Automation Monitoring, no. 62, vol. 63, ISSN 2450-2855, pp. 49-51, 2017.

[11] M. Strakowska, R. Strąkowski, M. Strzelecki, G. De Mey, B. Wiecek, "Thermal modelling and screening method for skin pathologies using active thermography", Biocybernetics and Biomedical Engineering, 10.1016/j.bbe.2018.03.009.

[12] M. Strakowska, P. Chatzipanagiotou, G. De Mey, V. Chatziathanasiou, B. Więcek, "Novel software for medical and technical Thermal Object Identification (TOI) using dynamic temperature measurements by fast IR cameras", 14th Quantitative InfraRed Thermography Conference, QIRT 2018, June 25-29, 2018, Berlin, DOI: 10.21611/qirt.2018.053 http://qirt.gel.ulaval.ca/archives/qirt2018/papers/053.pdf

[13] M. Strakowska, B. Wiecek, M. Strzelecki, A. Kaszuba, "Screening procedure based on cold provocation and thermal tissue modeling", 13th Quantitative Infrared Thermography Conference, 04-08 July 2016, Gdansk, Poland.

[14] M. Strąkowska, R. Strąkowski, M. Strzelecki, G. De Mey, B. Więcek, "Evaluation of Perfusion and Thermal Parameters of Skin Tissue Using Cold Provocation and Thermographic Measurements", Metrology and Measurement Systems, The Journal of Committee on Metrology and Scientific Instrumentation of Polish Academy of Sciences, vol. 23, issue 3, 2016.

[15] M. Strakowska, G. De Mey, B. Więcek, M. Strzelecki, "A Three Layer Model for The Thermal Impedance of The Human Skin: Modeling And Experimental Measurements", Journal of Mechanics in Medicine and Biology, 15(3), 2015, DOI: 10.1142/S021951941550044X .

[16] P. Chatzipanagiotou, V. Chatziathanasiou, G. De Mey, B. Wiecek, "Influence of soil humidity on the thermal impedance, time constant and structure function of underground cables: A laboratory experiment", App Therm Engin, vol. 113, pp. 1444–1451, 2017.

[17] V. Chatziathanasiou, P. Chatzipanagiotou, I. Papagiannopoulos, G. De Mey, B. Wiecek, "Dynamic thermal analysis of underground medium power cables using thermal impedance, time constant distribution and structure function", Applied Thermal Engineering, vol. 60, no. 1-2, pp. 256–260, 2013.

[18] H. Garnier, M. Mensler, A. Richard, "Continuous-time Model Identification from Sampled Data: Implementation Issues and Performance Evaluation", International Journal of Control, vol. 76, issue 13, pp. 1337–1357, 2003.

[19] L. Ljung, "Experiments with Identification of Continuous-Time Models", Proceedings of the 15th IFAC Symposium on System Identification, 2009.

[20] A.U. Jibia, M-J Salami, "An Appraisal of Gardner Transform-Based Method of Transient Multiexponential Signal Analysis", International Journal of Computer theory and Engineering, vol.4, pp. 16-24, 2012.

[21] B. S. Yarman, A. Kilinc, A. Aksen, "Immitance Data Modelling via Linear Interpolation Techniques: A Classical Circuit Theory Approach", Int. J. Circ. Theory. Appl., 32: 537-563, 2004.

[22] K. Wang, M. Z. Q. Chen, G. Chen, "Realization of a transfer function as a passive two port RC ladder network with a specified gain", Int. J. Circ. Theory. Appl., 45: 1467– 1481. doi: 10.1002/cta.2328, 2017.

[23] CAPTAIN - Computer-Aided Program for Time-Series Analysis and Identification of Noisy Systems. (http://www.es.lancs.ac.uk/cres/captain/)

[24] https://vigo.com.pl/

[25] Savitzky A., Golay M.J.E.: Smoothing and differentiation of data by simplified least squares procedures. Anal. Chem., vol. 36, pp. 1627–1639, 1964.

[26] https://www.comsol.com

[27] https://nl.mathworks.com/help/ident/ref/tfest.html

[28] Infrared detectors Room Temperature and TE-Cooled. [07.05.2019]. Available: https://www.boselec.com/wp-content/uploads/Linear/Vigo/ VigoLiterature/BEC-Vigo-IR-Detector-Catalog-03-08-19.pdf

[29] https://vigo.com.pl/

Analysis and Modelling
of ICs and Microsystems

Proceedings of the 27th International Conference *"Mixed Design of Integrated Circuits and Systems"*
June 25-27, 2020, Łódź, Poland

A Process, Voltage and Temperature Dependent Modeling Methodology for Industrial Requirements

Ivan Sejc, Robert Kappel

ams AG

Premstaetten, Austria

ivan.sejc@ams.com, robert.kappel@ams.com

Abstract—**Time efficient and accurate design verification is a key procedure in an industrial environment. In order to achieve both, a behavioral modeling can be used instead of more complex SPICE model options. This, however, requires a significant model complexity. This paper proposes a PVT and Monte Carlo modeling methodology based on the generation of polynomial equations with Matlab environment. These expressions are then used to describe the model behavior of its outputs and inputs. The use of behavioral model significantly reduces simulation time and still provides relatively high degree of accuracy while considering the variation in temperature, supply voltage and process corner.**

Keywords—**Behavioral modeling, PVT variations, Matlab**

I. Introduction

Design verification (DV) of mixed-signal systems has become more time consuming process because of its increasing complexity. In order to simulate the overall behavior of the system considering the PVT (Process Voltage Temperature) variation and MC (Monte Carlo) variations in SPICE based simulations, one has to consider and plan the necessary time effort and engineering resources accordingly. This is often in conflict with very tight project schedules leading to a design phase duration of several months in an industrial environment. Therefore, a time efficient DV process is necessary. A good candidate to optimize a DV process by reducing the simulation time is the use of Behavioral Models (BM). BM can effectively describe the behavior of the complex systems while significantly reducing the simulation time [1]. However, the BM must contain enough information for a relatively accurate DV. From the industrial point of view, the BM needs to show the exact functionality, including startup and power down behavior, but also needs to be dependent on the power supply, temperature, process corner and mismatch variations.

The possible BM implementations were already published in [1] and [2] by using response surface models (RSM) and distribution functions. Those allowed to model a circuit behavior under different PVT conditions while considering the aging effect [1]. However, in [1] and [2], significantly large amount of simulation data was necessary to cover MC variations and aging. The amount of simulation data was even bigger in [3], where no parameter identification was performed. This work presents an improved methodology for the BM development for industrial requirements by considering PVT and MC variations. RSM was used to model the analog behavior of the bandgap voltage reference with a help of Matlab Curve and Surface Fitting Toolbox. Due to its simple and user friendly implementation of

the algorithm it is able to reduce the overall time to prepare the necessary scripts which generate a polynomial approximation to the SPICE simulation results.

Section II. describes the modeling methodology and RSM generation by using the Matlab Fitting Curve Toolbox. In section III., the proposed methodology is applied on the example of the bandgap voltage reference circuit, considering the model accuracy, simulation time and complexity of the data pre-processing. Section IV. concludes the achieved results.

II. Modeling Methodology

The modeling flow is divided into several steps as shown in the figure 1.

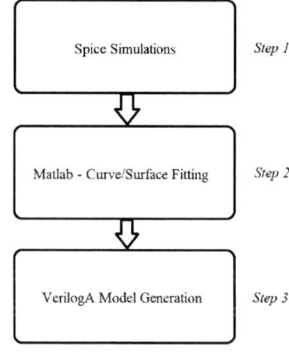

Figure 1. Modeling flow

A. Spice Simulations

The first step is to perform all the necessary SPICE simulations to obtain the output parameter y as a function of required input variables e.g. temperature, supply voltage or process corner. The output data are then saved as a *.csv* file and prepared for post-processing. In the case of MC simulations, Gaussian distribution of the parameter y is obtained. The mean value and standard deviation are extracted from the simulation.

The amount of the necessary simulations rises based on the number of variables. Temperature dependency of the output parameter y requires one variable sweep which results in two dimensional representation of $y(T)$ as shown in figure 2 for the example of bandgap reference voltage where $V_{ref} = y(T)$.

147

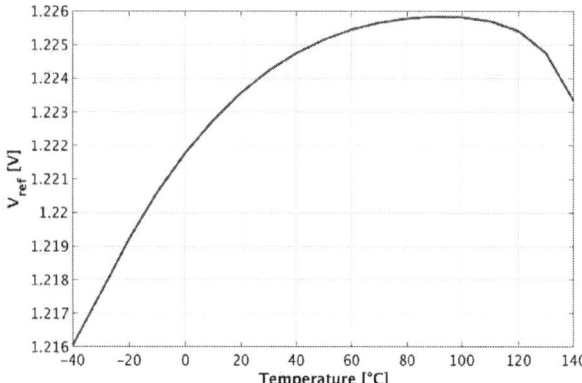

Figure 2. V_{ref} as function of temperature

Additionally, another variable can be added, in this case a supply voltage – V_{DD}. The output parameter V_{ref} is now function of two variables – $V_{ref}(T, V_{DD})$ leading to the three dimensional graphical representation, shown in figure 3.

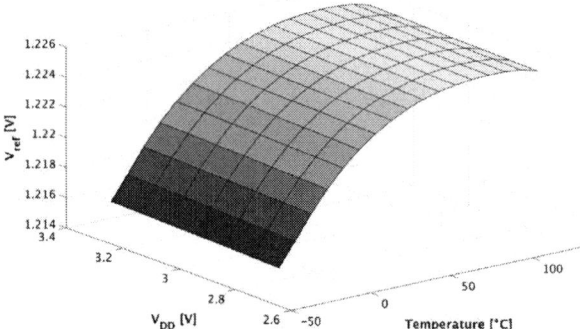

Figure 3. V_{ref} as function of temperature and supply voltage V_{DD}

The information about the process corner dependency can be also added as shown in figure 4. In this case, multiple surface representations of $V_{ref}(T, V_{DD})$ are obtained for different process corners.

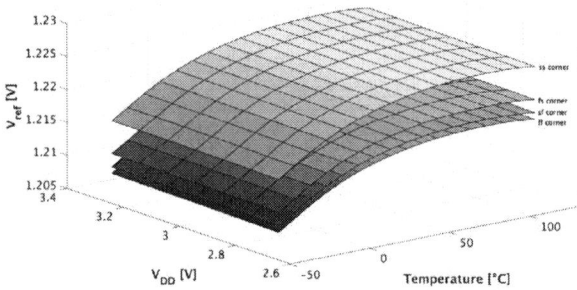

Figure 4. V_{ref} as function of temperature and supply voltage V_{DD} and process corner

B. Equation Generation in Matlab

The second step of the modeling methodology is generation of all the necessary mathematical equations in order to describe the behavior of the output V_{ref} by using Matlab Curve and Surface Fitting toolbox, since it provides relatively fast and user-friendly approach. First of all, curve fitting was used to approximate the V_{ref} with a polynomial equation which was chosen due to the character of the V_{ref}. Different types of approximation functions are available e.g. exponential, linear etc. The fitting toolbox generates the polynomial coefficients $c_0 \ldots c_n$ as shown in the general equation (1), where y refers to an output parameter of the system and x is a variable which can be temperature, supply voltage, etc.

$$y \approx c_0 x^0 + c_1 x^1 + c_2 x^2 + \cdots + c_n x^n \quad (1)$$

For the purpose of the bandgap voltage reference, the equation (2) is shown as more specific example where V_{ref} is an output parameter and T is the temperature.

$$V_{ref}(T) \approx c_0 T^0 + c_1 T^1 + c_2 T^2 + \cdots + c_n T^n \quad (2)$$

The accuracy of the function depends on its complexity, respectively on the number of coefficients used in equation. In order to increase the number of variables, surface fitting can be used. This approach allows now two independent variables (x_1, x_2) as shown in the general equation (3).

$$y \approx c_0 + c_1 x_1^1 x_2^1 + c_2 x_1^1 +$$
$$+ c_3 x_2^1 + c_4 x_1^1 x_2^2 + c_5 x_1^2 x_2^1 + \cdots + c_n x_1^m x_2^p \quad (3)$$

Again, for the purpose of the bandgap voltage reference, the equation (4) describes V_{ref} as a function of temperature (T) and supply voltage (V_{DD}).

$$V_{ref}(T, V_{DD}) \approx c_0 + c_1 T^1 V_{DD}^{\ 1} + c_2 T^1 +$$
$$+ c_3 V_{DD}^{\ 1} + c_4 T^1 V_{DD}^{\ 2} + c_5 T^2 V_{DD}^{\ 1} + \cdots + c_n T^m V_{DD}^{\ p} \quad (4)$$

In the case of process corner dependency, every corner case needs to have a separate equation as shown in the equation (5).

$$V_{ref}(T, V_{DD}, process) \approx \begin{matrix} V_{ref}(T, V_{DD}, ff\ corner) \\ \ldots \\ V_{ref}(T, V_{DD}, ss\ corner) \end{matrix} \quad (5)$$

C. VerilogA model generation

The third step is generation of a VerilogA model based on the derived equations in step 2. The output parameter quantity, either voltage or current can be simply considered as a function $V_{ref}(T, V_{DD}, process\ corner)$. This is shown in listing 1.

```
`include "constants.vams"
`include "disciplines.vams"

module bandgap(Vref,Vdd);

inout Vref,Vdd;

electrical Vref,Vdd;
thermal temp;

(* cds_inherited_parameter *) parameter real corners = 0; // parameter specified in model library

parameter real bg_out = 0;

analog begin

Temp(temp) <+ $temperature - 273.15;

if (corners == 1) // tt corner
V(Vref) <+ 1.2209+3.517e-5*(Temp(temp)*V(Vdd))+3.517e-5*(V(Vdd)*V(Vdd))+4.14e-
7*(Temp(temp)*Temp(temp))-3.4646e-7*(Temp(temp)*Temp(temp)*V(Vdd));
.
.
.

end
endmodule
```

Listing 1. Example of bandgap BM with polynomial expression describing V_{ref}

The *process corner* variable must be additionally specified in the technology model file as shown in the listing 2.

```
section ttmacro_mos_moscap
setscale options scalefactor=0.9
.
.
.
parameters corners = 1

endsection ttmacro_mos_moscap
```

Listing 2. Specification of the parameter 'corners' in the technology model file

The integer value of the parameter is then passed to a VerilogA model which can now distinguish different process options. The similar approach is applied for the case of the MC analysis. Here, additional information about a distribution type, mean and standard deviation value is specified. Mean and standard deviation are set to 1, as shown in the Listing 3., and can be normalized in the model later on.

```
section globalmc_localmc_mos_moscap
setscale options scalefactor=0.9
.
.
.
statistics {
  process {
    vary bg_voltage_mc dist=gauss std=1/1
  }
}

endsection globalmc_localmc_mos_moscap
```

Listing 3. MC parameter specification in the technology model file

III. EXAMPLE OF PVT DEPENDENT MODEL AND COMPARISON WITH SPICE SIMULATIONS

A good candidate for PVT BM generation is a bandgap voltage reference circuit which is shown in figure 5.

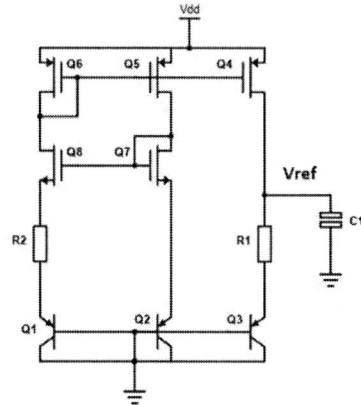

Figure 5. Bandgap voltage reference circuit

PTAT and CTAT currents ensure relatively small temperature dependency of the output reference voltage V_{ref}. The BM which describes the behavior of V_{ref}, contains a set of polynomial equations which are dependent on temperature (T) and supply voltage (V_{DD}). Every process corner option includes a polynomial equation with two variables. The simulation result of the BM for different PVT values is shown in figure 6. The accuracy of BM is defined by the number of polynomial coefficients. More coefficients result in more accurate BM, however, the simulation time will be longer. The accuracy results of the BM in comparison to the SPICE based simulation are shown in the table I.

TABLE I.
THE ACCURACY RESULTS OF THE BANDGAP BM

	Average Error	Maximum Error
BM - 2 coefficients	3.72%	5.83%
BM - 3 coefficients	1.83%	2.20%

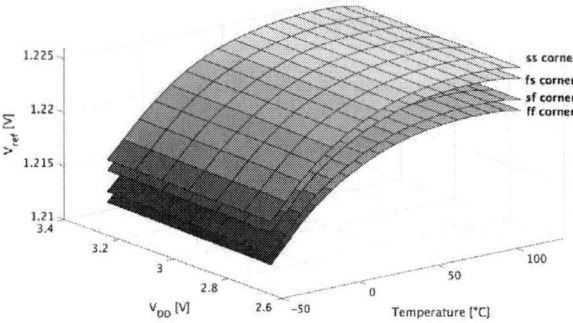

Figure 6. Bandgap voltage behavioral model simulation result

The table considers an absolute maximum error and an average error for two polynomial equations, the first one with two coefficients and the latter with three. The simulation time of the BM is compared with SPICE simulations in table II.

TABLE II.
SIMULATION TIME COMPARISON

	DC simulation	Transient simulation (1us)
SPICE netlist	511ms	830ms
BM	168us	44ms

The simulation time of the BM is almost 20 times shorter in the case of transient simulation than the SPICE model. The MC simulation results and comparison with SPICE netlist are shown in table III. As shown, the simulation time with BM shows almost 6 times improvement compared to SPICE based simulation.

TABLE III.
MONTE CARLO SIMULATION TIME COMPARISON

	MC simulation time (200 runs)
SPICE netlist	1193s
BM	200s

IV. CONCLUSION

This work presented a PVT dependent modeling methodology with a help of Matlab environment. The BM based on this methodology was compared in terms of accuracy and simulation time with SPICE based modeling of bandgap reference voltage circuit. The comparison showed significant simulation time enhancement with relatively good accuracy. The proposed modeling methodology also allows BM to be used in MC simulations and considers different process corner options which can provide more useful information in a DV process for relatively small amount of effort. This approach can be used to describe relatively large designs since only input and output behavior of the system is modelled with very low granularity.

REFERENCES

[1] M. Taddiken, T. Hillebrand, K. Tscherkaschin, S. Paul, and D. Peters-Drolshagen, "Parameter identification for behavioral modeling of analog components including degradation," in *Mixed Design of Integrated Circuits and Systems, 2016 MIXDES-23rd International Conference*, 2016, pp. 336–340.

[2] M. Taddiken, T. Hillebrand, S. Paul and D. Peters-Drolshagen, "Variation- and degradation-aware stochastic behavioral modeling of analog circuit components," 2017 14th International Conference on Synthesis, Modeling, Analysis and Simulation Methods and Applications to Circuit Design (SMACD), Giardini Naxos, 2017, pp. 1-4.

[3] N. Heidmann, N. Hellwege, M. Taddiken, D. Peters-Drolshagen, and S. Paul, "Analog behavioral modeling for age-dependent degradation of complex analog circuits," in *Mixed Design of Integrated Circuits & Systems (MIXDES), 2014 Proceedings of the 21st International Conference*. IEEE, 2014, pp. 317–322.

Proceedings of the 27th International Conference *"Mixed Design of Integrated Circuits and Systems"*
June 25-27, 2020, Łódź, Poland

Capacitance Deviation Caused by Mechanical Deformation of MEMS Inertial Structure

Jacek Nazdrowicz, Adam Stawiński, Andrzej Napieralski
Department of Microelectronics and Computer Science
Lodz University of Technology
Lodz, Poland
jnazdrowicz@dmcs.pl, astawinski@dmcs.pl, napier@dmcs.pl

Abstract—**Authors presents a very important problem of estimation capacitance structures used in MEMS sensors. The importance comes from the fact that during operation, inertial sensors deform under influence of external forces. The result of the problem is visible in capacitance deviations what directly may be seen in accuracy sensor measurement.**

Keywords—**MEMS, accelerometer, vibratory gyroscope, thermal expansion, temperature effect, capacitance.**

I. INTRODUCTION

MEMS inertial sensors are well-known microdevices commonly used for measurement linear or rotational motion of objects. MEMS (Micro-Electro-Mechanical Systems), are considered as devices and systems including mechanical elements, sensors, actuators, and electronic circuits fabricated on a one silicon substrate through the same technology. Design process of device is the first, crucial step in whole proces preceeding fabrication. Design, in turn, is the stage which is repeated many times because of the optimization of structure. This optimization comes from simulation results and analysis of stresses, deformation – generally, mechanical and electrical response.

Nowadays, MEMS structures are very interesting areas for researchers and discoverers from academic centers like universities and from commercial companies which have their own Research and Development departments.

Inertial MEMS sensors like accelerometers and gyroscopes are built of solid material, and therefore these are devices, which are governed by mechanics rules. As classical spring-mass-damping systems they are vibratory systems with elements which assure returning mass to an equilibrium state. The measurement process is based on proof mass displacement from equilibrium state along one, specified axis and convert it to an electrical signal. The method of transformation is various, however the most willingly applied is microsensors capacitance change detection.

There are some publications which consider capacitance deviation influenced by external factors. The essence of the problems of capacitance deviations caused not only during device operation, but also coming from design and fabrication are discussed in [11-14]. It can be found also performance and ASIC considerations for signal processing coming from the device and compensation of potential deviation. Deformation of MEMS structures was described in details in publications [1-7]. Consideration in reference to modal analysis, fluctuations of eigenfrequencies and Q-factors are included there. Additionally, in [8] results of FEM studies on capacitance are considered for electrostatic comb-drive microstructures. Analysis of influence of the edge-effect on the electric and additional impact on

capacitance in inclined-plate capacitor system is analyzed in [9]. In [10] results of analyzing an effects of temperature change on solid structure deformation are discussed, moreover effect on total performance.

II. THEORETICAL BACKGROUND

In MEMS inertial sensors, input quantities like acceleration or rotational velocity generate directly (accelerometers) or indirectly (gyroscopes) inertial force to the system (fig. 1). Nowadays, inertial solid-based microdevices are complex structures with many various geometry shapes and details, however in most cases these shapes consist mainly of well-known simple geometrical figures like beams. Such structures simplify modeling a device and allow to obtain results in mathematical models very close to the results of FEM analysis.

In MEMS structures, groups of anchored and movable electrodes are systems of serially or parallelly combined variable capacitors. The most popular geometry is called comb structure (fig. 1).

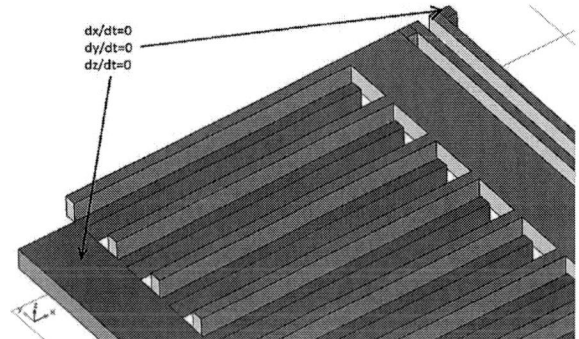

Fig. 1. Comb structure used in MEMS capacitive inertial sensors.

The MEMS comb capacitors consist of numbers of parallel electrodes. Capacitance estimation in a perfect situation (when all electrodes are in parallel) is very easy. These are just parallelly connected single capacitors and calculation of capacitance may be performed with use popular formula:

$$C_{comb} = \sum_{i=1}^{n} C_i = n\varepsilon \frac{A}{d}$$

The above formula assumes, that dimensions of all electrode fingers are the same and distances between stationary and movable electrodes are constant.

However, in case of solid inertial MEMS devices like sensors (where many details are combined with and these details are very susceptible to deformation), any displacement from equilibrium state, anchors, combination with springs potentially causes unnecessary structure stresses and deformations. Moreover, the source of these deformations may come also from other external sources like temperature. Those mentioned above factors influence on capacitance measurements and in turn, accuracy of sensors because of electrodes deflections, displacement and orientation.

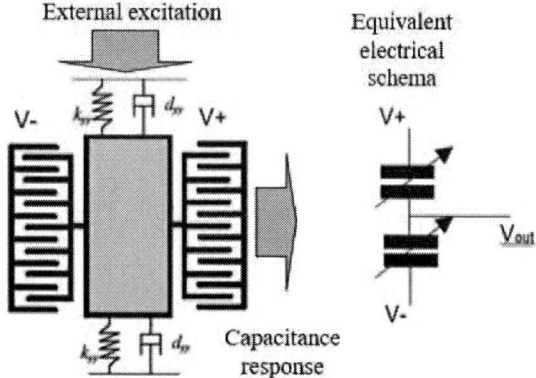

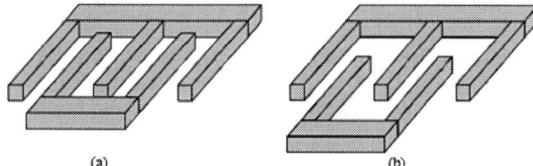

Fig. 2. Comb structure electrodes and equivalent electrical schemas.

Fig. 3. Comb structure electrodes before (a) and after (b) movement.

Fig. 2. shows that external excitation brings about deviation from the equilibrium state. This causes inertial mass motion (combined with electrodes), what naturally changes capacitances located on both sides of this mass (obviously in non-zero voltage presence) – fig. 3. Therefore, the external force causes signal changes in two different capacitances C_1 and C_2. Because we consider inertial sensors, the most famous form of external force is expressed by acceleration or rotational velocity. Inertial mass is in equilibrium state when acceleration or rotational velocity equals to 0. Because the output capacitance comes from comb structure (like shown in fig.1) the difference of C_1 and C_2 (C_1 is capacitance on the left, and C_2 – on the right) - output capacitance - equals to 0 or is negligible small. In such case, both capacitances C_1 and C_2 are equal and can be calculated then using following well-known formula:

$$C_1 = C_2 = C_0 = \frac{\varepsilon_0 \varepsilon A}{d_0} \qquad (1)$$

Force application in the form of acceleration, brings about a displacement of inertial mass from equilibrium state, what in turn, causes deviation of both capacitances. Consequently, capacitance on one side of the mass grows, and on the other side, drops and $C_1 \neq C_2$. Capacitance difference $\Delta C = |C_1 - C_2|$ becomes not equal to zero, moreover - what is important here – becomes proportional to mass displacement from equilibrium state. This proportion dependency is conditioned mostly on motion direction movable part, relative to each other and also on geometry of these movable parts.

Although, capacitance calculation is very simple, when deformations take place in such devices, they must be taken into consideration, because they may incredibly influence on many electrodes. In fig. 4. it is shown (large scale factor), that comb structures considered here, deform differently for different force vector applied. Supporting beam of comb structures adopts the shape of the "U" on the assumption, that the force vector is normal to supporting beam or "S" letter.

Fig. 4. Comb structures deformation for $\alpha \neq 0$ and particular top and bottom capacitances.

Although, seemingly deformation of supporting beam is not meaningful, it is important what happens to finger electrodes.

We see in fig. 5, that particular electrode location and orientation changes and it is in fact caused by supporting beam deformation. We may see, that each finger is oriented with different angle from the original position. Therefore, it is crucial to calculate capacitance separately.

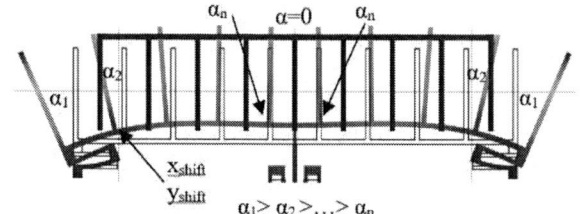

Fig. 5. Different angles of electrodes caused by structure deformation.

In case of "U" shape deformation the smallest (or none) deviation is in half of the length, the highest – at the ends. In case of "S" shape deformation we see that these are double "U" deformation and we have three maximum deviations – at the ends and in half of the length, whereas the smallest (or none) deviation – in 1/4 and ¾ of the length.

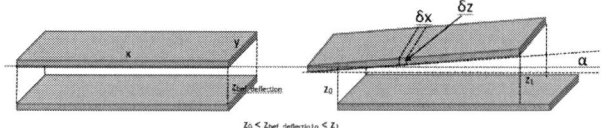

Fig. 6. Capacitor electrode orientation change caused by comb supporting beam deflection.

Obviously, perfect situation is when, all electrodes are oriented parallelly and their geometrical dimensions are constant, even in case of operating environment change. Such situation can take place, when comb structures are not clamped at all – or not directly. Vibratory MEMS devices operation principle is based on vibratory mass, therefore springs limit comb structure motion and causes additional negative capacitance deviations, and consequently causes fluctuations of the output signal – what is commonly present.

152

Let's consider non-parallel electrodes inclined at the α angle to reference angle (in fact 0 degree – angle with no deformation presence). To calculate capacitance, it is necessary to discretize finger electrodes onto very small capacitances like in fig. 6. Each small capacitor should be calculated independently. All these capacitances must be summed as parallel connection. According to the above calculation non-parallel single pair of electrodes may be performed with use following set of formulas:

$$\tan \alpha = \frac{z_1 - z_0}{x_1}$$
$$dC = \varepsilon_0 \frac{dA}{\delta z}$$
$$dA = y_1 dx$$
$$\delta z = z_0 + x \tan \alpha,$$
$$C = \int dC = \varepsilon_0 \int_A \frac{dA}{\delta z} = \varepsilon_0 \int_0^{x_1} \frac{y_1 dx}{z_0 + x \tan \alpha} =$$

$$= \varepsilon_0 \frac{y_1}{\tan \alpha} \ln \left(\frac{z_1}{z_0} \right) \xrightarrow[\tan \alpha = \alpha]{} C = \varepsilon_0 \frac{y_1}{\alpha} \ln \left(1 + \frac{x_1}{z_0} \right) \quad (3)$$

$$C_{total} = \sum_{i=1}^{n} C_i = \sum_{i=1}^{n} \varepsilon_0 \frac{y_i}{\alpha_i} \ln \left(1 + \frac{x_i}{z_{0_i}} \right)$$

where α – angle between plates (fig. 5), y, x – plate dimensions, z_0 – the smallest difference between plates, z_1-the largest differences between two plates.

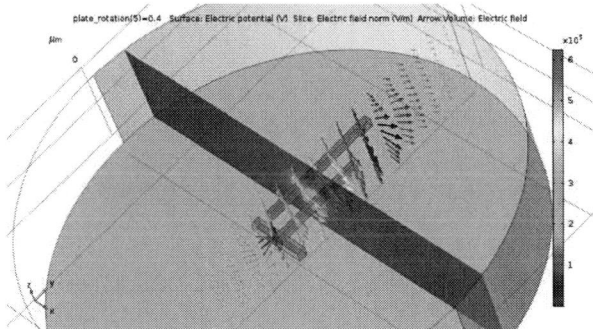

Fig. 7. Electric potential distribution for non-zero angle between electrodes.

Fig. 7. shows electric potential distribution for non-zero angle between electrodes, caused by supporting beam deformation. It shows here, that capacitances come not only from finger electrodes, but also from both stationary and movable supporting beams. It is crucial to take it into consideration, because it influences meaningfully on electric field in comb structures, caused by their direction vectors change. Electric field which is shown here comes from both electrodes, has additional functionality - returning mass to an equilibrium state. In case of electrode orientation change – according to presented here, vectors of electric field also change.

III. SIMULATIONS RESULTS

Both Matlab/SIMULINK and COMSOL model were used for capacitance calculation. (fig. 8). Geometrical dimensions of MEMS structure which were used for simulation performance are presented in table I.

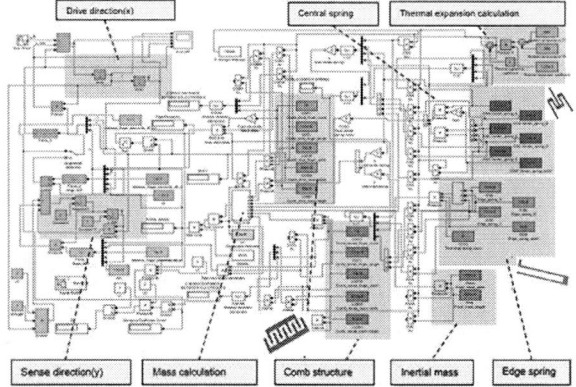

Fig. 8. Model of 2-DOF MEMS device based on 2nd order differential equations used for comb capacitance calculations.

Calculation of particular electrode angles were performed with the use of output Cartesian coordination obtained from simulations. Based on evaluating x and y coordinates displacement of electrodes caused by deformation is also calculated. Generally, a formula which may be used for that id following:

$$\alpha = \arcsin \frac{x_2 - x_1}{\sqrt{(x_2 - x_1)^2 + (y_2 - y_1)^2}}$$

In Table I geometrical details of sensing structure are presented. The material is polysilicon, with Young's modulus equals to 169GPa. Although all parameters in the table are constant, during simulation, modification of electrode length was also applied.

TABLE I.
GEOMETRICAL DETAILS OF SENSING STRUCTURES.

Quantity	Value
Width, length of inertial mass	$100*10^{-6}$m
Length of movable electrode	$200*10^{-6}$m
Length of static electrode	$10*10^{-6}$m
Device thickness	$5*10^{-6}$m
Width of movable electrode	$5*10^{-6}$m
Width of static electrode	$10*10^{-6}$m
d_0	0-15°
α	

Results of simulations for comb device parameter data, show that maximum angle that can be achieved is about 15 degrees and it concerns electrode dependently on location. Above this angle value – 15 degrees – some electrodes may contact with each other what consequently is short circuit.

Particular capacitances C_1 and C_2 in dependency of angle are presented in fig. 9 and 10. The stronger dependency on angle is in case of C_1 and it is strongly nonlinear. For angle close to maximum value capacitance drastically grows. Fig. 9. shows that capacitance C_1 grows exponentially as electrode angle grows from $1*10^{-14}$ to $6.5*10^{-14}$ F – thus, the growth is more than 6 times. Consequently, we can confirm that particular capacitances C_1 and C_2 are dependent on electrode angle and shift in referring to the initial situation. This shift is brought about by deformation change of supporting beam. Therefore, capacitance differences are $0.5-1*10^{-14}$F for angle 0 degrees and $0.5-6 *10^{-14}$F for maximum. Thus, we see, that value at maximum may be 6 times more in the whole range.

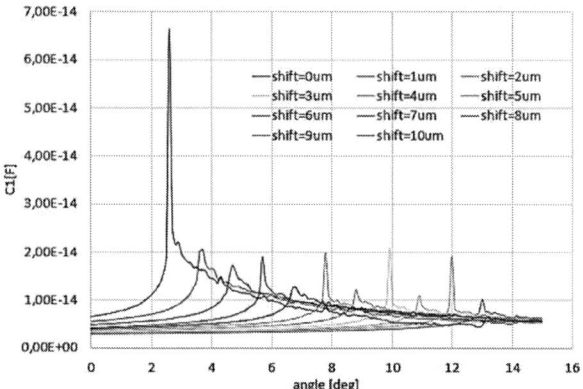

Fig. 9. C_1 capacitance dependency on angle for different electrode shift caused temperature expansion.

According to plots maximum capacitance peaks presented in fig. 9 it is clear that the amount of this maximum depends one side on angle and on the other side on the shift of electrode length. Therefore, there is a huge distribution of peak values for particular configurations. Although maximum capacitance decreases along with angle growth, for some specified cases, it grows. One may see as alternating occurring low and high peaks (for 1 μm maximum capacitance change is 2 times in some cases). The difference between alternating peaks drops along with shift drops. For angle 2.3 degrees capacitance significantly increases to $6.5*10^{-14}$ F.

In fig. 10. It can be seen that C_2 decreases for each shift along with angle growth, however the capacitance difference between particular values decreases along with angle grows. Consequently capacitance difference plot has the same shape like C_1 (fig. 9) but it is shifted along with capacitance axis.

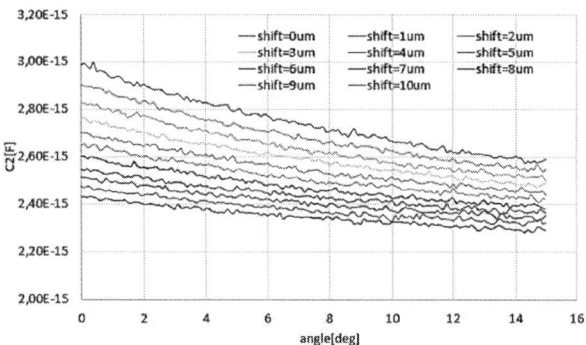

Fig. 10. C_2 capacitance dependency on angle for different electrode shift caused temperature expansion.

In fig. 11 there is capacitance dependency on angle for small angles (between 0 and 1 degree). It can be observed, that along with angle growth, the difference between both C_1 and C_2 increases and both capacitances fluctuate in this range.

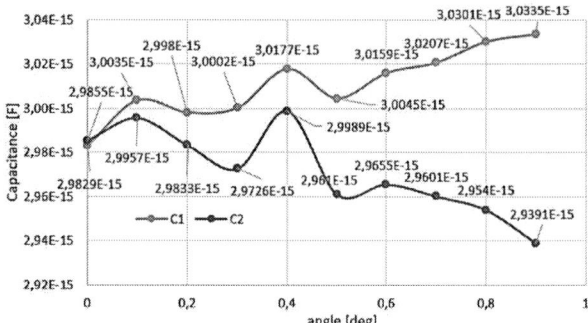

Fig. 11. C_2 capacitance dependency on angle for different electrode shift caused temperature expansion.

Comb structures are affected by some motion limitations. They come from distances between particular electrode as well as geometrical dimensions. Because of these limitations, the deformation may in turn limits capacitance, that device (comb structure) may achieve. Fig. 12 shows maximum capacitance which is allowed to achieve for given electrode length. It can be observed, that capacitance of a single pair of electrode (of comb) grows along with the length. For this given device capacitance slightly grows between $3.96*10^{-15}$F to $1*10^{-14}$F for length range $80-150*10^{-6}$m. Above length of $150*10^{-6}$m – capacitance dramatically grows. Fluctuation of maximum capacitance comes from the fact that during deformation of supporting beam, besides electrode orientation changes, it also moves, what may be seen in fig. 12.

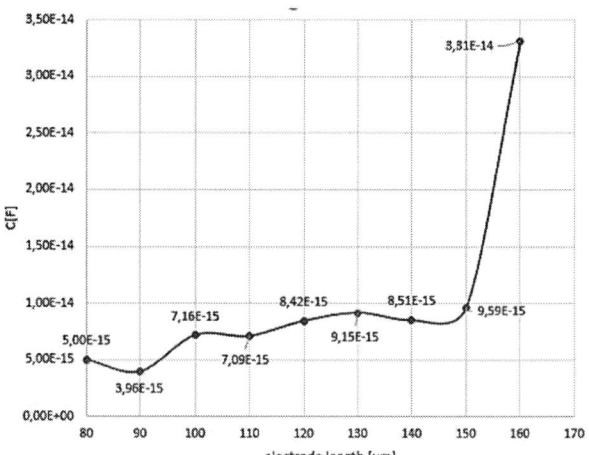

Fig. 12. The maximum capacitance allowable for specified electrode length.

IV. CONCLUSIONS

Comb structure is one of some types of capacitor solutions willingly applied in MEMS inertial sensors. Such complex structures are used because combination many capacitors for one specified functionality (actuating or sensing) multiply the instantaneous value of physical quantity used for mentioning purposes – in case of capacitance - Unfortunately, negative effects are also multiplied. Here, there is a good example of that.

Results of simulations of comb structures presented here show, that deformations of structure greatly influence on output signal and are the source of capacitance deviations. Moreover, it is clearly seen, that many nonlinearities, fluctuations may degrade accuracy of such device, therefore it is incredibly important to design comb structures with appropriate geometrical dimensions, especially considering supporting beam, which appeared to be part with the most influence on to MEMS output signal. This influence manifests through capacitance changes 10^{-15}-10^{-14} order (10 times difference) caused by mechanical deformation.

The analytical results calculated with Matlab/SIMULNK model based on presented formula gave similar results to FEM simulation, however, it is worth to underline that the values were a bit lower.

REFERENCES

[1] P. Chandradip, P. MCCluskey, "Simulation of the MEMS Vibratory Gyroscope through Simulink", Conference: Device Packaging, March 2012, https:// www.researchgate.net/publication/320170530_ Simulation_of_the_MEMS_Vibratory_Gyroscope_through_Simulink.

[2] D. Xia, S. Chen,S. Wang, H. Li, "Microgyroscope Temperature Effects and Compensation-Control Methods", Sensors, 9(10), 2009, pp. 8349-8376.

[3] J. C. Fang, J.L. Li, W. Sheng, "Improved temperature error model of silicon MEMS gyroscope with inside frame driving". J. Beijing Univ. Aeronaut. Astronaut. 32, 2006 pp.1277–1280.

[4] C. Patel, P. McCluskey, D. Lemus, "Performance and reliability of MEMS gyroscopes at high temperatures", 2010 12th IEEE Intersociety Conference on Thermal and Thermomechanical Phenomena in Electronic Systems, 2010, https://ieeexplore.ieee.org/document/ 5501319.

[5] C. Patel, A. Jones, J. Davis, P. McCluskey, D. Lemus, Temperature effects on the performance and reliability of MEMS Gyroscope Sensors, Proceedings of IPACK2009, InterPACK'09, July 19-23, 2009, San Francisco, California, USA, pp. 1-6, 2009, https://www.researchgate.net/publication/267600154_Temperature_Effe cts_on_the_Performance_and_Reliability_of_MEMS_Gyroscope_Senso rs.

[6] P. Lall, A. Abrol, L. Simpson, J. Glover, "Effect of Simultaneous High Temperature and Vibration on MEMS based Vibratory Gyroscope", IEEE ITHERM 2017, https://www.researchgate.net/ publication/317350570_Effect_of_Simultaneous_High_Temperature_an d_Vibration_on_MEMS_based_Vibratory_Gyroscope.

[7] D. Liu, X. Chi, J. Cui, L. Lin, Q. Zhao, Z. Yang, G. Yan, "Research on temperature dependent characteristics and compensation methods for digital gyroscope", 3rd International Conference on Sensing Technology, Nov. 30 – Dec. 3, Tainan, Taiwan, 2008 pp. 273-277.

[8] https://www.comsol.com/paper/download/295221/ hanasi_paper.pdf.

[9] https://pdfs.semanticscholar.org/bc17/b0d17b705716c5 2d1298102b4c721036e208.pdf.

[10] https://www.ncbi.nlm.nih.gov/pmc/articles/ PMC4435161/.

[11] P. Zajac, M. Jankowski, P. Amrozik, M. Szermer: Application of offset trimming circuit for reducing the impact of parasitics in capacitive sensor readout circuit, 26th International Conference "Mixed Design of Integrated Circuits and Systems" (MIXDES). Rzeszów, Poland 2019, pp. 178-181.

[12] M. Jankowski, P. Zajac, A. Napieralski, Fully differential read-out circuitry components for MEMS-based accelerometers, MEMSTECH 2018: XIV-th International Conference Perspective technologies and methods in MEMS design. Polyana, Ukraine 2018, pp. 86-90.

[13] C. Maj, M. Szermer, M. Jankowski, Influence of Fringing Field on Estimating of Comb Drive Accelerometer Performance, 2019 MEMSTECH XV-th International Conference on the Perspective Technologies and Methods in MEMS Design. Polyana, Ukraine 2019, pp. 67-70.

[14] M. Jankowski, M. Szermer, P. Amrozik, 2019 IEEE 15th International Conference on the Experience of Designing and Application of CAD Systems (CADSM), 26 Feb.-2 March 2019 Polyana, Ukraine 2019, pp. 23-26.

Proceedings of the 27th International Conference *"Mixed Design of Integrated Circuits and Systems"*
June 25-27, 2020, Łódź, Poland

Noise Resistance Estimation for a GaN JFET Using Small Signal Measurements for an X-band LNA

Evangelia A. Karagianni[1], Christina C. Lessi[2], Christos N. Vazouras[1], Athanasios D. Panagopoulos[2],
George Deligeorgis[3], George Stavrinidis[3], Athanasios Kostopoulos[3]

[1]Hellenic Naval Academy
Piraeus, Greece

[2]Division of Information Transmission Systems and Material Technology
School of Electrical and Computer Engineering
National Technical University of Athens
Zographou campus, Athens, Greece

[3]Foundation for Research and Technology Hellas
Creta, Greece

{evka, chvazour}@hna.gr, chrislessiee@gmail.com, thpanag@ece.ntua.gr, {deligeo, gstav, kosto}@physics.uoc.gr

Abstract—**Gallium Nitride technology is entering dynamically in the area of manufacturing integrated circuits. In this paper the design of a Low Noise Amplifier is presented. The transistor that is used is a bilateral, conditionally stable transistor and it has been built at the Foundation for Research and Technology Hellas. It is measured in order to get the Scattering parameters and the Noise Figure. The Noise Figure is additionally calculated, together with the noise resistance and the error between the calculated and the measured values is estimated for a single stage amplifier.**

Keywords—**GaN technology; coplanar technology; noise figure; noise resistance; low noise amplifier**

I. INTRODUCTION

Socially critical applications such as wireless and satellite communications, terrestrial and flying radar, wireless (5G) technologies, new generation active phase antenna radars, cable television and "intelligent" transport infrastructure are significantly affecting our daily lives. The cradle of all modern technology applications is nowadays the Internet of Things (IoT). These upcoming applications have upgraded requirements such as higher power, higher operating frequency, lower footprint and volume, enhanced functionality and lower cost [1]. They are based on the use of transceivers (T/R) (Figure 1) that allow the transmission / reception of electromagnetic waves. The central element of T/R is the final for the transmitter (or initial for the receiver) microwave part of the RF front end (or Low Noise Block), consisting, among others, of High Power Amplifiers (HPA) for signal transmission, Low Noise Amplifiers (LNA) to receive the signal, a common T/R antenna, and a switch allowing the transition from the transmitting circuit to the corresponding receiving.

The current industrial technology of HPA and LNA circuits is based on Gallium Arsenic Semiconductor (GaAs) devices that are hybridized to the antenna via a circulator with electromagnetic potential (T/R) isolation.

The predominant technology for replacing this GaAs is that of Gallium Nitride (GaN) due to a multitude of superior properties of the material such as a larger energy gap, higher charge density and high thermal conductivity, which allow its use in very high power density applications and consequently a smaller circuit surface area for self-performance [2]. Also, a higher electron impedance rate allows signal amplification at higher frequencies. GaN circuit devices can meet the increased demands of future applications for higher power (over 40 dBm), lower noise and higher operating frequency.

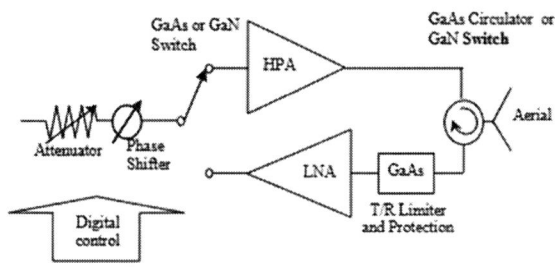

Fig. 1. Unit diagram of a T/R module architecture

II. GAN BASED LOW NOISE AMPLIFIERS

GaN-based LNA design has attracted particular interest because of its high input impedance and low noise resistance compared to GaAs and CMOS technologies [3]. Due to the larger bandwidth of GaN semiconductors, nearly three times larger than Si and GaAs, the technology-based LNAs withstand higher breakdown voltages as well as higher drain voltages. The high reception voltage results in a power density of 5W/mm from a single transistor and is a major design advantage compared to other technologies [4]. Furthermore, the passive circuits using these transistors can withstand higher power and voltage variations than other semiconductor materials [5], [6].

This research has been co-financed by Hellenic Navy, Hellenic Ministry of Defence.

Figure 1 illustrates the conventional architecture of a T/R, where we observe that the protection circuits (Limiter circuits) although they should be installed prior to the LNA circuit, they cannot be integrated into the same substrate as the LNA, increasing thus the cost and size of the system in GaAs technology. In addition, they limit the performance of the T/R by reducing the range and increasing the noise figure as they are positioned before the LNA circuit. By using a GaN transistor in the LNA, these protection circuits can be omitted since the transistors withstand higher input power without the use of an external protection circuit.

Offering comparable gain and noise ratio with other semiconductor technologies and the advantage of high-power input signal resistance, GaN technology is selected in this research for X-band LNA design, with a central operating frequency of 10GHz [7]-[10].

III. OVERVIEW OF THE RECENT TECHNOLOGY

Applications to both consumer goods and security and defense products since the early 2000s have been promoting GaN transistor technology (HEMT), more in the high power field (radar, telecommunications, etc.).

In 2012, the dual heterostructure (DHFET) AlN/GaN/AlGaN millimeter wave performance was presented and the minimum noise level was achieved at 36 GHz with a relative gain of 7.5 dB [11], while in 2013 an LNA with 2.3 dB noise figure at 7 GHz was presented [12]. In 2015, Optimum Existing Technology (SotA) for Noise Technologies for a Single-stage Low Noise Monolithic Amplifier (MMIC LNA) (NF = 3.3dB @ 30GHz) was achieved (as well as for three-stage LNA MMIC Amplifier with NF<3 dB above 14% of the frequency band) [13]. In 2016, it was published for the Ka band (27 GHz - 40 GHz) a MMIC LNA too, on a GaN substrate with NF<2 dB in a frequency range of 28 GHz- 39.2 GHz and with a wide range of useful DC polarized network conditions (Vd=0.6V - 4V, Pdc=5 mW - 310 mW) [14].

IV. GaN TRANSISTOR SELECTION AND MEASUREMENTS

Transistors can be completely characterized by their scattering parameters (S-prameters) that vary with frequency and bias level. With these parameters, it is possible to calculate potential instabilities (tendency to oscillation), maximum available gain, input and output impedances, and transducer gain. It is also possible to calculate optimum source and load impedances either for simultaneous conjugate matching or simply to help choose specific source and load impedances for a specified transducer gain. The source impedance, which results in the minimum (optimum) Noise Figure (NF) is called Zopt and when expressed in terms of reflection coefficients it is Sopt. Regarding noise, transistors can be completely characterized by their S-parameters, four (4) noise parameters and the source impedance. One common parameter is the minimum noise figure Fmin which is achieved at the specific optimum (complex) reflection coefficient (Sopt). Two of the other parameters are magnitude and angle of Sopt , with the fourth parameter being the equivalent noise resistance rn..

A Gallium Nitride (GaN) transistor, which operates on the X-band has been built at the Institute of Technology and Research of Foundation for Research and Technology Hellas

(FORTH) with the use of coplanar technology as shown in Figure 2. This JFET is a key building block of the design and implementation of the LNA, which will form part of a single-phase Active Phase Array System for shipping radar at the X-band. Small signal measurements have been made so far as well as NF measured values for the band 5-15 GHz, as it is shown in Table 1. Measurements have been made using an Anritsu MS4644B Vector Network Analyzer (VNA) with an input power of 0dBm. There are no measurements for the noise resistance nor for Sopt. Also, there are no large signal measurements. In addition strong signal metrics (with input power above 20dBm) are expected to be used for the design and implementation of an HPA at X-band with output power above 40W.

The transistor is biased with VGS (gate-source voltage) at 2.5Volts, VDS (drain-source voltage) at 15Volts and drain-source current at 32mA. Initially, it is assumed that the measured S11 is also the Sopt as the single LNA design for the testing will have the proper matching circuit both for the input and output.

In order to design an LNA, the noise parameters named Sopt, NFmin and rn, have to be known. Taking into account that we have only Sopt and NF, which are measured, it is possible to calculate the rest noise parameters and compare them with the measured ones. Measurements for the noise resistance will be made using the same VNA, with some additional connected components [15].

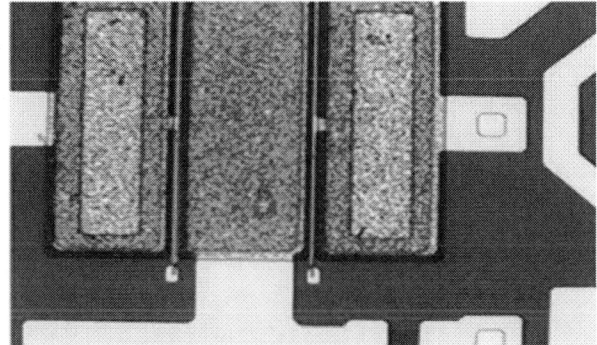

Fig. 2.   The measured JFET

freq	S(1,1)	S(1,2)	S(2,1)	S(2,2)
5.000 Hz	0.989 / -9.701	0.023 / 73.405	2.396 / 173.319	0.831 / -8.685
1.000 GHz	0.989 / -9.701	0.023 / 73.405	2.396 / 173.319	0.831 / -8.685
2.000 GHz	0.983 / -19.966	0.045 / 73.193	2.537 / 164.754	0.798 / -14.826
3.000 GHz	0.969 / -29.209	0.064 / 65.301	2.488 / 156.620	0.779 / -20.491
4.000 GHz	0.950 / -38.447	0.084 / 59.210	2.434 / 148.285	0.762 / -25.838
5.000 GHz	0.931 / -47.280	0.100 / 53.218	2.365 / 140.217	0.743 / -31.240
6.000 GHz	0.910 / -55.745	0.115 / 46.973	2.284 / 132.537	0.724 / -36.521
7.000 GHz	0.887 / -63.721	0.129 / 41.499	2.194 / 125.215	0.705 / -41.322
8.000 GHz	0.865 / -71.225	0.140 / 36.546	2.100 / 118.303	0.685 / -46.053
9.000 GHz	0.843 / -78.264	0.151 / 31.537	2.006 / 111.817	0.668 / -50.558
10.00 GHz	0.824 / -84.882	0.160 / 26.816	1.917 / 105.664	0.650 / -54.743
11.00 GHz	0.804 / -91.283	0.169 / 21.773	1.828 / 99.715	0.632 / -58.790
12.00 GHz	0.785 / -97.245	0.176 / 16.944	1.746 / 94.085	0.617 / -62.683
13.00 GHz	0.766 / -102.715	0.180 / 12.154	1.663 / 88.830	0.601 / -66.032
14.00 GHz	0.748 / -107.754	0.182 / 7.798	1.588 / 83.885	0.586 / -69.306
15.00 GHz	0.734 / -112.334	0.182 / 4.361	1.520 / 79.295	0.577 / -72.192

Fig. 3.   The measured S-parameters of the transistor

V. GaN NOISE PARAMETERS ESTIMATIONS

For the normalized noise resistance rn estimation we use the following formulas [16] where Rn is the noise resistance and F is the Noise Factor which is the measured Noise Figure in pure number. The transistor is assumed to be working at the minimum temperature Tmin=263K. The scattering parameters S21 and S11 have already been measured and Z0 is 50Ω.

$$F = 1 + \frac{a}{k \cdot T_0 \cdot |S_{21}|^2} \qquad (1)$$

$$a = |S_{21}|^2 \cdot \left(k \cdot T_{min} + \beta \cdot |S_{opt}|^2 \right) \qquad (2)$$

$$\beta = \frac{4 \cdot k \cdot T_0 \cdot R_n}{Z_0 \cdot |1 + S_{opt}|^2} \qquad (3)$$

For the NFmin=10log(Fmin) calculation we use the following formula

$$F = F_{min} + \frac{4 \cdot R_n \cdot |\Gamma_s - S_{opt}|^2}{Z_0 \cdot (1 - |\Gamma_s|^2) \cdot |1 + S_{opt}|^2} \qquad (4)$$

where Γ_s is the source reflection coefficient.

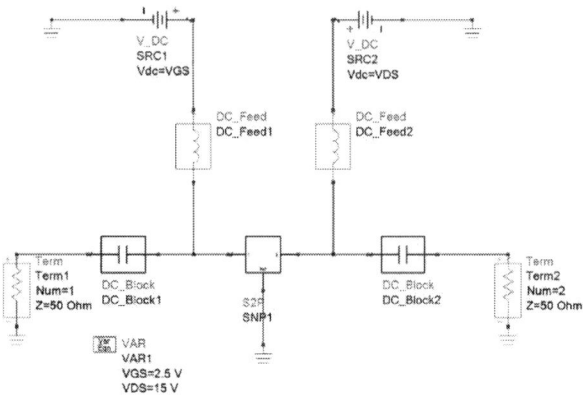

Fig. 4. Simulated circuit with the use of ADS (Advanced Design System)

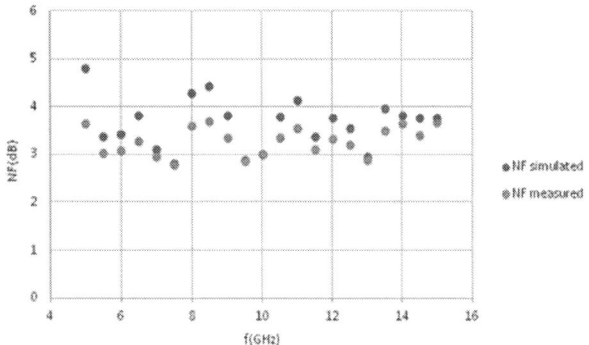

Fig. 5. Comparison between measured and simulated NF

In Figure 4 the design in Advanced Design System (ADS) is presented in order to test the performance of the JFET. Figure 5 gives the comparison between the simulated NF and the measured NF. The error entered varies from 0.15 % (at 9.5GHz) to 30% (at 5GHz). This is encouraging enough to start designing the Low Noise Amplifier without having the noise resistance measured, since the LNA is designed to operate at 10GHz, where the error is small enough to be accepted (0.4%).

VI. LNA DESIGN

For the first stage of the LNA the transistor is polarized with VGS at 2.5Volts, VDS at 15Volts and drain-source current at 32mA. The transistor is conditionally stable - as its shown in Figure 6 - and bilateral. Rollet's stability factors are K=0.436 and Δ=0.607 for the operating frequency f=10GHz. The maximum gain is 10.79 dB at 10GHz

A way to deal with the oscillations is to add a 90Ω resistance in series with the output and the simulated results give K=2.009 and Δ=0.488 which means that the transistor is unconditionally stable but S parameters as well as noise have been affected with a simulated value for the NF 3.694 dB (approximately 20% deviation). The maximum gain is 7.45 dB at 10GHz.

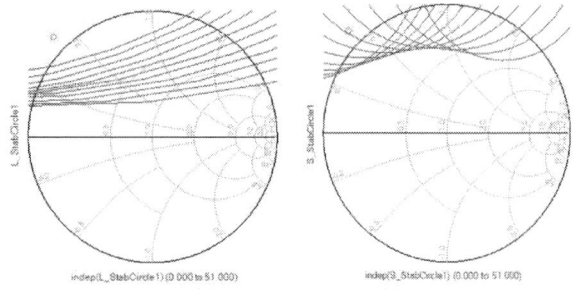

Fig. 6. (a) Load Stability circles and (b) Source stability circles for frequencies 5-15 GHz

Another method to deal with the problem of oscillations is accomplished through a coil at the source, with the aim of eliminating parasitic capacities increase gain [17], [18]. After connecting the proper matching circuits to the input and output of the first stage, the circuit of Figure 7 is obtained, with the results as shown in Figure 8. The LNA is stable at frequencies from 5 to 15GHz. The maximum gain is 6.94 dB at 10GHz and the NF is 3.2dB, almost 6% higher than the minimum of the transistor.

Regarding the design of the inductors, they must consist of as many coils as possible in order to avoid coupling between adjacent coils in the layout. However, the initial design that has taken place has taken into account to some extent coupling between adjacent coils to avoid problems with the amplifier layout.

Then, once the noise parameters for the transistor in the X-band are given by FORTH, the design will continue, adding more stages, so that the final design meets the initial requirements. Table I summarizes both designs.

TABLE I.
SUMMARY OF THE PERFORMANCE OF THE DESIGNS

	NF(dB)	Gmax (dB)	Stability
Transistor	3.002	10.789	No
Resistance in series	3.694	5.046	Yes
Coil in parallel	3.208	6.936	Yes

Με τη συγχρηματοδότηση της Ελλάδας και της Ευρωπαϊκής Ένωσης

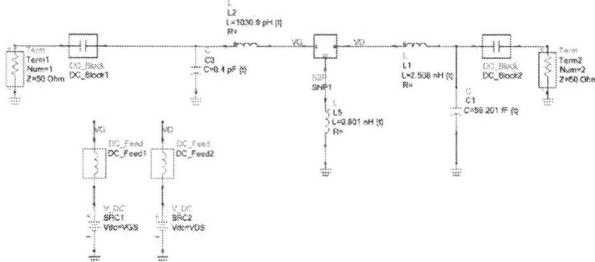

Fig. 7. Simulated circuit with the optimized matching circuits for the input and output, with the use of ADS (Advanced Design System)

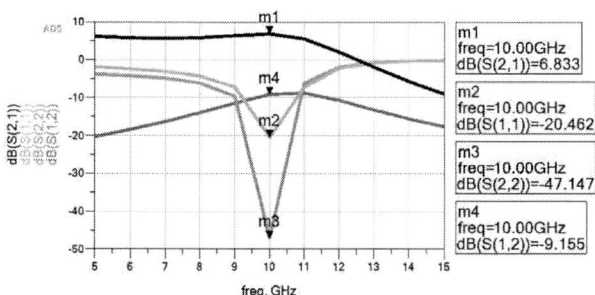

Fig. 8. Simulated results for the LNA's S-paremeters with the use of ADS

VII. CONCLUSION

The design of any kind of LNA demands, except S-parameters values, the noise parameters values for the operating frequency at least. This paper is presenting the method to design an LNA using only measured S-parameters and NF of a JFET conditionally stable and bilateral. The Rn is calculated and the error inserted to the NF is less than 0.4% at the operating frequency. Once the error between the original NF measured and the calculated one is acceptable, we will proceed to the LNA design. This design is ambitious and full of challenges, as it seems to be necessary to use the stability resistor, which worsens the transistor noise characteristics. The alternative method to deal with instabilities is the use of a coil at the source. This design give a NF=3.2 dB and a maximum gain approximately 7 dB. As the design is successful, the amplifier will be manufactured and measured.

ACKNOWLEDGMENT

This research has been co-financed by the European Union and Greek national funds through the Operational Program Competitiveness, Entrepreneurship and Innovation, under the call RESEARCH – CREATE – INNOVATE (project code: T1EDK-00329) and Hellenic Navy, Hellenic Ministry of Defence.

REFERENCES

[1] Sumit Saha, "RF front-end design for X-band using 0.15μm GaN Hemt Technology", thesis presented to Ottawa-Carleton Institute, Canada, 2016.

[2] David Schnaufer and Bror Peterson, "Gallium Nitride – A critical Technology for 5G", Qorvo, Dec. 2016

[3] C. F. Campbell, "Gallium Nitride Power MMICs Promise and Problems", Integrated Nonlinear Microwave and Milimeter-wave Circuits Workshop (INMMiC), November 2015

[4] A. Katz, M. Kubak ve G. DeSalvo, "A 6 to 16 GHz Linearized GaN Power Amplifier," 2006 IEEE MTT-S International Microwave Symposium Digest, San Francisco, California, 2006, p. 1364-1367

[5] C. F. Campbell, D. C. Dumka, "Wideband high power GaN on SiC SPDT switch MMICs," IEEE MTT-S Int. Microwave Symp. Dig., Anaheim, California, U.S.A, p. 145- 148, May 2010.

[6] O. Memioglu et al., "Design and Implementation of an Encapsulated GaN X-Band Power Amplifier Family," 2018 13th European Microwave Integrated Circuits Conference (EuMIC), Madrid, 2018, pp. 89-92

[7] O. Kazan et al., "An X-Band Robust GaN Low Noise Amplifier MMIC with Sub 2 dB Noise Figure," 2018 13th European Microwave Integrated Circuits Conference (EuMIC), Madrid, 2018, pp. 234-236.

[8] . Memioglu et al., "A High Power GaN, Quarter Wave Length Switch for X-Band Applications," 2018 IEEE 18th Mediterranean Microwave Symposium (MMS), Istanbul, 2018, pp. 195-197.

[9] P. Parikh, Y. Wu, M. Moore, P. Chavarkar, Mishra, R. Neidhard, L. Kehias, T. Jenkins, "High linearity, robust, AlGaN-GaN HEMTs for LNA and receiver ICs," Proceedings of IEEE Lester Eastman Conference on High Performance Devices, p. 415-421, August 2002

[10] U. Mishra, S. Likun, T. Kazior, and Y.-F. Wu, "GaNbased RF power devices and amplifiers," Proc. IEEE, vol. 96, no. 2, s. 287-305, February. 2008

[11] F. Medjdoub, et.al., "Sub-1-dB Minimum-Noise-Figure Performance of GaN-on-Si Transistors Up to 40 GHz", IEEE Electron Device Letters, VOL. 33, NO. 9, Sept. 2012

[12] Sergio Colangeli,, et.al., "GaN-Based Robust Low-Noise Amplifiers", IEEE Transactions on Electron Devices, VOL. 60, NO. 10, Oct. 2013

[13] S. D. Nsele, et.al., "Ka-band low noise amplifiers based on InAlN/GaN technologies", 2015 International Conference on Noise and Fluctuations

[14] M. Micovic, et.al., "Ka-Band LNA MMIC's Realized in Fmax > 580 GHz GaN HEMT Technology", Compound Semiconductor Integrated Circuit Symposium (CSICS), 2016 IEEE

[15] D. F. Wait, J. Randa, "Amplifier Noise Measurements at NIST", IEEE Transactions on Instrumentation and Measurement, vol. 46, No. 2, April 1997

[16] M. E. Mokari, W. Patience, "A New Method of Noise Parameter Calculation Using Direct Matrix Analysis", IEEE Transactions on Circuits and Systems-I: Fundamental Theory and Applications, Vol. 39, No. 9, September 1992

[17] T. Yao et al., "Algorithmic Design of CMOS LNAs and PAs for 60-GHz Radio," IEEE J. Solid-State Circuits, vol. 42, no. 5, pp. 1044–1057, May 2007.

[18] D. Pepe and D. Zito, "32 dB Gain 28 nm Bulk CMOS W-Band LNA," IEEE Microw. Wirel. Compon. Lett., vol. 25, no. 1, pp. 55–57, Jan. 2015.

On Applications of Fractional Derivatives in Circuit Theory

Jacek Gulgowski
University of Gdansk
80-308 Gdansk, Poland
Email: jacek.gulgowski@mat.ug.edu.pl

Tomasz P. Stefański
Gdansk University of Technology
80-233 Gdansk, Poland
Email: tomasz.stefanski@pg.edu.pl

Damian Trofimowicz
SpaceForest Ltd.
81-451 Gdynia, Poland
Email: d.trofimowicz@gmail.com

Abstract—In this paper, concepts of fractional-order (FO) derivatives are discussed from the point of view of applications in the circuit theory. The properties of FO derivatives required for the circuit-level modelling are formulated. Potential problems related to the generalization of transmission line equations with the use of FO derivatives are presented. It is demonstrated that some of formulations of the FO derivatives have limited applicability in the circuit theory. That is, the Riemann-Liouville and Caputo derivatives with finite base point have a limited applicability whereas the Grünwald-Letnikov and Marchaud derivatives lead to reasonable results of the circuit-level modelling.

Keywords—Circuit theory, circuit simulation, transmission lines, fractional calculus.

I. INTRODUCTION

The fractional-order (FO) calculus is a branch of mathematics investigating formulations of the derivative operator D^α with the order α being a real number ($\alpha \in \mathbb{R}$). Hence, the FO derivative operator D^α is a generalization of the standard integer-order (IO) concept of the n-fold differentiation D^n where n is an integer number ($n \in \mathbb{Z}$). The FO calculus has been applied in the circuit theory for many years [1]–[4]. Reviews of numerous formulations of FO derivatives can be found in classical monographs [5]–[7].

Some definitions of FO derivatives are well established and already applied in the circuit theory, whilst some definitions have been introduced recently. Hence, we discuss the applicability of four important derivative definitions, i.e., Riemann-Liouville, Caputo, Grünwald-Letnikov and Marchaud, from the circuit theory point of view. The ambiguity of definitions of the FO derivative, whose properties sometimes exclude them from applications in the circuit theory, is the motivation for our research. Recently, opinions appear that questionize applicability of FO derivatives and models in electrical sciences and engineering [8]–[10]. Such a discussion in literature suggests that the proposed analysis of properties of FO derivatives is currently necessary.

II. FO DERIVATIVE PROPERTIES NECESSARY FOR APPLICATIONS IN CIRCUIT THEORY

Already, several attempts have been made to specify the conditions that constitute the FO derivative [11]–[13]. Furthermore, these properties have also been discussed from the

point of view of possible applications in the electromagnetism [8], [14]. In this section, we would like to formulate the set of properties which allow for applications of the FO derivative concept in the circuit theory. It is assumed that considered functions are smooth enough for the FO derivative operator D^α ($\alpha > 0$) to be applied. Moreover, we should assume that considered functions are smooth enough that the sequence of derivatives D_x and D_t^α may be exchanged, i.e., $D_x D_t^\alpha = D_t^\alpha D_x$ (practically we may require that $v(x,t)$ and $i(x,t)$ are class C^2 functions). Hence, the FO derivative employed in the circuit theory should satisfy the following properties:

1) *Identity*
$$D^0 f(t) = f(t). \tag{1}$$

2) *Compatibility with IO Derivative*
$$D^\alpha f(t) = \frac{d^\alpha}{dt^\alpha} f(t), \qquad \alpha \in \mathbb{N}. \tag{2}$$

3) *Compatibility with IO Integral*
$$D^{-\alpha} f(t) = \int \cdots \int f(t) d^\alpha t, \qquad \alpha \in \mathbb{N}. \tag{3}$$

4) *Linearity*
$$D^\alpha (af(t) + bg(t)) = aD^\alpha f(t) + bD^\alpha g(t). \tag{4}$$

5) *Semigroup Property (also called the index law)*
$$D^\alpha D^\beta f(t) = D^\beta D^\alpha f(t) = D^{\alpha+\beta} f(t), \quad \alpha, \beta \in \mathbb{R}. \tag{5}$$

This property is sometimes validated under additional assumptions that, e.g., either $\alpha, \beta < 0$ or $0 < \alpha, \beta, \alpha + \beta < 1$. However, as it is noticed in [9], the condition (5) may not be satisfied for widely applied definitions of FO derivative.

6) *Trigonometric Functions Invariance*
$$D^\alpha e^{j\omega t} = (j\omega)^\alpha e^{j\omega t} \tag{6}$$

where $j = \sqrt{-1}$. This property is a generalization of the fundamental formula taken from the IO calculus. It is required for the representation of signals in the phasor analysis of circuits.

7) *Constant Function Derivative*

$$D^\alpha C = 0 \qquad (7)$$

where $C = Const$. This property results from the trigonometric functions invariance because one obtains (7) from (6) for $\omega = 0$.

Let us consider the function $f(t)$ defined in the entire real line for which the bilateral Laplace transform can be defined

$$\hat{f} = \mathcal{L}\{f(t)\} = \int_{-\infty}^{+\infty} f(t)e^{-st}dt. \qquad (8)$$

It is identical to the unilateral Laplace transform given by

$$\hat{f} = \mathcal{L}\{f(t)\} = \int_{0}^{+\infty} f(t)e^{-st}dt \qquad (9)$$

for causal functions (i.e., $f(t) = 0$ for $t \in (-\infty, 0)$). Let us assume that $\hat{f}$ is a solution of the circuit analysis obtained with the use of the Laplace transform. Then, substituting $s = j\omega$ into $\hat{f}$ should give the frequency-domain solution.

The circuit analysis can be executed in either time or frequency domain. The solutions obtained in both domains should be equivalent which means that results of time- and frequency-domain circuit analyses should be related by the Fourier transform. The properties (1)–(7) are required when using the classical methods of the circuit theory. For instance, in the next section, we demonstrate inconsistencies which can arise when the semigroup property is not valid for the FO derivative. Then, we review popular definitions of fractional derivatives from this perspective.

III. PROBLEMS RELATED TO MODELLING OF FO TRANSMISSION LINES

The IO model is able to describe characteristics of the transmission line in a limited frequency range. It stems from the fact that for THz frequencies, a conductor exhibits both frequency and spatial dispersion [15], [16]. Hence, the loss term in the traditional RLGC model with IO elements (see Fig. 1) becomes insufficient to describe the dispersion and non-quasi-static effects. It may further result in causality problems in the time-domain analysis. However, the FO RLGC model of the transmission line (see Fig. 2) can describe these effects also in the THz frequency range. In [15], [16], a causal and compact FO transmission line model for THz frequencies is developed for CMOS on-chip conductor. Due to inclusion of frequency-dependent dispersion loss and non-quasi-static

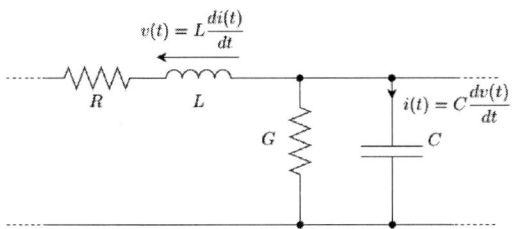

Fig. 1. Unit-cell equivalent circuit of IO model of transmission line.

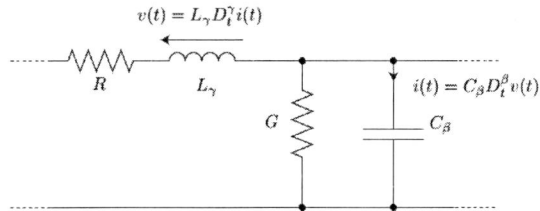

Fig. 2. Unit-cell equivalent circuit of FO model of transmission line.

effects, a good agreement of the characteristic impedance is observed between the FO model and measurements up to 110 GHz. On the other hand, the traditional IO RLGC model provides agreement between the model and measurements only up to 10 GHz. These results clearly demonstrate advantages of the FO transmission line modelling. However, as demonstrated below, the circuit-level modelling of FO transmission lines may be problematic if the definition of the FO derivative does not satisfy the semigroup property (5).

Let us consider the transmission line as in Fig. 2. Then, the voltage $v = v(x, t)$ and the current $i = i(x, t)$ along the line in the x direction are solutions to the following set of equations:

$$\frac{\partial v}{\partial x} = -Ri - L_\gamma D_t^\gamma i \qquad (10)$$

$$\frac{\partial i}{\partial x} = -Gv - C_\beta D_t^\beta v. \qquad (11)$$

Differentiating both equations with respect to x and substituting (10) into (11) and vice versa (with the use of $D_x D_t^\alpha = D_t^\alpha D_x$ property), the FO telegraph equations [17] are obtained as follows:

$$\frac{\partial^2 v}{\partial x^2} = RGv + RC_\beta D_t^\beta v + L_\gamma G D_t^\gamma v + L_\gamma C_\beta D_t^\gamma D_t^\beta v \qquad (12)$$

$$\frac{\partial^2 i}{\partial x^2} = RGi + RC_\beta D_t^\beta i + L_\gamma G D_t^\gamma i + L_\gamma C_\beta D_t^\beta D_t^\gamma i. \qquad (13)$$

One can notice that the voltage and the current satisfy the same telegraph equation if and only if $\beta = \gamma$ or the FO operators commute (i.e., $D_t^\gamma D_t^\beta = D_t^\beta D_t^\gamma$). For $\beta = \gamma = 1$, one obtains the telegraph equation [18] for the traditional RLGC model of the transmission line with IO elements.

Let us consider the phasor representation of the voltage and the current along the line

$$v = \Re(Ve^{j\omega t}) \qquad (14)$$

$$i = \Re(Ie^{j\omega t}) \qquad (15)$$

where $V = V(x)$ and $I = I(x)$ are respectively voltage and current phasors that are functions of the spatial variable x only and ω denotes the angular frequency. From (10)–(11), one obtains the following equations describing the FO transmission line in the frequency domain:

$$\frac{\partial V}{\partial x} = -[R + (j\omega)^\gamma L_\gamma]I \qquad (16)$$

$$\frac{\partial I}{\partial x} = -[G + (j\omega)^\beta C_\beta]V. \qquad (17)$$

161

Substituting (16) into (17) and vice versa, the telegraph equations are obtained in the frequency domain as follows:

$$\frac{\partial^2 V}{\partial^2 x} = [R + (j\omega)^\gamma L_\gamma][G + (j\omega)^\beta C_\beta]V \qquad (18)$$

$$\frac{\partial^2 I}{\partial^2 x} = [R + (j\omega)^\gamma L_\gamma][G + (j\omega)^\beta C_\beta]I. \qquad (19)$$

It occurs that the voltage and current phasors satisfy the same equation in the frequency domain. However, the problem considered in the time domain results in different equations (12)–(13) for the voltage and the current along the line. For definitions of the fractional derivative that satisfy both (5) and (6), the time-domain FO telegraph equations (12)–(13) are consistent with the frequency-domain formulation (18)–(19). Hence, the semigroup property (5) is required for the circuit analysis of the FO transmission lines. From this point of view, we consider popular definitions of FO derivatives in the next section.

IV. FO DERIVATIVES

Let us assume that considered functions, i.e., $f \colon \mathbb{R} \to \mathbb{R}$, are defined on the real line. Presented definitions of FO derivatives have *left-* and *right-sided* versions. In our investigations, only the *left-sided* version is considered. These approaches are symmetrical and the *left-sided* version operates on past times, hence, it is closer to the concept of causality. It is worth noticing that both Riemann-Liouville and Caputo derivatives are defined with the use of the *base point*. The classical approach assumes that the base point $a \in \mathbb{R}$ is a finite number, but the definitions may also be extended to $a = +\infty$. Usually, due to the causality assumption, $a = 0$ is selected and the domain of the function f can be restricted to the interval $[0, +\infty)$.

A. Riemann-Liouville

The Riemann-Liouville integral is defined as

$$J^\alpha f(t) = \frac{1}{\Gamma(\alpha)} \int_a^t f(\tau)(t-\tau)^{\alpha-1} d\tau \qquad (20)$$

where $\alpha > 0$ is an order of integration, Γ is the Gamma function and a is the fixed base point. Hence, the Riemann-Liouville derivative of order $\alpha \in (n-1, n)$ is defined as

$$D^\alpha f(t) = D^n J^{n-\alpha} f(t). \qquad (21)$$

The Riemann-Liouville derivative does not satisfy neither (5) nor (6). The semigroup property for the Riemann-Liouville intergral is discussed in [11, Section 3.2]. Some counterexamples (actually related to fractional differential equations) are presented in [19] (see Example 6.2 and Remark 5). For instance, the base point $a = 0$ leads to the derivatives of the sine and cosine functions being nonelementary functions (refer to e.g. [20, Propositions 11, 12])

$$D^\alpha \sin(\omega t) = \frac{t^{-\alpha}}{2} \left(E_{1,1-\alpha}(j\omega t) + E_{1,1-\alpha}(-j\omega t) \right) \qquad (22)$$

$$D^\alpha \cos(\omega t) = \frac{t^{-\alpha}}{2} \left(E_{1,1-\alpha}(j\omega t) - E_{1,1-\alpha}(-j\omega t) \right) \qquad (23)$$

where $E_{\alpha,\beta}(z)$ is the generalized Mittag-Leffler function. However, the Riemann-Liouville integral of neither sine nor cosine exists when $a = -\infty$ due to the divergent integral in the unbounded domain for $\alpha > 0$.

The Laplace transform of the FO derivative of the order α ($\alpha \in [n-1, n)$ where $n \in \mathbb{N}$) for the function $f \colon [0, +\infty) \to \mathbb{R}$ can be calculated as [7, Formula (1.85)]

$$\mathcal{L}\{D^\alpha f(t)\} = s^\alpha \mathcal{L}\{f(t)\} - \sum_{k=0}^{n-1} s^k [D^{\alpha-k-1} f(t)]_{t=0}. \qquad (24)$$

B. Caputo

The Caputo derivative is defined similarly to the Rimenann-Liouville derivative (21) but with the opposite order of FO integration and IO derivative operators

$$D^\alpha f(t) = J^{n-\alpha} D^n f(t) \qquad (25)$$

where $\alpha \in (n-1, n)$. It is also possible to define it based on the Riemann-Liovuille integral (see [7, Formula (1.12)] or [6, Formula (2.4.1)])

$${}_C D^\alpha f(t) = {}_{RL} D^\alpha \left[f(t) - \sum_{k=0}^{n-1} \frac{t^k}{k!} f^{(k)}(0) \right] \qquad (26)$$

where ${}_C D^\alpha$ is the Caputo derivative and ${}_{RL} D^\alpha$ is the Riemann-Liouville derivative with the same base point $a = 0$. These two definitions agree for a function f of the class C^n, although the definition (26) formally requires only the existence of $(n-1)$-th derivative in the neighbourhood of the base point $a = 0$. The formula (26) applied to the function $f(t) = \sin t$ for $\alpha \in (0,1)$ shows that (22) is also valid for the Caputo derivative with the base point $a = 0$. Therefore, the Caputo derivative does not satisfy (6) – for exact formulas see [20, Propositions 11, 12]. Moreover, the semigroup property (5) is not generally valid for this definition, refer to [19] and especially to Example 6.1 therein.

The Laplace transform of the FO derivative of the order $\alpha \in (n-1, n]$ (where $n \in \mathbb{N}$) of the function $f \colon [0, +\infty) \to \mathbb{R}$ can be calculated as [7, Formula (1.88)]

$$\mathcal{L}\{D^\alpha f(t)\} = s^\alpha \mathcal{L}\{f(t)\} - \sum_{k=0}^{n-1} s^{\alpha-k-1} f^{(k)}(0). \qquad (27)$$

C. Grünwald-Letnikov

The Grünwald-Letnikov derivative of the order $\alpha > 0$ is given by the discrete formula (refer to [5, Formula (20.7)])

$$D^\alpha f(x) = \lim_{h \to 0^+} \frac{1}{h^\alpha} \sum_{m=0}^\infty (-1)^m \binom{\alpha}{m} f(x - mh) \qquad (28)$$

where $\binom{\alpha}{m} = \frac{\alpha(\alpha-1)\dots(\alpha-m+1)}{m!}$. This definition satisfies both (5) (refer to [21, Section 2.6.1]) and (6) (refer to [21, Formula (2.65)]).

The bilateral Laplace transform of the derivative of the order $\alpha \in (n-1, n)$ (where $n \in \mathbb{N}$) of the function $f \colon \mathbb{R} \to \mathbb{R}$ can be calculated as [21, Sections 2.7.3 and 2.8]

$$\mathcal{L}\{D^\alpha f(t)\} = s^\alpha \mathcal{L}\{f(t)\}, \qquad \Re s > 0. \qquad (29)$$

D. Marchaud

The definition of the Marchaud derivative is similar to the Riemann-Liouville derivative with a base point $a = -\infty$. For a broad class of functions, these two definitions are equivalent, i.e., for a class of sufficiently smooth functions with an appropriate behaviour at $-\infty$ as discussed in [5, Section 5.4]. There are also very important differences between both definitions. The first one is that the Marchaud derivative can be calculated for a broader class of functions than the Riemann-Liouville derivative, including the sine and cosine functions. Furthermore, the Marchaud defintion is equivalent to the Grünwald-Letnikov definition for a broad class of functions, covering periodic functions and $L^p(\mathbb{R})$ functions for $p \in [1, +\infty)$, refer to [5, Theorems 20.2 and 20.4]. The recent survey paper [22] discusses both approaches in details.

In general, the Marchaud derivative is defined for $\alpha > 0$ (refer to [5, Section 5.5], [7, Section 1.3.1], and [23] for a historical perspective). When $\alpha \in (n-1, n)$ and $n \in \mathbb{N}$, then

$$D^\alpha f(t) = \frac{\alpha - n + 1}{\Gamma(n-\alpha)} \int_0^{+\infty} \frac{f^{(n-1)}(t) - f^{(n-1)}(t-\tau)}{\tau^{2+\alpha-n}} d\tau \tag{30}$$

where f is assumed to be smooth enough, e.g., $f \in C^{n-1}(\mathbb{R})$ with $f^{(n-1)}$ bounded. Because it is equivalent to the Grünwald-Letnikov derivative, it satisfies both (5) and (6). Anyway, (6) can be directly verified from the Formula 5 in Table 9.2 in [5]. The formulas

$$I_+^{(1-\alpha)}(\sin \omega t) = \omega^{\alpha-1} \sin(\omega t - (1-\alpha)\frac{\pi}{2}) \tag{31}$$

$$I_+^{(1-\alpha)}(\cos \omega t) = \omega^{\alpha-1} \cos(\omega t - (1-\alpha)\frac{\pi}{2}) \tag{32}$$

are satisfied for $\omega > 0$ and $\alpha \in (0,1)$. Then, (31)–(32) should be differentiated to get the fractional Marchaud derivative of the order α, i.e.

$$D^\alpha(\cos \omega t + j \sin \omega t) = \tag{33}$$

$$\omega^\alpha(-\sin(\omega t - (1-\alpha)\frac{\pi}{2}) + j \cos(\omega t - (1-\alpha)\frac{\pi}{2}) =$$

$$\omega^\alpha(-\sin(\omega t + \alpha\frac{\pi}{2} - \frac{\pi}{2}) + j \cos(\omega t + \alpha\frac{\pi}{2} - \frac{\pi}{2}) =$$

$$\omega^\alpha e^{j\omega t} e^{j\alpha\frac{\pi}{2}} = (j\omega)^\alpha e^{j\omega t}$$

for $\alpha \in (0,1)$ and $\omega > 0$. Similar derivations can be obtained for $\omega < 0$. For $\alpha > 1$, it is sufficient to refer to the semigroup property (5) with $\alpha = n - 1 + \{\alpha\}$ where $\{\alpha\} \in (0,1)$.

V. Conclusion

The time- and frequency-domain methods of circuit analysis should return equivalent results. It is demonstrated that in order to obtain the equivalence between results in the time and frequency domains, the FO derivative operator should satisfy the semigroup condition and be representable in the phasor domain. Out of four of the most popular approaches considered in this paper, only two of them are looking at the entire time-history of an input function and are appropriate choices for the circuit theory. The Riemann-Liouville and Caputo derivatives with finite base point have a limited applicability, whereas the Grünwald-Letnikov and Marchaud definitions (which are actually equivalent) lead to reasonable results.

References

[1] M. D. Ortigueira, "An introduction to the fractional continuous-time linear systems: the 21st century systems," *IEEE Circuits Syst. Mag.*, vol. 8, no. 3, pp. 19–26, 2008.

[2] A. S. Elwakil, "Fractional-order circuits and systems: An emerging interdisciplinary research area," *IEEE Circuits Syst. Mag.*, vol. 10, no. 4, pp. 40–50, 2010.

[3] T. J. Freeborn, "A survey of fractional-order circuit models for biology and biomedicine," *IEEE Trans. Emerg. Sel. Topics Circuits Syst.*, vol. 3, no. 3, pp. 416–424, 2013.

[4] T. P. Stefanski and J. Gulgowski, "Electromagnetic-based derivation of fractional-order circuit theory," *Comm. Nonlinear Sci. Numer. Simulat.*, vol. 79, p. 104897, 2019.

[5] S. G. Samko, A. A. Kilbas, and O. I. Marichev, *Fractional Integrals and Derivatives: Theory and Applications.* Gordon and Breach, New York, 1993.

[6] A. A. Kilbas, H. M. Srivastava, and J. J. Trujillo, *Theory and Applications of Fractional Differential Equations.* Elsevier Science, 2006.

[7] C. Li and F. Zeng, *Numerical Methods for Fractional Calculus.* Chapman and Hall/CRC, 2015.

[8] R. Sikora and S. Pawłowski. "Fractional derivatives and the laws of electrical engineering," *COMPEL*, vol. 37, no. 4, pp. 1384–1391, 2018.

[9] R. Sikora and S. Pawłowski, "On certain aspects of application of fractional derivatives in the electromagnetism," *Przeglad Elektrotechniczny*, vol. 94, no. 1, pp. 101–104, 2018.

[10] K. J. Latawiec, R. Stanisławski, M. Łukaniszyn, W. Czuczwara, and M. Rydel, "Fractional-order modeling of electric circuits: modern empiricism vs. classical science," in *2017 Progress in Applied Electrical Engineering (PAEE)*, June 2017, pp. 1–4.

[11] M. D. Ortigueira and J. Tenreiro Machado, "What is a fractional derivative?" *J. Comput. Phys.*, vol. 293, no. C, pp. 4–13, 2015.

[12] G. S. Teodoro, J. T. Machado, and E. C. de Oliveira, "A review of definitions of fractional derivatives and other operators." *J. Comput. Phys.*, vol. 388, pp. 195–208, 2019.

[13] E. C. de Oliveira and J. A. T. Machado, "A review of definitions for fractional derivatives and integral," *Math. Probl. Eng.*, vol. 2014, 2014.

[14] J. Gulgowski and T. P. Stefanski, "On applications of fractional derivatives in electromagnetic theory," *submitted for publication*, 2020.

[15] Y. Shang, W. Fei, and H. Yu, "A fractional-order rlgc model for terahertz transmission line," in *2013 IEEE MTT-S International Microwave Symposium Digest (MTT)*, June 2013, pp. 1–3.

[16] Y. Shang, H. Yu, and W. Fei, "Design and analysis of cmos-based terahertz integrated circuits by causal fractional-order rlgc transmission line model," *IEEE Trans. Emerg. Sel. Topics Circuits Syst.*, vol. 3, no. 3, pp. 355–366, 2013.

[17] S. M. Cvetićanin, D. Zorica, and M. R. Rapaić, "Generalized time-fractional telegrapher's equation in transmission line modeling." *Nonlinear Dyn.*, vol. 88, no. 2, pp. 1453–1472, 2017.

[18] E. W. Weisstein. (2020) "telegraph equation." from mathworld–a wolfram web resource. [Online]. Available: http://mathworld.wolfram.com/TelegraphEquation.html

[19] S. Bhalekar and M. Patil. "Can we split fractional derivative while analyzing fractional differential equations?" *Comm. Nonlinear Sci. Numer. Simulat.*, vol. 76, pp. 12–24, 2019.

[20] R. Garrappa, E. Kaslik, and M. Popolizio, "Evaluation of fractional integrals and derivatives of elementary functions: Overview and tutorial," *Mathematics*, vol. 7, no. 5, 2019. [Online]. Available: https://www.mdpi.com/2227-7390/7/5/407

[21] M. D. Ortigueira, *Fractional Calculus for Scientists and Engineers.* Berlin, Heidelberg: Lecture Notes in Electrical Engineering, Springer, 2011.

[22] S. Rogosin and M. Dubatovskaya, "Letnikov vs. Marchaud: A survey on two prominent constructions of fractional derivatives," *Mathematics*, vol. 6, no. 1, 2018. [Online]. Available: https://www.mdpi.com/2227-7390/6/1/3

[23] F. Ferrari, "Weyl and Marchaud derivatives: A forgotten history," *Mathematics*, vol. 6, no. 6, 2018. [Online]. Available: https://www.mdpi.com/2227-7390/6/1/6

Simulation of Signal Propagation Along Fractional-Order Transmission Lines

Tomasz P. Stefański
Gdansk University of Technology
80-233 Gdansk, Poland
Email: tomasz.stefanski@pg.edu.pl

Damian Trofimowicz
SpaceForest Ltd.
81-451 Gdynia, Poland
Email: d.trofimowicz@gmail.com

Jacek Gulgowski
University of Gdansk
80-308 Gdansk, Poland
Email: jacek.gulgowski@mat.ug.edu.pl

Abstract—In this paper, the simulation method of signal propagation along fractional-order (FO) transmission lines is presented. Initially, fractional calculus and the model of FO transmission line are introduced. Then, the algorithm allowing for simulation of the nonmonochromatic wave propagation along FO transmission lines is presented. It employs computations in the frequency domain, i.e., an analytical excitation is transformed to the frequency domain, multiplications with phase factors are executed, and finally the result is transformed back to the time domain. This algorithm involves elementary functions only and the fast Fourier transformation, hence, computations are numerically efficient and accurate. However, applicability of the method is limited by the sampling theorem. Numerical results are presented allowing for the evaluation of the method.

Keywords—Circuit simulation, transmission lines, fractional calculus, signal processing.

I. Introduction

The fractional-order (FO) modelling of transmission lines demonstrates advantages over classical approach for THz frequencies. The classical integer-order (IO) RLGC model allows for accurate modelling of transmission-line characteristics in a limited frequency range, because a conductor exhibits both frequency and spatial dispersion for high frequencies [1], [2]. Therefore, the dispersion and non-quasi-static effects cannot be accurately modelled by the classical model of the transmission line with IO elements. As a result, causality problems can appear in time-domain simulations employing the classical model. Fortunately, the FO RLGC model of the transmission line can describe these effects also in the THz frequency range and has already been successfully applied in engineering. In [1], [2], a causal and compact FO transmission-line model for THz frequencies is developed for CMOS on-chip conductor. For this model, a good agreement of the characteristic impedance is observed with measurements up to 110 GHz. However, the traditional IO model agrees with measurements only up to 10 GHz. These results clearly demonstrate advantages of the FO transmission-line modelling based on the time-fractional telegraph equations.

In mathematical literature, this type of equations has already been analysed. In [3]–[6], the time-fractional telegraph equation without the term linear in unknown function is considered with the use of both analytical and numerical methods on either bounded and unbounded domains. In [7], [8], the forcing term is introduced as well. The time-fractional telegraph equa-

tion with orders α and β of fractional differentiation ranging from zero to two is considered in [9], [10]. Furthermore, several numerical methods of simulation of the FO transmission lines have been recently proposed [11]–[16]. These methods are based on numerical computations of the inverse Laplace transformation. Significant analytical preprocessing on the complex s-plane is therefore required before these methods can be applied. Hence, we decided to propose the fast and efficient algorithm of the FO transmission-line simulation which is already successfully applied for simulations of the wave propagation in media described by FO models [17].

II. Fractional Calculus

Fractional calculus studies an extension of derivatives and integrals to non-integer orders [18]. It generalizes concepts of the classical calculus, hence, similar methods and tools are available within this framework, but with wider generality and applicability. In fractional calculus, several definitions of the derivative operator D^α with the order α being a real number ($\alpha \in \mathbb{R}$) are formulated [19]–[21]. The fractional calculus is widely applied for modelling of electrical circuits [22]–[25].

In our investigations, the Marchaud derivative concept is applied. In general, the Marchaud derivative is defined for $\alpha > 0$ (refer to [19, Section 5.5], [21, Section 1.3.1], and [26] for a historical perspective). When $\alpha \in (n-1, n)$ and $n \in \mathbb{N}$, then

$$D^\alpha f(t) = \frac{\alpha - n + 1}{\Gamma(n - \alpha)} \int_0^{+\infty} \frac{f^{(n-1)}(t) - f^{(n-1)}(t - \tau)}{\tau^{2+\alpha-n}} \mathrm{d}\tau \tag{1}$$

where f is assumed to be smooth enough, e.g., $f \in C^{n-1}(\mathbb{R})$ with $f^{(n-1)}$ bounded. The Marchaud derivative satisfies the condition

$$D^\alpha e^{j\omega t} = (j\omega)^\alpha e^{j\omega t} \tag{2}$$

where $j = \sqrt{-1}$. It is a generalization of the fundamental formula taken from the IO calculus, which is required for the representation of signals in the phasor analysis of circuits. Furthermore, the Marchaud derivative satisfies the semigroup property

$$D^\alpha D^\beta f(t) = D^\beta D^\alpha f(t) = D^{\alpha+\beta} f(t), \quad \alpha, \beta \in \mathbb{R}. \tag{3}$$

III. FO Model of Transmission Line

Let us consider the FO RLGC model of the transmission line proposed in [11]. Its elementary circuit of the length Δx is presented in Fig. 1. Using Kirchhoff's current and voltage

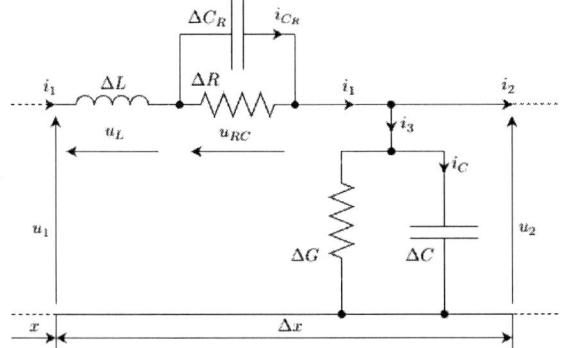

Fig. 1. Elementary circuit of FO RLGC model of transmission line.

laws, one obtains the following set of equations:

$$i_1 = \frac{u_{RC}}{\Delta R} + i_{C_R} \tag{4}$$

$$i_1 = i_2 + i_3 \tag{5}$$

$$i_3 = i_C + \Delta G u_2 \tag{6}$$

$$u_L + u_{RC} + u_2 - u_1 = 0. \tag{7}$$

For the circuit elements in Fig. 1, the following FO relations between voltages and currents are assumed:

$$u_L = \Delta L D_t^\alpha i_1 \tag{8}$$

$$i_{C_R} = \Delta C_R D_t^\beta u_{RC} \tag{9}$$

$$i_C = \Delta C D_t^\gamma u_2. \tag{10}$$

The voltages and currents at the edges of the elementary circuit in Fig. 1 are given by

$$i_1 = i(x, t) \tag{11}$$

$$i_2 = i(x + \Delta x, t) \tag{12}$$

$$u_1 = u(x, t) \tag{13}$$

$$u_2 = u(x + \Delta x, t). \tag{14}$$

The set of FO differential equations [11] can be formulated based on (4)–(14) assuming that $\Delta x \to 0$, i.e.

$$Ri(x, t) = u'(x, t) + \tau D_t^\beta u'(x, t) \tag{15}$$

$$\frac{\partial}{\partial x} i(x, t) = -C D_t^\gamma u(x, t) - G u(x, t) \tag{16}$$

$$\frac{\partial}{\partial x} u(x, t) = -L D_t^\alpha i(x, t) - u'(x, t). \tag{17}$$

In (15)–(17), the parameters are defined as $L = \Delta L / \Delta x$, $R = \Delta R / \Delta x$, $C = \Delta C / \Delta x$, $G = \Delta G / \Delta x$, $\tau = \Delta R \Delta C_R$,

$u'(x, t) = u_{RC}(x, t) / \Delta x$, all for $\Delta x \to 0$. This set of equations is solved under the initial-boundary conditions, i.e.

$$u'(x, 0) = 0, \quad i(x, 0) = 0, \quad u(x, 0) = 0, \quad x \geq 0 \tag{18}$$

$$u(0, t) = u_0(t), \quad \lim_{x \to +\infty} u(x, t) = 0, \quad t > 0. \tag{19}$$

In our investigations, we consider (15)–(19) after the normalization [11], i.e.

$$\frac{R}{L} \left(\frac{C}{G} \right)^{\frac{\alpha}{\gamma}} \bar{i}(\bar{x}, \bar{t}) = \bar{u}'(\bar{x}, \bar{t}) + \bar{\tau} D_{\bar{t}}^\beta \bar{u}'(\bar{x}, \bar{t}) \tag{20}$$

$$\frac{\partial}{\partial \bar{x}} \bar{i}(\bar{x}, \bar{t}) = -D_{\bar{t}}^\gamma \bar{u}(\bar{x}, \bar{t}) - \bar{u}(\bar{x}, \bar{t}) \tag{21}$$

$$\frac{\partial}{\partial \bar{x}} \bar{u}(\bar{x}, \bar{t}) = -D_{\bar{t}}^\alpha \bar{i}(\bar{x}, \bar{t}) - \bar{u}'(\bar{x}, \bar{t}) \tag{22}$$

$$\bar{u}'(\bar{x}, 0) = 0, \quad \bar{i}(\bar{x}, 0) = 0, \quad \bar{u}(\bar{x}, 0) = 0, \quad \bar{x} \geq 0 \tag{23}$$

$$\bar{u}(0, \bar{t}) = \bar{u}_0(\bar{t}), \quad \lim_{\bar{x} \to +\infty} \bar{u}(\bar{x}, \bar{t}) = 0, \quad \bar{t} > 0 \tag{24}$$

where $T = (C/G)^{\frac{1}{\gamma}}$, $l = T^{\frac{\alpha}{2}} / \sqrt{LG}$, $\bar{t} = t/T$, $\bar{x} = x/l$, $\bar{\tau} = \tau/T^\beta$, $\bar{u} = u/U$, $\bar{u}' = u'l/U$, $I = UT^{\frac{\alpha}{2}} \sqrt{G/L}$, $\bar{i} = i/I$.

IV. Simulation Method

Let us omit the bars over variables and parameters in (20)–(24). This set of equations is transformed to the frequency domain. Let us consider the phasor representation of the voltage and the current along the line

$$u(x, t) = \Re(U(x) e^{j\omega t}) \tag{25}$$

$$i(x, t) = \Re(I(x) e^{j\omega t}). \tag{26}$$

In (25)–(26), $U(x)$ and $I(x)$ are respectively voltage and current phasors that are functions of the spatial variable x only and ω denotes the angular frequency. Then, one obtains from (20)–(22)

$$\frac{R}{L} \left(\frac{C}{G} \right)^{\frac{\alpha}{\gamma}} I(x) = U'(x) + \tau (j\omega)^\beta U'(x) \tag{27}$$

$$\frac{\partial}{\partial x} I(x) = -(j\omega)^\gamma U(x) - U(x) \tag{28}$$

$$\frac{\partial}{\partial x} U(x) = -(j\omega)^\alpha I(x) - U'(x). \tag{29}$$

This set of equations can be reduced to a single equation which $U(x)$ satisfies [11]

$$\frac{\partial^2}{\partial x^2} U(x) - \psi(j\omega) U(x) = 0 \tag{30}$$

where

$$\psi(j\omega) = \frac{\left((j\omega)^{\alpha+\beta} + a(j\omega)^\alpha + b \right) \left((j\omega)^\gamma + 1 \right)}{\left((j\omega)^\beta + a \right)}. \tag{31}$$

In (31), $a = 1/\tau$ and $b = R/(\tau L)(C/G)^{\alpha/\gamma}$. Hence, the solution to (30) is given by

$$U(x) = U_0(j\omega) e^{-k(j\omega)x} \tag{32}$$

165

where

$$k(j\omega) = \sqrt{\psi(j\omega)}. \tag{33}$$

We select the propagation constant $k(j\omega)$ as the one with a positive real part. The frequency-domain solution (32)–(33) allows one to find the time-domain solution to the system (20)–(24). For this purpose, the computations are executed according to the scheme presented in Fig. 2. From the real data sequence $u_0(t)$, the analytic signal $u_a(t) = u_0(t) + j\mathcal{H}\{u_0(t)\}$ is obtained with the use of the Hilbert transformation [27]. The excitation $u_a(t)$ is transformed to the frequency domain, then its analytic representation is multiplied by the phase factors $e^{-k(j\omega)x}$, and finally the real part of the inverse Fourier transformation is taken as the solution $u(t)$. The Hilbert transformation can be executed in the frequency domain according to the formula

$$\mathcal{F}\{\mathcal{H}\{u_0\}\}(j\omega) = -jsgn(\omega)\mathcal{F}\{u_0\}(j\omega) \tag{34}$$

where $sgn(\omega)$ denotes the signum function. Hence, the algorithm steps can be combined together before the multiplication with the phase factors. The fast Fourier transformation is employed in the developed code to reduce the computational overhead of the method [27].

With the use of the proposed algorithm, one can avoid computations of the time-domain Green's function or numerical inversion of the Laplace transformation. However, the proposed simulation method has limited applicability due to the implementation of the method in the discrete-time domain. The phase factors $e^{-k(j\omega)x}$ involve attenuation and phase delay of the propagating wave. From this point of view, the phase delay must be grater or equal to the phase shift corresponding to the sampling time T_s of the complex signal $u_a(t)$. Hence, one obtains the condition allowing for the application of the method

$$\Im(k(j\omega)x) \geq \omega T_s. \tag{35}$$

For the complex discrete-time signal, the maximum frequency within the discrete spectrum is equal to $\frac{1}{T_s}$. With the use of (35), one can check if assumed values of the parameters α, β and γ allow for the application of the proposed simulation method.

If a linear time-invariant system is driven by harmonic excitations, or excitations that can be expanded using Fourier series or transform, then the proposed method can be applied. Hence, it is implicitly assumed that excitations are repeated after the total simulation time, which must be sufficiently long to allow simulated signals to decrease to zero.

V. NUMERICAL RESULTS

The numerical results presented below are obtained with the use of standard functions available in Matlab. The non-monochromatic signal propagation is analysed along the FO RLGC model of the transmission line. For easy verification of the method, let us consider the excitation which is Dirac's delta function. In this case, the algorithm reduces to the computations of the inverse Fourier transformation of the

phase factors $e^{-k(j\omega)x}$. The results for varying parameters of the line model are presented in Fig. 3. The obtained results are the same as the reference results in [11, Figs 7-10], hence, it proves the applicability of the proposed method. Furthermore, the implementation of the method does not require any significant analytical preprocessing on the complex s-plane for the application of the method.

VI. CONCLUSION

The simulation method of signal propagation along FO transmission lines is developed. It employs computations in the frequency domain, i.e., an analytical excitation is transformed to the frequency domain, multiplications with phase factors are executed, and finally the result is transformed back to the time domain. The method is numerically efficient and accurate. However, the applicability of the method is limited by the sampling theorem.

REFERENCES

[1] Y. Shang, W. Fei, and H. Yu, "A fractional-order rlgc model for terahertz transmission line," in *2013 IEEE MTT-S International Microwave Symposium Digest (MTT)*, June 2013, pp. 1–3.

[2] Y. Shang, H. Yu, and W. Fei, "Design and analysis of cmos-based terahertz integrated circuits by causal fractional-order rlgc transmission line model," *IEEE Trans. Emerg. Sel. Topics Circuits Syst.*, vol. 3, no. 3, pp. 355–366, 2013.

[3] R. C. Cascaval, E. C. Eckstein, C. L. Frota, and J. A. Goldstein, "Fractional telegraph equations," *J. Math. Anal. Appl.*, vol. 276, no. 1, pp. 145–159, 2002.

[4] J. Chen, F. Liu, and V. Anh, "Analytical solution for the time-fractional telegraph equation by the method of separating variables," *J. Math. Anal. Appl.*, vol. 338, no. 2, pp. 1364–1377, 2008.

[5] S. Momani, "Analytic and approximate solutions of the space- and time-fractional telegraph equations," *Appl. Math. Comput.*, vol. 170, no. 2, pp. 1126–1134, 2005.

[6] E. Orsingher and L. Beghin, "Time-fractional telegraph equations and telegraph processes with brownian time," *Probab. Theory Relat. Fields*, vol. 128, no. 1, pp. 141–160, 2004.

[7] F. Huang, "Analytical solution for the time-fractional telegraph equation," *J. Appl. Math.*, p. 890158, 2009.

[8] W. Jiang and Y. Lin, "Representation of exact solution for the time-fractional telegraph equation in the reproducing kernel space," *Comm. Nonlinear Sci. Numer. Simulat.*, vol. 16, no. 9, pp. 3639–3645, 2011.

[9] T. M. Atanackovic, S. Pilipovic, and D. Zorica, "A diffusion wave equation with two fractional derivatives of different order," *J. Phys. A Math. Theor.*, vol. 40, no. 20, pp. 5319–5333, 2007.

[10] S.-q. Zhang, "Solution of semi-boundless mixed problem for time-fractional telegraph equation," *Acta Math. Appl. Sin.*, vol. 23, no. 4, pp. 611–618, 2007.

[11] S. M. Cvetićanin, D. Zorica, and M. R. Rapaić, "Generalized time-fractional telegrapher's equation in transmission line modeling," *Nonlinear Dyn.*, vol. 88, no. 2, pp. 1453–1472, 2017.

[12] S. M. Cvetićanin, D. Zorica, and M. R. Rapaić, "Frequency characteristics of two topologies representing fractional order transmission line model," *Circuits, Syst. Signal Process.*, vol. 39, no. 1, pp. 456–473, 2020.

[13] N. A. R-Smith, A. Kartci, and L. Brancik, "Fractional-order lossy transmission line with skin effect using nilt method," in *2017 40th International Conference on Telecommunications and Signal Processing (TSP)*, July 2017, pp. 730–734.

[14] L. Brancik, A. Kartci, and N. A. R-Smith, "Matlab simulation of transmission lines with skin effect via fractional telegraph equations and nilt," in *2017 27th EAEEIE Annual Conference (EAEEIE)*, June 2017, pp. 1–5.

[15] L. Brancik, N. A. R-Smith, and A. Kartci, "Numerical simulation of nonuniform multiconductor transmission lines with hf losses in matlab: Laplace-domain and time-domain approaches," in *2018 28th International Conference Radioelektronika (RADIOELEKTRONIKA)*, April 2018, pp. 1–5.

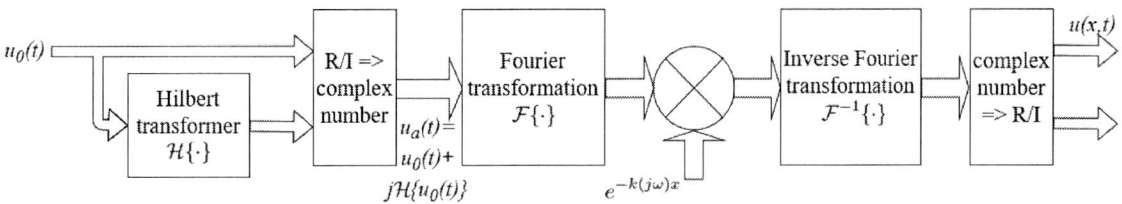

Fig. 2. Simulation scheme of the signal propagation along FO RLGC model of transmission line.

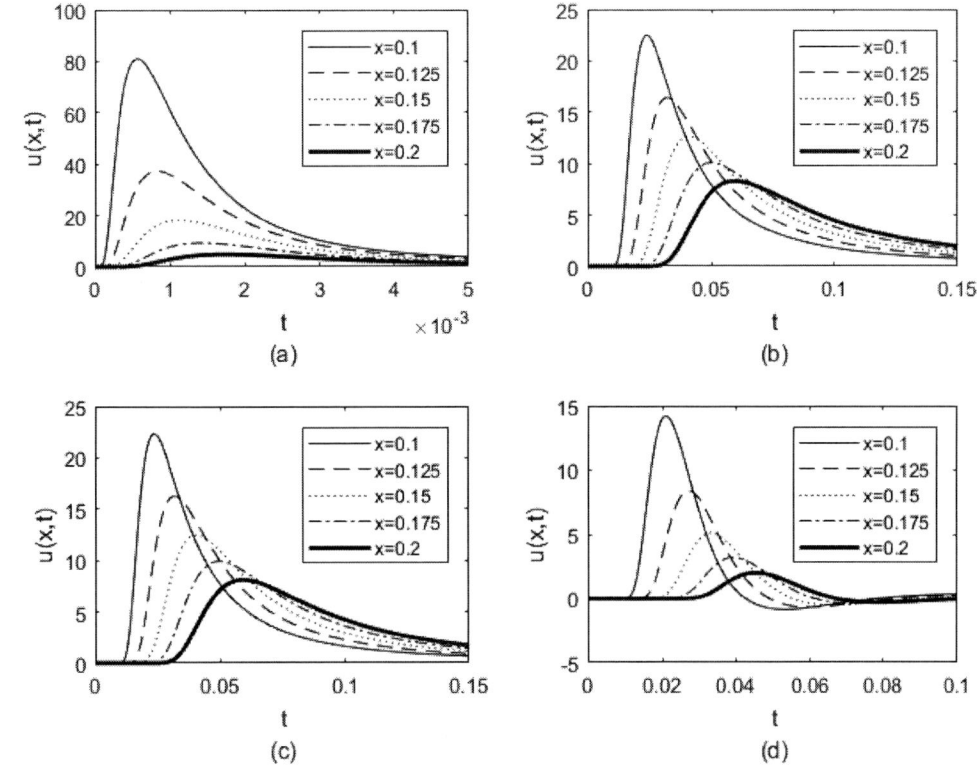

Fig. 3. Impulse response of FO RLGC model of transmission line for varying position x. (a) $\alpha = 1/4$, $\beta = \gamma = 2/3$, $a = 2\sqrt[3]{9}$, $b = 450$. (b) $\alpha = 5/6$, $\beta = \gamma = 2/3$, $a = 2\sqrt[3]{9}$, $b = 4.5$. (c) $\alpha = 5/6$, $\beta = \gamma = 2/3$, $a = 2\sqrt[3]{9}$, $b = 9$. (d) $\alpha = 5/6$, $\beta = \gamma = 2/3$, $a = 2\sqrt[3]{9}$, $b = 450$.

[16] N. A.-Z. R-Smith, A. Kartci, and L. Brancik, "Application of numerical inverse laplace transform methods for simulation of distributed systems with fractional-order elements," *J. Circuits, Syst. Comput.*, vol. 27, no. 11, p. 1850172, 2018.

[17] T. P. Stefanski and J. Gulgowski, "Signal propagation in electromagnetic media described by fractional-order models," *Comm. Nonlinear Sci. Numer. Simulat.*, vol. 82, p. 105029, 2020.

[18] E. W. Weisstein. (2020) "fractional calculus." from mathworld–a wolfram web resource. [Online]. Available: http://mathworld.wolfram.com/FractionalCalculus.html

[19] S. G. Samko, A. A. Kilbas, and O. I. Marichev. *Fractional Integrals and Derivatives: Theory and Applications.* Gordon and Breach, New York, 1993.

[20] A. A. Kilbas, H. M. Srivastava, and J. J. Trujillo, *Theory and Applications of Fractional Differential Equations.* Elsevier Science, 2006.

[21] C. Li and F. Zeng, *Numerical Methods for Fractional Calculus.* Chapman and Hall/CRC, 2015.

[22] M. D. Ortigueira, "An introduction to the fractional continuous-time linear systems: the 21st century systems," *IEEE Circuits Syst. Mag.*, vol. 8, no. 3, pp. 19–26, 2008.

[23] A. S. Elwakil, "Fractional-order circuits and systems: An emerging interdisciplinary research area," *IEEE Circuits Syst. Mag.*, vol. 10, no. 4, pp. 40–50, 2010.

[24] T. J. Freeborn, "A survey of fractional-order circuit models for biology and biomedicine," *IEEE Trans. Emerg. Sel. Topics Circuits Syst.*, vol. 3, no. 3, pp. 416–424, Sep. 2013.

[25] T. P. Stefanski and J. Gulgowski, "Electromagnetic-based derivation of fractional-order circuit theory," *Comm. Nonlinear Sci. Numer. Simulat.*, vol. 79, p. 104897, 2019.

[26] F. Ferrari, "Weyl and Marchaud derivatives: A forgotten history," *Mathematics*, vol. 6, no. 6, 2018. [Online]. Available: https://www.mdpi.com/2227-7390/6/1/6

[27] A. V. Oppenheim and R. W. Schafer. *Discrete-Time Signal Processing*, 3rd ed. Upper Saddle River, NJ, USA: Prentice Hall Press, 2009.

Proceedings of the 27th International Conference *"Mixed Design of Integrated Circuits and Systems"*
June 25-27, 2020, Łódź, Poland

Subband Structure and Ballistic Conductance of a Molybdenum Disulfide Nanoribbon in Topological 1T' Phase: A k·p Study

Viktor Sverdlov
Christian Doppler Laboratory
for Nonvolatile Magnetoresistive Memory and Logic
at the Insitute for Microelectronics
Technische Universität Wien
Vienna, Austria
sverdlov@iue.tuwien.ac.at

Al-Moatasem Bellah El-Sayed, Siegfried Selberherr
Institute for Microelectronics
Technische Universität Wien
Vienna, Austria
el-sayed@iue.tuwien.ac.at, Selberherr@TUWien.ac.at

Abstract—**We evaluate the subband structure in a narrow nanoribbon of 1T' molybdenum disulfide by employing an effective k·p Hamiltonian. Highly conductive topologically protected edge states whose energies lie within the bulk band gap are investigated. Due to the interaction of the edge modes located at the opposite edges, a small gap in their linear spectrum opens in a narrow nanoribbon. This gap is shown to sharply increase with the perpendicular out-of-plane electric field, in contrast to the behavior in a wide nanoribbon. The gaps between the electron and hole bulk subbands also increase with the electric field. The increase of the gaps between the subbands leads to a rapid decrease of the ballistic nanoribbon conductance and current with the gate voltage, which can be used for designing molybdenum disulfide nanoribbon-based current switches.**

Keywords—**topological insulators, topologically protected edge states, nanoribbons, subbands, k.p Hamiltonian, ballistic conductance**

I. INTRODUCTION

Edge states in two-dimensional (2D) topological insulators (TI) propagate without backscattering, making them attractive for designing highly conductive channels [1]. Recently it was discovered that the 1T' phase of MoS$_2$, a well-known 2D material with a high promise for future microelectronic devices [2], is a TI [3]. The inverted band structure is well approximated by parabolas, with the conduction and valence bands having masses of $m_{y(x)}^{d(p)}$ [3]. The spin-orbit interaction opens a gap at the intersection of the valence and conduction bands, which appears at a finite value of the momentum k_y along the OY axis. A topologically protected highly conductive edge state with a linear Dirac-like energy dispersion on the momentum k_x parallel to the OX axis must exist within this spin-orbit gap [3].

However, possessing robust conductive channels is only one requirement. To make a good switch it is necessary to suppress the current through the channel as a function of a perpendicular electric field induced by a gate. A standard approach is to close the gap in the bulk host material. In this case scattering between the protected edge and the non-protected electron-hole bulk states results in strong scattering, which effectively reduces the current through the edge states [4].

By applying an electric field E_z along the OZ axis perpendicular to a MoS$_2$ sheet in the 1T' phase, the bulk spin-orbit gap can be reduced, closed, and opened again as a "negative" gap at large electric fields [3]. The traditional band

order is restored from the inverted band structure, the gap becomes a direct gap, and no edge states are allowed in the gap.

This transition between the topological and conventional insulator phases in a wide 1T'-MoS$_2$ controlled by the electric field orthogonal to the sheet eliminates the edge states completely and can be used to further suppress the current [5]. In order to enhance the current through the channel it is beneficiary to have many edges by stacking several narrow nanoribbons. Here, we evaluate the subband structure in a narrow nanoribbon of 1T'-MoS$_2$ by using an effective k·p Hamiltonian [3]. In contrast to a wide channel, we find that a small gap in the spectrum of edge states in a nanoribbon [6] increases with the electric field. It results in a rapid decrease in the nanoribbon conductance with the field, which is potentially suitable for switching.

II. METHOD AND RESULTS

In order to investigate transport through a nanoribbon, the subband structure and the wave functions must be evaluated first. We parametrize the energy in units of the band inversion gap 2δ at k_y=0, while $k_{y(x)}$ in units of $k_0 = \left(\frac{4\delta}{\hbar^2} \frac{m_y^d m_y^p}{m_y^d + m_y^p} \right)^{1/2}$. By applying a unitary transformation [7], the 4×4 Hamiltonian [3] is cast in a block-diagonal form similar to the one in [6].

$$\begin{pmatrix} H(\mathbf{k}) & 0 \\ 0 & H^*(-\mathbf{k}) \end{pmatrix} \qquad (1)$$

The 2×2 Hamiltonian $H(\mathbf{k})$,$\mathbf{k}$=(k_x, k_y) in dimensionless units has the form

$$H(\mathbf{k}) = \begin{pmatrix} \frac{1}{2} - k_y^2 \frac{m}{m_y^p} - k_x^2 \frac{m}{m_x^p} & v_2 k_y - \alpha E_z + i\, v_1 k_x \\ v_2 k_y - \alpha E_z - i\, v_1 k_x & -\frac{1}{2} + k_y^2 \frac{m}{m_y^d} + k_x^2 \frac{m}{m_x^d} \end{pmatrix}, \quad (2)$$

where $m = \frac{m_y^d m_y^p}{m_y^d + m_y^p}$ and $v_{1(2)}$ are the dimensionless velocities. The parameters used in (2) are from [3] and listed in Table I.

The bulk energy dispersion obtained with the Hamiltonian (2) with an offset of $\Delta E = \frac{1}{2} \frac{m_y^d - m_y^p}{m_y^d + m_y^p}$ is shown in Fig. 1 for several k_x and E_z=0. The spin-orbit gap opened at $k_y = \pm k_0$ is increasing with k_x. Indeed, the gap is determined by the off-diagonal terms in (2). Since the off-diagonal terms in (2) can be written as $\sigma_y v_1 k_x$ and $\sigma_x v_2 k_y$, where $\sigma_{x(y)}$ are the Pauli matrices, the gap Δ is defined by $\Delta = (v_1^2 k_x^2 + v_2^2 k_y^2)^{1/2}$.

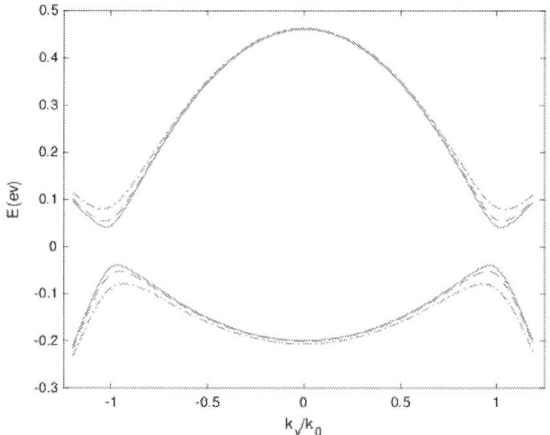

Fig. 1. Bulk energy dispersion in 1T'-MoS$_2$ two-dimensional material, E_z=0, for k_x=0 (solid line), k_x=0.1 k_0 (dashed line), and k_x=0.2 k_0 (dot-dashed line).

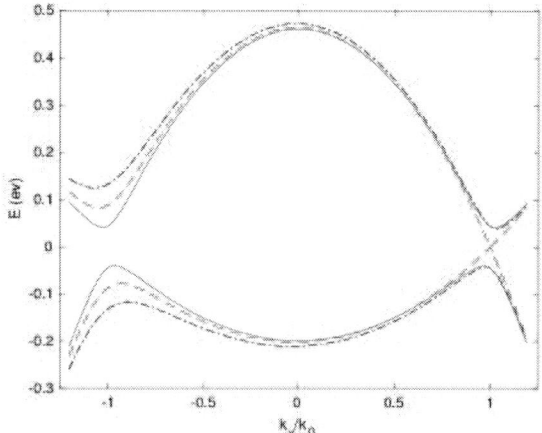

Fig. 2. Bulk energy dispersion in 1T'-MoS$_2$ two-dimensional material, k_x=0, for E_z=0 (solid line), αE_z=v$_2$ (dashed line) and αE_z=2v$_2$ (dot-dashed line).

By applying an electric field E_z along the OZ axis the gap at one of the minima (Fig.2, solid line) can be reduced, completely closed (Fig.2, dashed line), or even be opened again (Fig.1, dot-dashed line) at large electric fields. The gap at large electric fields becomes direct.

Let us consider a nanoribbon with a width in the OY direction of d=40/k_0. Only quantized values of the momentum k_y along the quantization axis OY are allowed. In addition, it is expected that at E_z =0 two topologically protected highly conductive edge states localized at opposite interfaces of a nanoribbon exist at any particular energy E within the gap opened by the spin-orbit interaction at k_y =± k_0.

A general form of the subband wave function $\psi_{k_x}(y)$ in the quantization OY direction is written as

$$\psi_{k_x}(y) = \sum_{j=1}^{4} A_j \begin{pmatrix} 1 \\ a(k_j, E) \end{pmatrix} \exp(ik_j y), \quad (3)$$

where k_j, j=1,…,4 are the roots of $E(k_x, k_j)$= E, A_j are constants, and

$$a(k_j, E) = \frac{-\frac{1}{2} + k_j^2 \frac{m}{m_y^p} + k_x^2 \frac{m}{m_x^p} + E}{v_2 k_j - \alpha E_z + i\, v_1 k_x} . \quad (4)$$

The subband energies are obtained by setting the wave function to zero at both edges. The characteristic equation

$$\det (\mathbf{M}) = 0, \quad (5)$$

where the matrix $\mathbf{M} = (M_1 \quad M_2 \quad M_3 \quad M_4)$ is composed of the columns M_j, j=1,..,4.

$$M_j = \begin{pmatrix} 1 \\ a(k_j, E) \\ \exp(ik_j d) \\ a(k_j, E) \exp(ik_j d) \end{pmatrix}, \quad (6)$$

is solved numerically, in complete analogy to the problem of finding the eigenenergies and eigenfunctions of a 2-band **k·p** Hamiltonian in silicon films [8]. Fig.3 displays the behaviour of the real part (the imaginary part is zero for E_z=0) of the determinant as a function of energy, for k_x=0. We are interested

TABLE I.
PARAMETERS [3,5] USED IN THE MODEL. m_e IS THE ELECTRON MASS, e IS THE ELECTRON CHARGE, AND d IS THE WIDTH IN OY DIRECTION.

Variable	Value
2δ	0.66 eV
v_1	3.87 10^5 m/s
v_2	0.46 10^5 m/s
m_x^p	0.5 m_e
m_y^p	0.16 m_e
m_x^d	2.48 m_e
m_y^d	0.37 m_e
α	0.03 e nm
k_0	1.485 nm^{-1}
d	$40k_0^{-1}$ = 26.94 nm

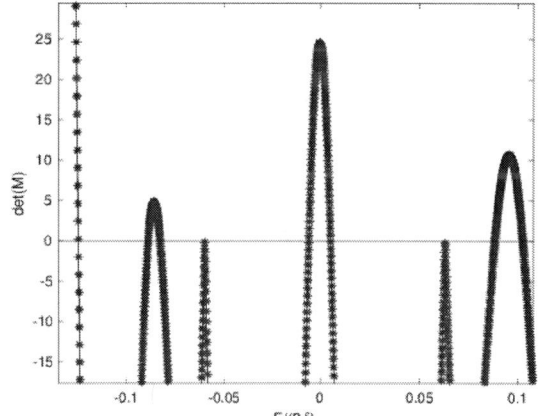

Fig. 3. Real (stars) and imaginary (line) parts of det(**M**) computed at k_x =0, E_z = 0, d=40k_0^{-1}. The bulk gap is seen at at $E \approx \pm0.065$, where the real part touches the OX axis from below. The subband energies are obtained from det(**M**)=0. Topological edge states are seen in the bulk gap ($E \approx \pm0.005$).

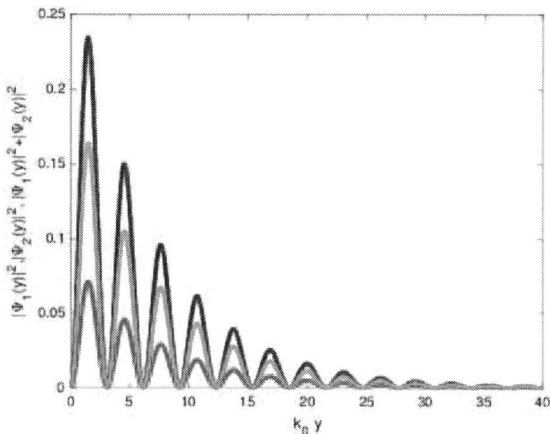

Fig. 4. The wave function square (blue) and its two spinor components (green and red) of the topological edge state evaluated at $\alpha E_z = 0.1v_2$, $k_x = 0.1k_0$, and $E \approx 0.005$. The subband wave function shows both oscillation and an exponential decay.

in the crossings of the curve with the axis OX. The bulk gap due to the spin-orbit interaction occurs at $E \approx \pm 0.065$. The value of the determinant approaches zero from negative values and touches it at a single point, when the energy E touches the minimum (maximum) of the dispersion curve. Therefore, $k_1 = k_2$ and $k_3 = k_4$ and the determinant (5) is zero.

All other intersections with the OX axis correspond to subband energies. We clearly observe two roots in the gap at $E \approx \pm 0.005$ A close inspection shows that the wave vectors k_j corresponding to these solutions are complex numbers. The wave functions corresponding to these solutions in the gap are located at an edge of the nanoribbon as shown in Fig.4 and Fig.5 for $k_x = 0.1k_0$ and $E \approx \pm 0.005$, respectively. In contrast to [6], where only an exponential decay was predicted, the wave functions display both oscillations and decay. Although the structure of the Hamiltonians considered here and in [6] are similar, the actual parameters differ. In particular, here the spin-orbit gaps are open at the finite values of $k_y = \pm k_0$. This displacement of the bulk band's minima from the Gamma-point at $k_y = 0$ is reflected in the oscillations of subband wave functions superimposed on the exponential decay.

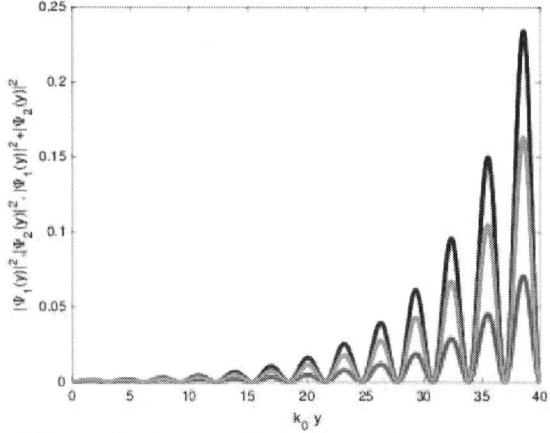

Fig. 5. The wave function square (blue) and its two spinor components (green and red) of the topological edge state evaluated at $\alpha E_z = 0.1v_2$, $k_x = 0.1k_0$, but $E \approx -0.005$. The wave function is localized at the opposite edge.

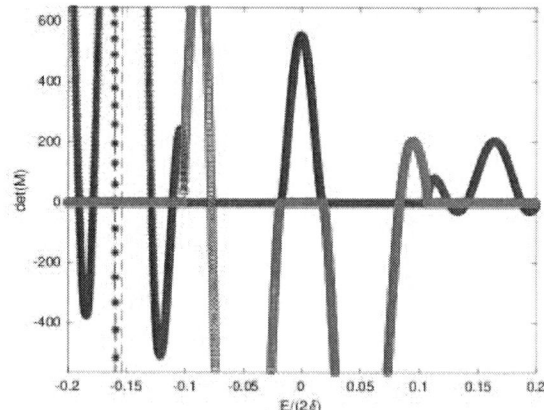

Fig.6 Real (blue stars) and imaginary (red circles) parts of det(**M**) computed at $k_x = 0$, $\alpha E_z = 0.7 v_2$, $d = 40k_0^{-1}$. At $E \approx \pm 0.02$ the imaginary part of the two k_j ensuring the localization at the edges is approaching zero. New type of roots with imaginary part equals to zero at $E \approx \pm 0.075$ appear.

The roots of the determinant for $|E| > 0.065$ correspond to the subbands with all k_j real. The wave functions are delocalised through the width of the nanoribbon. Due to the strong non-parabolicity of the bulk dispersion, the positions of the subband minima and the subband dispersions can only be found by solving (5) numerically.

Fig. 6 shows the behavior of the determinant at $\alpha E_z = 0.7v_2\hbar k_0$. In this case the gap at $k = k_0$ is reduced but not completely closed. However, due to the finite width of the nanoribbon, the edge modes seen at $E \approx \pm 0.02$ are already delocalized as the imaginary parts of two k_j responsible for the localization at the edges are becoming zero. At the same time, the two solutions at $E \approx \pm 0.075$ split off from the traditional subbands set as two of their k_j acquire an imaginary part. This happens due to the fact that, while the gap at $k_y = k_0$ shrinks with increasing E_z, the gap at $k_y = -k_0$ displays an opposite trend and becomes wider. Therefore, the lowest traditional subband initially outside of the gap enters the gap at $k_y = -k_0$ thus forcing the two roots to become complex.

The behaviour of the determinant at even higher electric field $\alpha E_z = 1.4v_2\hbar k_0$ is shown in Fig.7. As the field is larger than the

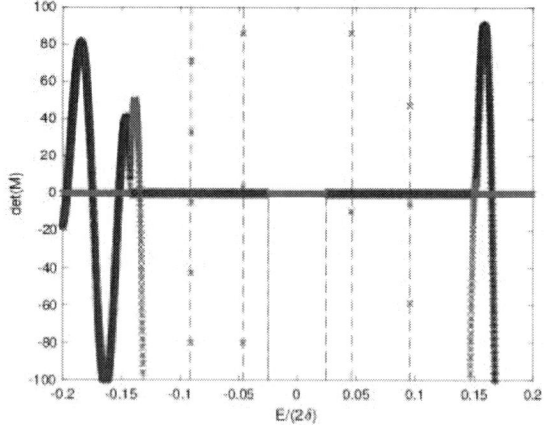

Fig.7. Real (blue stars) and imaginary (red crosses) parts of det(**M**) computed at $k_x = 0$, $\alpha E_z = 1.7v_2$, $d = 40k_0^{-1}$. No solution within the direct gap at $|E| < 0.03$ is allowed. Two solutions of with imaginary part equal to zero at $E \approx \pm 0.05$ and $E \approx \pm 0.08$ are now observed.

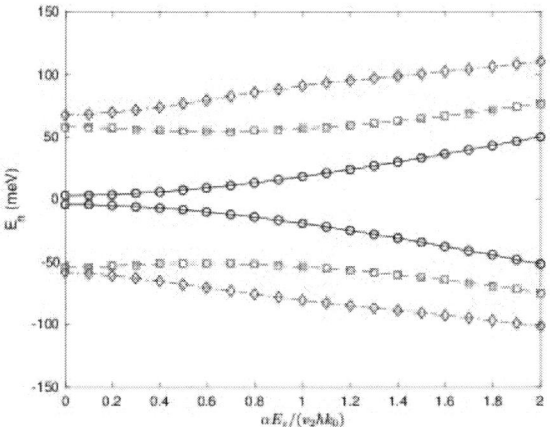

Fig. 8. Dependence of electron (hole) subband minima (maxima) on the electric field E_z for the first three subbands. In contrast to the bulk case, the gap never closes and keeps increasing with E_z

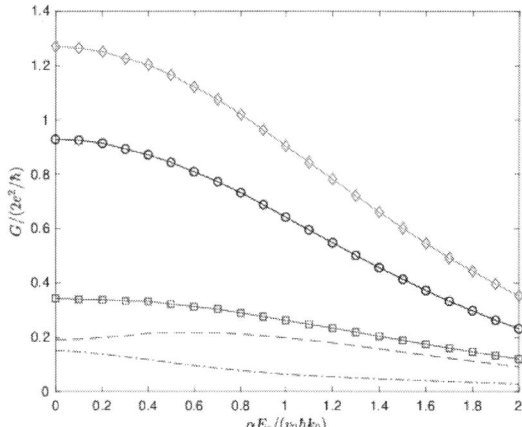

Fig. 9. Ballistic conductance (diamonds) of a 1T'-MoS$_2$ nanoribbon, with the contributions from the edge-like states (circles), and the remaining bulk-like subbands (squares). Dashed line from subbands shown in Fig.8 by squares; dot-dashed line from Fig.8, diamonds. Temperature T=300K, E_F=0

critical value field $\alpha E_z = v_2 \hbar k_0$ at which there is no gap at $k_y = k_0$, the gap seen at $|E| < 0.03$ is a direct gap. Therefore, as confirmed by the absence of zeroes of the determinant, no edge states are allowed in the gap. At the same time, there are already four subbands lying outside the direct gap at $k_y = k_0$ but still within the gap at $k_y = -k_0$. While the solutions at $E \approx \pm 0.09$ split off from the traditional subbands set in complete analogy to the situation in Fig. 6, the subbands at $E \approx \pm 0.05$ originate in the edge states, which were continuously pushed outside of the gap at $k_y = -k_0$.

Fig.8 shows the dependence of the electron (hole) subband minima (maxima) $E_n^{e(h)}$ on the electric field E_z. First, we note that the "bulk" gap in the nanoribbon defined as the difference between the extrema of the first traditional subbands (Fig.8, squares) shows signs of reduction, when the field is increased till $E_z \approx 0.7 v_2 \hbar k_0 / \alpha$, after which the trend is inverted. This behavior is in sharp contrast to that in wide ribbons in which the bulk gap closes at $E_z = v_2 \hbar k_0 / \alpha$ (Fig.1, dashed line).

Second, with increasing E_z the gap between the lowest electron and the highest edge-like subbands grows (Fig. 8, circles). As indicated in Fig.5 and Fig.6, all four k_j contain an imaginary part for $E_z < 0.7 v_2 \hbar k_0 / \alpha$. This value is lower than the value $E_z < v_2 \hbar k_0 / \alpha$ corresponding to the bands' inversion in the bulk. The increasing gap between the edge-like subbands is reflected in the decrease of the corresponding nanoribbon ballistic conductance shown in Fig.9 (circles). Although the edge-like subbands give the leading contribution in the conductance G (Fig.9, diamonds) computed as

$$ G = \frac{2e^2}{h} \sum_j \left[\frac{1}{\exp\left\{\frac{E_j^e - E_F}{k_B T}\right\} + 1} - \frac{1}{\exp\left\{\frac{E_j^h - E_F}{k_B T}\right\} + 1} + 1 \right], \quad (7) $$

where T is the temperature and E_F is the Fermi energy, the role of the other subbands shown in Fig.9 by squares is non-negligible. The first two electron (hole) bulk-like subbands give similar contributions to the ballistic conductance totaling to 30%. However, all contributions to the total conductance G

rapidly decrease as a function of E_z (Fig.9). This makes 1T'-MoS$_2$ potentially suitable for switching applications.

ACKNOWLEDGMENT

Financial support by the Austrian Federal Ministry for Digital and Economic Affairs and the National Foundation for Research, Technology and Development is gratefully acknowledged. A.-M.B.E.-S. was supported in part by project No. IN 23/2018 'Atom-to-Circuit modeling technique for exploration of Topological Insulator based ultra-low power electronics' by the Centre for International Cooperation & Mobility (ICM) of the Austrian Agency for International Cooperation in Education and Research (OeAD).

REFERENCES

[1] L. Kou, Y. Ma, Z. Sun, T. Heine, and C. Chen, "Two-dimensional topological insulators: Progress and prospects", J.Phys.Chem.Lett. vol.8, pp.1905-1919, 2017.

[2] Yu.Yu. Illarionov, A.G. Banshchikov, D.K. Polyushkin, S. Wachter, T. Knobloch, M. Thesberg, L. Mennel, M. Paur, M. Stöger-Pollach, A. Steiger-Thirsfeld, M.I. Vexler, M. Waltl, N.S. Sokolov, T. Mueller, and T. Grasser, "Ultrathin calcium fluoride insulators for two-dimensional field-effect transistors", Nature Electronics, vol.2, pp.230-235, 2019.

[3] X. Qian, J. Liu, L. Fu, and Ju Li., "Quantum spin Hall effect in two-dimensional transition metal dichalcogenides", Science, vol. 346, issue 6215, pp.1344-1347, 2014.

[4] W.G.Vandenberghe and M. V. Fischetti, "Imperfect two-dimensional topological insulator field-effect transistors", Nature Communications. Vol.8, art.14184 (pp. 1-8), 2017.

[5] L. Liu and J. Guo, "Assessment of performance potential of MoS$_2$-based topological insulator field-effect transistors", J.Appl.Phys., vol.118, art.124502 (pp.1-5), 2015.

[6] B. Zhou, H.-Z. Lu, R.-L. Chu, S.-Q. Shen, and Q. Niu, "Finite size effects on helical edge states in a quantum spin-Hall system" Phys.Rev.Lett., vol.101, art.246807 (pp.1-4), 2008.

[7] V. Sverdlov, A.-M.B. El-Sayed, H. Kosina, and S. Selberherr, "Topologically protected and conventional subbands in a 1T'-MoS$_2$ nanoribbon channel", EUROSOI-ULIS 2020, accepted.

[8] V. Sverdlov and S. Selberherr, "Silicon spintronics: Progress and challenges", Phys.Rep., vol.585, pp.1-40, 2015.

Microelectronics Technology and Packaging

Proceedings of the 27th International Conference *"Mixed Design of Integrated Circuits and Systems"*
June 25-27, 2020, Łódź, Poland

Challenges in Performance Improvement of Silicon Systems on Chip in Advanced Nanoelectronics Technology Nodes

Arkadiusz Malinowski
Department of Integration and Yield
GLOBALFOUNDRIES
Dresden, Germany
arek.malinowski@globalfoundries.com

Shiv Kumar Mishra
Department of Product Integration
GLOBALFOUNDRIES
Malta, 12020 NY, USA

Abstract—**Speed or clock rate of the first microprocessor released to the market in 1971 was 740 kHz. This microprocessor was intended for calculator application. Continuing increase of microprocessor speed and computing power led to explosion of numerous applications. Five decades later microprocessors speed reached 5 GHz and they have enough computing power leading to such wonders as an artificial intelligence, virtual reality and self-driving autonomous cars which were before only in a science fiction domain. However, an increase of a chip speed is very challenging and it comes with high price. The most straightforward chip speed improvement based on transistor physical dimensions scaling eventually ran out of steam. This led to stress (1990s) followed by strain (2000s) techniques development. When this became insufficient new device structure, FinFET, has been introduced into main stream manufacturing in 2011. However, similarly to the previous approaches, increasing computing power of microprocessors based on FinFET is running now out of steam due to difficult technological barriers and integration challenges. Difficulties and challenges outlined in this paper may end era of microprocessor computing power improvement based on classical silicon technology.**

Keywords—**Moore's Law, CMOS Technology, Technology Node; FinFET, Integration, Scalling; Performance Boost, System on Chip, CPU, driver current, leakage current, Fmax/Iddq.**

I. INTRODUCTION

In 1965 Gordon Moore claimed that the future of integrated electronics is the future of electronics itself and that the advantages of integration will bring about a proliferation of electronics, pushing this technology into many new areas. He predicted that integrated circuits will lead to such wonders as more powerful home computers, automatic controls for automobiles, and personal portable communications equipment [1]. Moore was quite sure about integrated electronics proliferation because he noticed it had a very critical factor to make it happen, namely: a cost reduction potential. He explained that the cost advantage will be a continuing process resulting from making smaller chips with increased computing power. A key to achieve this was to keep cramming more and more transistors onto integrated circuit by reducing their physical dimensions. Additionally to the area scaling and reducing the chip size, it will be explained in the next section

there is a strict relation between transistor physical dimensions and its performance. However, reducing transistor physical dimensions for its performance improvement is a tremendous technological task and it has reached limits. Due to undesired so called short channel effects reducing transistor physical dimensions started to harm its performance. From this reason in order to improve transistor performance additional technological steps and processes, so called performance boosters, are needed [2, 3]. Those steps usually increase costs, cycle time and reduce throughput. On top of that the transistor architecture change was needed as well. The CMOS technology scaling for the chip area downsizing is a very broad topic with a number of specific critical challenges which won't be discussed in this paper. More details related to CMOS technology scaling for the next generations can be found in [4, 5]. This paper focuses only on technological processes aimed at microprocessor performance improvement. Due to the limited space only key concepts will be presented, some will be only mentioned while skipping the other ones with secondary effects.

A. Microprocessor's Performance

Microprocessor's performance is usually advertised by its maximum frequency (F_{max}) or clock rate i.e. 5 GHz. It means a number of simple operations per second (5 billion simple operations per second in this case). Obviously, the more operations per second, the faster and more efficient microprocessor is[1]. And this is how personal computer (PC) era market was driven – it was all about speed and computing power for high performance (HP) applications. The clock rate of the first generation of computers was measured in kilohertz (kHz), the first PCs to arrive throughout the 1970s and 1980s had clock rates measured in megahertz (MHz), and in the 21st century the speed of modern CPUs is commonly advertised in gigahertz (GHz). Continuous computing power increase involved increased energy consumption which led to directly proportional heat generation. Thus, the heat dissipation had to be introduced. Cooling techniques where changing along with increased energy consumption, starting firstly with simple radiators (passive cooling) followed by fans (active cooling) and water cooling (liquid cooling) for the most advanced

---

[1] Microprocessor architecture as well as number of computing units (cores) play a significant role as well but it is beyond scope of this paper.

applications (including liquid nitrogen and helium in extreme cases [6]). Limitation of the power density (W/cm^2) than can be dissipated from microprocessor, has been also reached, so the power consumption has become a significant concern [7].

A problem became even more critical with the second wave of integrated circuit (IC) manufacturing revolution, namely low power (LP) mobile applications. Now, the power consumption has come into main play and has become the even more critical parameter. The low power chips including 5G and RF application for such devices as mobile phones, laptops, pads, and smartwatches are powered by batteries, so the power consumption determines how long those mobile devices can run on a battery. So, the chip behavior, namely its frequency as a function of the power consumption has become a key issue. The power consumption is directly related to the current drawn by the chip during operation. Especially critical for battery lifetime is the current drawn in quiescent state, so called I_{ddq}. The I_{ddq} testing is a very broad topic and typical plot correlating maximum frequency (F_{max}) with I_{ddq} leakage current for the chip performance estimation is shown in Fig. 1^2 [8, 9]. It is worthwhile to mention, that I_{ddq} measurements are also used for the chip defects analysis and more details can be found elsewhere. Improving the microprocessor performance means either improving its maximum frequency for a fixed I_{ddq} leakage (a), or reducing its leakage current I_{ddq} for a fixed maximum frequency (b).

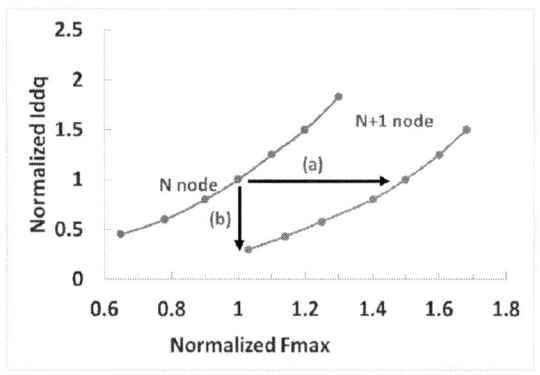

Fig. 1. F_{max} vs. I_{ddq} plot for two technology nodes.

The chip performance is determined by performances of the transistors and of the passive elements in the chip, e.g. parasitic resistances of the access regions and metallization lines at different levels and parasitic capacitances between semiconductor and metal areas. Therefore, improving the chip performance includes improving the transistor intrinsic performance.

B. Transistor Performance

Transistor performance is characterized by a so-called Universal Curve (U-curve) that plots I_{on} current (drain current in the saturation mode – I_{dsat}) versus I_{off} current (drain current with the gate voltage set to 0V) (Fig. 2). The transistor performance is improved when for a fixed I_{off} current I_{on} is

---

^{2} Other alternative for chip performance estimation is plot F_{max} vs V_{min} but this case won't be discussed in this paper.

improved (a), or alternatively for a fixed I_{on} current I_{off} is reduced (b). Typically, a new technology node N+1 offers somewhere between 15% - 40% I_{on} improvement at a fixed I_{off}, comparing to N-th technology node [2].

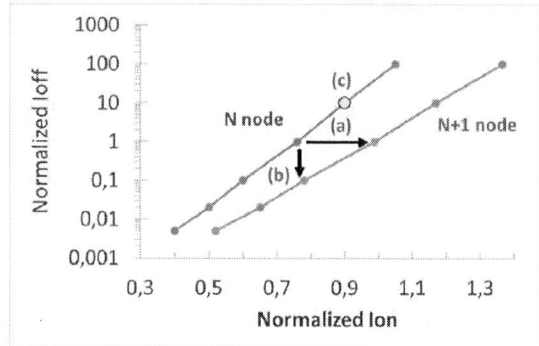

Fig. 2. Transistor I_{on} vs. I_{off} plot for two technology nodes.

Worth to highlight is that moving along the Universal Curve and increasing transistor I_{on} current (c) does not indicate performance transistor improvement since I_{on} increase is associated with an increase of I_{off}. Such situation can be obtained by saturation threshold voltage (V_{tsat}) shift (the most common knob to do so is a "halo" dose change) and is quite common practice. In a given technology node this allows for manufacturing chips for low as well as for high performance applications (Fig. 3).

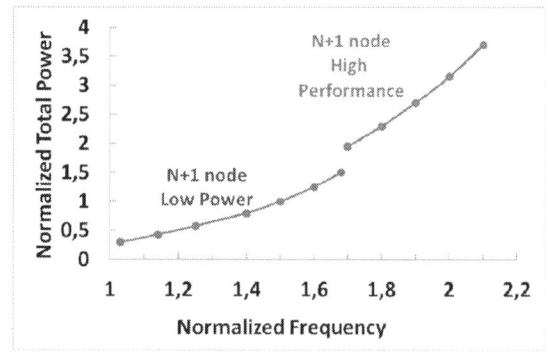

Fig. 3. Impact of V_{tsat} shif on chip performance in N+1 technology node.

Despite being very attractive because of easy realization, such an approach has a limitation, namely it saturates. Eventually making devices hotter does not bring benefits in terms of chip maximum frequency but the same time causes I_{ddq} leakage substantial surge. And this is why further chip performance enhancement requires the transistor performance improvement and methods how to achieve it will be discussed in the following sections.

II. PERFORMANCE IMPROVEMENT KNOBS

As it has been explained in the previous section, moving along U-curve by V_{tsat} shift does not cause transistor performance improvement. Instead, it is observed when for a fixed V_{tsat} either I_{on} is increased (Fig. 4a) or I_{off} is reduced (Fig. 4b).

a)

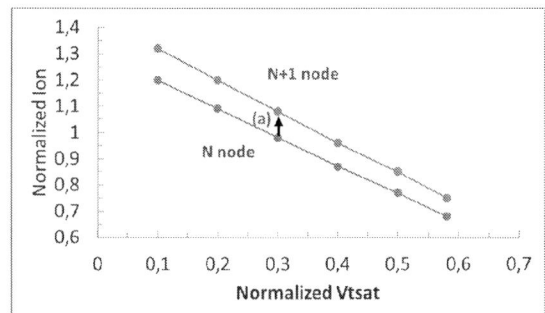

b)

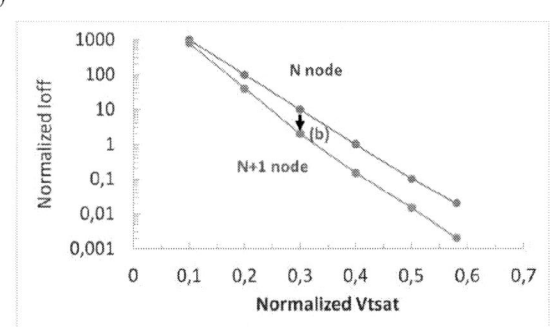

Fig. 4. Illustration of transistor performance improvement by a) increase of I_{on} @ V_{tsat}, or b) reduce I_{off} @ V_{tsat}.

Improving one current (i.e. I_{on}) without degrading the other one (i.e. I_{off}) is not a trivial tasks and in the next section it will be explained how this task has been achieved in early CMOS technology development.

A. Classical performance improvement

Despite Moore noticed that integrated electronics cost advantage will be a continuing process resulting from making smaller chips with an increased computing power, he never mentioned how this scaling might be done. His prediction was based on extrapolation of only a few points corresponding to early MOS technology nodes and taking into account mainly economical aspects of IC manufacturing [1]. First paper proposing a consistent set of scaling relationships that showed how a conventional device could be reduced in size was by Dennard in 1974 [10]. Additionally, in his paper Dennard also projected the sizable performance improvement expected from using in the integrated circuits very small MOSFETs of comparable dimensions and provided a table with the changes in integrated circuit performance which follow from scaling the circuit dimensions. The MOSFET drain current (I_D) is related to the device dimensions as show in equation (1):

$$I_D \propto \mu_n \, \varepsilon_0 \, \varepsilon_{ox} \, t_{ox}^{-1} \, W \, L_g^{-1} \qquad (1)$$

where: μ_n – carrier mobility, ε_0 – vacuum permittivity, ε_{ox} – dielectric permittivity, t_{ox} – gate oxide thickness, W – gate width, and L_g – gate length.

In the early CMOS technology era scaling and performance improvement were working hand in hand. Firstly, scaling of CMOS technology for following nodes was done by reducing

gate length L_g and thinning gate oxide t_{ox} [11]. Later on the carrier mobility μ_n enhancement techniques came into play based on tensile and compressive stress liners as well as on introducing strained silicon (SiGe) in 90 nm technology [12].

Eventually the classical scaling started facing significant challenges and potential roadblocks. The first one came from the gate oxide thickness. In order to keep up with scaling trends following Moore's Law the gate oxide thickness became so thin that led to a substantial leakage current through the gate and became a dominant leakage mechanism in the transistor. To solve this issue there was a need to use thicker gate oxide in order to reduce the gate leakage. However, in order to avoid the transistor performance degradation and to compensate a thickness increase the only solution was to use dielectrics of a higher permittivity (so called high-*k* materials). Starting from 45 nm technology node in 2007 this has led to replacement of SiO_2, the gate insulator of choice that has served in the MOS devices for more than 40 years, with high-*k* dielectric based on hafnium oxide (HfO_2) combined with metal gate [13].

When the gate leakage problem was solved in the following technology nodes the other leakage mechanism started to dangerously arise and started dominating transistor leakage along the gate length scaling. When the gate length is scaled to values which are in the same order of magnitude as depletion width in junctions formed between S/D and well, undesired short channel effects (SCE) arise. The main problem related to the SCEs is a drain-induced-barrier-lowering (DIBL) that causes significant drain current leakage increase along the drain voltage in a weak inversion mode. Despite extensive junction engineering works employing shallow highly doped source/drain extension (SDE), halo/pocket ion implants as well as retrograde well it was not possible to eliminate the SCEs and this could stop CMOS scaling at 22 nm technology (Fig. 5).

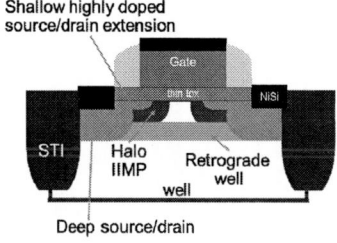

Fig. 5. Junction engineering in advanced planar technology nodes [11].

The only solution was to change the transistor architecture to the one with enhanced gate control over the channel. This is possible in 3D structures were the gate is wrapping the channel from multiple sides. For this reason Intel has introduced into high volume manufacturing 3D transistor structure (due to presence of fin named FinFET) in May 2011 (Fig. 6) [14]. Switching to high-*k* and metal gate was very challenging technologically but changing transistor structure to 3D was truly revolutionary and resulted in unprecedented performance improvement. Compared to last planar technology 32 nm, Intel reported 37% performance increase at low voltage and > 50% power reduction at constant performance for new 22 nm FinFET based technology.

177

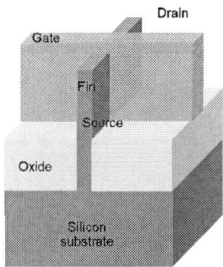

Fig. 6. FinFET 3D view illustration.

Since 2011 the era of FinFET has started and dominated CMOS main stream manufacturing. However, performance enhancement of FinFET started facing very critical challenges. These challenges may actually end the Moore's Law and will be discussed in the next section.

B. Critical challenges of FinFET Transistors Performance Improvement

Currently, in mainstream manufacturing are FinFETs with fin pitch 34 nm and gate pitch 54 nm [2, 15]. This allows to stuff as much as 100 million transistors in 1 square millimeter. Unlikely how it has been before such hyper scaling causes tremendous difficulties for FinFET performance improvement.

First of all, the architecture of the fin is very problematic, namely it is very thin (Fig. 7). Turns out that what is good for SCE control causes significant issues with performance improvement. A thin fin denotes a high channel resistance. In order to reduce the source/drain resistance raised source/drain (RSD) process is carried out by selective epitaxial growth (SEG) technique.

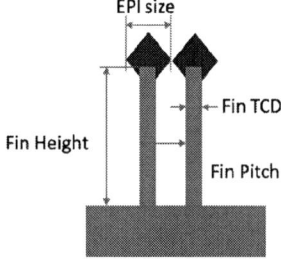

Fig. 7. Key features of the fin.

However, scaling the fin pitch causes less epi volume. Smaller epi volume leads to significantly higher source/drain resistance. S/D doping levels are reaching dopant solubility levels in Silicon ($\sim 10^{21}$ cm^{-3}) [16]. This also involves more complicated processes for dopant activation such as laser spike anneal (LSA) or dynamic surface anneal (DSA). What has become especially critical for high doping levels is a dopant de-activation. The process temperature as low as 500 C (in following gate module processes even with higher temperature) can cause significant dopant de-activation leading to significant S/D resistance increase [17]. Besides, a smaller fin pitch poses a higher risk for epi to epi shorts and thinner fin (especially top CD) causes significant DC parameters degradation due to

quantum confinement effect. This effect is observed for feature size smaller than 7 nm [18]. Another DC degradation comes from SiGe diminishing stress when the fin thickness and epi volume are reduced. This is partially being compensated by increase of fins height (increase of effective width in (1)). However, a tall fin with a very tight pitch results in a very challenging material fill for spacer formation (fill material pinch-off). Additionally, a tall fin also leads to significant increase of parasitic effective capacitance (C_{eff}) [19, 20]. To obtain high performance microprocessors it is necessary to provide transistors not only with high drain saturation current but also with low parasitic capacitances according to switching time formula:

$$t = CV/I \qquad (2)$$

where: C – parasitic capacitances, V – supply voltage, and I – drain saturation current. This requires spacer material replacement and switching to materials with low-k which integration is challenging and also has impact on transistors reliability (time-dependent dielectric breakdown - TDDB).

Following classical scaling / performance improvement by gate length (L_g) reduction in FinFET manufacturing is approaching to critical roadblock. When L_g is approaching and going below 17 nm, a replacement metal gate (RMG) recess process becomes extremely difficult (Fig. 8). A typical SoC (system-on-chip) requires multiple threshold voltages (mVT) i.e. regular and low for NFETs and PFETs. A volume-based mVT integration process requires multiple deposition / etch steps as well as fill and recess of sacrificial material which become very hard to control [21].

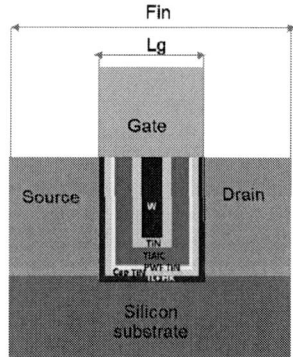

Fig. 8. Schematic view of gate in FinFET.

Tighter pitch leaves also less space for contacts (CA). In the middle-end-of-line (MOL) contact resistance (R_c) is a paramount problem.

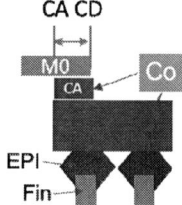

Fig. 9. Cobalt local interconnect [15].

178

Low R_c is a key issue to boost device performance for sub-10nm nodes. A key point is to achieve $Rc < 10^{-9}$ Ωcm^2 which requires, as it has been mentioned earlier, extremely high active dopant concentration ($> 5\times10^{20}$ cm^{-3}) in the source/drain epi regions. In order to further reduce the contact resistance also new materials are needed. In the latest technology nodes tungsten has been replaced with cobalt [15]. New innovative solutions are needed to further reduce contact resistance for the next technology nodes [22–24].

C. RC delay – Back End of Line

Till now a discussion about microprocessor performance was related to the transistor performance. However, critical impact on microprocessor maximum frequency has its wiring connections as well. Metal lines and a dielectric between them form so called RC delay which has a critical impact on final chip F_{max}. Scaling and decreasing the metal pitch leaves less volume for metal fill. A constant barrier/liner (B/L) thickness has the biggest impact on the copper volume and therefore resistance (Fig. 10).

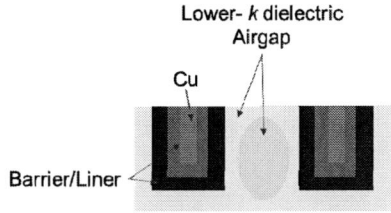

Fig. 10. Interconnect metallization.

Solutions with thinner B/L are needed and an ideal barrier-less solution offers the highest volume for metal fill. Big game changer, copper metallization which replaced aluminum, soon may be as well replaced with barrier-less ruthenium metallization [25]. In order to reduce the capacitance between metal lines, air-gapped interconnects have been introduced in 2014 [26]. Air-gaps offer lower capacitance but they are associated with reliability (TDDB) and mechanical integrity concerns.

D. Impact of process variation on performance

Last issue that needs to be highlighted and which is has critical impact on microprocessor performance is process variation. Minor technology process dispersion causes significant impact on FinFET electrical parameters fluctuations and its performance [27-29]. In many cases, maximum process variations have reached manufacturing tool capabilities and are getting tighter than equipment control limits. Multiple processes such as lithography and overlay, etch, deposition, epitaxy, chemical-mechanical polishing (CMP) and thermal processing cause very hard to control across-wafer and pitch-dependent process variability impacting not only microprocessor performance but yield and reliability as well. This problem is additionally enhanced when combined with local layout effects (LLE). Next generations of FinFETs with aggressively scaled pitches and improved performance as well as power efficiency won't be possible without new tools and processes such as super fine plasma etch [30, 31].

III.   SUMMARY

Intel's first 80286 chips released to market in 1982 were specified for a maximum clock rate of 4 MHz. In year 2000 AMD released first microprocessor with speed 1 GHz. This means that during roughly 20 years microprocessor clock rate increased 250 times. Another 20 years down the road to the year 2020, the best microprocessor reached top boost frequency 4.9 GHz [32, 33]. This means that during last 20 years microprocessor speed increased only 5 times. And this the best illustrates how problematic is increasing microprocessor speed by increasing clock rate due to difficult challenges described in this paper. For this reason other solutions have been implemented in order to increase the microprocessor speed. The first one was switching the architecture from 32-bit to 64-bit. A 64-bit processor is more capable (faster) than a 32-bit processor, because it can handle more data at once. The first AMD64-based processor, the Opteron, was released in April 2003 [34]. The other way to increase microprocessor computing speed was to multiple its cores. Multi-core processor, firstly demonstrated by IBM in 2001, can run instructions on separate cores at the same time, increasing overall program speed (supporting multithreading or other parallel computing techniques) [35].

What will happen next? The CMOS technology scaling will be extended most likely to 3 nm node. New device structure possibly Nano-sheet transistor (NSFET) will be introduced. Very interesting worth to mention fact is that according to International Roadmap for Devices and Systems predictions in year 2033 should be 0.7 nm technology in production based on vertical gate all around (VGAA) architecture with VGAA diameter 6 nm and pitch 14 nm [36]. Will it happen or not it is very hard to predict at the moment. Nevertheless sometime after 2025 year classical CMOS technology scaling as well as Moore's Law will end. Beyond Moore era will begin and for that new paradigm is required. Surprisingly graphene-based or carbon nanotube-based solutions, despite extensive research and large publicity, are not planned in CMOS mainstream manufacturing in foreseeable future [37]. Despite a lack of continuing compliance with Moore's Law there are many ways to improve performance without squeezing more transistors onto each chip. One way is a Heterogeneous Integration (HI) where small pieces of silicon (chiplets) containing IP and performing specific functions will be integrated with other chiplets through package-level integration [38-40]. SiP technology enables delivering cost effective functional diversification (equivalent to scaling) to maintain the pace of progress.

REFERENCES

[1] G. Moore, "Cramming more components onto integrated circuits", Electronics, April 19, 1965, pp. 114-117.

[2] S. Narasimha et al., "A 7nm CMOS technology platform for mobile and high performance compute application", 2017 IEEE International Electron Devices Meeting (IEDM), DOI: 10.1109/IEDM.2017.8268476

[3] A. Malinowski, J. Singh, US Patent 10,546,943

[4] A. Malinowski, C. F. Tan, N. Sassiat, M. Wiatr, US Patent 9,812,573

[5] A. Malinowski, J. Chen, S. K. Mishra, S. Samavedam and D. Sohn, "What is Killing Moore's Law? Challenges in Advanced FinFET Technology Integration," 2019 MIXDES - 26th International Conference "Mixed Design of Integrated Circuits and Systems", Rzeszów, Poland, 2019, pp. 46-51.

[6] ZDNet, "World's fastest processor is an overclocked beast", Sept. 14, 2011, https://www.zdnet.com/article/worlds-fastest-processor-is-an-overclocked-beast-video/

[7] S. Sen, V. Natarajan, and Ch. Abhijit., "Low-Power Adaptive Mixed Signal/RF Circuits and Systems and Self-Healing Solutions." (2011), 10.1007/978-1-4419-7418-1_9.

[8] P. Su and Y. Li, "Process technological analysis for dynamic characteristic improvement of 16-nm HKMG bulk FinFET CMOS circuits," 2016 IEEE 16th International Conference on Nanotechnology (IEEE-NANO), Sendai, 2016, pp. 812-815.

[9] A. Nazakat, L. Yungui, R. Liu and V. Chew, "Novel Techniques of FIB Edit on VDD Routing in Internal Circuit for IDDQ Leakage Failure Analysis," 2018 IEEE International Symposium on the Physical and Failure Analysis of Integrated Circuits (IPFA), Singapore, 2018, pp. 1-5.

[10] R.H. Dennard, F.H. Gaensslen, V.L. Rideout, E. Bassous, A.R. LeBlanc, "Design of ion-implanted MOSFET's with very small physical dimensions", IEEE Journal of Solid-State Circuits, Vol. 9 , Issue 5, 1974, pp. 256 – 268.

[11] Intel Technology Journal, Vol. 6, Issue 2, May 16, 2002 https://www.intel.com/content/dam/www/public/us/en/documents/research/2002-vol06-iss-2-intel-technology-journal.pdf

[12] T. Ghani et al., "A 90nm high volume manufacturing logic technology featuring novel 45nm gate length strained silicon CMOS transistors," IEEE International Electron Devices Meeting 2003, Washington, DC, USA, 2003, pp. 11.6.1-11.6.3.

[13] K. Mistry et al., "A 45nm Logic Technology with High-k+Metal Gate Transistors, Strained Silicon, 9 Cu Interconnect Layers, 193nm Dry Patterning, and 100% Pb-free Packaging," 2007 IEEE International Electron Devices Meeting, Washington, DC, 2007, pp. 247-250.

[14] Intel 22 nm Technology, https://simplecore.intel.com/newsroom/wp-content/uploads/sites/11/2011/05/22nm-Details_Presentation.pdf, May 2011.

[15] C. Auth et al., "A 10nm high performance and low-power CMOS technology featuring 3rd generation FinFET transistors, Self-Aligned Quad Patterning, contact over active gate and cobalt local interconnects," 2017 IEEE International Electron Devices Meeting (IEDM), San Francisco, CA, 2017, pp. 29.1.1-29.1.4.

[16] S. Mochizuki et al., "Advanced Arsenic Doped Epitaxial Growth for Source Drain Extension Formation in Scaled FinFET Devices," 2018 IEEE International Electron Devices Meeting (IEDM), San Francisco, CA, 2018, pp. 35.2.1-35.2.4.

[17] M. Schaekers, E. Rosseel, A. Schulze, H. Tielens, "Dopant activation and De-activation in HiP Si:P", IMEC / Globalfoundries, Malta, NY.

[18] E. Sperling, "Quantum Effects At 7/5nm And Beyond", Semiconductor Egineering, May 23, 2018, https://semiengineering.com/quantum-effects-at-7-5nm/

[19] S. Y. Mun et al., "Quantitative model of CMOS inverter chain ring oscillator's effective capacitance and its improvements in 14nm FinFET technology," 2018 IEEE International Conference on Microelectronic Test Structures (ICMTS), Austin, TX, 2018, pp. 153-156.

[20] S. Kim, M. Guillorn, I. Lauer, P. Oldiges, T. Hook and M. Na, "Performance trade-offs in FinFET and gate-all-around device architectures for 7nm-node and beyond," 2015 IEEE SOI-3D-Subthreshold Microelectronics Technology Unified Conference (S3S), Rohnert Park, CA, 2015, pp. 1-3.

[21] S. Hung, Gate Stack Module, Cross Business Unit, Applied Materials, "Multi-Vt Engineering and Gate Performance Control for Advanced FinFET Archetecture", IEDM2017 short course.

[22] S. K. Mishra et al., US Patent 10,084,093

[23] S. K. Mishra et al., US Patent 9,419,082

[24] S. K. Mishra et al., "Middle of Line: Challenges and Their Resolution for FinFET Technology" submitted to Advanced Semiconductor Manufacturing Conference – AMSC2020, May 4-7, 2020, Saratoga Springs, NY, USA.

[25] Z.Tökei, IMEC, "Sub-5nm Interconnect Trends and Opportunities", IEDM2017 short course.

[26] S. Natarajan et al., "A 14nm logic technology featuring 2nd-generation FinFET, air-gapped interconnects, self-aligned double patterning and a 0.0588 µm2 SRAM cell size," 2014 IEEE International Electron Devices Meeting, San Francisco, CA, 2014, pp. 3.7.1-3.7.3.

[27] D. Tomaszewski, A. Malinowski, M. Zaborowski, P. Salek, L. Lukasiak and A. Jakubowski, "Fluctuations of electrical characteristics of FinFET devices," 2009 MIXDES-16th International Conference Mixed Design of Integrated Circuits & Systems, Lodz, 2009, pp. 61-66.

[28] A Malinowski, A Kociubinski, P Salek, L Lukasiak, M Zaborowski, D Tomaszewski, A Jakubowski, "Electrical characterization of FinFETs", 2008 15th International Conference on Mixed Design of Integrated Circuits and Systems, pp. 65-69.

[29] A. Malinowski, PhD thesis "Analysis of Dispersion of Electrical Parameters of FinFETs", Warsaw University of Technology, The Institute of Microelectronics and Optoelectronics, May 24, 2016.

[30] A. Malinowski et al., "A novel fast and flexible technique of radical kinetic behaviour investigation based on pallet for plasma evaluation structure and numerical analysis", 2013 J. Phys. D: Appl. Phys. 46 265201, doi: 10.1088/0022-3727/46/26/265201

[31] A. Malinowski, PhD thesis "Study of radical kinetic behavior investigation technique and its application in ultimate CMOS TCAD topography simulation", Nagoya University, Japan, March 26, 2012.

[32] ZDNet, "It's official: AMD hits 1,000MHz first", https://www.zdnet.com/article/its-official-amd-hits-1000mhz-first-5000096067/

[33] Tom's Hardware, "Best Gaming CPUs for 2020", https://www.tomshardware.com/reviews/best-cpus,3986.html

[34] Digital Trends, "32-bit vs. 64-bit: What it really means", https://www.digitaltrends.com/computing/32-bit-vs-64-bit-operating-systems/

[35] IBM, "Power 4 The First Multi-Core, 1GHz Processor", https://www.ibm.com/ibm/history/ibm100/us/en/icons/power4/

[36] IEEE International Roadmap for Devices and Systems, Metrology, 2017 edition, https://irds.ieee.org/images/files/pdf/2017/2017IRDS_MET.pdf.

[37] A. Malinowski et al., "Modeling Considerations and Performance Estimation of Single Carbon Nano Wall based Field Effect Transistor by 3D TCAD Simulation Study", Transactions of the Materials Research Society of Japan, Vol. 35, Issue 3, pp. 669-674.

[38] T. Bieniek et al., "Multi-Domain Modeling and Simulations of the Heterogeneous Systems", Journal of Telecommunications and Information Technology, 2010, pp. 34-39.

[39] S. K. Moore, "Intel's View of the Chiplet Revolution", IEEE Spectrum, https://spectrum.ieee.org/tech-talk/semiconductors/processors/intels-view-of-the-chiplet-revolution

[40] International Technology Roadmap for Semiconductors 2.0, 2015 Edition, Heterogeneous Integration.

Proceedings of the 27th International Conference *"Mixed Design of Integrated Circuits and Systems"*
June 25-27, 2020, Łódź, Poland

Recessed and P-GaN Regrowth Gate Development for Normally-off AlGaN/GaN HEMTs

Chaymaa Haloui[1,2], Gaëtan Toulon[3], Josiane Tasselli[1], Yvon Cordier[4], Éric Frayssinet[4],
Karine Isoird[1], Frédéric Morancho[1], Mathieu Gavelle[2]

[1]LAAS-CNRS, Toulouse University, CNRS, UPS, Toulouse, France
[2]CEA Tech Occitanie, Toulouse, France
[3]EXAGAN, Toulouse, France
[4]CRHEA-CNRS, Valbonne, France
chaloui@laas.fr

Abstract—**A new normally-off AlGaN/GaN HEMT structure is proposed. The regrowth of a P-GaN layer on the AlGaN/GaN heterostructure after the gate recess allows the achievement of the enhancement mode. A shift in the threshold voltage to positive values has been proved through simulation results. A precise control of the etch depth for the gate recess is detailed.**

Keywords—**HEMT; normally-off; AlGaN/GaN; gate recess; RIE; P-GaN regrowth.**

I. INTRODUCTION

AlGaN/GaN high-electron mobility transistors (HEMTs) have attracted worldwide attention in power electronics as candidates for next-generation of high-speed switching devices. Thanks to the large electric field of GaN and the high carrier mobility and density in the two-dimensional electron gas (2DEG), AlGaN/GaN HEMTs can achieve high breakdown voltage and realize ultrahigh power density operation with low power losses.

While most of the demonstrated AlGaN/GaN HEMTs are inherently normally-on with a negative gate threshold voltage, normally-off mode is strongly demanded to fulfill the requirements of power electronics applications; normally-off devices are inherently secure and suitable for energy converters requiring specifically high system reliability. Several approaches, each with its own limitations, have been proposed to convert the inherent depletion mode (normally-on) into an enhancement one (normally-off). Fluorine plasma ion implantation [1], oxygen treatment [2], gate injection transistor (GIT) [3] and P-GaN gate [4] are the most developed ones. In this paper a structure that combines two approaches (recessed and P-GaN regrowth gate) is presented. The use of a P-GaN layer on the AlGaN/GaN heterostructure under the gate contact region lifts up the band diagram, which causes the depletion of the 2DEG channel, even in the absence of external bias. First, the simulated performance of the new device will be presented, then an overview of the technological process mandatory to realize such a device will be proposed, with a particular focus on the gate recess step.

II. THE SIMULATED STRUCTURE

Numerical simulations were performed with Sentaurus TCAD tools in order to have an insight of the HEMT parameter sensitivity. The designed structure has source, drain and gate contact lengths of 1 μm, a gate-source distance of 2 μm and a gate-drain distance of 15 μm. A gate field plate of 3 μm is added at drain side. Contrary to other studies [5] [6], the deep ionization energy of the Mg dopant is not considered, which means that this region presents an "equivalent" uniform doping profile. Comparative simulations (not presented) with the use of the "incomplete ionization" model present similar results in terms of threshold voltage. Two cases were analyzed: the one with the gate metal directly on top of the P-GaN region, and the other with an insulator between them (Fig. 1).

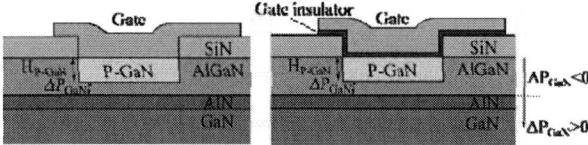

Fig. 1. Gate detail of the proposed structure with metal on P-GaN (left) and with gate insulator on top of P-GaN (right)

First simulations were performed on structures with metal on P-GaN. Band diagram through the gate at 0 V for different AlGaN thicknesses below the gate is presented in

Fig. 2. The etched AlGaN effect can be seen on the conduction band profile, especially at GaN side. Negative ΔP_{GaN} values means that the remaining AlGaN layer under the gate provides higher polarization charge to deplete, that finnaly results to lower threshold voltage.

Simulation results of the threshold voltage variations as a function of the P-GaN doping are represented in Fig. 3. When considering ohmic contact, no depletion appears on P-GaN, that can be confirmed by the absence of band bending on

Fig. 2, so the threshold voltage is independent on P-GaN doping concentration. On the contrary, the Schottky contact puts the energy bands downward for lower values of doping concentration, while only high doping values will pin the Fermi

181

level close to the valence band. Therefore, the threshold voltage variation is related to the depletion thickness on P-GaN.

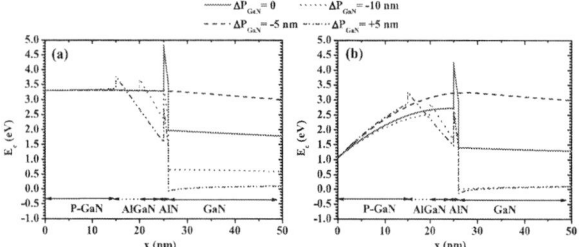

Fig. 2. Conduction band energy profile through the gate for different ΔP_{GaN} values for ohmic (a) and Schottky (b) gate contact.

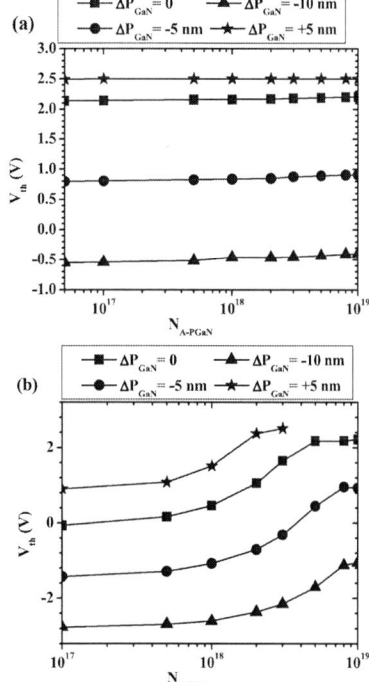

Fig. 3. Threshold voltage as a function of P-GaN doping concentration for different ΔP_{GaN} values when the gate contact is defined ohmic (a) and Schottky (b).

For the analysis of the structure with a gate insulator, the material used as insulator as well as its thickness have to be considered. Materials such as SiO_2 or Al_2O_3 are promising due to their higher bandgap and electron affinity [7], which means an increase in the conduction band continuity with the P-GaN. However, the lower relative permittivity of SiO_2 is not favourable in terms of electrical characteristics since it gives lower transconductance as represented in Fig. 4. Moreover, it has been demonstrated that SiO_2 induces high density of surface states with GaN [8].

Threshold voltage variations with P-GaN doping concentration for different insulator materials and thicknesses are presented in Fig. 5. Interface states density has been inserted, based on extracted values for structures with the Si_3N_4

layer deposited on P-GaN by LPCVD. These values are expected to change with the insulator material and its deposition method.

Results from structures with insulator show similar trends that previously observed. The additional layer of dielectric absorbs more or less electric field when a gate voltage is applied, depending on its thickness and permittivity, enhancing by the way the threshold voltage variation with doping concentration, which means that the threshold voltage presents stronger voltage variations with P-GaN doping. If high P-GaN doping concentration can be demonstrated, high positive threshold voltage is achievable.

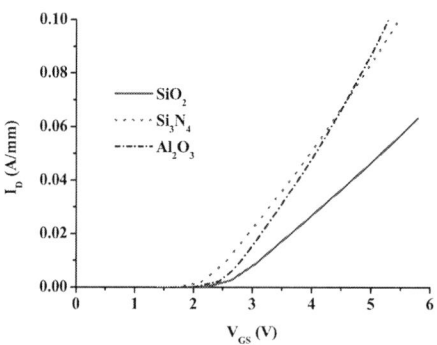

Fig. 4. Transfer characteristics I_D (V_{GS}) of HEMT with different gate insulators of 30 nm thick.

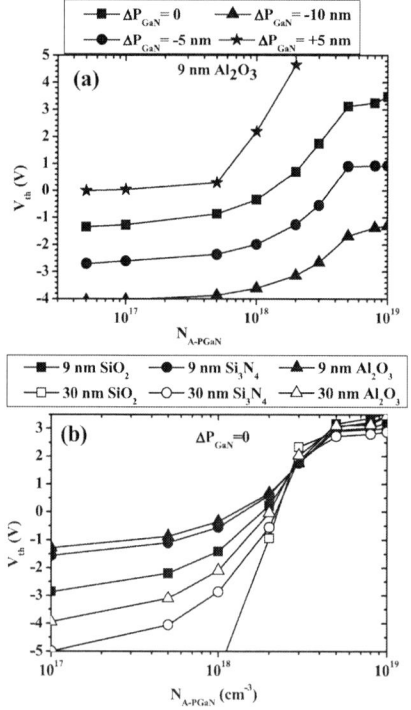

Fig. 5. Threshold voltage as a function of doping concentration for different parameters: (a) P-GaN distance(ΔP_{GaN}) from AlN surface for 9 nm of Al_2O_3 and (b) different insulator materials and thicknesses.

III. GATE RECESS PROCESS

A fabrication process of the presented device was conducted in order to validate the simulation results previously detailed. The AlGaN/GaN HEMT layers were grown by metal organic chemical vapor deposition (MOCVD) on Si: the epilayer stacking is composed of an alternation of AlN/GaN layers, followed by a 1.5 μm GaN, 1.5 nm AlN interface enhancement layer, a 25 nm $Al_{0.3}GaN_{0.7}$ layer and a 10 nm SiN cap layer.

Before the gate recess, 50 nm of Si_3N_4 and 100 nm of SiO_2 were deposited on the epitaxial layers, by LPCVD and ICPECVD respectively (Fig. 6.a) The role of Si_3N_4 is to conserve the 2DEG density and the SiO_2 one is to prevent the growth of the P-GaN outside the gate region. The gate is then opened through the etching of three materials: SiO_2 and Si_3N_4 were both removed by CHF_3/O_2 ICP plasma and AlGaN by Cl_2 RIE dry etching (Fig. 6.b). This step is pursued by localized MBE epitaxy of a P-GaN layer into the gate region (Fig. 6.c). The SiO_2 masking layer is removed thereafter by wet etching (Fig. 6.d).

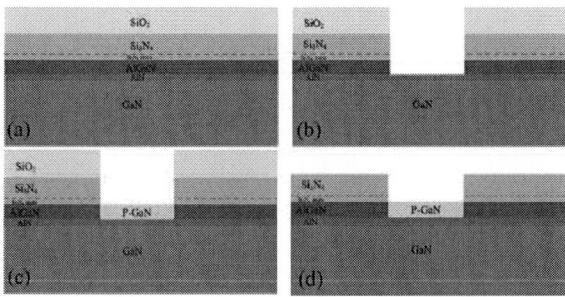

Fig. 6. Schematic process flow of the localized regrowth of p-GaN on the gate region. (a) Succession of the epitaxial layers on the Si substrate; (b) gate recess by dry plasma etching (CHF_3/O_2 for Si_3N_4 and SiO_2, Cl_2 for AlGaN); (c) P-GaN regrowth by MBE with a Mg doping; (d) SiO_2 removal by wet etching.

A. AlGaN etching

For the gate recess, inductively coupled plasma or reactive ion etching (ICP-RIE) are the most used techniques owing to their anisotropy etching and efficiency. However, several requirements must be fulfilled when using these techniques such as a smooth surface morphology and a low damage. As mentioned before, a normally-off P-GaN technology requires the use of a thin AlGaN barrier layer (25 nm), therefore the etching rate has to be well-controlled in order to preserve the 2DEG properties. Selectivity is a very important factor in mastering etching rate: it depends on several parameters such as gas chemistry, chamber pressure and temperature. To the best of our knowledge, there is no selective recipe for AlGaN etching and the etch rate strongly depends on plasma conditions. The majority of AlGaN etching processes are based on chlorine as a principal etching agent: the chlorine-containing reactant may be boron trichloride (BCl_3), chlorine (Cl_2), or a mixture of the two gases [9].

In the present work, a Cl_2-based etching was carried out in a RIE plasma etch system for the AlGaN gate recess. The etching conditions were the same as reported by D. Buttari et al. [10]: RF power = 60 W, pressure = 5 mTorr and

Cl_2 flow = 10 sccm. The wafer was patterned using ECI 3012 photoresist of 1.1 μm thick. The removal of the photoresist mask, after etching, was performed by acetone and isopropanol (IPA), followed by DI water and O_2 plasma (800 W). However, further cleaning was necessary for removing the photoresist post-etch residues remaining on the AlGaN layer: it will be discussed later. Etch depths were measured by transmission electron microscopy (TEM) and determined to be about 6 nm, 19 nm and 21 nm for 25 s, 35 s and 45 s etching times respectively. A FIB sectional view of the gate after etching is presented in Fig. 7 for 35 s AlGaN etching.

The use of a medium RIE power (< 100 W) and the consequent slow etch rate allowed an accurate control of AlGaN thinning. Times lower than 25 s have not been tested because of the eventual presence of a thin surface oxide layer formed before the etch, that can give rise to a dead time by inhibiting etch at the beginning [11]. Above 45 s, the AlGaN is completely etched as well as the underlying layers, AlN and part of GaN.

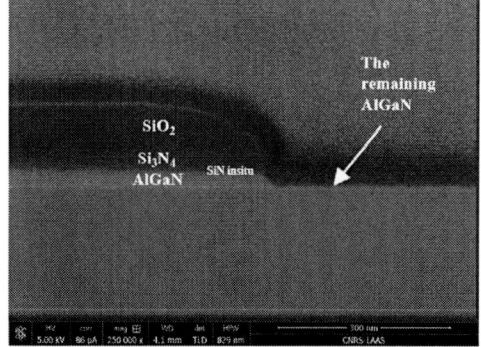

Fig. 7. Picture of FIB cut after the gate recess for RIE AlGaN etching of 35 s.

B. Surface roughness

The root means square roughness RMS measured by AFM on a 2 x 2 μm2 windows before and after etching are respectively 0.43 nm and 1.22 nm as shown in Fig. 8. The AlGaN surface observed after a partial etching of 25 s exhibited a RMS three times higher than the non-etched AlGaN one. It is very difficult to compare these values to those of literature because the roughness after etching depends on the as-grown surface roughness and the etching parameters. However, the obtained values remain in the range of values reported in literature [12].

It has been proved that the RMS roughness of the etched AlGaN can be strongly affected by oxidation during plasma etching; the in-situ produced aluminum oxides may provide a self-mask effect [13]. Given that the bond energy of Al-O (21.2 eV/atom) is higher than that of Al-N (11.52 eV) and Ga-N (8.92 eV), AlGaN etch would be limited by the formation of the aluminum oxide. Therefore, the non-uniformity of the aluminum oxide distribution will cause roughness on the etched surface. The addition of an appropriate quantity of BCl_3 to the plasma mixture could help to improve the smoothness of the etched surface. Indeed, BCl_3 reacts with oxygen to form some BCl_xO_y gases which could facilitate the removal of the oxygen remaining into the chamber [14].

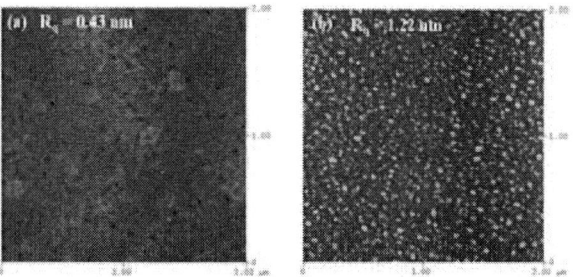

Fig. 8. AFM surface images of the AlGaN (a) before etching, (b) after Cl_2-RIE etching for 25 s: RF power = 60 W, pressure = 5 mTorr and Cl_2 flow = 10 sccm.

C. Post-etch residues

After the resist removal, we observed some residues on the etched surface that were very difficult to eliminate by conventional wet stripping methods. The SEM image in Fig. 9.a shows a veil residue on the etched region. These residual impurities are inherent by-products of the Cl_2-based etch process. They are probably formed by a mixture of species stemming from the plasma ions, the photo-resist mask and the etched materials (SiO_2, Si_3N_4 and AlGaN), which prevents dissolution by solvents. Several experiments conducted to remove these tenacious post-etch residues by common chemical strippers were ineffective [15]. The selected solution consists of UV insolation of the wafer followed by a developing step. The efficiency of the after-etching cleaning can be observed in Fig. 9.b.

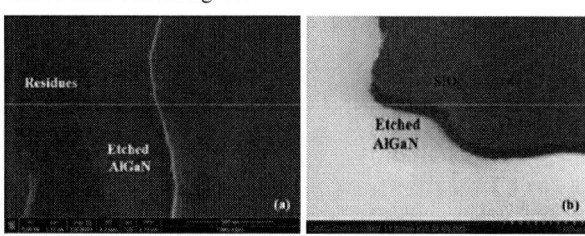

Fig. 9. SEM image showing (a) post-etch residues after the resist removal by acetone, isopropanol and DI water followed by O_2 plasma (5 min, 800 W); (b) the removal of the post-etch residues by UV insolation and development after etching.

IV. CONCLUSION

In this work, a new design of a normally-off HEMT was proposed. The introduction of a P-GaN layer on the AlGaN/GaN heterostructure under the gate contact allows the normally-off functionality. The simulations were carried out in two cases: metal on P-GaN (ohmic and Schottky) and gate with insulator for different recess depths and P-GaN doping concentrations. High threshold voltages can be achieved by reducing the thickness of the remaining AlGaN. With the ohmic contact, the threshold voltage is independent on P-GaN doping concentration, contrary to the Schottky one. With the gate insulator structure, the electrical performance depends on the type of insulator and its thickness. Then, a focus on the gate recess process, mandatory for the realization of such devices, was also detailed in this work. A precise AlGaN etching was achieved by adjusting Cl_2-RIE etching parameters. The roughness surface after etching was three times higher than the non-etched one but remains acceptable for a RIE mode. Some

post-etch residues were found on the etched AlGaN surface and were removed by UV insolating and developing steps. The work of realization oh the whole fabrication process of the AlGaN /GaN HEMT is underway in order to demonstrate the normally-off behavior of such a device.

ACKNOWLEDGMENT

This work was partly supported by LAAS-CNRS micro and nanotechnologies platform member of the French RENATECH network.

REFERENCES

[1] K. J. Chen et al., "Physics of fluorine plasma ion implantation for GaN normally-off HEMT technology," 2011, pp. 19.4.1-19.4.4, doi: 10.1109/IEDM.2011.6131585.

[2] Y.-L. He et al., "Recessed-gate quasi-enhancement-mode AlGaN/GaN high electron mobility transistors with oxygen plasma treatment," Chin. Phys. B, vol. 25, no. 11, p. 117305, Nov. 2016, doi: 10.1088/1674-1056/25/11/117305.

[3] H. Okita et al., "Through recessed and regrowth gate technology for realizing process stability of GaN-GITs," in 2016 28th International Symposium on Power Semiconductor Devices and ICs (ISPSD), Prague, Czech Republic, 2016, pp. 23–26, doi: 10.1109/ISPSD.2016.7520768.

[4] Y. Zhong et al., "Effect of Thermal Cleaning Prior to p-GaN Gate Regrowth for Normally Off High-Electron-Mobility Transistors," ACS Appl. Mater. Interfaces, vol. 11, no. 24, pp. 21982–21987, Jun. 2019, doi: 10.1021/acsami.9b03130.

[5] L. Efthymiou, G. Longobardi, G. Camuso, T. Chien, M. Chen, and F. Udrea, "On the physical operation and optimization of the p-GaN gate in normally-off GaN HEMT devices," Appl. Phys. Lett., vol. 110, no. 12, p. 123502, Mar. 2017, doi: 10.1063/1.4978690.

[6] I. Hwang et al., "p-GaN Gate HEMTs With Tungsten Gate Metal for High Threshold Voltage and Low Gate Current," IEEE Electron Device Lett., vol. 34, no. 2, pp. 202–204, Feb. 2013, doi: 10.1109/LED.2012.2230312.

[7] J. Robertson and B. Falabretti, "Band offsets of high K gate oxides on III-V semiconductors," J. Appl. Phys., vol. 100, no. 1, p. 014111, Jul. 2006, doi: 10.1063/1.2213170.

[8] T. Hashizume, S. Ootomo, T. Inagaki, and H. Hasegawa, "Surface passivation of GaN and GaN/AlGaN heterostructures by dielectric films and its application to insulated-gate heterostructure transistors," J. Vac. Sci. Technol. B Microelectron. Nanometer Struct., vol. 21, no. 4, p. 1828, 2003, doi: 10.1116/1.1585077.

[9] W. Yang, T. Ohba, S. Tan, K. J. Kanarik, J. Marks, and K. Nojiri, "Atomic Layer Etching of GaN and Other III-V Materials," US 2016/0358782 A1, 2016.

[10] D. Buttari et al., "Systematic characterization of Cl_2 reactive ion etching for improved ohmics in AlGaN/GaN HEMTs," IEEE Electron Device Lett., vol. 23, no. 2, pp. 76–78, Feb. 2002, doi: 10.1109/55.981311.

[11] D. Buttari et al., "Origin of etch delay time in Cl2 dry etching of AlGaN/GaN structures," Appl. Phys. Lett., vol. 83, no. 23, pp. 4779–4781, Dec. 2003, doi: 10.1063/1.1632035.

[12] G. Greco, F. Iucolano, and F. Roccaforte, "Review of technology for normally-off HEMTs with p-GaN gate," Mater. Sci. Semicond. Process., vol. 78, pp. 96–106, May 2018, doi: 10.1016/j.mssp.2017.09.027.

[13] Z. Gao, M. F. Romero, and F. Calle, "Etching of AlGaN/GaN HEMT structures by Cl2-based ICP," p. 4.

[14] Y. Han et al., "Nonselective and smooth etching of GaN/AlGaN heterostructures by Cl2/Ar/BCl3 inductively coupled plasmas," J. Vac. Sci. Technol. Vac. Surf. Films, vol. 22, no. 2, pp. 407–412, Mar. 2004, doi: 10.1116/1.1641054.

[15] G. Levitin, C. Timmons, and D. W. Hess, "Photoresist and Etch Residue Removal," J. Electrochem. Soc., vol. 153, no. 7, p. G712, 2006, doi: 10.1149/1.2203096.

Testing and Reliability

Proceedings of the 27th International Conference *"Mixed Design of Integrated Circuits and Systems"*
June 25-27, 2020, Łódź, Poland

A Human Immunity Inspired Intrusion Detection System to Search for Infections in an Operating System

Patryk Widulinski, Krzysztof Wawryn

Faculty of Electronics and Computer Science

Koszalin University of Technology

Koszalin, Poland

patryk.widulinski@tu.koszalin.pl, wawryn@tu.koszalin.pl

Abstract—In the paper, an intrusion detection system to safeguard computer software is proposed. The detection is based on negative selection algorithm, inspired by the human immunity mechanism. It is composed of two stages, generation of receptors and anomaly detection. Experimental results of the proposed system are presented, analyzed, and concluded.

Keywords—artificial immune system; intrusion detection system; negative selection algorithm; computer security, operating system

I. INTRODUCTION

During several decades, Intrusion Detection Systems (IDS) have been developed to protect computer software and computer networks from virus attacks. Traditional IDS designed as computer security programs are not able to cope with an increasing complexity of computer software and a detection of previously unknown intrusions. There are several analogies between tasks executed by the Intrusion Detection Systems and Natural Immune Systems, so immune mechanisms of the human defense system become a source of inspiration for IDS designers. Exploration and adoption of these mechanisms in IDS solutions have been developed to meet listed challenges. Many approaches to IDS based on Artificial Immune Systems have been presented in the literature. Most of them are focused on anomaly intrusion detection systems and negative selection method to detect infections (anomalies) in the computer software. The negative selection method is used to recognize self and nonself patterns in the computer software and allows to detect occurred anomalies. Early, fundamental works on IDS based on Artificial Immune Systems may be found in [1]–[6]. They initiate plenty of more sophisticated systems. A few examples of those systems are presented in [7]–[14].

Our IDS is devoted to detection of infections in an operating system. It is based on the negative selection algorithm consisting of two steps, receptor generation and anomaly detection. Code sections of the program and receptors are both represented by binary strings. Receptors are generated randomly and receptors recognizing self patterns of the code sections are discarded. In this way, a set of receptors recognizing nonself patterns only, is created. These receptors are used to search for the infections in the operating system by the anomaly detection

algorithm. The idea of the algorithm relies on a comparison between the receptor and read fragments of the program. If a consecutive given number of bits is identical, an anomaly is detected which means that the read fragment of the program is infected. The paper is organized as follows. The proposed IDS is described in Section II. Analysis of the experimental results is presented in Section III, and finally conclusions about the work are presented in Section IV.

II. OUR SOLUTION

We propose an AIS-based system which may be used to detect irregularities within compiled computer programs. A diagram of the system is shown in Fig. 1.

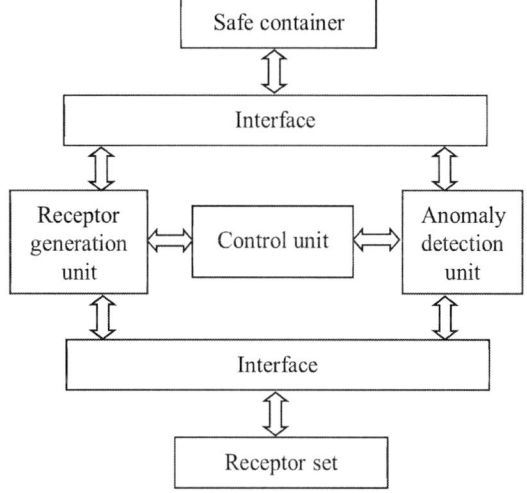

Fig. 1. A diagram of the proposed IDS.

The IDS monitors an area prone to infections in the operating system. The monitored area is called the safe container. The safe container houses programs the integrity of which is required. IDS operation is possible only if programs in the safe container are in a valid, supported format. The program format that is supported by the IDS is Win32 PE (Portable Executable). A typical valid PE file contains sections, which have specific characteristics such as their purpose (code sections, initialized data, uninitialized data) as well as indications whether or not they can be read, written to and executed by the CPU.

The control unit is a block in the IDS which controls receptor generation and anomaly detection units. Proposed IDS scans given PE files for executable code sections using the files' headers and generates special binary strings, called receptors for each file. Generated receptors are stored in a separate container and are used to detect anomalies within the files in the safe container.

A. Receptor generation

In order to detect anomalies within the program code, the IDS needs to construct a set with receptors. Receptors are binary strings of length l. If used within specific formulae, they have the ability to recognize nonself code patterns. A flowchart of the receptor generation algorithm is shown in Fig. 2.

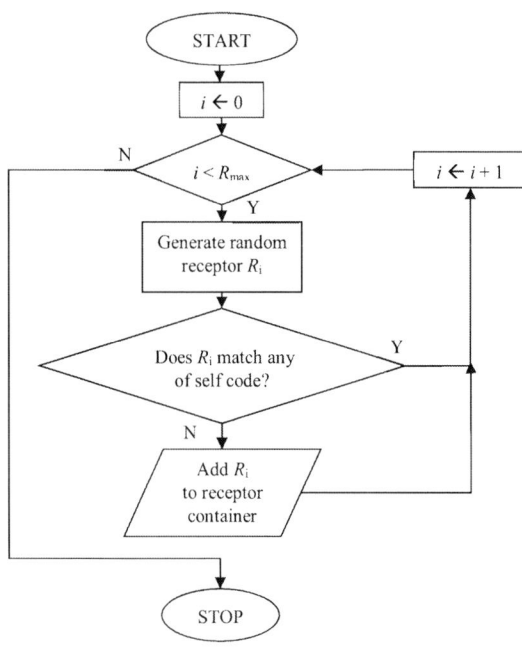

Fig. 2. A flowchart of the receptor generation algorithm.

The number of receptors to be generated by the algorithm is specified by the user and is denoted R_{max}. The maximum quantity of receptors that can be generated by the system for a single file is defined by (1):

$$max(R_{max}) = 2^l, \qquad (1)$$

where l denotes the length of the receptor in bits. Various methods to construct the receptor set exist. The approach

of the proposed IDS is to generate the receptors randomly. In this receptor generation algorithm, a pseudorandom 8-bit number generator is used to create l-sized receptors. The generator produces pseudorandom 8-bit numbers from range [0; 255]. The number of 8-bit fragments to randomly generate for a single receptor is denoted f and is specified by (2):

$$f = ceil(div(l, 8)), \qquad (2)$$

where l is the length of the receptor in bits, $div(a, b)$ is the result of arithmetic division of a by b and $ceil(a)$ is the ceiling function. Once the 8-bit fragments are generated, they are concatenated together in a string sense, creating a random receptor. If l is not divisible by 8, redundant bits are present in the receptor. The redundant bits are stored in memory, but are not used further in the generation or detection algorithms. Examples of random receptor generation are shown in Fig. 3 and Fig. 4.

01101001	11011000	10000101	11101101
69	D8	85	ED

Fig. 3. An example of a random receptor generation for $l = 28$, $f = 4$ (above: binary representation, below: hexadecimal representation). Bits not marked in bold are redundant.

11100011	10111011
E3	BB

Fig. 4. An example of a random receptor generation for $l = 16$, $f = 2$ (above: binary representation, below: hexadecimal representation). Note absence of redundant bits.

In the figures above, two examples of random receptor generation are listed. In Fig. 3, for $l = 28$ it can be seen that the random number generator produced four bytes: 0x69, 0xD8, 0x85 and 0xED. These bytes were then concatenated together to create a factually 32-bit receptor 0x69D885ED. However, due to $l = 28$, four bits will not be used further by the proposed algorithms. In Fig. 4, the number $l = 16$, and because it is a number divisible by 8, no redundant bits are present. Therefore, 8-bit numbers 0xE3 and 0xBB are joined together to create a receptor, 0xE3BB, and the entirety of it will be used further – if it passes the false positive checks.

A distinctive feature of an AIS-based approach using receptors is the activation threshold, denoted e. The activation threshold is a mandatory number of consecutive matching bits between a receptor and a read code fragment to consider that receptor matched, or "activated". If there are no at least e consecutive bits between the structure and the receptor, then the receptor is not considered matched/activated. Because the purpose of a receptor is to detect nonself structures and not the self patterns, it is imperative that every generated receptor does not cause false positives due to self code detection. To assure this, the IDS checks every newly generated receptor against the file that has to be safeguarded. The algorithm opens the PE file through a file system interface and navigates its PE header to locate information about code sections. The IDS then compares the newly generated receptor against all code in the code sections of the PE program. To achieve proper comparison, the algorithm reads file data in chunks of length l (the same as receptor size), starting at the beginning of every code section.

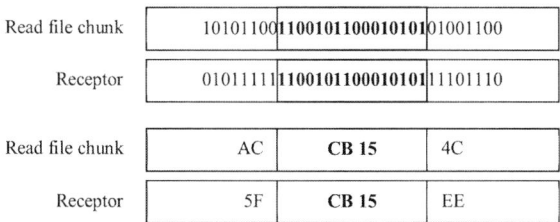

Fig. 5. An example of receptor matching a file fragment for $l = 32$, $e = 16$. The fragment was matched with a window shift of $k = 8$.

If at any point during the scan the newly generated receptor matches at least e consecutive bits anywhere in the code sections, then it is ruled out as a possible receptor. Consecutive e bits are called a window. The current position of the window with regard to a receptor is called the window shift and is denoted k. In case the receptor doesn't match e consecutive bits anywhere in code sections, it is added to the receptor set through an interface. The final number of receptors added to the receptor set is denoted R_n.

B. Anomaly detection

Proposed system detects anomalies in the code sections of programs in the safe container using receptors generated earlier, when the files were in a trusted state. A flowchart of the anomaly detection algorithm is shown in Fig. 6.

Before the algorithm starts scanning PE programs for irregularities, the current number of receptors is retrieved from the receptor set and is denoted R_n. For every receptor present in the set, the algorithm reads PE file code section fragments in sizes of l and performs a comparison between the receptor and retrieved fragments.

Whenever a consecutive e count of bits is identical, the receptor is said to be activated, and an anomaly is detected. The information about the detection of this anomaly is then forwarded to the user, including the location of occurrence.

III. EXPERIMENTAL RESULTS

Presented algorithms have been tested experimentally. A Windows implementation of the proposed IDS has been written in Microsoft Visual C# 2017. The implementation was tested on the Windows 7 64-bit operating system with an Intel i7-7700K CPU.

A file directory has been set up on the test computer's hard drive to serve as the self container for the detection system.

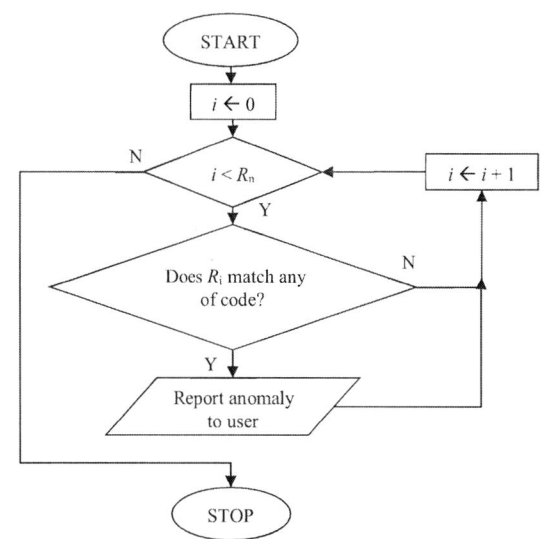

Fig. 6. A flowchart of the anomaly detection algorithm.

The directory contained one file and its properties are shown in Fig. 7.

File name	sample.exe
File size (bytes)	9216
Code section count	1
Code section range(s)	[0x100; 0x400]
Total code section size (bytes)	768

Fig. 7. Properties of the clean PE file in the safe container.

The IDS has been tested using various receptor bit sizes l and activation thresholds e. Each full l and e combination test is called a run. A single run consisted of 10 rounds of 1000 receptors to be generated. Therefore, $R_{max} = 10000$ in the testing procedure. The receptors were generated using a clean, unmodified PE file. After receptor generation, an anomaly was artificially introduced into the PE file to simulate a virus infection. The anomaly was introduced at the same location in the file every time and consisted of 0x90 machine code bytes (x86 NOP instruction). For each l and e combination, three anomaly sizes were tested. The tested anomaly sizes were 1 byte, 2 bytes and 4 bytes. Five runs were done for every l, e and anomaly size combination.

First tests were done using $l = 16$ and $e = 8$. The results of the tests are presented in Tables I, II and III, where *Run #* = number of run, S_{avg} = average discarded receptor count per 1000 generated receptors, Gtr_{avg} = average time of receptor generation per round in milliseconds, R_n = final receptor count in the receptor set, Gt_{avg} = total time for receptor generation in milliseconds, M = number of receptors that matched the introduced anomaly and At = total time for anomaly detection algorithm to finish execution.

TABLE I
EXPERIMENTAL RESULTS FOR $l = 16$, $e = 8$, ANOMALY SIZE = 1 BYTE.

Run #	S_{avg}	Gtr_{avg}	R_n	Gt_{avg}	M	At
1	997	327	24	3279	0	26
2	997	311	28	3123	0	30
3	996	291	32	2922	0	40
4	996	278	39	2788	0	43
5	996	295	32	2963	0	36

TABLE II
EXPERIMENTAL RESULTS FOR $l = 16$, $e = 8$, ANOMALY SIZE = 2 BYTES.

Run #	S_{avg}	Gtr_{avg}	R_n	Gt_{avg}	M	At
1	996	268	37	2690	0	40
2	997	260	25	2611	0	27
3	996	312	35	3137	0	39
4	996	326	34	3270	1	36
5	996	321	31	3218	1	33

TABLE III
EXPERIMENTAL RESULTS FOR $l = 16$, $e = 8$, ANOMALY SIZE = 4 BYTES.

Run #	S_{avg}	Gtr_{avg}	R_n	Gt_{avg}	M	At
1	995	319	46	3203	2	49
2	996	314	34	3146	0	37
3	996	364	39	3648	0	44
4	997	311	26	3121	1	27
5	997	225	28	2269	2	29

TABLE IV
EXPERIMENTAL RESULTS FOR $l = 32$, $e = 16$, ANOMALY SIZE = 1 BYTE.

Run #	S_{avg}	Gtr_{avg}	R_n	Gt_{avg}	M	At
1	57	1002	9427	10029	0	9895
2	64	991	9358	9922	0	9832
3	63	990	9366	9912	1	9851
4	56	995	9440	9958	0	9911
5	63	987	9363	9880	1	9815

TABLE V
EXPERIMENTAL RESULTS FOR $l = 32$, $e = 16$, ANOMALY SIZE = 2 BYTES.

Run #	S_{avg}	Gtr_{avg}	R_n	Gt_{avg}	M	At
1	61	1000	9388	10010	2	9871
2	64	986	9357	9868	3	9839
3	60	984	9395	9856	1	9921
4	61	989	9389	9899	2	9882
5	64	987	9357	9879	0	9841

TABLE VI
EXPERIMENTAL RESULTS FOR $l = 32$, $e = 16$, ANOMALY SIZE = 4 BYTES.

Run #	S_{avg}	Gtr_{avg}	R_n	Gt_{avg}	M	At
1	65	1008	9348	10090	2	9812
2	66	995	9338	9966	0	9820
3	57	995	9424	9962	2	9881
4	61	998	9390	9993	2	9814
5	63	989	9366	9906	3	9772

For $l = 16$ and $e = 8$, in case of anomaly size of 1 byte there were no detections at all for the introduced anomaly. Across all five test runs, for every 1000 receptors generated randomly by the IDS an average count of 996 receptors was discarded due to their generation of false positives, which means that 4 receptors were accepted into the receptor set per one round of generation. All receptors were generated and ready to use in 3015 milliseconds on average, and the full anomaly check took 35 milliseconds on average.

The test case of a 2-byte anomaly saw successful detection by the IDS during runs 4 and 5. Proposed system detected the introduced anomaly with a single receptor in 36 milliseconds during the fourth run, and in 33 milliseconds during the fifth test run. Across all runs, the average activated receptor count was 0.4 and detection rate was 40%. The 4-byte anomaly test case saw better results than the 2-byte case. In a total of three runs the anomaly has been detected, and in two runs the anomaly has been matched by at least two receptors. Across all five runs, the detection rate was 60%.

The next experimental tests were performed for $l = 32$ and $e = 16$. The results of these tests are shown in Tables IV, V and VI.

For $l = 32$ and $e = 16$, in the case of a 1-byte anomaly which was previously undetected by 16-bit receptors, an improvement can be observed. The anomaly was detected by single receptors in two runs. The average count of receptors present in the set after generation was 9390.8 across five

runs and the detection rate was 40%. Introduction of a 2-byte anomaly in the PE file was detected with a solid 80% success rate across five test runs, compared to the previous result of 40%.

In the case of anomaly size of 4 bytes, detection rate was 80%, which is an improvement of 20 percentage points over the case where $l = 16$. However, it can be observed that an improvement of detection rates came at a price of much longer receptor generation times and anomaly detection times. The memory footprint of receptor storage also dramatically increased. For example, in the case of $l = 16$, $e = 8$ and anomaly size of 4 bytes, only 26 16-bit receptors were present in the receptor set when a detection occurred during run 4, compared to the average receptor count of 9373.2 in results shown in Table VI.

IV. CONCLUDING REMARKS

In the paper, an intrusion detection system to detect anomalies within an operating system was proposed. The algorithms in the proposed IDS are based on a negative selection algorithm inspired by the human immune system. The IDS detects anomalies within code sections of Win32 PE programs in a specific location on the hard drive. Anomalies are detected by binary strings called receptors, which are generated randomly. Randomly generated receptors do not produce false positives due to the implementation of a self verification process.

Proposed IDS has been implemented in C# language and tested on a computer running the Windows 7 operating system. During testing, the implemented solution successfully detected infected files. For a receptor size of 16 bits and an activation threshold of 8 bits, the system's detection rate was 60% for 4 byte anomalies. The detection rate can be improved by increasing the receptor size. For a receptor size of 32 bits and an activation threshold of 16 bits, the system's detection rate was 80%. However, a trade-off between detecting the anomalies quickly with a small memory footprint less accurately and detecting the anomalies more reliably at a greater memory expense, was observed. The research may continue with testing the IDS with files of greatly varying code section sizes.

REFERENCES

[1] A. Somayaji, S. Forrest, S. Hofmeyr, and T. Longstaff, "A sense of self for unix processes," in IEEE Symposium on Security and Privacy, 1996, pp. 120-128.

[2] A. Somayaji, S. Hofmeyr, and S. Forrest, "Principles of a computer immune system," in New Security Workshop, 1997, pp. 75-82.

[3] S. Forrest, A.S. Perelson, L. Allen, and R. Cherukuri, "Self-nonself discrimination in a computer," in IEEE Symposium on Security and Privacy, IEEE Computer Society, 1994, p. 202.

[4] J. Kephart, "A biologically inspired immune system for computers," in Fourth International Workshop on Synthesis and Simulation of Living Systems, Artificial Life IV, 1994, pp. 130-139.

[5] D. Dasgupta, "Immunity-based intrusion detection systems: a general framework," in 22nd National Information Systems Security Conference (NISSC), 1999.

[6] D. Dasgupta, and F. Gonzalez, "An immunity-based technique to characterize intrusions in computer networks," IEEE Transactions on Evolutionary Computation, vol. 6(3), 2002, pp. 281-291.

[7] P.S. Andrews, and J. Timmis, "Tunable detectors for artificial immune systems: from model to algorithm," Bioinformatics for Immunomics, Springer, New York, NY, USA, vol. 3, 2010, pp. 103-127.

[8] T.S. Sobh, and W.M. Mostafa, "A cooperative immunological approach for detecting network anomaly," Applied Soft Computing, vol. 11(1), 2011, pp. 1275-1283.

[9] D. Wang, F. Zhang, and L. Xi, "Evolving boundary detector for anomaly detection," Expert Systems with Applications, vol. 38(3), 2011, pp. 2412-2420.

[10] S.T. Powers, and J. He, "A hybrid artificial immune system and self organizing map for network intrusion detection," Information Sciences, vol. 78(15), 2008, pp. 3024-3042.

[11] G.Y. Li, and T. Guo, "Receptor editing-inspired real negative selection algorithm," Computer Science, vol. 39, 2012, pp. 246–251.

[12] C.A. Laurentys, G. Ronacher, R.M. Palhares, and W.M. Caminhas, "Design of an artificial immune system for fault detection: a negative selection approach," Expert Systems with Applications, vol. 37(7), 2010, pp. 5507–5513.

[13] R. Fanelli, "A hybrid model for immune inspired network intrusion detection," Springer-Verlag, Phuket, Thailand, 2008.

[14] P. Mostardinha, B.F. Faria, A. Zúquete, and F. Vistulo de Abreu: A negative selection approach to intrusion detection. In: C.A.C. Coello, J. Greensmith, N. Krasnogor, P. Liò, G. Nicosia, M. Pavone (eds.) Artificial Immune Systems, Lecture Notes in Computer Science, vol. 7597, 2012, pp. 178-190.

The Application of NIR Spectrometer for Average Temperature Measurement in Optical Fibers Based on Spontaneous Raman Scattering for DTS Applications

Iyad S. M. Shatarah, Bogusław Więcek

Institute of Electronics
Lodz University of Technology
Lodz, Poland
iyad.safwat.shatarah@dokt.p.lodz.pl, boguslaw.wiecek@p.lodz.pl

Abstract—Continuously excited Raman scattering in optical fibers is proposed for temperature remote sensing in a Distributed Temperature Sensing (DTS) system. Such an approach is suitable for average temperature measurements over the entire optical fiber or in the chosen set points. The system is operating at 1550 nm to achieve long distance temperature applications. This paper proposes the use of sensitive NIR spectrometer instead on WDM splitter. It allows controlling and choosing the appropriate wavelength of the Raman Anti-Stokes and Raman Stokes backscattered radiation. Moreover, two different types of optical fibers were tested in order to verify the DTS system capabilities, and to present the different impact of temperature upon different optical fibers types. The obtained results were satisfying and promising.

Keywords—Raman scattering, Raman spectrometry, Raman Stokes, Raman Anti-Stokes, optical fibers, Distributed Optical Fiber Sensor, Distributed Temperature Sensing.

I. INTRODUCTION

Distributed optical fiber sensors (DOFS) have been implemented in several applications in the past decades [1]. Nowadays DOFS are gaining more attention and are being applied in several new fields, whereby researchers and companies are discovering new opportunities in saving lives, time and costs [2]. This evolution is a result of the huge improvement in optical fibers' parameters and production techniques. Furthermore, DOFS system construction and data analyzing methods have improved significantly, providing faster and more accurate results of temperature [3] and strain [4] along the fiber.

The general principle of DOFS systems is the measurement of the changes along the optical fiber, which in this case is the actual sensor [5]. Fiber optics are smaller, cheaper and can be easily implemented in tight and harsh environments, thus provide several measurement points along pipeline [6], oil wells [7], structures [8], soil and water [9]. Fiber sensors can detect temperature changes, strain and cracks, fire and oil leakage [10].

DOFS systems are based on the measurement and analysis of Rayleigh, Brillouin and Raman scatterings phenomena [5].

The analysis is being held in the optical time domain reflectometry (OTDR) [11] or the optical frequency domain reflectometry (OFDR) [12]. This paper presents a DOFS system for the measurement of temperature changes based on the analysis of Raman scattering.

II. THEORETICAL BACKGROUND

Light injected in optical fibers scatters in every direction. Backscattered light is often studied in order to obtain information about changes along the optical fiber [13]. In order to obtain the backscattered light, an optical circulator should be implemented in the set-up. A light source for DOFS system can be a laser or a laser diode. A pulse laser with known power and pulse width is used in OTDR based system [14], whereas a continuous wave laser is used in OFDR based systems, with a tunable function generator or frequency modulator to set the source frequency [15]. The receiver can be a photodiode (usually an avalanche photodiode), spectrum analyzer or spectrometer. Distributed temperature sensing (DTS) systems are the type of DOFS systems that is used for temperature measurement [16]. DTS systems are commonly developed in single – ended, double – ended and loop configurations [10]. Figures 1 and figure 2 present a general single-ended and loop set-up, respectively. The advantage of the loop configuration is covering a wider space around the monitored structure (such as wells), thus providing several testing points at various positions.

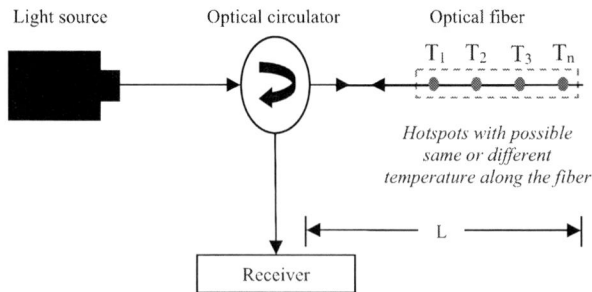

Fig. 1. Single-ended configuration for DTS systems

Depending on the wavelength of the scattered light, there are three types of light scattering: Rayleigh, Brillouin and Raman scatterings (figure 3) [5]. DTS systems measure the intensity of Raman Anti-Stokes in order to determine temperature changes due to its high dependency on temperature [17]. In order to enhance DTS systems' capabilities, demodulation algorithms were suggested, using Rayleigh or Raman Stokes to normalize the results [18]. DTS systems rely on the measurement of spontaneous Raman scattering [10]. DOFS systems for strain measurements are based on Brillouin scattering [19].

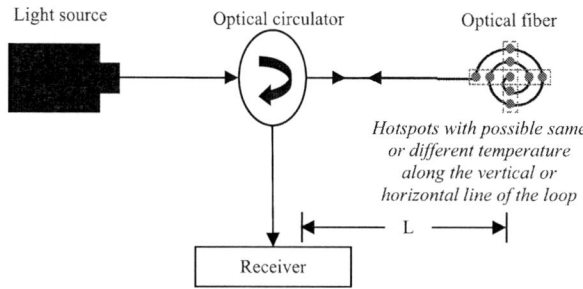

Fig. 2. Loop configuration for DTS systems

Light scattering in optical fibers is one of the most influencing factors on fiber's attenuation, which describes the losses in signal intensity along the fiber. Rayleigh scattering is the major source of attenuation [13], thus it emerges in equations that describe the power losses along the fiber and the equations of backscattered Raman Stokes and Raman Anti-Stokes powers [20].

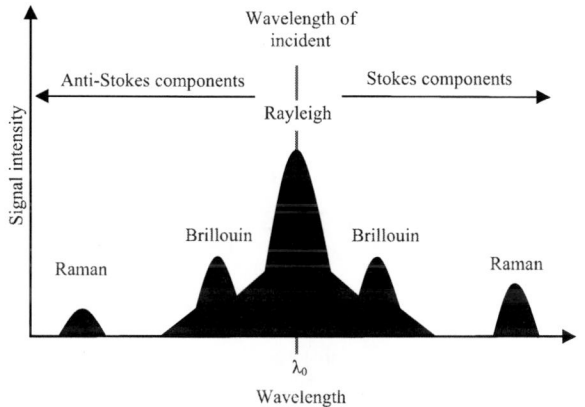

Fig. 3. Diagram of Rayleigh, Brillouin and Raman backscattering intensity below (Anti-Stokes) and above (Stokes) the wavelength of injected light

III. EXPERIMENTAL SET-UP

Figure 4 presents the set-up of the DTS system used in the measurements. This setup is based on the loop configuration. The author's laser diode driver board was introduced in [15]. The operating wavelength of the used laser diode is 1550 nm in order to achieve long distance temperature measurements. The optical circulator leads light from the laser diode through port 1 to the optical fiber through port 2, and the backscattered light is

led through port 2 to the WDM filter through port 3. This guarantees the obtainment of the whole backscattered light. The backscattered light is typically filtered into Rayleigh, Raman Stokes and Raman Anti-Stokes by using WDM filters. Alternatively, in the set-up presented in this research, the spectrometer NIRQUEST256-2.5 [21] was used in order to observe the backscattered light. This spectrometer works in the NIR range (900 – 2500 nm). The implementation of the spectrometer allows the elimination of WDM filter. Backscattered Rayleigh radiation has a very high intensity and covers a larger range of the wavelength spectrum than theoretically should be, thus WDM filters in most cases are not able to precisely filter the backscattered radiation to pure Raman Stokes and Anti-Stokes components, which leads to incorrect measurements. The whole spectrum of the backscattered light can be obtained using the software OceanView provided by OceanOptics company. The author's MATLAB calculation script help avoiding the inability of the WDM filter through choosing the desired range for each the backscattered radiation.

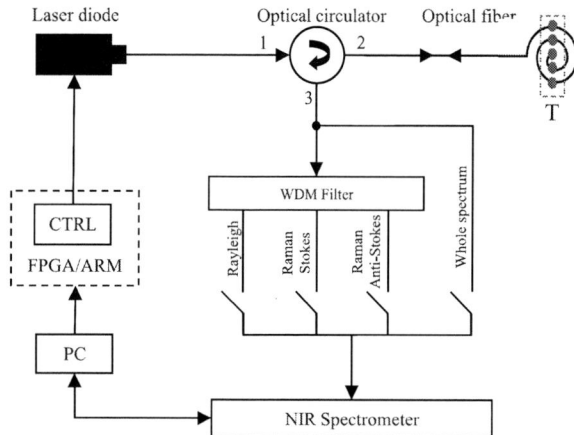

Fig. 4. Block diagram of the proposed DTS system

A. Spectrometer parameters

The process of the measurements starts with setting the parameters of the spectrometer [22]:

- Integration time: This parameter defines the period of time, that the CCD (Charge-coupled device) will take in order to sum the detected light.

- Scans to average: This parameter determines the number of spectral acquisitions the spectrometer driver will collect before averaging the results. This option averages the data from a single point. Due to lag this parameter causes to the software, it is chosen to be turned off in this set-up. Instead, a calculation script in MATLAB was created in order to average the recorded files.

- Boxcar Width: This parameter is used to smoothen the plot through by averaging the data of multiple, adjacent points on the CCD.

The parameters of the spectrometer were determined after a set of testing measurements in order to confirm the optimal parameters for the proposed set-up.

B. DTS system parameters

- Laser board driver: a low power, low cost and reliable driver for the laser diode was specially designed in order to control the operating mode (continuous wave, frequency modulated wave). Safety LED red light was also implemented to signal the beginning of laser diode radiation [15].

- Laser diode: 20 mW single-mode (SMF-28e) pigtailed laser diode, and the operating wavelength is 1550 nm.

- Optical circulator: The operating wavelength of the optical circulator is 1550 nm. The optical fiber of the optical circulator's ports is single-mode (SMF-28e).

- Optical fiber: 1 km bare single-mode optical fiber (SMF-28e) and 100 m loose tube single-mode optical fiber (SMF-28e).

- WDM filter: 3 port Wavelength divide multiplexing filter, with 1450 nm, 1550 nm and 1660 nm passband channels.

- NIR spectrometer: NIRQUEST256-2.5 spectrometer form OceanOptics, combined with OceanView software. The spectrum coverage range of this spectrometer is 900 – 2500 nm [21].

C. DTS system optical parameters

The optical parameters of the DTS system are defined based on the laser diode and the optical fiber. The wavelength and power of the laser diode depend on the desired range of the DTS system and the reflectometry domain for analysis (OTDR, OFDR) [10]. The choice of the optical fiber's mode, doping, tube and isolation depend on the application environment [23]. Table I shows the optical parameters for the DTS system used in this research. These parameters are the fundament for modelling and essential calculations for the DTS system [20]. The optical fiber used in this DTS system's experiments is a 1000 m single-mode SMF-28e bare fiber. Preliminary measurements for a 100 m loose tube single-mode fiber have been realized in order to verify the DTS system operation with different optical fiber types.

TABLE I.
OPTICAL PARAMETERS OF THE PRESENTED DTS SYSTEM

Laser diode	Optical power	mW	20
	Wavelength	nm	1550
Raman Stokes wavelength		nm	1660
Raman Anti-Stokes wavelength		nm	1450
Raman Stokes effective power attenuation coefficient		dB/km	0.22
Raman Anti-Stokes effective power attenuation coefficient		dB/km	0.25
Raman Stokes capture coefficient		m^{-1}	3.04E-10
Raman Anti-Stokes capture coefficient		m^{-1}	4.00E-10

D. Measurements process and data analysis

The process of measurements starts with checking the status of every optical connector. The backscattered radiation is weak, any scratch or dirtiness may cause optical losses at the connectors. After connecting the DTS system, one must pay attention to avoid moving or touching the optical fibers and optical connectors due to its possible effect on measurements' accuracy.

After setting the desired spectrometer parameters, measurements for the DTS system could be started. The first set of measurements was held for a DTS system with WDM filter implemented in the system, for the single-mode bare optical fiber. The first measurement was held in room temperature, then the optical fiber was put in a controlled heating chamber, where the temperature was changed and controlled. For each measurement, 512 files were saved for each Raman Stokes channel and Raman Anti-Stokes channel, then each file was read and filtered using the MATLAB calculation script by applying Savitzky – Golay filter in order to reduce noise. Then the area under the spectral curve was then calculated for each recording, and finally they were averaged. The Raman Anti-Stokes to Raman Stokes ratio was then calculated in order to obtain much reliable results of temperature impact on the Raman Anti-Stokes intensity.

The second set of measurements was held for a DTS system without the implementation of the WDM filter, also for the single-mode bare optical fiber. The author's calculation script replaced the WDM filter. For this purpose, a series of calibrating measurements were held, in order to determine the optimal ranges to divide the backscattered spectrum. The final version of the script is fully automatic, and has a respectful capabilities of finding the optimal ranges, filtering and calculating the areas under the curve. The second set measurements were done similarly to the first measurements set, and after dividing the spectrum of the backscattered light to 3 ranges: Raman Anti-Stokes, Rayleigh and Raman Stokes, the areas under the curve for the chosen Raman Stokes and Raman Anti-Stokes ranges were calculated, then the ratio was calculated.

Preliminary measurements were also held for a loose tube single-mode optical fiber, with the WDM filter implemented in the DTS system set-up. The aim of these measurements is to test the DTS systems capabilities with different types of optical fibers. These measurements also prove the different temperature impact on different optical fibers types and constructions. The measurements for this set-up could not be held in the heating chamber due to the size of the bare optical fiber. Therefore, the DTS system was installed in a room with an air conditioning system, that is capable of temperature setting in the range of 16 – 31°C.

IV. RESULTS

A. Results for the single-mode bare optical fiber

Table II and figure 5 present the results obtained for the first set of measurements: with the WDM filter implemented in the DTS system. The impact of the wide Rayleigh backscattered radiation is clearly visible in figure 5. However, the DTS system still manages to obtain reliable results of the sensitivity of Raman Anti-Stokes backscattered radiation to temperature changes.

TABLE II.
THE RATIO OF RAMAN ANTI-STOKES TO RAMAN STOKES FOR DIFFERENT TEMPERATURES OF THE OPTICAL FIBER, WITH THE IMPLEMENTATION OF WDM FILTER, FOR THE BARE OPTICAL FIBER

Temperature °C	Anti-Stokes to Stokes ratio %
23	21.85
30	22.10
45	24.48
60	27.08

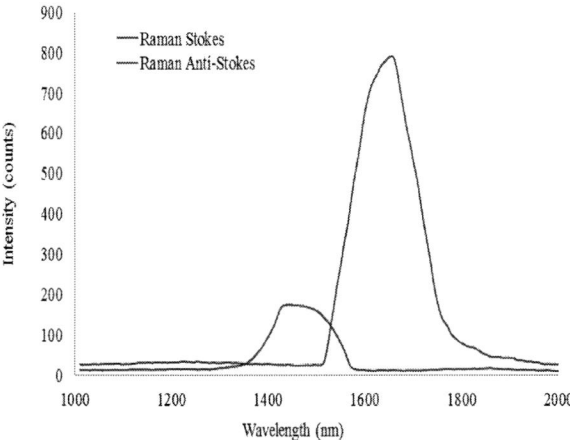

Fig. 5. Raman Stokes and Anti-Stokes backscattered radiations as seen using WDM filter, at fiber's temperature 30°C

The results for the second set of measurements are presented in table III and figure 6: without WDM filter implemented in the DTS system. The DTS system still handles the measurements well, and the use of the author's MATLAB calculation script even improves the resolution of the system. The time needed for this set of measurements is relatively shorter that the time needed for the first set, due to the lack of disconnecting and reconnecting WDM filter channels to the spectrometer. This also leads to less changes in the setup. DTS systems are sensitive and the less the changes in the setup, the more effective and reliable are the obtained results.

TABLE III.
THE RATIO OF RAMAN ANTI-STOKES TO RAMAN STOKES FOR DIFFERENT TEMPERATURES OF THE OPTICAL FIBER, WITHOUT THE IMPLEMENTATION OF WDM FILTER, FOR THE BARE OPTICAL FIBER.

Temperature °C	Anti-Stokes to Stokes ratio %
30	17.80
45	21.53
60	26.25

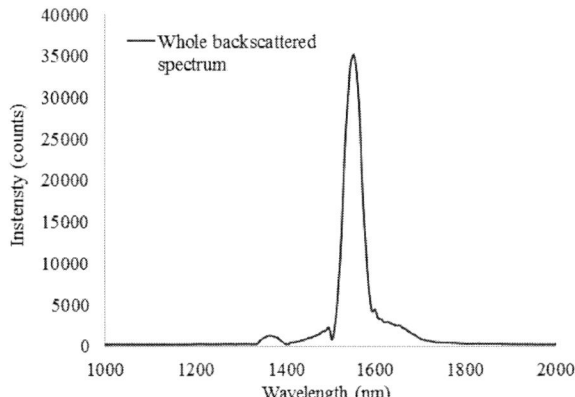

Fig. 6. The backscattered radiations as seen without using WDM filter, at fiber's temperature 30°C

B. Preliminary results for the single-mode loose tube optical fiber

Table IV and figure 7 present the results obtained for the DTS system with the single-mode loose tube fiber. These results are only for the set-up with a WDM filter. As mentioned previously, the optical fiber was placed in a room, which temperature was controlled by air conditioning. However, the temperature of the room could be set only in the range of 16 – 31°C. For this optical fiber type, the DTS also produces promising results. Further research is being held on this optical fiber and graded-index multi-mode fibers.

TABLE IV.
THE RATIO OF RAMAN ANTI-STOKES TO RAMAN STOKES FOR DIFFERENT TEMPERATURES OF THE OPTICAL FIBER, WITH THE IMPLEMENTATION OF WDM FILTER, FOR THE LOOSE TUBE OPTICAL FIBER

Temperature °C	Anti-Stokes to Stokes ratio %
16	15.48
24	18.00
31	21.00

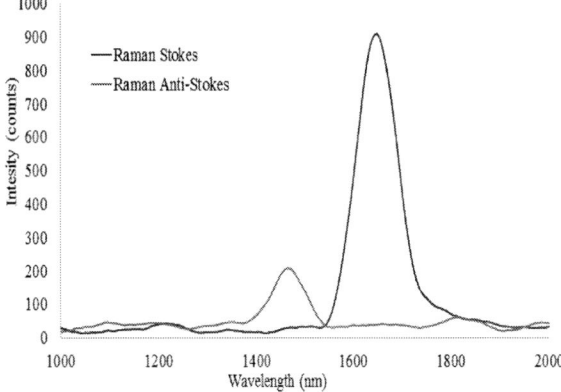

Fig. 7. The backscattered radiations as seen using WDM filter, at fiber's temperature 31°C

V. CONCLUSION

This paper provided a theoretical background for light scattering in optical fibers, and the potential applications of this phenomenon. In this paper, a full description of DTS systems was presented, with respect to the most important parameters that should be taken into consideration while building such systems.

The DTS system presented in this paper allows obtaining reliable results regarding the impact of temperature on Raman Anti-Stokes intensity. The results presented in this paper show the impact of temperature over two different types of optical fibers: bare optical fiber and loose tube optical fiber. Both optical fibers are single-mode (SMF28-e), however the construction, isolation and protection layers are different, thus the impact of temperature on them is different. This system also helps reducing the costs and the ability to work without a WDM filter. The author's calculation script improves the potentials of achieving more reliable results.

A series of further experiments are planned to be held. Several types of optical fibers will be tested. Moreover, experiments will be extended through the implementation of photodiodes as a receiver for the DTS system. Model, alongside results will soon be published.

REFERENCES

[1] L. Thévenaz, M. Facchini, A. Fellay, M. Niklès and P. Robert, "Evaluation of local birefringence along fibres using Brillouin analysis" in Reprint from Conference Digest OFMC'97, Teddington UK, NPL Publication, 1997, pp.82-85.

[2] D. A. Krohn, T. MacDougall and A. Mendez, Fiber optic sensors: Fundamentals and applications, 4th ed., SPIE, ISBN: 9781628411805, 2015.

[3] A. H. Hartog and A. P. Leach, "Distributed temperature sensing in solid-core fibres" in Electronics Letters, vol. 21, no. 23, 1985, pp.1061-1062.

[4] A. Masoudi and T. P. Newson, "Contributed review: Distributed optical fibre dynamic strain sensing" in Review of Scientific Instruments 87, 011501 (10 pages), 2016, https://doi.org/10.1063/1.4939482

[5] A. H. Hartog, An Introduction to Distributed Optical Fibre Sensors, CRC Press, Boca Raton, ISBN 9781138082694, 2017.

[6] D. Inaudi and B. Glisic, "Long-Range Pipeline Monitoring by Distributed Fiber Optic Sensing" in J. Pressure Vessel Technol., 132(1): 011701 (9 pages), 2010, https://doi.org/10.1115/1.3062942

[7] S. Adachi, "Distributed Optical Fiber Sensors and Their Applications" in SICE Annual Conference, Japan, 2008, pp.329-333.

[8] W. Li and X. Boa, " High Spatial Resolution Distributed Fiber Optic Technique for Strain and Temperature Measurements in Concrete Structures" In Proceedings of the International Workshop on Smart Materials, Structures NDT in Canada 2013 Conference & NDT for the Energy Industry, Canada, 2013.

[9] L. Schenato, "A Review of Distributed Fibre Optic Sensors for Geo-Hydrological Applications" in Applied Science, 7, 896 (42 pages), 2017, https://doi.org/10.3390/app7090896

[10] I.S.M Shatarah and R. Olbrycht, "Distributed temperature sensing in optical fibers based on Raman scattering: theory and applications" in Measurement Automation Monitoring, vol. 63, no. 02, 2017, pp.41-44.

[11] M. Tateda and T. Horiguchi, "Advances in Optical Time-Domain Reflectometry" in Journal Of Lightwave Technology, vol. 7. no. 8, 1989, pp.1217-1224.

[12] Z. Ding, C. Wang, K. Liu, J. Jiang, D. Yang, G. Pan, Z. Pu and T. Liu, "Distributed Optical Fiber Sensors Based on Optical Frequency Domain Reflectometry: A review" in Sensors, 18(4), 1072 (31 pages), 2018, https://doi.org/10.3390/s18041072

[13] M. Iten, Novel Applications of Distributed Fiber-optic Sensing in Geotechnical Engineering, vdf Hochschulverlag AG at ETH Zurich, ISBN: 978-3-7281-3454-7, 2011.

[14] A. Zrelli, M. Bouyahi and T. Ezzedine, "Measurements of Temperature Through Raman Scattering" in Procedia Computer Science, vol. 73, 2015, pp.350-357.

[15] I.S.M Shatarah and B. Więcek, "Application of Software-Defined Radio for Rayleigh and Raman Scattering Measurement in Optical Fibers" in Measurement Automation Monitoring, vol. 64, no. 04, 2018, pp.112-115.

[16] A. Ukil, H. Braendle and P. Krippner, "Distributed Temperature Sensing: Review of Technology and Applications" in IEEE Sensors Journal, vol. 12, no. 5, 2012, pp.885-892.

[17] F. Suarez, J. Dozier, M. B. Hausner and J. S. Selker, "Heat Transfer in the Environment: Development and Use of Fiber-Optic Distributed Temperature Sensing" in Developments in Heat Transfer, Dr. Marco Aurelio Dos Santos Bernardes (Ed.), ISBN: 978-953-307-569-3, InTech, Available from: http://www.intechopen.com/books/developments-in-heat-transfer/heat-transfer-in-the-environmentdevelopment-and-use-of-fiber-optic-distributed-temperature-sensing

[18] I. S. M. Shatarah and R. Olbrycht, "Distributed temperature sensing in optical fibers based on Raman scattering: demodulation algorythms" in Measurement Automation Monitoring, vol. 63, no. 02, 2017, pp. 45-47.

[19] S.. H Kim, J. J. Lee and I. B. Kwon, "Structural monitoring of a bending beam using Brillouin distributed optical fiber sensors" in Smart Materials and Structures 11 396 , 2002, pp.396-403.

[20] I.S.M Shatarah, R. Olbrycht and B. Więcek, "Modeling of Spontaneous Raman Scattering in silica light guides for Distributed Temperature Sensing" in 14th Quantitative Infrared Thermography Conference (QIRT), Berlin 2018, pp. 209-220.

[21] Datasheet NIRQuest: Installation and Operation Manual, https://www.oceaninsight.com/globalassets/catalog-blocks-and-images/manuals--instruction-old-logo/spectrometer/nirquest.pdf

[22] Datasheet OceanView: Installation and Operation Manual, https://www.oceaninsight.com/globalassets/catalog-blocks-and-images/manuals--instruction-old-logo/software/oceanviewio.pdf

[23] M. A. Farahani and T. Gogolla, "Spontaneous Raman Scattering in Optical Fibers with Modulated Probe Light for Distributed Temperature Raman Remote Sensing" in Journal of Lightwave Technology, vol. 17, no. 8, 1999, pp.1379-1391.

Power Electronics

1MHz Gate Driver in Power Technology for Fast Switching Applications

Roberto Di Lorenzo, Andrea Baschirotto
Physics Department "G.Occhialini"
University of Milan - Bicocca
Milan, Italy
r.dilorenzo3@campus.unimib.it

Albino Pidutti, Paolo Del Croce
IFAT DC ATV BP PD ADL
Infineon Technologies
Villach, Austria
Albino.Pidutti@infineon.com

Abstract—**The demand for low-cost integrated circuits for automotive applications is increasing, while their cost must remain low to maintain product competitiveness. In this scenario, to guarantee DC-DC Buck converters high-efficiency and low cost (in terms of external components) increasing switching frequency is mandatory. The main problems are inherent the parasitic inductances and the parasitic capacitance of power MOSFET. This paper deals with the main critical aspects of increasing such switching frequency and show how to replace the external Schottky diode with an integrated structure. The case of a high-speed monolithic integrated circuit to control a load current is here proposed. Proper design allows to achieve switching frequency up to 1MHz with 94.4% efficiency.**

Keywords—**gate driver, DC/DC converter, Buck**

I. INTRODUCTION

Day-by-day the demand for switching DC-DC converters is increasing over conventional converters because of their higher efficiency. Getting from the voltage battery to a few volts with power electronic circuits requires very large voltage conversion ratio. Primary converters, such as buck converters, use the large duty cycle to achieve large step-down ratios. The efficiency can be kept high by reducing losses, reducing the number of external components and reducing the dimensions of parasitic components (like Equivalent Series Resistor). Inductive DC-DC converters can accomplish this task but require large magnetic components (inductors), which cannot be integrated and, typically, feature low power efficiency at light or no load. This drawback can be mitigated by increasing the switching frequency from the typical ≈300kHz up to 1MHz and beyond, but this appears critical when high switching frequency has to be implemented in high-power nodes with very large devices and, then, very large parasitic capacitance to be driven. The trade-off between switching frequency and conversion efficiency will be part of the power transistor design. In [1] are presented the simulations results about the gate driver at 1MHz.

In many DC-DC converters applications [2]-[4] there is an external diode to charge the bootstrap capacitor and an external power transistor to supply the load. These external components increase the PCB size and cost and reduce efficiency due to the associated parasitic impedance. In this paper the on-chip implementation of such components is studied. The power transistor [5] and the block used to charge the bootstrap capacitor have been integrated on silicon, allowing minimizing capacities, inductances and resistances in the PCB. In this way it is possible increasing the gate driver frequency up 1MHz, with

high dV/dt for V_{DS} on power transistor. The proposed gate driver, bootstrap circuit and level shifter are designed for a load that requires a constant current of 3A with an input voltage range of 4.5V to 27V.

The paper is organized as follows. The gate driver architecture and the block designs are presented in Section II. The experimental results are given in Section III, and Sections IV concludes the paper.

II. GATE DRIVER ARCHITECTURE AND BLOCK DESIGN

A. Gate driver Architecture

Fig. 1 shows the structure of the proposed on-chip HV gate driver in an asynchronous buck DC-DC converter. The main blocks are: the dead time generator, the level shifter, the inherent bootstrap charge, and the driver. $V_{bootstrap}$ is a floating voltage that can range from below ground to above battery voltage. In the next section, the operation of each block will be explained.

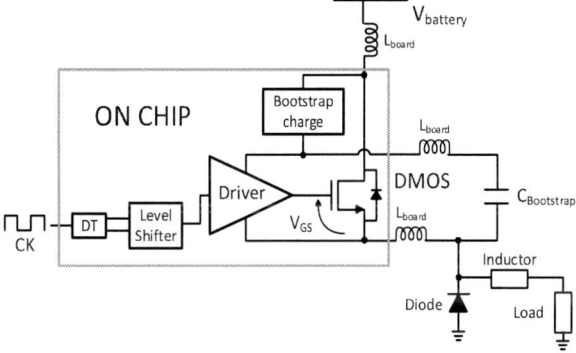

Fig. 1. Block diagram of gate driver

B. Dead Time Generator

The Dead Time Generator is the Timing Control (TC) block that provides an appropriate dead-time inserted between the switching on-phase and off-phase of the power transistor to avoid overlapping of the two states. The Dead Time Generator supports the level shifter to produce two signals: one signal turns on the switch (ck_{low}) and the second one turns it off (ckq_{low}).

The adopted 'analog' solution is shown Fig. 2. The non-overlapping time is defined by means of two capacitors of the same value (2pF) to obtain ≈20ns non-overlapping time. To calculate the optimal dead-time in a given application, the fall time in the actual circuit needs to be taken into account. In addition, variations in temperature and device parameters could also affect the effective dead-time in the actual circuit. Therefore, the nominal 20ns dead-time is chosen to avoid any PVT variations to produce shoot-through current. This analog solution, for this technology, reduces the die area compared to the digital version.

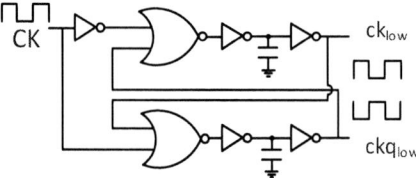

Fig. 2. Dead time generator

C. Level shifter

A fast and robust level shifter is necessary to transfer the signals to switch on and switch off the high-side driver. The scheme of Fig. 3 is based on low and high voltage transistor. When the ck_{low} is high, the V_{ck_high} is a digital zero and vice versa when ck_{low} is low. The transistors M_1 and M_2 are high-voltage devices (HV), while other transistors are low-voltage devices, because the difference between $V_{bootstrap}$ and Vs (voltage source on power switch) will be properly limited by design.

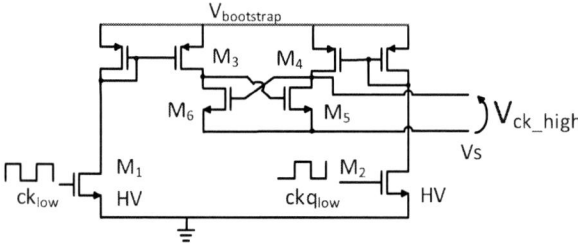

Fig. 3. Level shifter

This level shifter operates with a floating supply voltage, indeed $V_{bootstrap}$ is above the battery voltage when the power transistor turns on. Two complementary clocks (ck_{low}, and ckq_{low}) generated by the Dead Time Generator drive this structure. Between the two signals, there is a dead time to stabilize the level shifter, preventing a non-switching due to excessively high frequency. When clock ck_{low} is high, M1 is turned on and M2 is off, in the first branch there is a current and, therefore, M3 turns on and the voltage on $V_{D(M3)}$ is:

$$V_{D(M3)} \approx V_{bootstrap} - V_{DS(M3)} \qquad (1)$$

Therefore, transistor M5 turns on, because

$$V_{D(M3)} > V_{th(M5)} \qquad (2)$$

while M6 turns off. The delay between input and output can be kept in order of 6.4ns enabling high speed operations. The transistors must be large enough to withstand voltage fluctuations without changing state. For this case, the current consumption is around 100uA.

D. Power switch

The power transistor design depends on the switching frequency and the associated losses. Fig. 4 shows the losses related to a power transistor with an $R_{ds(on)}$ equal to 150mOhm and that of a transistor with an $R_{ds(on)}$ equal to 50mOhm. As the switching frequency increases, the capacitive losses are more relevant than the conduction losses. On the other hand, for low frequency the power losses are proportional to the switch resistance. Therefore, a small power transistor 150mOhm is preferable when the frequency increases. As a consequence, also the occupied area reduces.

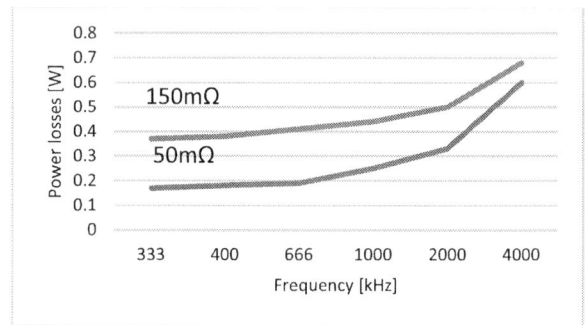

Fig. 4. Power losses on power switch vs frequency

E. Gate Driver

The gate driver in Fig. 5 determines the state of the power MOSFET. The gate driver is composed by a pair of low voltage MOSFET and it is designed to pilot a power transistor.

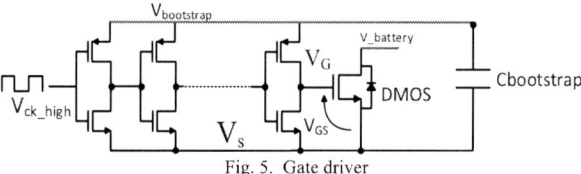

Fig. 5. Gate driver

The delay between the input signals and the output signal for such a driver is negligible with respect to other delays.

Since NMOS offers higher mobility than PMOS, it is convenient to use the NMOS as high side power in terms of implementation area (while providing the same on-resistance during conductions). Because of the high frequency operation requirement, the higher ohmic power transistor of Fig. 4 has been selected. To design a fast switching on-chip gate driver, the equivalent gate capacitance of the power transistor has been calculated (C=1nF). From the basic equations below:

$$Q = C * \Delta V \qquad (3)$$

$$Q = I * \Delta t \qquad (4)$$

where Q is the equivalent charge, ΔV is the maximum V_{GS} on power transistor, Δt is the rise time of V_{DS} and I is the average gate current to charge the equivalent capacitor. Once the rising time is set, the above equations are exploited to obtain the average current to switch:

$$I = \frac{C * \Delta V}{\Delta t} = 300mA \qquad (5)$$

Moreover, to design the gate driver last branch the following approximation is assumed:

$$\Delta t = 3 * \tau = R_{on_{pmos}} * C \qquad (6)$$

$$R_{on_{pmos}} < 10\Omega$$

To guarantee a secure margin, $R_{on_{pmos}}$ is designed to be 2Ω. Moreover, during simulations W/L=48 with multiplicity equal to 100 has been used (Fig. 5) and for the $R_{on_{nmos}}$ the ratio is 2.5 times smaller.

F. Bootstrapping technique for N-type high-side switch

This circuit recharges the external bootstrap capacitor, used to generate the voltage for turning on the power switch. This system is preferred to the charge pump since it is faster and this is mandatory for the target high switching frequency.

The bootstrapping scheme is illustrated in Fig. 6. During the switching operations, when the power switch is off, the electric charge is stored in the capacitor from $V_{battery}$, resulting in the voltage across of the capacitor to be $V_{bootstrap} = I_{dc} * R_{ref}$. The switch M2 inside the level shifter (Fig. 3) turns off, while M1 must be turned on, and the voltage across the bootstrap capacitor acts as a temporary supply voltage for the level-shifter and as a buffer for power transistor. As soon as the power transistor is switched on, its source voltage is pulled up as well as the auxiliary supply voltage level $V_{bootstrap}$ because of the bootstrap capacitor. Note that the parasitic diode of the HV NMOS becomes reversely biased since the bulk terminal is always biased such that its voltage is below the input one. Afterwards, the bootstrap capacitor slowly discharges through the level-shifter and the buffers. However, its capacitance is large enough to maintain a sufficient charge until the switch turns off again. Furthermore, Fig. 3 shows that the V_{ck_high} may exceed of few volts the battery voltage, which is approximately the value reached by the out_1 node minus the voltage drop on power transistor. This scheme allows to charge the bootstrap capacitor with switching frequencies >1MHz, because the capacitor starts to charge when the voltage bootstrap is one Vth_{MHV} lower than the voltage battery.

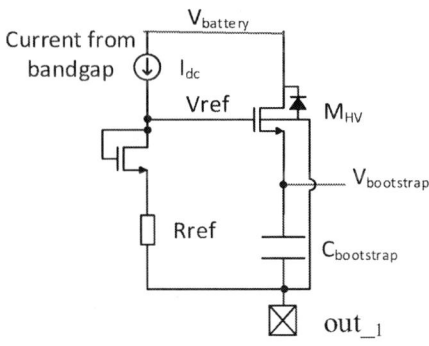

Fig. 6. Bootstrap circuit

To charge the bootstrap the current coming from a bandgap reference is used. This bandgap current makes the bootstrap voltage insensitive to temperature variation. Due to the M_{HV} body effect, its threshold increases leading to $V_{bootstrap}$ inaccuracy (the reference voltage will be the sum of $V_{bootstrap}$ and

Vth_{MHV}). To have the same voltage between the $V_{bootstrap}$ and the node above R_{ref}, a diode-connected transistor it is used to compensate the V_{GS} drop over M_{HV}. The proper bootstrap capacitance is selected according to the application as:

$$C_{bootstrap} \geq \frac{Q_{tot}}{\Delta V_{bootstrap}} = \frac{3nC}{0.4V} = 7.5nF \qquad (7)$$

III. EXPERIMENTAL RESULTS

The proposed device has been realized in a low-cost BCD technology. Fig. 7 and Fig. 8 shows the measured falling time (6.35V/ns) and rising time (3.55V/ns) behavior. In reference [6] for this time was 5ns for rise time e 10ns for fall time.

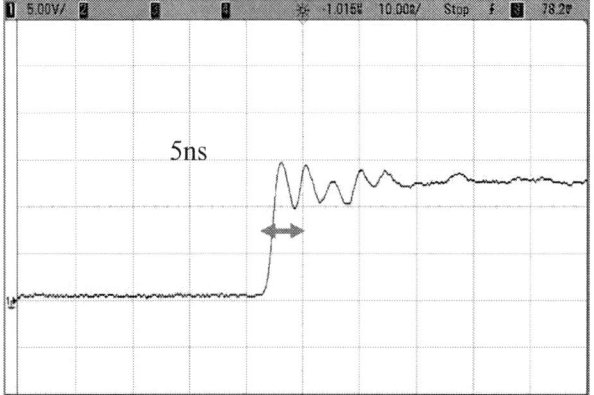

Fig. 7. V_{DS} rise time Vin = 12V

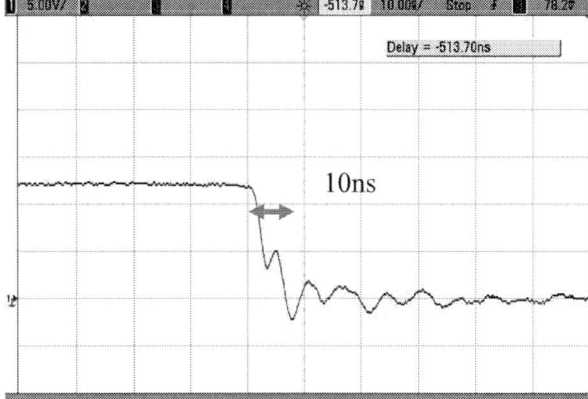

Fig. 8. V_{DS} fall time V_{in} = 12V

The circuit has been tested with three load (5, 10, 50Ω), at 12V of input voltage. The ringing are mainly caused by the parasitic inductances present in the bonding wires.

The chip has been tested at the edges of the possible duty cycle values, i.e. for the minimum duty cycle below 5% (see Fig. 9) and the maximum duty cycle above 90% (see Fig.10). This structure allows to reach a very low and high duty cycle up to 1MHz (Fig. 11), as a consequence the input voltage can take values very close to the Vout or over 27V, and inductor can be reduces up to 36% (from 47uH to 30uH).

201

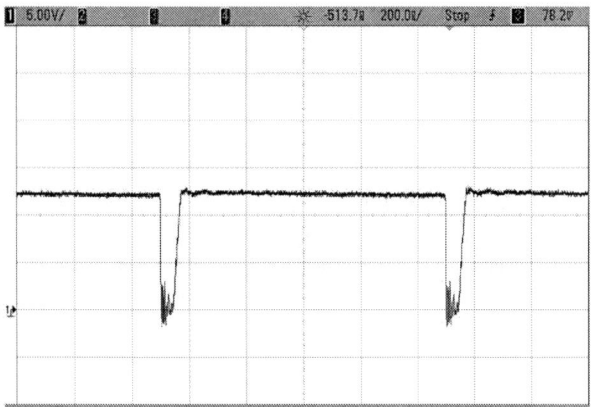

Fig. 9. V_{DS}, duty cycle 5%

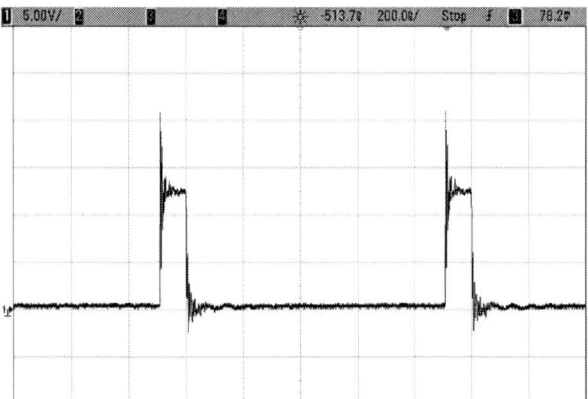

Fig. 10. V_{DS}, duty cycle 90%

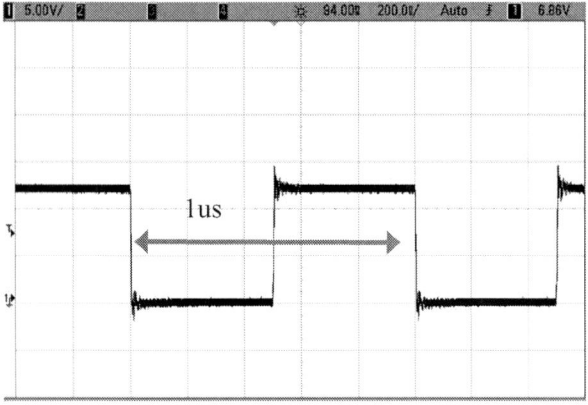

Fig. 11. V_{DS} on power switch

A. Efficiency

Tab. I presents results about the efficiency for standard conditions. To verify the efficiency, duty cycle values of 30% and 50% has been used. This configuration allows to check that at the same load and the same input voltage, the case with a duty cycle of 30% will be worse than that with D = 50%. Because with 30% of duty cycle, the Schottky diode leads for 70% of the time and this reduces the overall efficiency. Tab. II presents the efficiency of this work compared to the other literature DC-DC.

TABLE I.
EFFICIENCY

Efficiency			
Vin [V]	Duty Cycle [%]	Load [Ω]	η [%]
12	50	50	90,04
12	50	10	92,28
12	50	5	89,01
12	30	50	88,54
12	30	10	92,00
12	30	5	90,45

TABLE II.
PERFORMANCE COMPARISONS OF DIFFERENT BUCK CONVERTERS

	Comparison			
	LTC3630 [6]	LM5007 [7]	LM5017 [8]	This Work
Converter topology	Synch. Buck	Async. Buck	Sync. Buck	Asynch. Buck
Input Voltage (V)	12.5 - 76	12 - 75	12.5 - 95	4.5 - 27
Fsw (kHz)	N. A.	400	200	1000
Output voltage (V)	12	10	10	6
Max Output Current (A)	0.5	0.4	0.6	3.5
Efficiency	92% at Vin = 24V	93.7% at Vin = 15V	92.5% at Vin = 24V	92.28% at Vin = 12V

IV. CONCLUSIONS

In this paper an on-chip gate driver has been presented. The implemented high speed design solutions allow rapid state transitions of the high-side switch with an efficiency of 92.28%. An additional advantage is in the usage of a low-cost BCD technology in the design. As consequence of that, the circuit reaches high efficiency even in a high frequency regime without giving up on a competitive cost for the final product.

ACKNOWLEDGMENT

A part of the work has been performed in the project iDev40. The iDev40 project has received funding from the ECSEL Joint Undertaking (JU) under grant agreement No 783163. The JU receives support from the European Union's Horizon 2020 research and innovation programme. It is co-funded by the consortium members, grants from Austria, Germany, Belgium, Italy, Spain and Romania.

REFERENCES

[1] Di Lorenzo R., Gasparri O., Pidutti A, Del Croce Paolo, Baschirotto A., On-Chip Power Stage and Gate Driver for Fast Switching Applications, In *2019 26th International Conference on Electronics Circuits ans Systems (ICECS)*. IEEE, DOI: 10.1109/ICECS46596.2019.8965103.

[2] Bau, P., Cousineau, M., Cougo, B., Richardeau, F., Colin, D., & Rouger, N. (2018, July). A CMOS gate driver with ultra-fast dV/dt embedded control dedicated to optimum EMI and turn-on losses management for GaN power transistors. In *2018 14th Conference on Ph. D. Research in Microelectronics and Electronics (PRIME)* (pp. 105-108). IEEE, DOI: 10.1109/PRIME.2018.8430331.

[3] Subotskaya, V., Mihal, V., Tulupov, M., & Deutschmann, B. (2018). Optimized gate driver for high-frequency buck converter. *e & i Elektrotechnik und Informationstechnik*, *135*(1), 40-47. DOI: 10.1007/s00502-017-0570-7.

[4] Mednik, A. (2005). Automotive LED lighting needs special drivers. *Power Electronics Technology Magazine*.

[5] Elmoznine, Abdellatif, Buxo, Jean, Bafleur, Marise, & Rossel, Pierre (1990). The smart power high-side switch: description of a specific technology, its basic devices, and monitoring circuitries. *IEEE Transactions on Electron Devices*, *37*(4), 1154-1161, DOI: 10.1109/16.52454.

[6] Liu, Z., Cong, L., & Lee, H. (2015). Design of on-chip gate drivers with power-efficient high-speed level shifting and dynamic timing control for high-voltage synchronous switching power converters. *IEEE Journal of Solid-State Circuits*, *50*(6), 1463-1477, DOI: 10.1109/JSSC.2015.2422075.

[7] Texas Instruments Inc., LM5007 Datasheet: High Voltage (80 V) Step Down Switching Regulator, Mar. 2013.

[8] Texas Instruments Inc., LM5017 Datasheet: 100 V, 600 mA Constant On-Time Synchronous Buck Regulator, Dec. 2013.

An Influence of the Operation Mode of a LED Lamp of the HUE Type on Its Electrical and Optical Parameters

Przemysław Ptak, Krzysztof Górecki

Department of Marine Electronics
Gdynia Maritime University, Gdynia, Poland
k.gorecki@we.umg.edu.pl, p.ptak@we.umg.edu.pl

Jakub Heleniak

Faculty of Electrical Engineering
Gdynia Maritime University, Gdynia, Poland
kuba.97-1997@o2.pl

Abstract—In the paper electrical and optical properties of a LED lamp of the HUE type are analysed. This lamp has a possibility to remote control the mode of its operation. Properties of this class of lamps given by the producer was characterised and the measuring set-up to measure electrical and optical parameters of the considered lamps are proposed. The obtained results of measurements of the investigated lamps emitting light of different colours are shown and discussed. Attention was paid on parameters characterising influence of the LED lamp on the electroenergy network. Additionally, influence of feeding voltage, the fixed value of illuminance and colour of the emitted light on THD and PF parameters is analysed and discussed.

Keywords—LED lamps; wireless lighting; total harmonic distortion; power factor; measurements

I. Introduction

Solid-state light sources are commonly used in lighting technique [1-5]. Both LED lamps emitting white light (lighting of rooms) and LED lamps emitting light of the fixed colour (accent lighting, backlight) are used [2, 3, 6]. LED lamps consist of three components [7]: LED modules emitting light, feeders and the case with a heat-sink.

The important aim in lighting systems is to obtain a desirable colour of the emitted light and its illuminance. Classical LED lamps do not show such functionality and the mentioned above parameters are fixed in the stage of production of these lamps. On the other hand, from the literature [2, 3, 5, 8] it is known that illuminance emitted by power LEDs can be easily regulated by changing the value of current feeding these diodes, and the colour of the emitted light can be regulated using RGB diodes and regulating the value of current flowing through each diode.

Development of solid-state light sources and systems of intelligent lighting of buildings resulted in production of wireless controlled LED lamps which operate using mobile devices [9-11]. This control is realised with the use of radio waves and such systems as Bluetooth, Wi-Fi or ZigBee [10]. The radio signal is transmitted between the controller and the lamp contains the address of the lamp and the required parameters describing illuminance and the colour of the emitted light [9, 10].

One of the main advantages of LED lamps is high luminous efficiency [5, 12, 13] which causes that these lamps are characterised by low consumption of electrical energy. Unfortunately, quality of these feeders typically included in such lamps is low [7, 12, 14-22] and they cause essential noise of current received from the electroenergy network, which can lead to worsening of coefficients characterising quality of electrical energy [15, 16, 20, 21]. To these coefficients belong e.g.: total harmonic distortion THD and the power factor PF [11, 12, 19, 22, 23].

Philips is one of the main producers of wireless controlled LED lamps, offering among other things LED lamps of the type HUE [24]. The first lamps belonging to this group were introduced to the market in October 2012. They make possible among other things, a change of colour of the emitted light. At present, three generations of HUE lamps are accessible on the market. Such lamps are fully compatible with one other [24].

In the literature there is a lot of research on electrical and optical properties [7, 12, 14, 15, 25, 26] of selected types of LED lamps. However, no lamps that have a function of remote control have not yet been examined.

In the present paper the measurements results of parameters of a LED lamp of the type HUE emitting light of three basic colours: red, green and blue are presented. The second section contains a description of wireless control of LED lamps of the type HUE. The third section contains a description of the used measuring set-up. In the fourth section the results of measurements of selected characteristics of the examined lamp are shown and discussed.

II. Tested Lamp

The LED lamp of the type HUE which emits optical radiation in three different colours: green, red and blue was selected for experimental investigations. Colour of the emitted light was selected by means of the application Philips HUE 3.0.1 dedicated to mobile devices with the operating systems Windows, iOS and Android. This application makes it possible to configure up to 50 LED lamps of the type HUE and to choose a colour from the palette of sixteen million colours, and also to regulate intensity of illuminance within the range from 1 to 100 % with the step 1%. The examined lamp of the type White and Colour Ambiance model 9CK is characterised with the maximum value of the emitted luminous flux equal to 806 lm. Electrical power consumed from the electroenergy network is equal to 9.5 W [24].

Colour of the emitted optical radiation in the case of the LED lamp White and Color Ambience is obtained by means of RGB LEDs. These diodes are controlled by means of three separate PWM (Pulse Width Modulation) signals. Colour of illuminance is regulated by changing a value of the suitable level of the duty cycle for every signal PWM. This is one of the methods to obtain a suitable colour of the emitted light, widely described in the literature [7, 15]. Communication between the selected LED lamp and the mobile device takes place by means of the protocol ZigBee, described in the standard IEEE 802.15.4 [10, 24].

An indispensable component of the system of the remote control is a communication bridge – the HUE Control Bridge. It is used for remote communication with end devices of the system HUE (with LED lamps) [24]. The communication bridge HUE operates in the waveband from 2.4 to 2.484 GHz. The bridge uses this frequency range to communicate with the examined LED lamp and with the mobile device. The router Wi-Fi and the communication bridge HUE are connected by means of LAN (Local Area Network) by the RJ -45 cable. These devices are fed from the electroenergy network by means of the feeder producing dc voltage of the value equal to 5V.

III. MEASUREMENT SET-UP

In order to measure characteristics of the examined LED lamps of the type HUE the measuring set-up was designed and constructed. This set-up makes it possible to measure optical and electrical parameters of the examined LED lamps. In order to measure optical parameters of the selected LED lamp the light-tight measuring-track containing the pipe PCV of the length of 50 cm was built. In Fig. 1 the block scheme of the measuring set-up is presented.

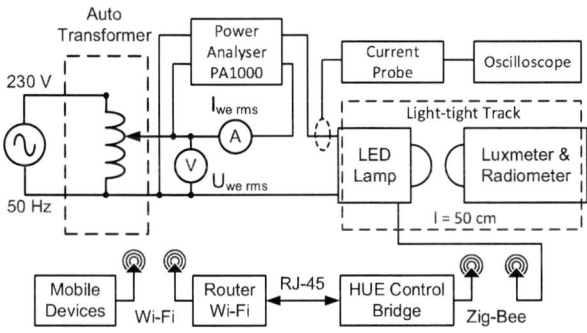

Fig. 1. Diagram of the measurement set-up

This set-up contains an autotransformer ATS-REG1.2 [27], two multimeters Unit 804 [28] realising the function of the ammeter and the voltmeter, the current probe Tektronix TCPA300 [29], the single-phase power analyser Tektronix PA1000 [30] and the oscilloscope Gw Instek GDS 2104a [31]. Feeding voltage of the examined LED lamp is regulated with the use of the autotransformer connected to the electroenergy network. This voltage is regulated within the range from 100 V to 230 V, making it possible to obtain different values of illuminance and power density of the emitted light for the examined LED lamps of the type HUE.

Illuminance of the emitted light by the investigated lamp is registered by means of the illuminometer L 200 [32], whereas power density of the emitted light is registered by means of the radiometer HD2302 [33]. Measuring probe of these devices was placed at the end of the optical path. In order to measure spectral characteristics of the examined LED lamp the spectrometer of the type USB 650 by Ocean Optic [34] was used. Measurements of optical parameters were controlled by the PC.

In order to examine quality of electrical energy received from the electroenergy network the oscilloscope and the current probe were used. These instruments enabled measurements of spectrum of current received from the electroenergy network. Additionally, spectrum of this current was measured by means of the single-phase power analyser of the type PA1000 by Tektronix [35].

IV. RESULTS

In order to test properties of the selected LED lamp measurements of some parameters of this lamp at different RMS values of sinusoidal feeding voltage of frequency equal to 50 Hz were performed. By means of the measuring set-up described in the third section optical power density, illuminance and RMS values of input current at the steady state were measured. Additionally, waveforms of illuminance and spectral characteristics of the examined LED lamp were measured. Measurements of the coefficient of total harmonic distortion THD and the power factor PF were performed for different RMS values of feeding voltage.

In Fig. 2 measured dependences of the RMS value of feeding current or the fixed value of illuminance d for two different RMS values of feeding voltage equal to 120 V and 230 V, respectively, are presented.

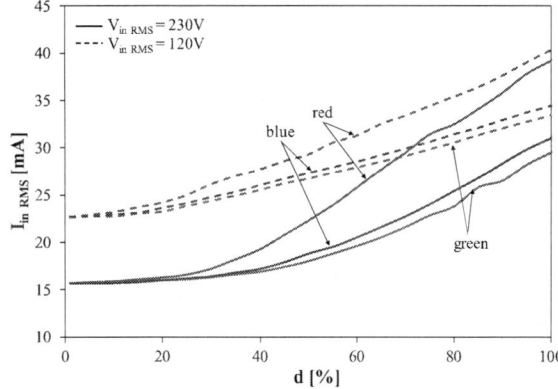

Fig. 2. Measured dependences of the RMS value of feeding current on the fixed value of illuminance

As it could be expected, an increase in the fixed value of illuminance caused an increase in the RMS value of feeding current. This current decreased with an increase in feeding voltage. It is proper to notice that the greatest consumption of current from the power source corresponds to light emission of red colour, and the least – at light emission of green colour. This is connected with spectral sensitivity of the human eye.

In Fig. 3 measured dependences of power density of light emitted by the investigated lamp on the fixed value of illuminance d are shown.

The considered dependence is a monotonically increasing function. It is proper to notice that at the maximum fixed value of illuminance power density of the emitted light is the greatest for red light, and the smallest for green light. Values of power density corresponding to these colours of the emitted light differ between each other even three times.

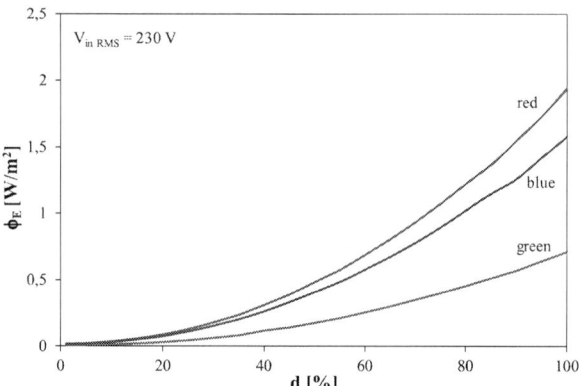

Fig. 3. Measured dependences of power density of the emitted light on the fixed value of illuminance

Measurements of waveforms of parameters of illuminance emitted by the examined LED lamp were performed, too. These measurements show constancy of this parameter, which testifies to efficient removal of heat generated in the examined LED lamp [8, 13] or suitable compensation of changes of internal temperature of light sources by a change in the value of current flowing through these sources.

In Fig. 4 spectral characteristics of the examined LED lamp operating in the mode of emission of red, blue and green light are presented.

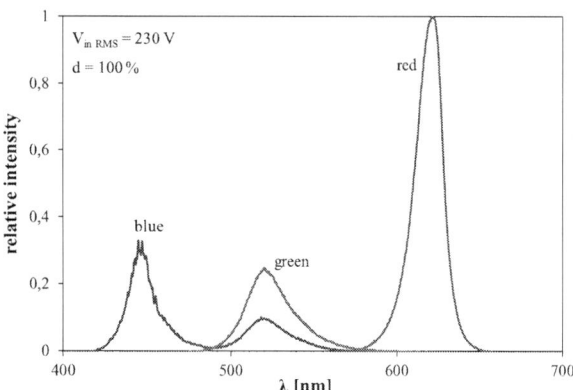

Fig. 4. Measured spectral characteristics of the tested LED lamp operating in the mode of emission of red, green and blue light

It can be noticed that when the lamp operates in the mode of emission of red light, its spectral characteristic is the narrowest. In turn, in the mode of emission of blue light two maximums are visible on spectral characteristics corresponding to blue and green colours.

Fig. 5 illustrates waveforms of feeding current of the examined lamp operating in the mode of emission of green light.

Visible deviations of the measured current from the desirable sinusoidal waveform are observed. It is worth paying attention to large current impulses occurring in the regions of extrema of waveforms of the network voltage. The shape of these waveforms shows capacitive character of input impedance of the feeder contracted in the examined lamp and the use in its full wave rectifier. Impulses of feeding current reach even 130 mA.

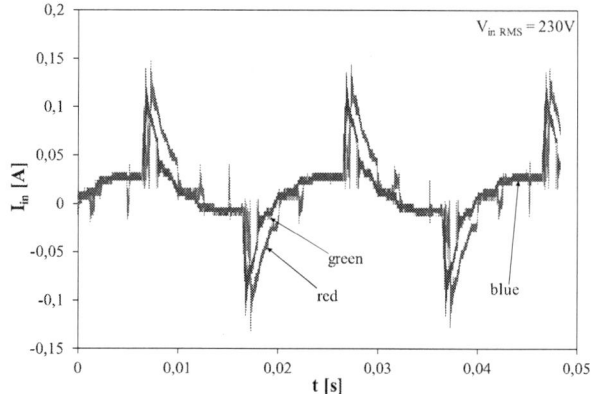

Fig. 5. Measured waveforms of feeding current of the tested lamp obtained in the mode of emission of red, green and blue light

In Fig. 6 spectrum of feeding current of the examined LED lamp emitting light of three basic colours is presented.

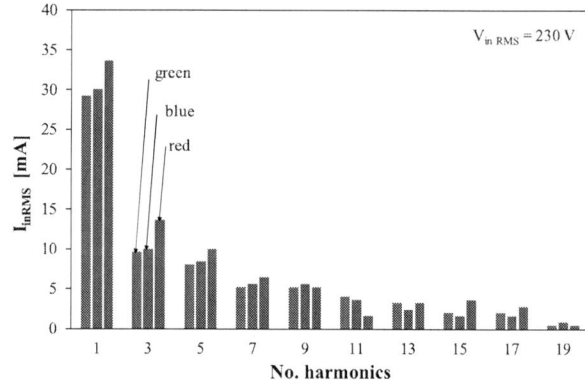

Fig. 6. Measured spectrum of feeding current of the tested lamp obtained at selected modes of light emission

In spectrum of feeding current only odd harmonics are visible. The prevailing role is played by the first harmonic, but in the measured spectra harmonics to the number 19 are visible. It is proper to notice that colour of the emitted light influences relations between values of each harmonic in the spectrum of feeding current.

Total deformation of the waveform of feeding current can be characterised by means of the coefficient of total harmonic distortion THD. The value of this coefficient is described by means of the definitional example of the form

$$THD = \frac{\sqrt{\sum_{k=2}^{n} I_k^2}}{\sqrt{\sum_{k=1}^{n} I_k^2}} \cdot 100\% \qquad (1)$$

where I_k denotes k-th harmonic of current, whereas n is the number of the used harmonics.

Using values of each harmonic of feeding current measured by means of the current probe and the oscilloscope, the THD values are calculated for feeding current of the lamp operating in modes of emission of red, green and blue light. Values of the THD calculated using the formula (1) for different values of feeding voltage are marked in Fig. 7 by means of dashed lines. In this figure also the results of measurements of the THD value of feeding current of the examined lamp performed with the use of Power Analyzer PA1000 are shown. These results are marked with solid lines.

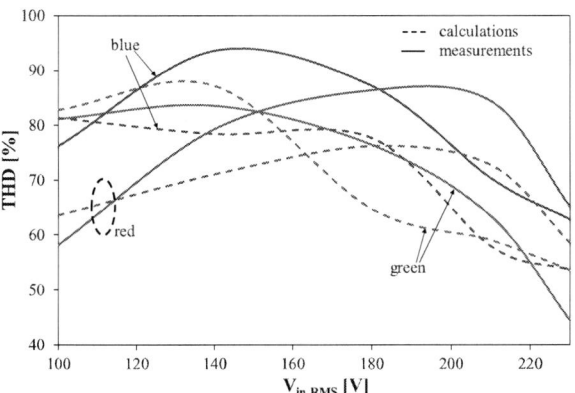

Fig. 7. Measured dependence of the THD of feeding current of the tested lamp on input voltage

As can be observed, the THD values of feeding current strongly depend on colour of the emitted light and on feeding voltage. Analysing the results obtained by means of the power analyser is visible that in the considered range of changes of feeding voltage the maximum of dependence THD(V_{in}) appears. For each of the considered colours of the emitted light the smallest THD value at the nominal value of feeding voltage equal to 230 V is obtained. At this value of feeding voltage the greatest THD value of current of the power supply is obtained at red light emission. It amounts to 65%. The smallest THD value equal to 45% was obtained at production of green light. In the case of dependence THD(V_{in}) calculated with the use of the formula (1) and measurements performed by means of the oscilloscope it is clearly visible that these values considerably are different from the values obtained by means of the power analyser. The observed differences between the THD values obtained with the help of different instruments result probably from the difference within the range of harmonics numbers of current used to calculate the THD value. Differences in the THD values obtained by means of both methods reach even 15%.

Unfavourable influence of the device fed from the electroenergy network on this network is typically characterised by the power factor PF by means of the power analyser. In Fig. 8 dependences of the power factor on feeding

voltage are presented. In this figure solid lines mark the results of measurements obtained by means of the power analyser, whereas dashed lines – the results obtained with the use of calculations based on measurements performed with the oscilloscope. The applied power analyser measures the value of the power factor based on the definitional example of the form [30]

$$PF = \frac{\frac{1}{T} \int_{0}^{T} v_{in}(t) \cdot i_{in}(t) dt}{V_{inRMS} \cdot I_{inRMS}} \qquad (2)$$

where T denotes the period of waveforms of current $i_{in}(t)$ and voltage $v_{in}(t)$, whereas V_{inRMS} and I_{inRMS} are RMS values of feeding voltage and current, respectively. Calculations were performed with the use of the formula of the form [35]

$$PF = \frac{1}{\sqrt{1 + \left(\frac{THD}{100\%}\right)^2}} \qquad (3)$$

where the THD is estimated with the use of the formula (1) and the results of measurements performed with the use of the oscilloscope.

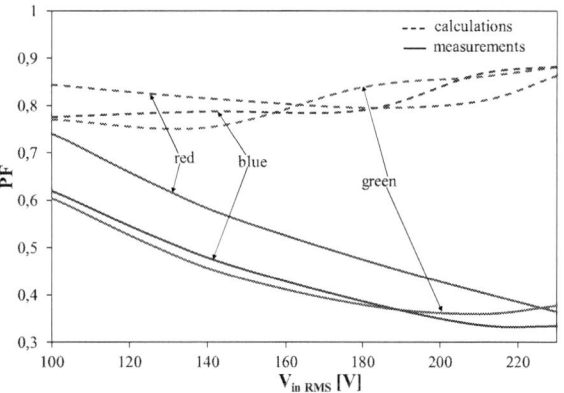

Fig. 8. Measured dependence of the PF of the tested lamp on input voltage

It is visible that the value of the power factor is a decreasing function of feeding voltage. At the nominal value of feeding voltage the value of the power factor is contained in the range from 0.35 to 0.4. This is a very small value, which shows that the considered lamp very unfavourably influences the electroenergy network. The obtained results of measurements show that any power factor correction circuit was not applied in the feeder of the examined lamp.

It is also worth noticing that by means of the formula (3) overestimated values of the power factor, exceeding even 0.85 are obtained. This means that in the considered case this formula should not be used. Correctness of dependence (3) is limited only to small values of the THD [35].

Comparing the obtained results of measurements of the power factor to the values of this parameter obtained for other light sources and given among other things in the paper [14] it can be stated that the use of the examined lamp, to a considerably higher degree, result in worsening quality of electrical energy in the electroenergy network than classical bulbs and popular LED lamps. The obtained values PF for the examined lamp are nearing the values corresponding to a fluorescent lamp.

V. CONCLUSIONS

In the paper experimental findings illustrating properties of electrical and optical LED lamps of the type HUE emitting light of selected colours are presented. A manner of regulating useful parameters of the considered lamp is presented and characteristics illustrating relation between values of the fixed optical parameters and the measured values of the emitted luminous flux and spectral characteristics of the examined lamp are shown.

Attention was paid to properties of the system feeding lighting elements of the examined lamp. Influence of both: feeding voltage and colour of the emitted light and illuminance of the lamp on the efficient value of current of the power supply, its waveform, the coefficient of the total harmonic distortion and the power factor is discussed.

Measurements were performed with the use of the authors' measuring set-up. Measurements of the RMS value of the feeding current show that it is an increasing function of the fixed value of illuminance and a decreasing function of feeding voltage. Waveforms of feeding current registered during experimental research confirm strong deformation and they can constitute a proof that in the feeding system a block of the power factor correction was not applied. The performed measurements and calculations of the coefficients THD and PF also show that the examined lamp unfavourably influences quality of energy in the electroenergy network. It is proper to pay attention that values of parameters THD and PF obtained by means of the power analyser accept far less favourable values than in the case of using oscilloscopic measurements and simplified analytic formulas given in the literature. The observed differences result probably from the fact that the waveband, wherein harmonics values of feeding current are measured, is considerably wider in the case of the power analyser. When the oscilloscope is used, fast Fourier transformation taking into account only about 20 harmonics frequencies of the network is realised. Therefore, disturbances of frequency higher than 1 kHz were not taken into account while determining the THD value.

Findings presented in this paper can be useful for designers of lighting systems with LED lamps. It is also proper to take into account in investigations influence of electronic devices on the electroenergy network to make allowance for harmonics of higher order which correspond to current impulses connected with the use of switch-mode power supplies of LED lamps. The objective of further investigations will be a proposal of systemic solutions making it possible to determine the THD and an increase of the PF value characterising waveforms of feeding current of the examined LED lamps.

ACKNOWLEDGMENT

The project financed within the program of the Ministry of Science and Higher Education called "Regionalna Inicjatywa Doskonałości" in the years 2019-2022, the project number 006/RID/2018/19, the sum of financing 11 870 000 PLN.

REFERENCES

[1] Weir, B. Driving the 21st Century's Lights. IEEE Spectrum 2012, Vol. 49, No. 3, pp. 42-47.

[2] Martin, P.S. High power white LED technology for solid state lighting. Lumileds. Availaible online: https://www.ele.uva.es/~pedro/optoele/LEDs/LEDilumination.pdf, (accessed 24 Jan 2019).

[3] Krames, M. Progress and future direction of LED technology. Available online: https://www.netl.doe.gov/ssl/PDFs/Krames.pdf (accessed 24 Jan 2019).

[4] Czyżewski D.: LED substitutes of conventional incandescent lamps. Przegląd Elektrotechniczny, Vol. 88, No. 11a, 2012, pp.123-127.

[5] Schubert, E.F. Light emitting diodes, 2nd ed.; Cambridge University Press: New York, USA, 2008.

[6] Chang, M.H.; Das, D.; Varde, P.V.; Pecht, M. Light emitting diodes reliability review. Microelectronics Reliability 2012, Vol. 52, pp. 762-782.

[7] P. Ptak, K. Górecki: Modelling power supplies of LED lamps. International Journal of Circuit Theory and Applications, Vol. 46, No. 3, 2018, pp. 629-636.

[8] K. Górecki, P. Ptak: Modelling LED lamps in SPICE with thermal phenomena taken into account. Microelectronics Reliability, Vol. 79, 2017, pp. 440-447.

[9] P. Apse-Apsitis, A.Avotins, L. Ribickis.: Wirelessly controlled led lighting system. 2nd IEEE Energycon Conference and Exhibition, 2012, pp. 952-955

[10] B.C. Mishra, A.S. Panda, N.K. Rout, S.K. Mohapatra.: A novel efficient design of Intelligent Street Lighting monitoring system using ZigBee network of devices and sensors on Embedded Internet Technology. International Conference on Information Technology, 2015, pp. 201-205.

[11] L. Svilainis.: Comparison of the EMI Performance of LED PWM Dimming Techniques for LED Video Display Application. Journal of display technology, Vol. 8, No. 3, 2012, pp. 162-165.

[12] Górecki K.: The influence of power supply voltage on exploitive parameters of the selected lamps. Informacije MIDEM - Journal of Microelectronics, Electronic Components and Materials, Vol. 43, No. 3, 2013, pp. 193-198.

[13] Lasance C.J.M., Poppe A.: Thermal management for LED applications. Springer Science+Business Media, New York, 2014.

[14] Górecki K., Górecka K., Górecki P.: Porównanie właściwości eksploatacyjnych wybranych typów lamp LED. Przegląd Elektrotechniczny, Vol. 88, No. 11a, 2012, pp. 111-114.

[15] Uddin S., Shareef H., Mohamed A., Hannan M. A.: Harmonics and thermal characteristics of low wattage LED lamps. Przegląd Elektrotechniczny, Vol. 88, No. 11a, 2012, pp. 266-271.

[16] A.Ndokaj , A. Di Napoli.: LED Power Supply and EMC Compliance. 2nd IEEE Energycon Conference and Exhibition, 2012, pp. 254-258

[17] M. F. Pinto, T. R. F. Mendonça, C. A. Duque and H. A. C. Braga.: Power Quality Measurements Embedded in Smart Lighting Systems, 2015, pp. 1202-1206

[18] R. Bloudíček, Š. Lužica, S. Rydlo, L. Hon.: Power Supply in LED Airport Lighting Systems. International Conference on Military Technologies, 2017, Brno, pp. 588-592.

[19] S. Uddin, H. Shareef, A. Mohamed, M. Hannan: An Analysis of Harmonics from DimmableLED Lamps. IEEE International Power Engineering and Optimization Conference, 2012, Malesia, pp. 182-185.

[20] Yaung Hwan Lho: A Study on methodology to improve the power factor of high power LED modul. 14th International Conference on Control, Automation and Systems, 2014, Gyeonggi-do, Korea, pp. 1404-1406.

[21] A.Pollock, H. Pollock, C. Pollock.: High Efficiency LED Power Supply. IEEE Journal of Emerging and Selected Topics in Power Electronics, Vol. 3, No. 3, 2015, pp. 617-622.

[22] G. Rata, M. Rata.: The Study of Harmonics from Dimmable LED Lamps, using CompactRIO. 13th International Conference on Development and Application Systems, Romania, 2016, pp. 180-183.

[23] C. Ionescu, M. Dima, D. Bonfert.: Flicker Distortion Power Factor Analysis in Lighting LED's, 23rd International Symposium for Design and Technology in Electronic Packaging, 2017, Romania, pp. 280-285.

[24] HUE system website. Online availaible: https://www2.meethue.com/pl-pl/p/hue-mostek/8718696511800, (accesed 20 February 2020).

[25] Chung Y. C., Lee K. M., Choe H. J., Sung C. H., Kang B.: Low-cost drive circuit for AC-direct LED lamps. IEEE Transactions on Power Electronics, Vol. 30, No. 10, 2015, pp. 5776-5782.

[26] V. T. Sreedevi, V. Devi, A. A. Sunil.: Analysis and Simulation of A Single Stage Power Supply for LED Lighting. International Conference on Green Computing, Communication and Conservation of Energy, 2013, pp. 453 -456.

[27] Autotransformer ATS-REG1.2 https://www.sklep.cyfronika.com.pl/pl/p/ATS-REG1.2-Autotransformator-z-regulowanym-napieciem-moc-1.25-kW/15527, (accesed 29 January 2020)

[28] Multimeter Unit 804 https://www.gotronik.pl/ut804-cyfrowy-multimetr-laboratoryjny-p-326.html, (accesed 29 January 2020)

[29] Current probe Tektronix TCPA300 https://pl.rs-online.com/web/p/akcesoria-do-sond/7644391/, (accesed 29 January 2020)

[30] Power analyser Tektronix PA1000 https://pl.farnell.com/tektronix/pa1000/ac-power-analyzer-1ph-1v-to-600v/dp/2772544, (accesed 29 January 2020)

[31] Oscilloscope Gw Instek GDS 2104a https://www.tme.eu/pl/details/gds-2104a/oscyloskopy-cyfrowe/gw-instek/, (accesed 29 January 2020)

[32] Luxmeter L 200 https://www.sonopan.com.pl/pl/produkty/swiatlo/luksomierz-precyzyjny-l-200/, (accesed 29 January 2020)

[33] Radiometer HD2302 https://sklep.emd.net.pl/foto-radiometr-hd23020-delta-ohm-p-2724.html, (accesed 29 January 2020)

[34] Spectrometer USB 650 Ocean Optic https://www.oceaninsight.com/products/spectrometers/usb-series/, (accesed 29 January 2020)

[35] M. H. Rashid: Power Electronics Handbook, Elsevier, 2007.

Signal Processing

Combining ε-similar Fuzzy Rules for Efficient Classification of Cardiotocographic Signals

Michal Jezewski[1], Robert Czabanski[1], Jacek M. Leski[1], Adam Matonia[2], Radek Martinek[3]

[1]Department of Cybernetics, Nanotechnology and Data Processing
Silesian University of Technology, Gliwice, Poland
[2]Łukasiewicz Research Network – Institute of Medical Technology and Equipment, Zabrze, Poland
[3]VSB – Technical University of Ostrava, Ostrava, Czech Republic
Emails: mjezewski@polsl.pl, rczabanski@polsl.pl, jleski@polsl.pl, adamm@itam.zabrze.pl, radek.martinek@vsb.cz

Abstract—**CardioTocoGraphic (CTG) monitoring is the primary method of fetal condition assessment. Due to the inter- and intra-observer disagreement between experts when evaluating signals visually, a well established solution supporting the diagnostic decision is automated classification of CTG signals. The goal of this paper is to propose a method of simplifying the fuzzy classifier rule base by combining ε-similar rules, to achieve high quality of CTG signals classification, but with fewer conditional rules. The results of experiments performed using the benchmark CTG database confirm the efficiency of the introduced method.**

Keywords—**fuzzy classifier, rule base simplification, ε-insensitivity, fetal monitoring.**

I. Introduction

CardioTocoGraphic (CTG) monitoring, involving the analysis of fetal heart rate (FHR) and uterine contractions, is a widely used method to assess fetal condition. The FHR signal is commonly acquired using Doppler ultrasound method [1], [2]. The visual evaluation of the signals by clinicians is characterized by inter- and intra-observer disagreement [3], therefore methods to support clinical decision are needed. A well-established approach is the classification of CTG signals represented by their quantitative parameters [4], [5], [6], [7], [8], [9], [10], [11], [12], [13].

Classification consists in assigning objects to predefined classes. Among many classification algorithms [14], the support vector machine (SVM) [15] is considered one of the best, ensuring high quality of classification. However, it does not provide interpretability of decisions made, which is of particular importance in case of medical diagnosis support. Interpretability is ensured, for example, by fuzzy rule-based classifiers. To determine classifier rules, fuzzy clustering methods can be used [13], [16], [17].

The extracted rule base may include redundant rules, which reduce interpretability and increase the rule base complexity. Therefore, research on rule base simplification have been carried out. For example, extraction and simplification of the rule base by means of fuzzy clustering, partition validation, approximate similarity analysis and parameter learning was proposed in [18]. The work [19] presents simplification algorithm using similarity measure between trapezoidal fuzzy sets. Rule base

simplification by replacing each group of inconsistent rules with a single equivalent rule is described in [20]. The authors of [21] presented application of multidimensional scaling and non-linear constraint optimization to simplification of fuzzy models with Gaussian fuzzy sets. The research [22], [23] is focused on the rules selection method based on the evaluation of their fitness in relation to the training data, whereas the elimination of redundant rules using the least angle regression algorithm is described in [24].

The aim of this study is to introduce a rule base simplification method which is based on ε-insensitive distance between rule premises. The proposed method consists in reducing the number of rules by combining ε-similar rules into representative rules. The goal is to achieve efficient classification of CTG signals to support assessment of fetal condition with fewer classifier rules. The premises of the initial (not simplified) rule base were found using clustering with pairs of ε-hyperballs procedure [13]. Experiments were carried out using the benchmark dataset of CTG signals.

II. Extraction of the Initial Rule Base

The rule base of the applied classifier consists of the following R fuzzy rules [25]:

$$\left\{ \textbf{if} \bigwedge_{n=1}^{N} x_{kn} \text{ is } A_n^{(r)}, \textbf{ then } y^{(r)} = \delta_{y,y_r} \right\}_{r=1}^{R}, \quad (1)$$

where x_{kn} denotes the n-th component of the feature vector $\mathbf{x}_k$, $A_n^{(r)}$ is the fuzzy set (described by the Gaussian membership function) referring to this component in the premise of the r-th rule, and δ_{y,y_r} denotes the singleton in the rule conclusion. The vector $\mathbf{x}_k$ can be assigned to one of two classes based on the sign of the classifier output value [25]

$$y_{0k} = \frac{\sum_{r=1}^{R} \mu_{\mathbf{A}^{(r)}} (\mathbf{x}_k) \, y^{(r)}}{\sum_{r=1}^{R} \mu_{\mathbf{A}^{(r)}} (\mathbf{x}_k)}, \quad (2)$$

where

$$\forall_{1 \le r \le R} \quad \mu_{\mathbf{A}^{(r)}} (\mathbf{x}_k) = \exp \left[-\frac{1}{2} \sum_{n=1}^{N} \left(\frac{x_{kn} - v_{rn}}{\gamma s_{rn}} \right)^2 \right]. \quad (3)$$

In the above formula v_{rn} (s_{rn}) is the center (width, scaled by the parameter γ) of the Gaussian membership functions, which are determined using the $\text{CPP}_{\text{ST}}^{\varepsilon}$ procedure. The rule conclusions ($y^{(r)}$) are found based on the IRLS error minimization procedure (see [13] for details).

A. Clustering with Pairs of ε-Hyperballs Procedure ($\text{CPP}_{\text{ST}}^{\varepsilon}$)

The clustering with pairs of ε-hyperballs is a method dedicated for determining premises of fuzzy rules [13]. The idea of $\text{CPP}_{\text{ST}}^{\varepsilon}$ is to introduce additional "prototypes in-between" to "ordinary prototypes" obtained by the $\text{FC}\varepsilon\text{H}$ clustering separately in each of the two classes of objects. Details of $\text{FC}\varepsilon\text{H}$ are provided in Section II-B. The prototypes in-between should be located near the border between two classes and their role is to define the rules that ensure correct classification of objects located near the border. The additional prototypes ($\mathbf{v}_i^{(b)}$) are calculated on the basis of $c^{(p)}$ pairs of ordinary prototypes (one prototype from the first class, one from the second), which are found by the ant algorithm [13]

$$
\underset{1 \leq i \leq c^{(p)}}{\forall} \mathbf{v}_i^{(b)} = \frac{\sum\limits_{k \in \mathcal{K}_{\omega_1}} \left(u_{sk}^{(1)}\right)^m \mathbf{x}_k + \sum\limits_{k \in \mathcal{K}_{\omega_2}} \left(u_{tk}^{(2)}\right)^m \mathbf{x}_k}{\sum\limits_{k \in \mathcal{K}_{\omega_1}} \left(u_{sk}^{(1)}\right)^m + \sum\limits_{k \in \mathcal{K}_{\omega_2}} \left(u_{tk}^{(2)}\right)^m}, \quad (4)
$$

where s, t are the indices of ordinary prototypes forming a pair, $u_{sk}^{(1)}$, $u_{tk}^{(2)}$ are the elements of partition matrices in both classes, $\mathcal{K}_{\omega_1}$, $\mathcal{K}_{\omega_2}$ are the sets of indices of objects from both classes, and m denotes the degree of fuzziness of clusters. Both ordinary and in-between prototypes are used as centers (v_{rn}) of the Gaussian membership functions in the rule premises. The corresponding dispersions (s_{rn}) are determined separately for ordinary and in-between prototypes.

B. Fuzzy Clustering with ε-Hyperballs ($\text{FC}\varepsilon\text{H}$)

In traditional clustering, prototypes are represented as points in the feature space. In [13] we described a fuzzy clustering based on ε-insensitive distance [26]

$$
g_{ik} = \begin{cases} d_{ik} - \varepsilon, & d_{ik} > \varepsilon, \\ 0, & d_{ik} \leq \varepsilon, \end{cases} \quad (5)
$$

where d_{ik} is the Euclidean distance between prototype $\mathbf{v}_i$ and object $\mathbf{x}_k$. As a result, $\text{FC}\varepsilon\text{H}$ prototypes can be interpreted as hyperballs with the center $\mathbf{v}_i$ and radius ε. For objects located inside the hyperball, their distances to the prototype center are considered as equal to 0. The $\text{FC}\varepsilon\text{H}$ criterion function has the form [13]

$$
J^{(\varepsilon)}(\mathbf{U}, \mathbf{V}) = \sum_{i=1}^{c} \sum_{k=1}^{K} (u_{ik})^m g_{ik}^2, \quad (6)
$$

where: $\mathbf{U}$ ($\mathbf{V}$) is the partition (prototypes) matrix, c (K) denotes number of clusters (objects). The necessary conditions for minimization of (6) are:

$$
\underset{1 \leq i \leq c}{\forall} \underset{1 \leq k \leq N}{\forall} u_{ik} = \begin{cases} (g_{ik})^{\frac{2}{1-m}} \bigg/ \sum\limits_{j=1}^{c} (g_{jk})^{\frac{2}{1-m}}, & I_k = \emptyset, \\ \underset{i \in \tilde{I}_k}{\forall} 0, \quad \sum\limits_{i \in I_k} u_{ik} = 1, & I_k \neq \emptyset, \end{cases}
$$

$$
\quad (7)
$$

where: $I_k = \{i|\ 1 \leq i \leq c;\ g_{ik} = 0\}$, $\tilde{I}_k = \{1, 2, \cdots, c\} \setminus I_k$, and

$$
\underset{1 \leq i \leq c}{\forall} \mathbf{v}_i = \frac{\sum\limits_{k \in \Omega_k} (u_{ik})^m \left(\frac{\varepsilon}{d_{ik}} - 2\right) \mathbf{x}_k}{\sum\limits_{k \in \Omega_k} (u_{ik})^m \left(\frac{\varepsilon}{d_{ik}} - 2\right)}, \quad (8)
$$

where $\Omega_k = \{k\,|d_{ik} > \varepsilon\}$.

III. RULE BASE SIMPLIFICATION

The proposed algorithm for simplification of the rule base consists in reducing its size by combining ε-similar rules into representative rules. The ε-similarity is defined on the basis of ε-insensitive distance [26]. Consequently, two rules ith and jth are considered ε-similar (ε_S-similiar), if they indicate the same class of objects (have conclusions of the same sign), and the distances between their centers of Gaussian functions do not exceed the value of ε (ε_S)

$$
\underset{\substack{1 \leq n \leq N \\ i \neq j,\ y^{(i)} y^{(j)} > 0}}{\forall} \left| v_{in} - v_{jn} \right| \leq \varepsilon_S. \quad (9)
$$

In the algorithm, the ε-similarity is checked for all possible pairs of rules and the ε-similar rules are combined into groups. Each group is then represented by the "strongest" (representative) rule. The "strength" (F_i) of the ith rule is determined based on its conclusion and summarized activation levels (3):

- for rules indicating first class ($y^{(i)} > 0$)

$$
F_i = \frac{\mu_{\mathbf{A}^{(i)}}^{(\omega_1)}}{\mu_{\mathbf{A}^{(i)}}^{(\omega_2)}} y^{(i)}, \quad (10)
$$

- for rules indicating second class ($y^{(i)} < 0$)

$$
F_i = -\frac{\mu_{\mathbf{A}^{(i)}}^{(\omega_2)}}{\mu_{\mathbf{A}^{(i)}}^{(\omega_1)}} y^{(i)}, \quad (11)
$$

where

$$
\mu_{\mathbf{A}^{(i)}}^{(\omega_1)} = \frac{\sum\limits_{k \in \mathcal{K}_{\omega_1}} \mu_{\mathbf{A}^{(i)}}(\mathbf{x}_k)}{N^{(1)}}, \quad \mu_{\mathbf{A}^{(i)}}^{(\omega_2)} = \frac{\sum\limits_{k \in \mathcal{K}_{\omega_2}} \mu_{\mathbf{A}^{(i)}}(\mathbf{x}_k)}{N^{(2)}}, \quad (12)
$$

and $N^{(1)}$, $N^{(2)}$ denote numbers of objects in both classes.

The algorithm of combining ε-similar rules into representative rules can be explained using a simple example (see Figure 1) with a rule base consisting of three, single input rules ($R = 3$, $N = 1$). The rule no. 1 is similar to rules no. 2 and no. 3, and rule no. 2 is not similar to rule no. 3. There are possible four cases that should be considered to illustrate the algorithm:

1) $F_1 > F_2$ and $F_1 > F_3$ – in this case, rule no. 1 represents rules no. 1 and no. 2, also, rule no. 1 represents rules no. 1 and no. 3, and finally rule no. 1 represents all three rules,

2) $F_2 > F_1$ and $F_1 > F_3$ – here rule no. 2 represents rules no. 1 and no. 2, rule no. 1 represents rules no. 1 and no. 3, and finally rule no. 2 represents all three rules,

214

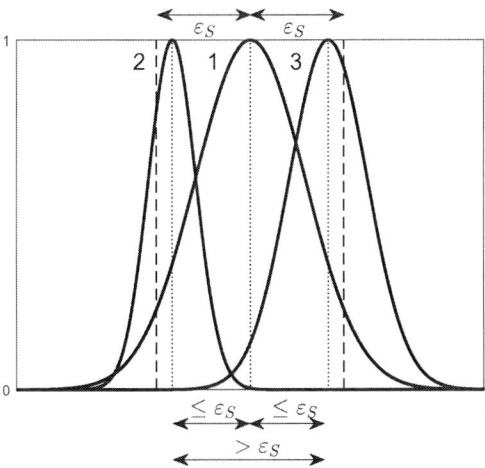

Fig. 1. Illustration of the algorithm of combining ε-similar (ε_S-similar) rules into representative rules. The figure shows a rule base consisting of three, single input rules ($R = 3$, $N = 1$). The rule no. 1 is similar to rules no. 2 and no. 3, and rule no. 2 is not similar to rule no. 3.

3) $F_2 > F_1$ and $F_3 > F_1$ – rule no. 2 represents rules no. 1 and no. 2, rule no. 3 represents rules no. 1 and no. 3, and finally rules no. 2 and no. 3 represent the rule base,

4) $F_1 > F_2$ and $F_3 > F_1$ – in this situation, rule no. 1 represents rules no. 1 and no. 2, rule no. 3 represents rules no. 1 and no. 3, and finally rule no. 3 represents all three rules.

The algorithm is run separately for both indicated classes, and stops if there are no ε-similar rules in the rule base.

IV. Research Material and Experiments

The experiments were carried out using CTU-UHB [27], [28] benchmark dataset of CTG signals. Since only raw signals were available, using the computerized fetal monitoring system [29] we calculated 12 parameters that quantitatively describe them in the time domain [13]. As a reference signal assessment, the result of the evaluation the newborn state can be used. There are three main attributes describing the newborn condition: Apgar score, percentile of the birth weight and measurement of blood pH. The Apgar score represents a visual assessment of the newborn five characteristics (appearance, pulse, grimace, activity, respiration). The birth weight determines whether the fetus has developed properly during pregnancy, and the blood pH in the neonatal umbilical artery describes the acid-base balance of newborn blood, providing information about possible fetal hypoxia. In this work, we used pH measurement, since it is most often used as an objective reference CTG signal evaluation. However, different threshold values are used to separate the normal and abnormal fetal condition [7], [11], [12]. In this work we assumed 7.15 as a threshold value, i.e. cases with the pH > 7.15 were defined as normal (438 cases). In addition, for comparison we also conducted experiments for pH threshold of 7.05.

Signals were randomly divided into 100 training and testing sets to apply the Monte-Carlo cross-validation, maintaining the proportion between classes. During our previous research we observed that the two-stage classification improves the results [13]. Hence, the two-stage scheme was used in this work. The classification quality was evaluated by sensitivity (Se), specificity (Sp), and quality index (QI) defined as $QI = \sqrt{Se * Sp}$. Fist, parameter values were found to ensure the highest mean QI value for the first ten testing sets. Next the final results for all 100 testing sets were calculated. For more implementation details see [13].

V. Results and Discussion

Table I compares the classification quality achieved using the proposed simplification method with the results obtained when using the full (not simplified) rule base. We can conclude that the proposed simplification is beneficial because it provides a higher QI value. Moreover, the increase in QI is accompanied by a significant increase of Se, which is of particular importance in medical applications. In addition, these results were obtained with a smaller number of rules (R_1 and R_2) of both component classifiers (in total 49.56 on average compared to 78.00).

TABLE I

Comparison of the Rule Base Simplification (CPP_{ST}^{ε}+RBS) with the Results Obtained Using the Full Rule Base (CPP_{ST}^{ε})

	CPP_{ST}^{ε} + RBS	CPP_{ST}^{ε}
QI	74.06 (3.28)[*]	72.30 (3.84)
Se	65.33 (6.08)	57.91 (6.52)
Sp	84.23 (3.82)	90.61 (2.50)
R_1	16.70 (0.84)	40
R_2	32.86 (0.87)	38

[*] mean value (std. dev.)
[#] no. of rules of the first (R_1) and the second (R_2) component classifiers

The classification of CTU-UHB signals with a pH threshold of 7.15 was also the subject of other studies. The comparison of various methods is given in Table II. The influence of image-based time-frequency (IBTF) features on least squares SVM (LS-SVM) classification was investigated by the authors of [7]. The SVM with IBTF features was analyzed in [8], the results included QI only. The study [9] concerned the SVM with features extracted based on empirical mode decomposition and discrete wavelet transform (SVM+EMD+DWT). The last column shows the results calculated for the Lagrangian SVM (LSVM) [30] with Gaussian kernel function. Table II shows the efficiency of our solution because it provides the highest values of performance measures. The exception is the highest Se for LSVM, but it is accompanied by a very low Sp and, consequently, the lowest QI.

The comparison of results using the pH threshold of 7.05 is shown in Table III. The results presented in [10] included the QI value, in the remaining cases (including this work) the QI was calculated based on the mean Se and Sp values. One

TABLE II
COMPARISON OF RESULTS WITH OTHER CLASSIFICATION METHODS

	QI	Se	Sp
CPP_{ST}^{ε}+RBS	74.06	65.33	84.23
LS-SVM+IBTF [7]	64.04	63.45	65.88
SVM+IBTF [8]	73.45	–	–
SVM+EMD+DWT [9]	63.44	57.42	70.11
LSVM	44.87	85.60	23.58

can notice that the highest QI (Se=70.27% and Sp=94.42%) was obtained with the proposed method.

TABLE III
COMPARISON OF RESULTS FOR pH THRESHOLD OF 7.05

	QI
CPP_{ST}^{ε}+RBS	81.45
Spilka et al., 2013 [4]	64.64
Spilka et al., 2016 [5]	68.99
Georgoulas et al., 2017 [6]	72.94
Comert and Kocamaz, 2019 [10]	72.48
Zarmehri et al., 2019, [11]	71.37

VI. CONCLUSIONS

In the presented work a method of simplifying the fuzzy classifier rule base was proposed. It consists in reducing the number of rules by combining ε-similar rules into representative rules. The ε-similarity is defined based on the ε-insensitive distance between premises of rules. The initial (not simplified) rule base is obtained by means of clustering with pairs of ε-hyperballs procedure.

The goal of the resulting fuzzy classifier was to achieve high quality classification of cardiotocographic (CTG) signals, but with smaller number of rules. A benchmark dataset of CTG signals was used in the experiments. The results confirmed the effectiveness of the proposed simplification method – the achieved classification quality was higher compared to literature results. Moreover, the total number of rules decreased by more than a third.

Both, the proposed rule base simplification method and the considered classifier (Takagi-Sugeno-Kang fuzzy system of the zero order) are defined using basic arithmetic operations and are characterized by low computational complexity. Therefore they can be easily implemented in microelectronic devices.

ACKNOWLEDGMENT

This research was partially supported by National Science Centre, Poland under grant 2017/27/B/ST6/01989, in part by statutory funds (BK-2020) of the Silesian University of Technology, Department of Cybernetics, Nanotechnology and Data Processing and in part by the Rector's research and development grant, Silesian University of Technology, grant no. 02/130/RGJ20/0001.

REFERENCES

[1] J. Jezewski, D. Roj, J. Wrobel, and K. Horoba, "A novel technique for fetal heart rate estimation from Doppler ultrasound signal," *BioMedical Engineering OnLine*, vol. 10, no. 1, pp. 1–17, 2011.

[2] J. Wrobel, D. Roj, J. Jezewski, K. Horoba, T. Kupka, and M. Jezewski, "Evaluation of the robustness of fetal heart rate variability measures to low signal quality," *Journal of Medical Imaging and Health Informatics*, vol. 5, no. 6, pp. 1311–1318, 2015.

[3] J. Spilka, V. Chudacek, P. Janku, L. Hruban, M. Bursa, M. Huptych, L. Zach, and L. Lhotska, "Analysis of obstetricians' decision making on CTG recordings," *Journal of Biomedical Informatics*, vol. 51, pp. 72–79, 2014.

[4] J. Spilka, G. Georgoulas, P. Karvelis, V. P. Oikonomou, V. Chudacek, C. Stylios, L. Lhotska, and P. Janku, "Automatic evaluation of FHR recordings from CTU-UHB CTG database," in *Information Technology in Bio- and Medical Informatics*, Bursa, M. et al., Ed. Berlin, Heidelberg: Springer, 2013, pp. 47–61.

[5] J. Spilka, V. Chudacek, M. Huptych, R. Leonarduzzi, P. Abry, and M. Doret, "Intrapartum fetal heart rate classification: cross-database evaluation," in *XIV Medit. Conf. on Med. and Biol. Eng. and Comp. 2016*, Kyriacou, E. et al., Ed. Cham: Springer Int. Pub., 2016, pp. 1199–1204.

[6] G. Georgoulas, P. Karvelis, J. Spilka, V. Chudacek, C. D. Stylios, and L. Lhotska, "Investigating pH based evaluation of fetal heart rate (FHR) recordings," *Health and Technology*, vol. 7, no. 2, pp. 241–254, 2017.

[7] Z. Comert, A. F. Kocamaz, and V. Subha, "Prognostic model based on image-based time-frequency features and genetic algorithm for fetal hypoxia assessment," *Computers in Biology and Medicine*, vol. 99, pp. 85 – 97, 2018.

[8] Z. Comert, A. M. Boopathi, S. Velappan, Z. Yang, and A. F. Kocamaz, "The influences of different window functions and lengths on image-based time-frequency features of fetal heart rate signals," in *2018 26th Sig. Proc. and Comm. App. Conf. (SIU)*, 2018. pp. 1–4.

[9] Z. Comert, Z. Yang, S. Velappan, A. M. Boopathi, and A. F. Kocamaz, "Performance evaluation of empirical mode decomposition and discrete wavelet transform for computerized hypoxia detection and prediction," in *2018 26th Sig. Proc. and Comm. App. Conf. (SIU)*, 2018, pp. 1–4.

[10] Z. Comert and A. F. Kocamaz, "Fetal hypoxia detection based on deep convolutional neural network with transfer learning approach," in *Software Engineering and Algorithms in Intelligent Systems*, R. Silhavy, Ed. Cham: Springer Int. Pub., 2019, pp. 239–248.

[11] M. N. Zarmehri, L. Castro, J. Santos, J. Bernardes, A. Costa, and C. C. Santos, "On the prediction of foetal acidaemia: A spectral analysis-based approach," *Computers in Biology and Medicine*. vol. 109, pp. 235 – 241, 2019.

[12] M. Jezewski, R. Czabanski, K. Horoba, and J. M. Leski, "Clustering with pairs of prototypes to support automated assessment of the fetal state," *Applied Artificial Intelligence*, vol. 30. no. 6, pp. 572–589, 2016.

[13] M. Jezewski, R. Czabanski, J. M. Leski, and J. Jezewski, "Fuzzy classifier based on clustering with pairs of ε-hyperballs and its application to support fetal state assessment," *Expert Systems with Applications*, vol. 118, pp. 109–126, 2019.

[14] M. N. Murty and V. S. Devi, *Pattern recognition. An algorithmic approach*. London: Springer-Verlag, 2011.

[15] I. Steinwart and A. Christmann, *Support vector machines*. New York: Springer-Verlag, 2008.

[16] S. Porebski and E. Straszecka, "Improving the quality of clustering-based diagnostic rules by lowering dimension of the cluster prototypes," in *Progress in Computer Recognition Systems*, Burduk, R. et al., Ed. Cham: Springer Int. Pub., 2020, pp. 47–56.

[17] M. Jezewski, J. M. Leski, and R. Czabanski, "Classification based on incremental fuzzy (1+p)-means clustering," in *Man-Machine Interactions 4*, Gruca, A. et al., Ed. Cham: Springer Int. Pub. Switzerland, 2016, pp. 563–572.

[18] M.-Y. Chen and D. A. Linkens, "Rule-base self-generation and simplification for data-driven fuzzy models," *Fuzzy Sets and Systems*, vol. 142, no. 2, pp. 243–265, 2004.

[19] B. Rezaee, "Rule base simplification by using a similarity measure of fuzzy sets," *Journal of Intelligent & Fuzzy Systems*, vol. 23, no. 5, pp. 193–201, 2012.

[20] A. Gegov, F. Arabikhan, and D. Sanders, "Rule base simplification in fuzzy systems by aggregation of inconsistent rules," *Journal of Intelligent & Fuzzy Systems*, vol. 28, no. 3, pp. 1331–1343, 2015.

[21] G. E. Tsekouras, "Fuzzy rule base simplification using multidimensional scaling and constrained optimization," *Fuzzy Sets and Systems*, vol. 297, pp. 46 – 72, 2016.

[22] S. Porebski and E. Straszecka, "Extracting easily interpreted diagnostic rules," *Information Sciences*, vol. 426, pp. 19 – 37, 2018.

[23] S. Porebski, P. Porwik, E. Straszecka, and T. Orczyk, "Liver fibrosis diagnosis support using the dempster–shafer theory extended for fuzzy focal elements," *Engineering Applications of Artificial Intelligence*, vol. 76, pp. 67 – 79, 2018.

[24] J. M. Leski, R. Czabanski, M. Jezewski, and J. Jezewski, "Fuzzy ordered c-means clustering and least angle regression for fuzzy rule-based classifier: Study for imbalanced data," *IEEE Transactions on Fuzzy Systems*, pp. 1–15, 2019.

[25] J. M. Leski, "Fuzzy (c+p)-means clustering and its application to a fuzzy rule-based classifier: toward good generalization and good interpretability," *IEEE Transactions on Fuzzy Systems*, vol. 23, no. 4, pp. 802–812, 2015.

[26] J. M. Leski, "An ε-insensitive approach to fuzzy clustering," *International Journal of Applied Mathematics and Computer Sciences*, vol. 11, no. 4, pp. 993–1007, 2001.

[27] V. Chudacek, J. Spilka, M. Bursa, P. Janku, L. Hruban, M. Huptych, and L. Lhotska, "Open access intrapartum CTG database," *BMC Pregnancy and Childbirth*, vol. 14, no. 1, pp. 1–12, 2014.

[28] A. L. Goldberger, L. A. N. Amaral, L. Glass, J. M. Hausdorff, P. C. Ivanov, R. G. Mark, J. E. Mietus, G. B. Moody, C.-K. Peng, and H. E. Stanley, "PhysioBank, PhysioToolkit, and PhysioNet: Components of a new research resource for complex physiologic signals," *Circulation*, vol. 101, no. 23, pp. e215–e220, 2000.

[29] J. Jezewski, A. Pawlak, K. Horoba, J. Wrobel, R. Czabanski, and M. Jezewski, "Selected design issues of the medical cyber-physical system for telemonitoring pregnancy at home," *Microprocessors and Microsystems*, vol. 46, pp. 35 – 43, 2016.

[30] O. L. Mangasarian and D. R. Musicant, "Lagrangian support vector machines," *Journal of Machine Learning Research*, vol. 1, pp. 161–177, 2001.

Fusion of Position Adjustment from Vision System and Wheels Odometry for Mobile Robot in Autonomous Driving

Jarosław Zwierzchowski*, Dawid Pietrala†, Jan Napieralski*
* Department of Microelectronics and Computer Science, Lodz University of Technology
Wolczanska 221/223, 90-924 Lodz, Poland
† Department of Automation and Robotics, Kielce University of Technology
Al. 1000-lecia P.P. 7, 25-314 Kielce, Poland
Emails: jzwie@dmcs.pl, dpietrala@tu.kielce.pl, jnapier@dmcs.pl

Abstract—Autonomous mobile vehicles need advanced systems to determine their exact position in a certain coordinate system. For this purpose, the GPS and the vision system are the most often used. These systems have some disadvantages, and so the GPS signal is unavailable in rooms and may be inaccurate, while the vision system is strongly dependent on the intensity of the recorded light. This work presents a system for determining the position of the robot base on information about the distance travelled coming from each vehicle wheel and the IMU sensor. However, wheels odometry introduces an additive measurement error (rise in every measure cycle). In this work there has been a vision correction system applied, where corrections are calculated by measuring the distance to artificial markers. This system precisely determines such correction. Each of these systems is described in the paper, in particular the vision system. There is also a description of a fusion algorithm of wheels odometry and vision correction system. The presented system was tested on an artificially built test track.

Keywords—autonomous driving, artag, odometry, signals fusion.

I. INTRODUCTION

The article describes the implementation of algorithms to determine the position and orientation of the mobile robot in a certain coordinate system without the participation of the GPS positioning system. Authors divided described task into two subsystems constantly sending messages to each other via UART bus. The first subsystem consists of programs located in the robot control system and determines the robot's trajectory calculated from vehicle wheels. This is called wheels odometry and it is widely used in the automotive industry. A general discussion of this solution has been described in [1], [2]. Currently, designers of modern solutions to increase the accuracy of determined robot position, very common are using a combination of many independent positioning systems. Therefore, the authors added a second subsystem, that was a vision system. In the literature, a vision system with an implemented SLAM algorithm is the most often described [3]. In this article, the vision system only provided position corrections, and was geared toward finding characteristic landmarks. The problem with landmarks was studied in the paper [5].

To check the correctness of the implemented subsystems, a mobile robot was designed and built. What is more, many tests were performed on the test track. This track was accurately measured and a map was made based on these measurements. Control points and position of graphic markers - ARTags were marked on the map. The experiment assumed that the robot autonomously was able to move from the starting point to the next and subsequent checkpoints with the greatest accuracy. Besides if the robot calculated that a checkpoint was reached the blue lamp mounted on its board stated to flash.

By design, numbers from 0 to 16 are encoded in ARTags. To accurately determine the position of the robot, the vision system needed to recognize three markers. While if two markers had been detected, two robot positions were determined, fortunately one of which could be easily rejected. If one marker was detected, the robot could be located with some accuracy, one was positioned on the circle indicated by the distance between the robot and the marker. To calculate the distance to the markers, a specially designed head with two cameras set under a 20-degree account was used. The appearance of the head is shown in (Fig. 6). Two cameras with FHD Logitech 920C resolution were used for testing.

The software that interpreted the robot's space was initially tested on a laptop with an Intel Core i7 processor and 32 GB RAM, which at 1024x760 resolution allowed to run the ARTag recognition algorithm at 15 frames per second. The vision system communicated with the robot's main processor via a serial bus and transmitted the numbers of founded markers and the distances to them. Information from the wheel odometry was also transmitted to the main processor. Ultimately, the important step of the whole task was to combine the subsystems described above into one decision-making robot control system, i.e. calculating signals fusion. Tests for the accuracy and speed of vision system were described in [4]. The general structure diagram of the robot system is shown in the figure (Fig. 1).

All algorithms regarded to wheels odometry were implemented in C language and are flashed in STM processor. The vision system was written in C++ language.

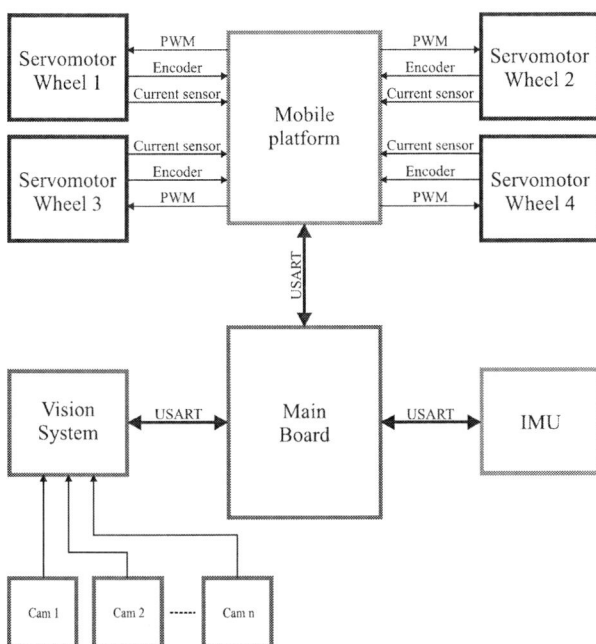

Fig. 1. Block diagram of the construction of a mobile vehicle

The general overview of the robot was shown in (Fig. 2).

Fig. 2. General view of the vehicle with the implemented autonomous system

II. DETERMINING VEHICLE POSITION

In this chapter, the description of the method for acquiring information about the position and orientation of the mobile robot, both from sensors determining travelled distance and the vision system, will be presented. At the end of the chapter, the method for fusing those signals will be described.

A. Position From Wheels and IMU

The current relative position of the mobile robot in two-dimensional space was determined by a vector of two co-ordinates $\mathbf{p}_w = [x, y]$. Those coordinates were determined according to equations (1)

$$x = \sin\theta \int v_x dt + x_0, \quad y = \cos\theta \int v_y dt + y_0, \quad (1)$$

where: θ - current vehicle yaw angle relative to the axis Y, v_x, v_y - vehicle speed components, x_0, y_0 - initial vehicle position. In order to determine the current relative position of the mobile robot, information about the current linear robot speed and its current angle relative to the Y axis (azimuth) in some coordinate system is necessary. For this purpose, each of the robot wheels is equipped with a DC motor with an incremental encoder and also for each wheel there was a PID controller algorithm implemented to control set angular velocity. The control system, at each calculation step, determines current linear vehicle speed based on known angular velocities of individual wheels and geometrical dimensions. While, azimuth is read from an inertial navigation system equipped with an accelerometer, gyroscope and magnetometer.

B. Position From ARTags

Let's define image I as a two-dimensional array containing grayscale intensity of pixel recorded, $I : \Omega \subset \mathbb{R}^2 \to \mathbb{R}_+$; $(x, y) \mapsto I(x, y)$, where Ω is the domain of the image. Point $P = (x, y, z)^T$ from 3D space is donate as $p = (u, v)^T \subset \Omega$. The vision system has been developed and will be presented in the 6 steps described below.

1) Edge detection - Canny algorithm: First, all edges of the image should be detected. The edge is a transition (change in the value of adjacent pixels). That means if $I(x_k, y_1) > I(x_k, y_2)$ then we have a vertical edge. In the simplest case the edge we will be gained using the equation

$$\frac{\partial I(x, y)}{\partial x} \approx \frac{I(x+1, y) - I(x, y)}{1} \quad (2)$$

However, showing edge in this manner can only result if the image is filtered with a low-pass filter, for example, a Gaussian filter $h(\tau, \sigma)$, because the edge will hide in the noise. This means that the final form of the edge detection filter will be equal to

$$\nabla I = \left[\frac{\partial(h(\tau, \sigma) * I(x, y))}{\partial x}, \frac{\partial(h(\tau, \sigma) * I(x, y))}{\partial y} \right] \quad (3)$$

wherein Canny edge detector threshold $\tau > 0$ and standard deviation $\sigma > 0$. Next step is as follows. If $\nabla I^T \nabla I$ is larger than the prefixed gradient threshold and is in local maximum along the gradient, set this pixel as edge. The final step of the algorithm is to determine the one-point edge from thehysteresis of 2 thresholds [6].

2) Contour detection and polygonal approximation: The second step is to determine objects contours, which edges were obtained in the previous step. For this purpose, the algorithm of the polygon contour description was used. Its main goal is to specify the vertices of the polygon according to the Ramer-Douglas-Peucker algorithm. Let's define vector

$D = [p_0 \; p_1...p_n]$ being points lying on the contour of the object. Determine the segment L_k with the beginning at p_0 point and ending at point p_n what is more, determine all lines perpendicular to this segment L_k passing through the points from vector D and determine the longest norm L^2 d_k for this straight line

$$\{\bigwedge p \in D : max\|p_i \perp L_k\|_2 = d_k\} \tag{4}$$

If $d_k > T$ then set a new segment L_{k+1} at the beginning at current L_k segment beginning and end at point p_k, for whose d_k has been determined, where T is arbitrarily selected threshold. Save point p_k as the vertex of the polygon in vector $\hat{D}$. If $d_k < T$ then calculate L_{k+1} such that the beginning is at point p_k, and the end at point p_n. Reassign d_{k+1} according to formula (4). Finish the algorithm if the L segment cannot be created.

3) Rejecting incorrect markers: The next step of the graphic markers recognition task is to choose appropriate candidates based on the following criteria. The first criterion is to check whether vector vertices $\hat{D}_i$ has 4 elements. Then we check if the vertices form closed contours, i.e whether they depict geometrical figure. The last two tests check if the geometric figure is convex and whether the distance between consecutive segments resulting from the fusion of points in vector $\hat{D}$ is big enough. Vector $\hat{D}_i$, which fulfil tests mentioned above is saved as candidate containing marker shape.

4) Removing perspective and warping algorithms: The candidate selected in the previous step may include a marker to recognize which we need to find such position of camera coordinate system to make vector $\hat{D}_i$ containing vertices of any quadrangle to be able to convert into vertices of the square. In the article, the warping perspective method was used. Next, knowing the transformation matrix we could use equation $\hat{I}(x,y) = A \cdot I(x,y)$ in order to distort the image.

5) Detecting marker frame and read its code: Having determined picture from the previous point it is possible to start reading the tag code. To do this, we divide the image into a grid of 9x9 squares and then we count the white pixels in each square. If their number will be higher than the defined threshold, we determine that the square represents the binary number 1, oppositely we assign the square value 0. The resulting matrix containing binary numbers is compared with the code patterns. Each marker has a frame with 2 squares thickness.

6) Calculation of the position and orientation of the marker: The perspective equation for a calibrated camera can be written as follows

$$\lambda[u, v, 1] = K[R|t][X, Y, Z, 1]^{\mathrm{T}} \tag{5}$$

where K is the matrix of internal camera parameters containing the focal length and the physical centre of the image, matrix $[R|t]$ is the matrix of the external camera's parameters containing rotations and translations of the camera coordinate system relative to some global coordinate system. The task is to determine Q from the equation $Q \cdot H = 0$, where $H = K[R|t]$. Using the homogeneity of the image points p_i

we can write the equations, $P_i^{\mathrm{T}} h_1^{\mathrm{T}} + 0 \cdot h_2^{\mathrm{T}} - u_i P_i^{\mathrm{T}} h_3^{\mathrm{T}} = 0$ and $0 \cdot h_1^{\mathrm{T}} + P_i^{\mathrm{T}} h_2^{\mathrm{T}} - v_i P_i^{\mathrm{T}} h_3^{\mathrm{T}} = 0$, where $h_i, i = 1, 2, 3$ are rows of the unknown matrix H. To determine the H matrix, Singular Value Decomposition (SVD) method shall be used. The minimum number of p points and their equivalents in the world space P needed to determine the matrix H equals 4. As a result of calculating SVD one of the matrices is the matrix $[R|t]$, where $t = [x, y, z]$, from which we will determine the distance between the vehicle and the marker $d_v = \lambda\sqrt{x^2 + y^2 + z^2}$.

C. Fusion Of Wheels And Vision Signals

This method of determining the position of a robot is exposed to errors, e.g. from wheel slip. What's more, the error of the currently determined position is the sum of all errors that have occurred since the beginning of the journey. To determine the relative position of correction determined according to the method presented above, a comparison of results from the vision system was used.

A set of markers were placed into workspace of the mobile robot, whose values and positions were known. The vision system sent information about codded numbers of currently detected markers and distance to them, to the robot controller. Then, based on the information received, an amendment was determined according to the following procedure. There were three cases. The first occurs when the vision system correctly recognizes one marker. This is shown schematically in the (Fig. 3),

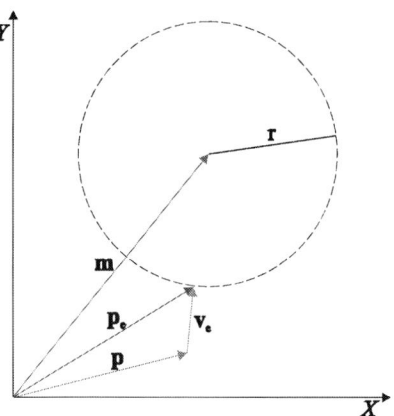

Fig. 3. Determination scheme of position correction vector when one marker is detected

where: **m** - vector determining the position of the marker, **p** - vector of the robot position determined from the odometry system, $\mathbf{p_c}$ - vector determining the corrected position of the robot, $\mathbf{v_c}$ - correction vector, r - distance from the marker obtained from the vision system. The corrected vehicle position is calculated as a point on a circle with a radius equal to the distance to the marker obtained from the vision system and the center point of the marker that is closest to the current vehicle position determined from the odometry system.

The second case takes into account the situation when the vision system correctly detects two markers simultaneously (or with a time interval of not more than 0.3s). This is schematically illustrated in (Fig. 4),

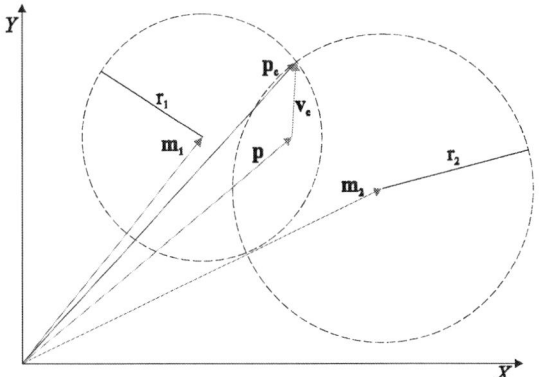

Fig. 4. Determination scheme of position correction vector when two markers are detected

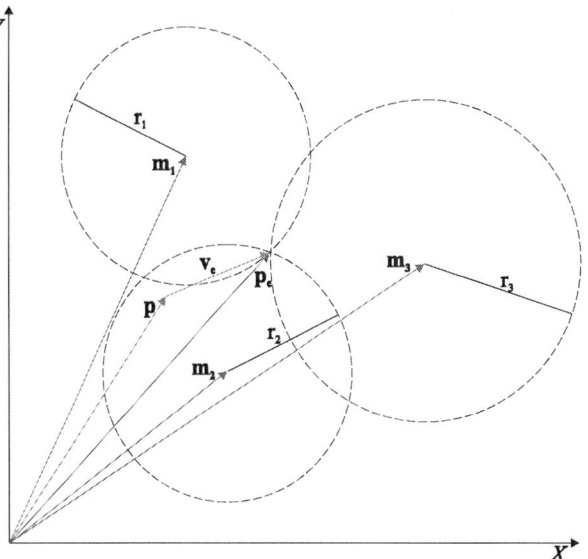

Fig. 5. Determination scheme of position correction vector when three markers are detected

where: m_i - vector defining the position of the i marker, p - vector of the robot position determined from the odometry system, p_c - vector defining corrected robot position, v_c - correction vector, r_i - distance from i marker obtained from the vision system. In this situation, the corrected robot position is in one of two places, which are the intersection of two circles with radii equal to vehicle distance from the markers and the centres lying in the markers point. The robot chose that solution, which is closer to the robot's currently determined odometry position.

The third case takes into account the situation when the vision system correctly detects three or more markers simultaneously (or with a time interval of not more than 0.3s). This is schematically illustrated in (Fig. 5), where the designations were adopted as for (Fig. 4). In this situation, three markers are selected. The corrected robot position is determined as the intersection of all three circles, with radii equal to the distance from the markers and with centres at the marker points.

Correctness data was immediately sent to the robot control system and the tentative robot position was updated. The position from vision system had the highest priority.

III. SYSTEM TESTS

System tests of automatic driving were divided into two stages. The first stage relates to testing the distance measuring accuracy from the vehicle to the marker. Tests on the correctness of estimating the distance were carried out in good lighting conditions, thanks to which the differences between the actual distance and the one determined by the algorithm were small and for a distance of up to approximately 3m, the relative errors were in most cases less than 1 percent. For worse conditions, e.g. underexposure, the maximum detection distance was significantly reduced. The strong light falling on the marker could even prevent detection. A detailed description was presented in [4] where it was noted that absolute

Fig. 6. Mobile vehicle with vision system on test track

measurement errors increase with distance. The most accurate measurements were obtained for distances up to 3 m and this corresponds to the distance for which the camera was calibrated.

The second stage of the test was performed in the laboratories of the Kielce University of Technology and at the European Rover Challenge international competition in 2018 and 2019. The (Fig. 7) presents a map of test track with

221

red marked control points (letter W), green marked points with graphic markers (letter L), where the point number was coded using ARTags. The places with obstacles were marked in black (letter R). The robot's position on the track was marked on the map with a red dotted line. The side of the map grid means 1m. In the (Fig. 8) the arrows show the corrections made during the driving. The black arrow indicates the correction made after detecting the markers number L4 and L5. While the red arrows indicate where the control system made corrections after detecting the L6 and L7 markers. As mentioned in the paper [4] the recognition range of the system was limited to about 6m, therefore corrections are calculated near the markers. During main drive at the competition, the vehicle reached the checkpoints with the following accuracy. Checkpoints W1 - 2cm, W2 - 5cm, W3 - 20cm, W4 - 60cm. The W4 checkpoint was very difficult to reach due to the large wheels slip of the vehicle, and the fact that vision system did not have two markers in its field of view. Vehicle control system did not make position correction and decided that the vehicle had reached the checkpoint without disruption. The same situation happened with the WX point, which was hidden behind the hill.

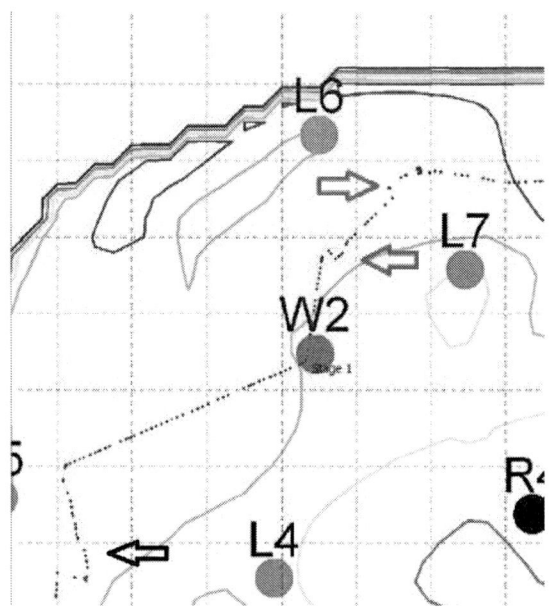

Fig. 8.   Mobile vehicle with vision system on test track

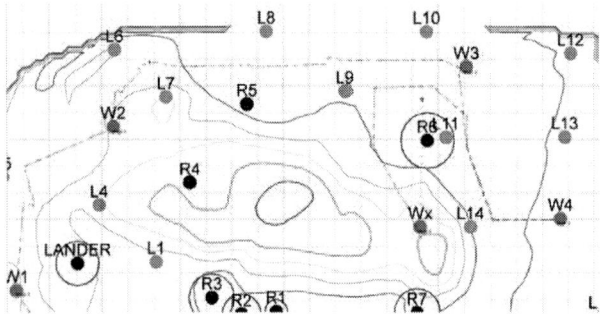

Fig. 7.   The screenshot of the software that shows view the position of the robot

IV. CONCLUSION

The article describes the automatic driving system of a mobile vehicle, which was equipped with two independent systems based on which the vehicle's position on the test track can be determined. The experiment assumes that GPS cannot be used. The first of the implemented systems calculated the robot positions from the robot wheels. This is the main system of the robot positioning in space, however, it is characterized by the fact that the more the distance traveled by the robot, the bigger positioning error increases. Therefore, the authors decided to develop a system that provides corrections from an independent system that gives an absolute measurement to known characteristic points with a well-known location on the track. The measurement to the markers was carried out using a vision system and if the marker or markers appear in the field of view of the camera, the system begins to use the detection algorithm and determining the correction. What is more, the vehicle informs about this fact by lighting the blue

lamp on board. The conducted tests of the system show that the accuracy of position determination for the vision system reached up to 2cm. Driving the entire test track gave over 60cm error at the last point and it was caused mainly by the lack of visibility of any marker. The place of making corrections by the system is shown in the screenshots and marked with arrows. The authors will check the working of the system with an additional GPS reference system in the future.

ACKNOWLEDGMENT

Authors would like to thank the Kielce University of Technology IMPUS Team, who won European Rover Challenge 2018, 2019 in Poland and University Rover Challenge 2019 in Utah USA.

REFERENCES

[1] D. Scaramuzza and F. Fraundorfer, Visual Odometry Part I: The First 30 Years and Fundamentals, IEEE Robotics and Automation Magazine 2011
[2] D. Scaramuzza and F. Fraundorfer, Visual Odometry Part II: Matching, Robustness, Optimization, and Applications, IEEE Robotics and Automation Magazine 2012
[3] C. Cadena, L. Carlone, H. Carrillo, Y. Latif, D. Scaramuzza, J. Neira, I. Reid and J. J. Leonard, Past, Present, and Future of Simultaneous Localization and Mapping: Toward the Robust-Perception Age IEEE TRANSACTIONS ON ROBOTICS, VOL. 32, NO. 6, 2016
[4] A. M. Annusewicz and J. Zwierzchowski, Zastosowanie algorytmu rozpoznawania znaczników graficznych do nawigacji robota mobilnego Automation Conference 2020 (in review process).
[5] X. Zhong, Y. Zhou, H. Liu, Design and recognition of artificial landmarks for reliable indoor self-localization of mobile robots. In: International Journal of Advanced Robotic Systems 14(1), 2017.
[6] J. Canny, A computational approach to edge detection. In: IEEE Transactions, Pattern Analysis and Machine Intelligence, 6, pp. 679-698 1986.

Proceedings of the 27th International Conference *"Mixed Design of Integrated Circuits and Systems"*
June 25-27, 2020, Łódź, Poland

Marker Detection Algorithm for the Navigation of a Mobile Robot

Anna Annusewicz
Kielce University of Technology
Aleja Tysiaclecia Panstwa Polskiego 7, 25-314 Kielce
Email: aannusewicz@tu.kielce.pl

Jarosław Zwierzchowski
Lodz University of Technology
ul. Żeromskiego 116, 90-924 Łódź
Email: jzwie@dmcs.pl

Abstract—**This article discusses a vision system for recognizing fiducial markers including the ARTag type. The system was developed to navigate a mobile robot. It is an alternative solution to replace algorithms based on GPS signals. The paper describes the algorithm for detecting markers in detail. The developed system was tested under various field and lighting conditions. The paper presents obtained results of tests checking the correctness of estimating the distance from the camera to the marker.**

Keywords—**mobile robot, fiducial marker, ARTag, vision system, navigation.**

I. INTRODUCTION

The navigation of mobile robots plays an important role in autonomous movement tasks. A GPS signal is commonly used for this purpose, but it is not always possible to use a GPS-based navigation system, e.g. underground or in buildings. An alternative way is to use vision systems. A robot that is equipped with a camera can use artificial markers located at known points in its environment to determine its position and orientation. Many types of markers have been developed so far. Artificial markers have been designed for the use in augmented reality. Billinghurst, Poupyrev and Kato presented an ARToolKit marker [1], and Fiala developed an ARTag marker [2]. These types of fiducial markers have been designed to calculate the position and orientation of the camera relative to the world coordinate system to properly set the perspective of artificially added virtual objects. Four vertices are necessary to determine the position of the camera relative to the marker. This is the minimum number of points to obtain unambiguous information about the translation and rotation of the camera coordinate system relative to the marker connected with the world coordinate system. The solution of the Perspective-Three-Points (P3P) problem was presented by Xiao-Shan Gao et al. [3]. They developed an algorithm that makes possible to determine the position of the camera relative to the marker based on image coordinates of characteristic points (marker vertices), coordinates of these points in the world system (2D to 3D corresponding points) and camera intrinsic parameters. Knowing the position of the camera relative to objects in the real world is also very useful in robotics. Augmented reality markers such as ARTag markers found application in robot navigation. Applications of different markers in the vision systems of mobile robots are discussed, for example, by Babinec et al. [4], Zhong et al. [5] and Kazala and

Straczynski [6]. The vision system described in this paper was developed for the navigation of a mobile robot. The marker detection algorithm was extended using additional criteria. The functioning of the algorithm was tested under various field and lighting conditions.

II. VISION SYSTEM

The vision system was developed for the mobile robot shown in Fig. 1. Its function is to enable accurate robot navigation without using a GPS signal. The position of the robot in space is determined by encoders. The location of the markers in the robot space is known; thus, the recognition of appropriate markers allows for precise navigation of the robot. Using only the odometry method could cause measurement errors resulting, e.g., from wheel slips. The vision system consists of two Logitech HD Pro Webcam C920 cameras and a Nvidia Jetson TX2 module performing digital image

Fig. 1. The mobile robot for which the vision system was developed at a test site

processing operations. Cameras are mounted on the sides of the robot in such a way that their frames do not overlap. Fiducial markers of the ARTag type were used to navigate the mobile robot. The vision system was designed to detect markers, recognize their indexes and determine the position of the robot relative to the ARTags.

A. Algorithm for recognizing graphic markers

The algorithm discussed here uses the marker recognition method presented in [7,8]. This method is also useful for detecting other types of markers, e.g. ARToolKit Plus, Matrix or BinARyID, shown in Fig. 2. The algorithm was implemented on the Nvidia module in C++ language using OpenCV and QT libraries. The application performs image processing operations for the two cameras and sends information to the main robot control system using a serial port.

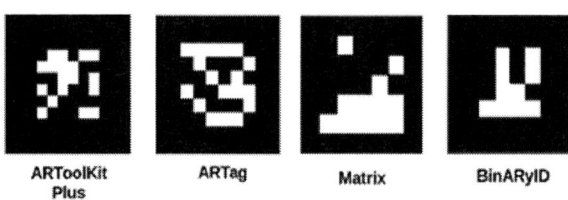

ARToolKit Plus **ARTag** **Matrix** **BinARyID**

Fig. 2. Examples of fiducial markers [7]

Fig. 3 shows a general block diagram of the algorithm for analyzing one frame. The algorithm works in real time and analyzes subsequent frames from both cameras in order to find markers.

The first stage of the image analysis involves converting the captured frame to a gray scale and performing adaptive thresholding. This process allows us to obtain an edge image and it is faster than Canny's edge detection algorithm [9]. The edge image is used to search for all contours using the method developed by Suzuki and Abe [10]. As most of the contours found are useless, they should be subjected to preliminary filtration to focus only on those that are likely to represent markers.

During the preliminary filtration, the algorithm rejects contours smaller than the declared minimum contour size. Others are approximated using a polygon [11]. All contours that are not quadrangles are rejected because the original shape of the marker is a square. In addition, more conditions were added to the algorithm, which cause the rejection of more contours that are not marker. As a result, fewer contours are subject to the internal code analysis. First, quadrilaterals that are not convex are discarded, then the algorithm verifies the criterion of the minimum length of the sides of the approximated quadrangle. After rejecting some candidates, the algorithm checks if the contours representing potential markers are not too close to each other. In the case of a pair of such markers, only one is selected based on the length of circumference criterion. Contours which satisfy these conditions are saved as potential markers and are further analyzed. The points representing the

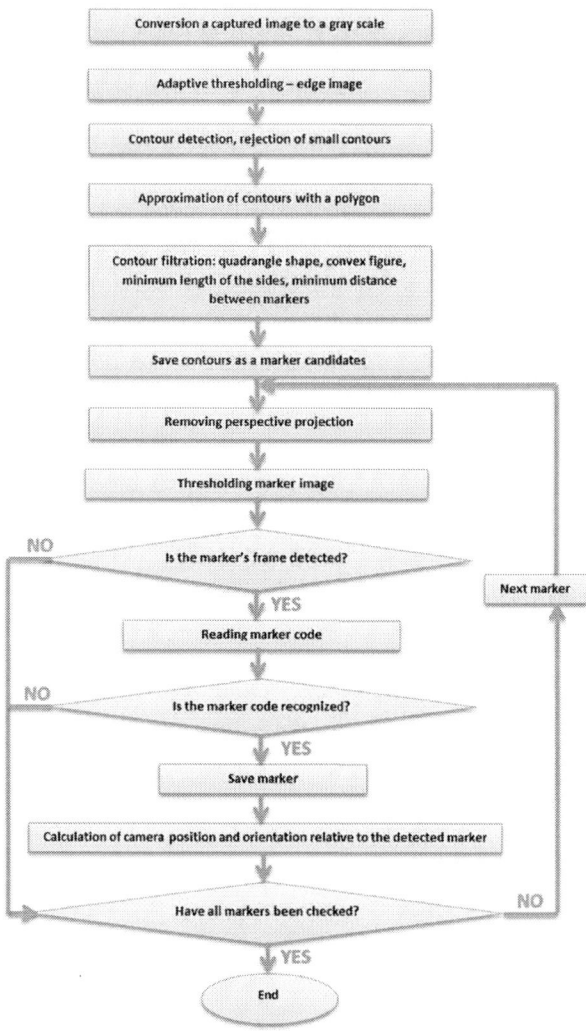

Fig. 3. Marker detection algorithm

corners of the figure are corrected using the method of setting sub-pixel accuracy. To interpret the center of the marker, the perspective view is removed so that individual cells can be extracted. The resulting image of the marker candidate is divided into cells whose number corresponds to the number of marker bits and frame thickness. This system uses the ARTag markers with five bits. It is assumed that the marker's frame thickness corresponds to two cells. The marker's image is subjected to thresholding operation using the Otsu algorithm [12], because the code verification should be performed on a binary image. First, the algorithm checks the cells belonging to the frame. If all of its elements are black, the marker code is verified. Black represents zero while white corresponds to one. The value of individual cells is determined based on most pixels of white or black color. The determined marker code is compared with the database of known codes saved in the program; based on the code, the appropriate index is assigned to the marker. The code is compared to the database for four different marker orientations. Using the detected ARTags, the position and orientation of a camera relative to the marker are estimated. The algorithm calculates the distance of the mobile robot to the detected marker from the translation vector. The world coordinate system is associated with the center of the marker, and the dimensions of the sides are known. This information allows us to determine the coordinates of the marker corners in the world coordinate system. Based on the corner coordinates in the image known from the approximation of the contour and the intrinsic parameters of the camera, it is possible to calculate the position of the camera coordinate system relative to the marker. For this purpose, the algorithm uses a method solving the Perspective-n-Point problem implemented in the OpenCV library based on the Levenberg-Marquardt optimization [13].

B. Camera calibration

In order to estimate the position of the camera relative to the marker based on the corresponding points from the image and the world coordinate system, it is necessary to know the camera intrinsic parameters. Each of the cameras used was calibrated using the method proposed by Zhang [14] in the MATLAB program.

III. ANALYSIS OF THE VISION SYSTEM OPERATION

The marker detection algorithm was tested for various camera resolution settings. The tests were carried out for Full HD resolution (1920x1080) and 1024x576. All the detection tests were performed on a marker with 142 mm long side. The calculated distance between the marker and the camera was compared with the distance measured using a measuring tape with an accuracy of 1 mm, which was treated as a reference measurement.

TABLE I
MEASUREMENT DATA - RESOLUTION 1920x1080

Distance marker-camera [mm]		Relative error [%]
Measurement	Vision system	
270	273	1.3704
430	442	2.8837
643	654	1.7263
885	887	0.2373
1043	1044	0.0959
1313	1312	0.0457
1524	1522	0.1312
1775	1772	0.1690
2013	2005	0.3974
2273	2260	0.5719
2602	2584	0.6918
2820	2811	0.3191
3061	3075	0.4574
3242	3281	1.2030
3364	3392	0.8323
3683	3722	1.0589
4071	4099	0.6878
4342	4419	1.7734
4790	4874	1.7537
5019	5118	1.9725
5215	5281	1.2656
5560	5638	1.4029

TABLE II
MEASUREMENT DATA - RESOLUTION 1024x576

Distance marker-camera [mm]		Relative error [%]
Measurement	Vision system	
304	306	0.7566
556	557	0.1799
758	761	0.4749
1115	1107	0.6906
1420	1415	0.3521
1670	1645	1.4970
1915	1896	0.9922
2083	2067	0.7681
2283	2262	0.9198
2635	2617	0.6831
2955	2947	0.2707
3260	3237	0.7055
3530	3506	0.6572
3895	3883	0.3081
4185	4181	0.0956
4585	4601	0.3490
4765	4697	1.4271

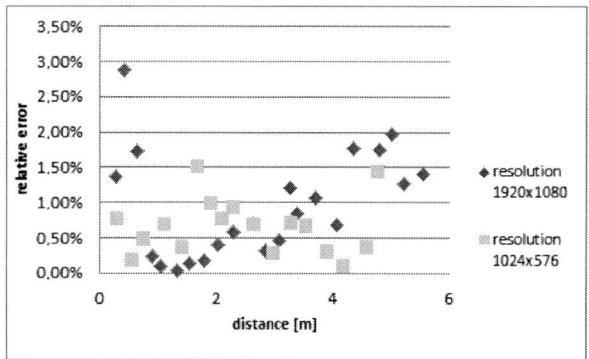

Fig. 4. Relative error versus the distance between the camera and the marker for measurements performed by the vision system for various resolutions

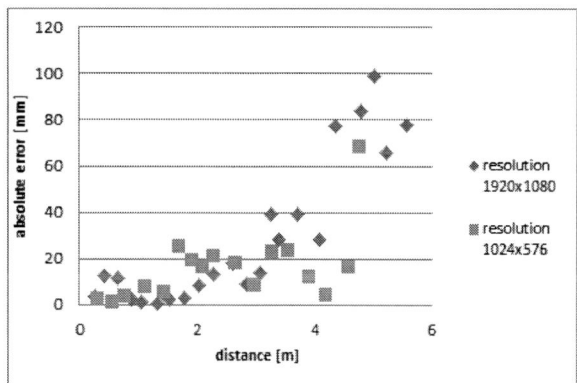

Fig. 5. Absolute error versus the distance between the camera and the marker for measurements performed by the vision system for different resolutions

For lower resolution, the operating speed of the vision system ranged from 12 to 15 fps, while for Full HD resolution the maximum image processing speed was up to 5 fps. In both cases, the delay was similar (about 1 s). The use of higher resolution allowed us to increase the distance to over 5 m (from which the marker could be recognized), while in the low resolution case the maximum distance was over 4 m. The absolute measurement error increased with distance. The results are provided in Table I, Table II and the graphs in Fig. 4 and Fig. 5. The measurement was carried out approximately every 200 mm.

IV. CONCLUSION

The vision system described in this paper was used to navigate a mobile robot. The algorithm is suitable for detecting ARTag type markers. The system was tested for various camera resolution settings: Full HD and 1024x576. For the lower resolution, the algorithm speed was about three times higher at the expense of reducing the distance from which ARTag can be detected. More frequent image analysis may be required if a robot moves at a higher speed or when changes in the lighting conditions are extreme, e.g. from deep shadows to intense light. For a lower frame rate, the camera might not be able to adapt to the light intensity. During the system testing, there were situations when the marker could not be detected because of this reason. However, if a robot operates in larger spaces, where the markers are located at large distances, it is necessary to choose a higher resolution, which allows us to significantly increase the detection distance. The selection of the appropriate resolution depends on the operating conditions of the algorithm. Experimental studies carried out on the presented mobile robot showed that better navigation was obtained using FullHd resolution. This allowed the recognition of distant markers. The speed of the robot was adjusted so that the number of frames per second was sufficient.

The tests to check the correctness of estimating the distance from the camera to the marker were carried out in good lighting conditions. As a result, the differences between the actual distance and the one estimated by the algorithm were small and for a distance of up to about 3 m, the relative errors were in most cases less than 1%. For worse conditions, e.g. underexposure, the maximum detection distance was significantly reduced, and the intense light shining on the marker could even prevent detection.

Absolute measurement errors increase with distance. The most accurate measure-ment results were obtained for distances of up to 3 m, and this corresponds to the distance for which the calibration was carried out.

The insertion of additional conditions rejecting more unnecessary contours coused that fewer contours are subject to internal code analysis, which uses complex image processing functions.

ARTag markers work well in mobile robot navigation applications. In the future, this vision system will also be used to support the autonomous operation of manipulators.

REFERENCES

[1] I. Poupyrev, H. Kato and M. Billinghurst, *ARToolkit User Manual, Version 2.33*. Human Interface Technology Lab, University of Washington, 2000.

[2] M. Fiala, *ARTag, a fiducial marker system using digital techniques*, in Proceedings / CVPR, IEEE Computer Society Conference on Computer Vision and Pattern Recognition, vol. 2, pp. 590 - 596, 2005.

[3] Xiao-Shan Gao, Xiao-Rong Hou, Jianliang Tang and Hang-Fei Cheng, *Complete Solution Classification for the Perspective-Three-Point Problem*, in IEEE Transactions on Pattern Analysis and Machine Intelligence, vol. 25, pp. 930- 943, 2003.

[4] A. Babinec, P.Hubinský, L.Jurišica. and F.Duchoň, *Visual Localization of Mobile Robot Using Artificial Markers*, in Procedia Engineering, vol. 96, pp. 1-9, 2014.

[5] X.Zhong, Y.Zhou, and H.Liu, *Design and recognition of artificial landmarks for reliable indoor self-localization of mobile robots*, in International Journal of Advanced Robotic Systems, vol. 14, 2017.

[6] R. Kazala, and P. Straczynski, *Increasing accuracy of the mobile robot positioning system by using ARTags*, in 24th International Conference Engineering Mechanics, Svratka, Czech Republic, 2018.

[7] S. Garrido-Jurado, R. Muñoz-Salinas, , F.J. Madrid-Cuevas, and M.J. Marín-Jiménez, *Automatic generation and detection of highly reliable fiducial markers under occlusion*, in Pattern Recognition, vol. 47, pp. 2280 - 2292, 2014.

[8] Y. Kim, S. H. Yang, K. W. Yang and N. G. Dagalakis, *Design of MEMS vision tracking system based on a micro fiducial marker*, in Sensors and Actuators A: Physical, vol. 234, pp. 48 - 56, 2015.

[9] J. Canny, *A computational approach to edge detection*, in IEEE Transactions, Pattern Analysis and Machine Intelligence, vol. 6, pp. 679-698, 1986.

[10] S. Suzuki, and K. Abe, *Topological structural analysis of digitized binary images by border following*, in Computer Vision, Graphics, and Image Processing, vol. 30, pp. 32–46, 1985.

[11] D.H. Douglas, and T. K. Peucker, *Algorithms for the reduction of the number of points required to represent a digitized line or its caricature*, in Cartographica: The International Journal for Geographic Information and Geovisualization, vol. 10, pp. 112–122, 1973.

[12] N. Otsu, *A threshold selection method from gray-level histograms*, in IEEE Transactions on Systems, Man, and Cybernetics, vol. 9, pp. 62–66, 1979.

[13] K. Levenberg, *A method for the solution of certain non-linear problems in least squares*, in: Quarterly of Applied Mathematics, vol. 2, pp. 164–168, 1994.

[14] Z. Zhang, *A Flexible New Technique for Camera Calibration*, in IEEE Transactions on Pattern Analysis and Machine Intelligence, vol. 22, pp. 1330–1334, 2000.

Modified Particle Swarm Optimization Algorithm Facilitating Its Hardware Implementation

Michal Rajewski, Zofia Długosz, Rafal Dlugosz, Tomasz Talaska

UTP University of Science and Technology
Faculty of Telecommunication, Computer Science
and Electrical Engineering
Kaliskiego 7, 85-796
Bydgoszcz, Poland
Email: michalrajon@gmail.com, zosia.dlugosz@gmail.com, rafal.dlugosz@gmail.com, talaska@utp.edu.pl

Abstract—**This paper presents various modifications and developments in the PSO algorithm. PSO is an algorithm based on the behavior of swarms. In this work we focus mainly on simplifying the algorithm by replacing random number generators, which can be a problem when implementing algorithms in hardware. We investigate how changing the methods for algorithm updates from random to more deterministic approach influences the results. This paper shows whether it is possible to achieve same as random or better results using a simplified algorithm, which uses simple mathematical operations in the algorithm optimization process.**

Keywords—**Particle Swarm Optimization; Optimization; Deterministic methods; Hardware implementation issues;**

I. INTRODUCTION

The PSO algorithm has been a well-known and widely used algorithm for many years, especially useful when searching for the function optimum. The algorithm is modeled on the behavior of clusters of animals such as ants or bees. It consists of searching the "area" of functions by many so-called agents who try to get to the extremum. The strength of the algorithm is the cooperation of agents with each other and them sharing experience regarding existing discoveries. Each agent is based on the knowledge acquired during exploration, taking into account the best value achieved, but also shares its knowledge with the swarm. The algorithm is especially useful when the searched function has many extrema, because in such situation searching the area by only one unit would not bring the desired results, limiting the effect of the operation to finding a local extremum, not a global one.

The way the agents move is similar to how animals move in nature, but there are also some differences. The space in which the agents move is fully abstract, and the movement of individuals is only a simplified mathematical description of what can be found in reality.

Every particle in the swarm is characterized by the following variables:

- position – position of the particle in space,
- velocity – speed at which the particle moves,
- local best – position of found personal local optimum,
- global best – position of current found global optimum.

An important parameter of each particle, which also describes local best and global best, is the fitness function (FF) output value. It determines how far the unit is from the target function optimum.

In the original version of the algorithm, both the initial position and the weight values, by which the position is updated in subsequent iterations, are random values. In this work, the goal was to simplify the algorithm and explore alternative ways to update the algorithm parameters. The scope of work included the analysis of updating units in a more orderly and deterministic manner, i.e. eliminating the element of randomness in the algorithm. Elimination of randomness is also an extremely important issue in the case of hardware implementation, because the hardware implementation of the randomization block is computationally expensive.

In the literature, the topic of the PSO algorithm has been raised and discussed many times. Various PSO usage scenarios have been tested. In addition to the traditional search for a function extremum, it has also been used to optimize weights in the neural network learning process, instead of using traditional backpropagation for this purpose [1].

Extensive literature addresses the issue of improving the PSO. Such actions were taken for e.g. in [2], [3], where imposing restrictions on the speed of particles was tested. Authors of [4] also conducted extensive analysis, which checked many aspects such as the selection of particle parameters or the influence of particles on each other.

There have also been attempts to simplify the PSO algorithm in terms of matching it to hardware [5], where the algorithm parameter settings were analyzed for fixed values.

II. AN OVERVIEW OF THE CONVENTIONAL PSO ALGORITHM

The first phase of the PSO algorithm is the initialization of the positions of particular particles in the swarm, X, and their velocities, V. In the conventional approach, the initial values of X parameters are selected randomly.

From the hardware implementation point of view, the drawing operation of the X and the V vectors is relatively complex. For this reason, in our works we investigated various ways of the initialization and their impact on the effectiveness of the optimization process of the swarm. For example, one of the tested options was setting the initial values of the velocities to zero. This aspect of the implementation of the PSO algorithm requires further investigations, and therefore it is not the main topic of the presented work.

The process of the optimization of the swarm is performed in iterations that in the following part of the paper are referred to as epochs. Each iteration, k, is divided into steps described below for a classical PSO algorithm.

A. Computation of the fitness function

In each iteration, the fitness function (FF) value is calculated for each particle based on the positions of these particles in the n-dimensional input data space. The FF determines the problem that is solved by the PSO algorithm. This function may be provided in the form of a mathematical expression, as for example in [6]–[10]. On the other hand, in a real world situation, its value may be determined by measuring of various physical quantities [11], [12]. However, independently on the approach, the values returned by the FF indicate how well particular particles match the optimal solution, i.e. how close they are to an optimal value.

B. Updating personal best values

At this stage, each particle verifies its own personal best value, p_{best}. If the newly calculated FF value for a given particle is better than its previous value, then the previous value is replaced by a new one. The position for which the best value was found is also remembered in the memory of a given particle.

C. Updating global best value

In this step, based on previously calculated FF values, the algorithm determines which of the particles has the best value, p_{best}, in relation to the sought optimum of the FF. Depending on the problem, the optimum may be either the minimum or the maximum value of the FF. The value that matches a given extreme of FF in the best way is stored in the memory of the overall swarm, as the so-called global best value, g_{best}.

Similarly to the p_{best} values, the position for which the g_{best} was found is also remembered in the memory of the swarm.

D. Updating positions and velocities of all particles

In this step of a given iteration, k, each particle updates first its velocity vector. It depends on three separate components, namely the inertial, the cognitive and the social ones. The are briefly characterized below.

1) Inertial component: The velocity vector from the previous iteration, $k-1$, is multiplied by a weight vector W. The elements of the W can be set to equal values, which means that each element of the velocity vector is multiplied by the same factor in a given iteration. In the conventional PSO algorithm, the values of the W vector are gradually reduced in subsequent iterations. The inertial component is computed as follows:

$$V_{ic,k} = W \cdot V_k \qquad (1)$$

2) Cognitive component: It depends on the positions of particular particles, for which their personal best values, p_{best}, of the FF were found. The cognitive component is computed as follows:

$$V_{cc,k} = c_1 \cdot r_1 \cdot (p_{best}.X - X_k) \qquad (2)$$

3) Social component: It is dependent on the position, for which the best FF value, g_{best}, for the overall swarm was found. The social velocity component is computed using following formula:

$$V_{sc,k} = c_2 \cdot r_2 \cdot (g_{best}.X - X_k) \qquad (3)$$

In the equations above, the $p_{best}.X$ and $g_{best}.X$ terms mean the position vectors, for which personal and global best values have been found. All these components allow to compute the resultant velocity, which is expressed as follows:

$$V_k = V_{ic,k} + V_{cc,k} + V_{sc,k} \qquad (4)$$

In the original version of the algorithm, the last two components are updated using random weight vectors r_1 and r_2.

On the basis of the computed velocities, the positions of particular particles in the swarm are updated, as follows:

$$X_{k+1} = X_k + V_k \qquad (5)$$

where:
- k – current iteration,
- X_k – position of a given particle in a given iteration,
- V_k – velocity of a given particle in a given iteration,
- W, c_1, c_1, r_1, r_2 – parameters that play the role of the weights.

E. Final step – stopping the optimization process

The optimization process may be completed based on various assumptions – conditions. One of them is to conduct a specific number of iterations. Alternatively, one can specify a number of iterations, for which the global best and personal best values do not change or change only insignificantly.

The optimization process should return the position, for which the FF reaches its extreme value. It is often desirable, that even all particles are located in the global FF extreme. However, it is also possible that particular particles oscillate around the extreme value of the FF. Under real conditions it can also happen that the FF reaches an extreme value or close to it at various points in the input data space. In this situation, particular particles can became representatives of these positions.

III. PROPOSED SOLUTIONS AT THE ALGORITHM LEVEL

In this work we present some modifications of the way how the velocities of particular particles are computed during the optimization process of the swarm. They lead to a simplification of the PSO algorithm when it comes to the hardware implementation. They also allow to increase a control over the changes occurring in subsequent iterations of the optimization process.

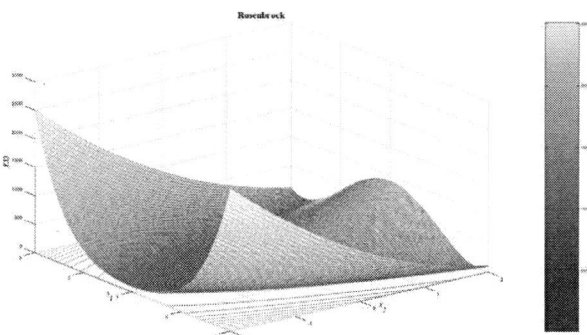

Fig. 1. Rosenbrock fitness function used to test proposed algorithms

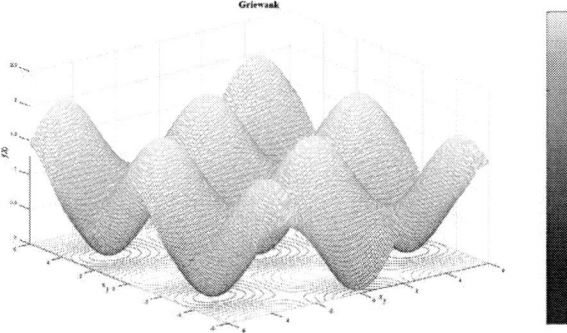

Fig. 2. Griewank fitness function used to test proposed algorithms

One of the main objectives of the carried out investigations is transistor level implementation of the PSO algorithm. For this reason, in the proposed solutions we try to eliminate the randomness of some coefficient appearing in the original algorithm.

The proposed modifications of the conventional PSO algorithm are presented below. The behavior of particular modified versions was verified for three selected FFs: the sphere, Rosenbrock and the Griewank ones. The Rosenbrock and the Griewank FFs are shown in Figs. 1 and 2, respectively. They offer different complexities. The simplest one is the sphere, in which there are no local minima and only a single global minimum. The remaining functions are much more complex, featuring multiple local minima.

Experiments were carried out for 64 particles in the swarm and for 100 iterations.

A. Proposed methods of modification of r parameters

Changes to the r_1 and r_2 parameters affect the velocity components described by 2 and 3. During the tests, the r_1 and r_2 parameters were updated as in the proposed methods, presented below.

1) Square root dependence on the iteration number : In this approach the values of the r parameters are equal to square root of a given epoch number, as follows:

$$r_{1,2} = \frac{r_{1,2}}{\sqrt{k}} \qquad (6)$$

This means that with each subsequent iteration, the rapidity of the particle position change is reduced, which in turn allows to avoid some oscillations at final stages of the optimization process. The solution boils down to a slow narrowing of the search area in order to indicate the optimum more precisely. In this approach, we define an η parameter, which is the radius of the proximity around the optimum, in which the algorithm finishes its operation.

2) Dependence on the identification number in the swarm: Unlike the previous method described in III-A1, in this approach the values of r_1 and r_2 for particular particles are constant, for the overall optimization process of the swarm. However, there are differences between particular particles in the swarm. To describe this method, suppose there is N_{pop} particles in the population (the swarm). Each of them has its unique identification number $i = 1, \ldots, N_{\text{pop}}$. Taking it into account, the values of the r_1 and r_2 parameters for particular particles are computed as follows:

$$r_{1,i} = r_{\text{base}} \cdot \frac{i}{N_{\text{pop}}} \qquad (7)$$

$$r_{2,i} = r_{\text{base}} \cdot \frac{i}{N_{\text{pop}}} \qquad (8)$$

where r_{base} is an additional base factor that may be used to control the course of the algorithm.

In this method, the range of motion of particular particles depends on their value i. Depending on this factor, the movements of some particles are more precise, while others more general. It results in a different impact of the p_{best} and the g_{best} factors on the trajectories of particular particles. Those of them whose indexes are high can cover greater distances in the input data space between subsequent iterations k. In this case, their search area largely depends on the current values of the p_{best} and the g_{best} factors. This is because the values of velocity components described with (2) and (3) are larger compared to their inertial components (1).

On the other hand, the particles with small identification numbers i have stronger inertial velocity components, and thus their movements are to higher extent dependent on the historical values of their velocities. Therefore, these particles do not change their movement directions abruptly. It can be said that this method introduces a type of specialization to the behavior of particular particles.

3) "Unblock" approach: In this approach, a series of historical FF output values, $y_{\text{f},k}$ is recorded in the memory of a given particle. For i^{th} particle in the swarm, this series is expressed as follows:

$$Y_{\text{f},i} = \left\{ y_{\text{f},i,k}, y_{\text{f},i,k-1}, \ldots, y_{\text{f},i,k-N_{\text{ws}}} \right\} \qquad (9)$$

where N_{ws} is the size of the window that determines the number of previously computed and recorded samples of the FF output. The N_{ws} factor is one the parameters that may be modified.

In contrary to p_{best} factors, which always keep historical optimum output values from the FF for particular particles, the $Y_{\text{f},i}$ signals may either increase or decrease between subsequent iterations.

In this approach, the algorithm detects stagnation periods in particular Y_{f} signals. The stagnation means a situation, in which the values of the FF does not change at all or changes slightly within a certain range. The stagnation period usually appears when the particle is stuck in the local optimum or its area. In order to get out of this situation, it is necessary to change the behavior of the particle, by modifying its r coefficients. In the presented case, the r parameters are halved in such a situation. However, this may lead to a situation, in which the cognitive and the social velocity components are suppressed. To avoid such a situation, after a given number of the halving operations, the values of r are increased by a selected factor.

B. Verification of the proposed solutions

To assess the effectiveness of the presented solutions the investigations were carried out for three cost functions, mentioned above, i.e. for Sphere, Rosenbrock and Griewank ones. Selected results are presented in Fig. 3. Particular diagrams present averaged results for each of the described approaches for ten iterations. The Sphere FF has been introduced as a baseline solution that allows to assess the algorithms used to solve simple problems. To evaluate more challenging examples, the Rosenbrock FF – for the problem with a single local minimum, and the Griewank FF that represents a problem with many local optima.

The best performance for the Sphere FF was the solution described in III-A1 above. It is shown in Fig. 3(a). This is due to a constantly decreasing value of r, which translates into decreasing distances traveled by particular particles in following iterations, causing a more accurate convergence to the optimum located at the zero point. Despite the promising efficiency for this FF, this method is not suitable for more complex problems.

The solution based on assigning different roles for particular particles, depending on their identification numbers, i, showed similar characteristics to the previously described approach. The values achieved for both these solutions, as well as the trend curve are similar. The "Unblock" solution, achieved one of the three best results regardless of the used FF.

IV. IMPACT OF THE PROPOSED SOLUTIONS ON HARDWARE IMPLEMENTATION

Details related to the hardware implementation of the PSO algorithm depend on the direction of such implementation. Sequential , parallel or mixed implementations can be considered here. We opt for a mixed approach, in which some operations are performed in series, but the main part of them is carried out in parallel. The assumption is that each particle in the swarm is implemented as a separate block in the integrated circuit. These algorithm steps that do not require interaction with other members of the swarm are performed independently

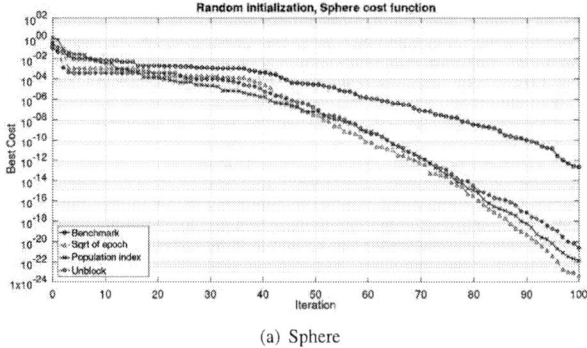

(a) Sphere

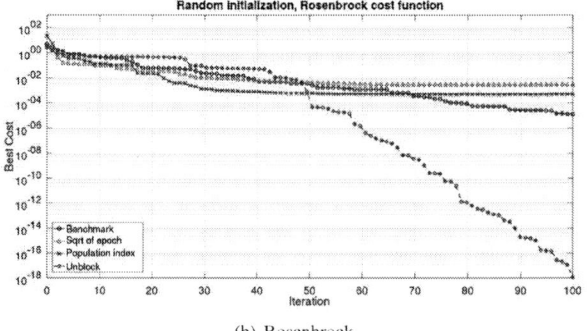

(b) Rosenbrock

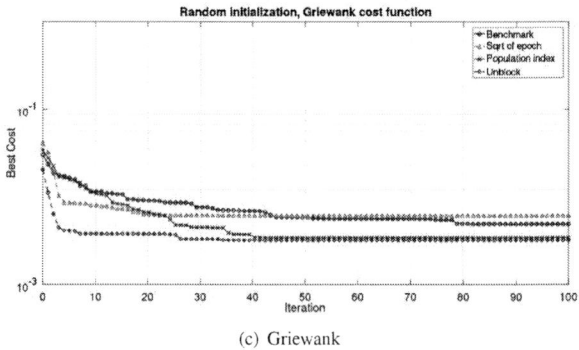

(c) Griewank

Fig. 3. Comparison of proposed methods for modifying r_1 and r_2 coefficients by means of the software model of the PSO algorithm

and at the same time in each of these blocks. In practice, the only operation that requires interaction between particles is the search for a global best value based on current p_{best} values. The system responsible for this operation is the subject of separate studies carried out by us.

For practical reasons, sequential processing applies to computing of particular components of the V and X vectors. In digital systems, particular components of these vectors are multi-bit signals. Their fully parallel processing would mean the necessity of processing from several dozen to even several hundred one-bit signals at the same time, depending on the number of dimensions of the input data space.

Considering this, we assumed that particular components of these vectors are processed sequentially, but simultaneously in each of the particles. In conventional PSO algorithms, separate r_1 and r_2 randomly generated coefficients are associated with each component of the V vector. Implementation of these coefficients is the main topic of the presented work.

Hardware implementations of the classic PSO algorithm require a dedicated component included in each particle, responsible for the randomization of the values of these coefficients for each iteration, k. In the literature one can find various hardware implementations of such blocks. One of the popular methods is to use ring oscillator-based true random number generators (TRNG) [13]–[15]. In a core block in such solutions the rings composed of an odd number of NOT gates are used, as well as other digital elements that include XOR gates and D-flip flops. Typically, these circuits have a fairly complex structure. Elementary TRNGs described in the literature usually allow to single bits of digital signals. This means that drawing the multi-bit r_1 and r_2 signals for each vector component V requires a strong serialization of the operations or using many of such circuits in parallel. For this reason, our objective is to avoid the necessity of using such circuit in the chip. From this reason, we proposed various deterministic methods as presented above. Deterministic methods are much faster and offer a simpler structure.

As the simulation investigations showed, the relatively best results were obtained for the "Unblock" approach, stable for different FFs. This solution does not require complex operations. To change the values of the r_1 and r_2 coefficients, it is needed to decrease or increase their value. The division operation by 2 is in this case performed by shifting the bits to the right by one position.

One of the key operations in this approach is also tracking the values recorded in the series Y_f for a given particle. Theoretically, this requires storing the overall block of the FF output signal samples. However, one can propose another method in which the instantaneous values of $Y_{\max}$ and $Y_{\min}$ values of this signal are stored. Both these signals are initially set equal to a first sample of the Y_f series, so a $\Delta Y_k = Y_{\max,k} - Y_{\min,k}$ term equals zero. For each subsequent sample Max and Min operation are performed with the previous values $Y_{\min,k-1}$ and $Y_{\max,k-1}$ and both these variables are updated accordingly. At the same time, the number of the performed Min and Max operations is counted. When the counter value reaches a certain level (N_{ws}), the circuit checks if the ΔY_k has exceeded an assumed level. If not, this means that the series is in stagnation phase. In this case the $Y_{\max}$ and $Y_{\min}$ are reset so that they are equal to the current sample, while the values of the r_1 and r_2 coefficients are modified.

The obtained results showed that not only it is possible, but it also can give better results than updating particle position

V. CONCLUSION

The goal of this work, was to test whether it is possible to replace random modules in PSO with more deterministic approaches. The study involved changing the particle update formula so that it uses only simple, mathematical expressions and therefore is simple to implement in hardware.
randomly. Using deterministic calculations in algorithm optimization process, gives more control over it and allows to search the fitness function space more precisely. It also gives more control over each particles movement, which means it is possible to program some of them to search further space, and other to keep in close range. The modifications allowed to avoid stopping at a local function extremum and thanks to the specific and precise particle movement the chances to reach global best position were much higher.

REFERENCES

[1] Grzesiak L., Erwiński K., Paprocki M., "Implementacja sieci neuronowej MLP FPNN w strukturze układu programowalnego FPGA", *Przeglad Elektrotechniczny*, January 2014.

[2] Shi Y., Eberhart R. C. "Parameter selection in particle swarm optimization", *Proceedings of Evolutionary Programming*, VII (EP98), pp.591-600, 1998.

[3] Eberhart R. C., Shi Y. "Comparing inertia weights and constriction factors in particle swarm optimization", *Proceedings of the 2000 Congress on Evolutionary Computation*, 2000.

[4] Carlisle A., Dozier G. "An off-the-shelf PSO", *Proceedings of the Particle Swarm Optimization Workshop*, pp. 1-6, 2001.

[5] Bratton D., Blackwell T. "A Simplified Recombinant PSO", *Hindawi Publishing Corporation Journal of Artificial Evolution and Applications*, 2007.

[6] Acharya, D., Goel, S., Asthana, R., Bhardwaj, A., "A Novel Fitness Function in Genetic Programming to Handle Unbalanced Emotion Recognition Data", *Pattern Recognition Letters*, Elsevier, 2020.

[7] Malhotra, R., Khanna, M., "Dynamic selection of fitness function for software change prediction using Particle Swarm Optimization", *Information and Software Technology*, Elsevier, Vol. 112, 2019.

[8] Liu, D., Jinling, D., Xiaohua, C., "A Genetic Algorithm Based on a New Fitness Function for Constrained Optimization Problem", *International Conference on Computational Intelligence and Security*, 2011, pp.6-9.

[9] Chen, Q., Worden, K., Peng, P., Leung, A.Y.T., "Genetic algorithm with an improved fitness function for (N)ARX modelling", *Mechanical Systems and Signal Processing*, Elsevier, Vol. 21, No. 2, 2007, pp. 994-1007.

[10] Acharya, D., Goel, S., Asthana, R., Bhardwaj, A., "A novel fitness function in genetic programming to handle unbalanced emotion recognition data", *Pattern Recognition Letters*, Elsevier, Vol. 133, 2020, pp. 272-279.

[11] Cao, J., Cui, H., Shi, H., Jiao, L., "Big Data: A Parallel Particle Swarm Optimization-Back-Propagation Neural Network Algorithm Based on MapReduce". *PLOS ONE*, 11(6), 2016.

[12] Cheng, S., Zhang, Q., Quande, Q., "Big Data Analytics with Swarm Intelligence", *Industrial Management & Data Systems*, Vol. 116, 2016.

[13] Vikram Belur Suresh, "On-Chip True Random Number Generation in Nanometer CMOS", *University of Massachusetts Amherst*, 2012.

[14] "Evaluation of VIA C3 Nehemiah Random Number Generator", *White paper by Cryptographic Research Inc.*, 2003.

[15] Benjamin Jun and Paul Kocher, "The Intel Random Number Generator", *Cryptography Research, Inc. White Paper Prepared For Intel Corporation*, 1999.

Testing Stability of Digital Filters Using Multimodal Particle Swarm Optimization with Phase Analysis

Damian Trofimowicz
SpaceForest Ltd.
81-451 Gdynia, Poland
Email: d.trofimowicz@gmail.com

Tomasz P. Stefański
Gdansk University of Technology
80-233 Gdansk, Poland
Email: tomasz.stefanski@pg.edu.pl

Abstract—In this paper, a novel meta-heuristic method for evaluation of digital filter stability is presented. The proposed method is very general because it allows one to evaluate stability of systems whose characteristic equations are not based on polynomials. The method combines an efficient evolutionary algorithm represented by the particle swarm optimization and the phase analysis of a complex function in the characteristic equation. The method generates randomly distributed particles (i.e., a swarm) within the unit circle on the complex plane and extracts the phase quadrant of function value in position of each particle. By determining the function phase quadrants, regions of immediate vicinity of unstable zeros, called candidate regions, are detected. In these regions, both real and imaginary parts of the complex function change signs. Then, the candidate regions are explored by subsequently generated swarms. When sizes of the candidate regions are reduced to a value of assumed accuracy, then the occurrence of unstable zero is verified with the use of discrete Cauchy's argument principle. The algorithm is evaluated in four benchmarks for integer- and fractional-order digital filters and systems. The numerical results show that the algorithm is able to evaluate the stability of digital filters very fast even with a small number of particles in subsequent swarms. However, the multimodal particle swarm optimization with phase analysis may not be computationally efficient in stability tests of systems with complicated phase portraits.

Keywords—Digital filters, Discrete-time systems, Stability analysis, Digital signal processing.

I. INTRODUCTION

The stability analysis is an important topic in almost every area of engineering. Most of electronic circuits must be stable to properly operate and execute tasks for which they are designed. It is also vitally important in the digital signal processing. In general, the discrete-time linear time-invariant (LTI) system is (asymptotically) stable if and only if all zeros of the characteristic equation ($f(z) = 0$) are within the unit circle at the complex z-plane [1]–[3]. The direct approach is to find all zeros of the charatcristic equation (e.g., a denominator of a transfer function). However, it might be a difficult task because for some systems, e.g., of fractional order [4], [5], the characteristic equation may not be based on a polynomial. Furthermore, the search space for zeros of the characteristic equation of the system is outside the unit circle ($|z| > 1$), hence, it is infinite.

We have already proposed numerical tests [6], [7] allowing for evaluation of the system stability by employing modern techniques of global root finding based on Delaunay's triangu-

lation and discrete Cauchy's argument principle (DCAP). The motivation behind this work is to benchmark a novel stability test based on the multimodal particle swarm optimization with phase analysis (MPSO-WPA) targeting discrete-time systems, especially digital filters. Its main advantage over other methods relies on stochastic space exploration by subsequent swarms, hence, the probability of detection of all zeros is increasing each time swarm positions are updated. Furthermore, the proposed method is based on the phase analysis, hence, it is not sensitive to numerical precision issues resulting from the overflow of arithmetic operations.

II. STABILITY OF DIGITAL FILTERS

The fundamental equations describing the stability of digital filters are presented in this section.

A. Integer-Order Digital Filters

The integer-order discrete-time LTI system (digital filter) can be represented in the Z-transform domain by the transfer function

$$H(z) = \frac{Y(z)}{X(z)} = \frac{Num(z)}{Den(z)} \qquad (1)$$

where X denotes the input signal and Y denotes the output signal. The characteristic equation

$$f(z) = 0 \qquad (2)$$

for (1) is based on the polynomial of the z variable

$$f(z) = Den(z) = \prod_{i=1}^{L}(z - p_i) \qquad (3)$$

where p_i denotes i-th zero of the function $Den(z)$.

B. Fractional-Order Digital Filters

The fractional-order discrete-time LTI system (digital filter) can be represented by the transfer function

$$H(z) = \frac{Y(z)}{X(z)} = \frac{Num(u)}{Den(u)} \qquad (4)$$

where

$$u = z(1 - z^{-1})^{\alpha}. \qquad (5)$$

In opposite to the integer-order systems, the characteristic equation of the fractional LTI system is not based on a polynomial of the z variable but on the complex function

$$f(z) = Den[z(1 - z^{-1})^\alpha] = 0. \tag{6}$$

Both integer- and fractional-order systems are stable if and only if all roots (i.e., zeros) of the characteristic equation on the complex z-plane are located inside the unit circle. In the developed test, the region outside the unit circle is transformed inside the unit circle by the following transformation:

$$z = w^{-1}. \tag{7}$$

MPSO-WPA allows one to explore complex functions with singularities. However, to reduce the computing time, the singularities can be extracted. Hence, the following equation is advantageous for the analysis:

$$F(w) = (w - p)^K f(w^{-1}) = 0. \tag{8}$$

In (8), p is an exemplary singularity and K is its multiplicity.

III. Procedure for Stability Evaluation

The numerical test for stability evaluation based on MPSO-WPA is executed in the following steps:

A. Algorithm Initialization and Generation of Initial Swarm

The algorithm initializes parameters such as the algorithm accuracy, the inertia weight, the initial velocity, the scaling factor, the number of particles in swarms, the number of swarm iterations. Next, a uniformly distributed random swarm (i.e., the set of particle coordinates) is generated within the unit circle $|w| < 1$ on the complex w-plane

$$W_{i,j} = \{w_{i,j,1}, w_{i,j,2}, ..., w_{i,j,n}\} \tag{9}$$

where i is the swarm number, j is the iteration number and n is the number of particles in the swarm population. Indices i and j are set to zero for the initial swarm. It is assumed that $W_{0,0}$ is not updated in subsequent iterations.

B. Generation of Subsequent Swarms

Next, i-th index is increased and subsequent uniformly distributed swarm $W_{i,j} = \{w_{i,j,1}, w_{i,j,2}, ..., w_{i,j,n}\}$ is generated on the complex plane. The second swarm ($i = 1$) is generated globally and its role is to perform a wide space search of the complex plane. If i-th index is greater than one, then subsequent swarms are generated locally only inside candidate regions.

C. Search for Regions in Vicinity of Zeros and Poles

Evaluate the function argument for each particle coordinates and compute the phase quadrant in which the corresponding function value is located

$$Q_n = \begin{cases} 1, & 0 \le \arg f(w_n) < \frac{\pi}{2} \\ 2, & \frac{\pi}{2} \le \arg f(w_n) < \pi \\ 3, & \pi \le \arg f(w_n) < \frac{3\pi}{2} \\ 4, & \frac{3\pi}{2} \le \arg f(w_n) < 2\pi \end{cases}. \tag{10}$$

Then, Delaunay's triangulation is applied to the coordinates of all particles, i.e., triangular connections between particles are generated. Finally, the quadrant distance along each of the connections is computed, i.e.,

$$\Delta Q_p = \Delta Q_{p2} - \Delta Q_{p1}. \tag{11}$$

The connections with the quadrant distance equal $|\Delta Q_p| = 2$ involve particles for which both real and imaginary parts of function values have different signs. These connections (called candidates) are considered as potential immediate vicinity of either zero or pole. If the candidate connections are not found, then, return that the system is stable and exit. Otherwise, the coordinates of the centre of each candidate connection and its length are computed. If the highest length of collected candidate connections within a total population is lower than the assumed initially accuracy, then proceed to the step E. If the number of current iteration is equal to the number of swarm iterations or its multiple, execute the loop to the step B. In other cases, proceed to the next step D.

D. Narrowing Candidate Regions

The distance between particle coordinates and coordinates of candidate centres is computed. For each particle, the lowest distance is determined and corresponding coordinates of the candidate centre are stored in the set

$$P_{k,j} = \{p_{k,j,1}, p_{k,j,2}, ..., p_{k,j,n}\} \tag{12}$$

where $k = 1, ..., i$. Next, the particles are accelerated in the direction of the closest p coordinates by updating the velocity of each particle, i.e.

$$V_{k,j+1} = gV_{k,j} + cu(P_{k,j} - W_{k,j}) \tag{13}$$

where $W_{k,j}$ denotes the coordinates of particles, $P_{k,j}$ denotes the coordinates of the closest candidate centre, $V_{k,j+1}$ denotes the particle velocities, c denotes the scaling factor towards the closest candidate centre, g denotes inertial weight which determines how much the particle keeps its original direction and u denotes a random number between 0 and 1. Next, the coordinates of the swarm particles $W_{k,j}$ are updated by

$$W_{k,j+1} = W_{k,j} + V_{k,j+1} \tag{14}$$

where $k = 1, ..., i$ and j denotes the iteration index. Finally, go to the searching for candidates step C.

E. Verification of Candidate Regions

In the final step, the existence of zero inside the unit circle is confirmed by DCAP and integration of quadrant differences along the path between particles around the candidate region

$$q = \frac{1}{4} \sum_{p=1}^{P} \Delta Q_p. \tag{15}$$

The parameter q is a positive integer when function zero is found inside the candidate region. When multiple zeros are located inside the candidate region, value of q indicates the number of zeros inside this region (calculated with their multiplicities). When one or more zeros are found, the system is unstable. The parameter q is a negative integer when pole is found. Finally, the parameter q is equal to zero when there is neither zero nor pole inside the unit circle.

IV. NUMERICAL RESULTS

Numerical benchmarks are executed on a personal computer equipped with Intel i7-4700MQ processor. The MPSO-WPA code is developed in Matlab. The characteristic equation, i.e., the function whose zeros are investigated, and algorithm parameters are the input data. The output data are values and multiplicities of zeros and poles found. Moreover, the code returns information if the system is either stable or unstable. In the first test, the stability of integer-order digital filter is evaluated. Then, fractional-order systems are evaluated in terms of stability.

A. Integer-Order Digital Filter

Let us consider the digital filter [7] described by

$$H(z) = \frac{G}{A(z)} \qquad (16)$$

where G is the gain factor and $A(z)$ is the characteristic-equation function defined by

$$
\begin{aligned}
A(z) = &+ 1.0000 - 2.5400z^{-1} + 3.0429z^{-2} - 2.9211z^{-3} \\
&+ 3.7088z^{-4} - 3.9740z^{-5} + 3.0221z^{-6} \\
&- 2.3163z^{-7} + 1.9791z^{-8} - 1.1265z^{-9} \\
&+ 0.3855z^{-10} - 0.2189z^{-11} + 0.1171z^{-12}.
\end{aligned}
\tag{17}
$$

For the filter stability, all zeros of $A(z)$ must be located inside the unit circle. By applying the transformation (7), one obtains

$$
\begin{aligned}
A(w) = &+ 1.0000 - 2.5400w + 3.0429w^2 - 2.9211w^3 \\
&+ 3.7088w^4 - 3.9740w^5 + 3.0221w^6 \\
&- 2.3163w^7 + 1.9791w^8 - 1.1265w^9 \\
&+ 0.3855w^{10} - 0.2189w^{11} + 0.1171w^{12}.
\end{aligned}
\tag{18}
$$

The benchmark is executed with a small number of particles in each swarm, set to only 200 particles. The swarm distribution is uniform inside the unit circle. The particle distribution as well as Delaunay's triangulation applied to the final iteration are presented in Fig. 1. Two conjugate zeros are found inside the unit circle, i.e.

$$w = 0.829238 \pm 0.536757j.$$

This stability test based on MPSO-WPA lasts about 0.16 second for the accuracy set to $\varepsilon = 10^{-5}$. The algorithm executes 11 iterations and generates 600 particles in total. The outer region of the unit circle is transformed into the inner one, thus, the filter has zeros outside the unit circle on the complex z-plane. Hence, the filter is unstable.

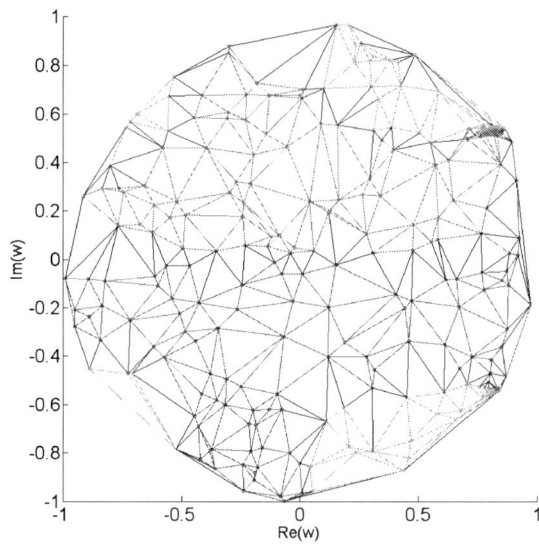

Fig. 1. Distribution of MPSO-WPA particles with Delaunay's triangulation for final iteration. Stability test of integer-order digital filter is executed in the transformed region $|w| < 1$. Phase quadrants of particles on the w-plane: • Q=1, ° Q=2, ° Q=3, • Q=4.

B. Fractional-Order Digital Filter

Let us consider the fractional-order digital filter which is described by

$$H(z) = \frac{1}{(\frac{2}{T}\frac{1-z^{-1}}{1+z^{-1}})^{\alpha+\beta} + a(\frac{1-z^{-1}}{1+z^{-1}})^{\alpha} + c} \qquad (19)$$

where a and c are the design parameters, α and β are the fractional-order parameters and T is the sampling period. The filter characteristic equation in the w-domain is given by

$$F(w) = s^{(\alpha+\beta)} + as^{\alpha} + c \qquad (20)$$

where

$$s = \frac{2(1-w)}{T(1+w)}. \qquad (21)$$

The pole $w = -1$ can be extracted, thus, the stability is tested based on the following equation:

$$F(w)(1+w)^{\alpha+\beta} = 0. \qquad (22)$$

The function is tested for $\alpha = \beta = 0.5$, $a = 1$, $c = -1000 + 50j$ and $T = 0.001$. The final distribution of particles as well as Delaunay's triangulation applied to the final swarm distribution are presented in Fig. 2. A single zero is found inside the unit circle, i.e.

$$w = 0.346947 + 0.022324j.$$

This stability test lasts about 0.16 second for the accuracy set to $\varepsilon = 10^{-5}$. The algorithm executes 9 iterations and generates 600 particles in total. The filter is unstable because zero is found inside the unit circle on the complex w-plane.

234

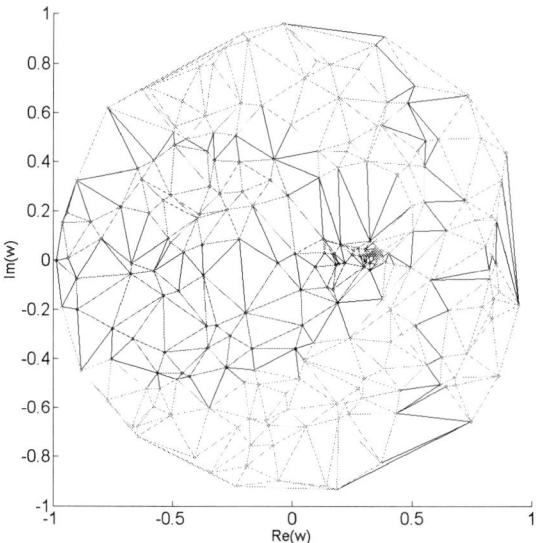

Fig. 2. Distribution of MPSO-WPA particles with Delaunay's triangulation for final iteration. Stability test of fractional-order digital filter is executed in transformed region $|w| < 1$. Phase quadrants of particles on the w-plane: • Q=1, °Q=2, °Q=3, °Q=4.

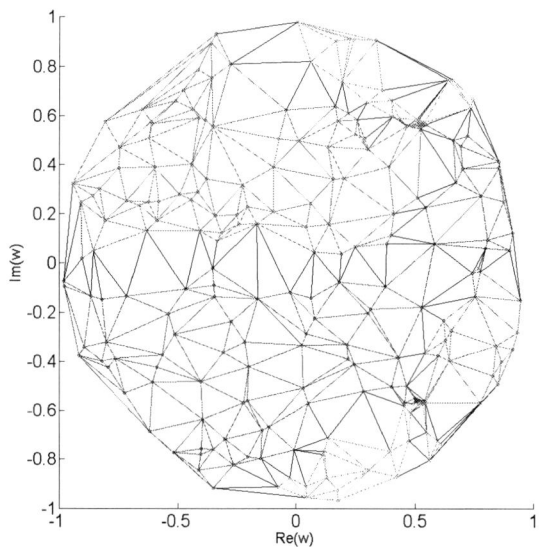

Fig. 3. Distribution of MPSO-WPA particles with Delaunay's triangulation for final iteration. Stability test of fractional-order system is executed in transformed region $|w| < 1$. Phase quadrants of particles on the w-plane: •Q=1, °Q=2, °Q=3, •Q=4.

C. Fractional-Order System

MPSO-WPA can be applied to test stability of various discrete-time systems, also those described by the fractional-order state-space equations [4], [5]

$$\Delta^\alpha \mathbf{x}(n+1) = \mathbf{A}_f \mathbf{x}(n) + \mathbf{B}\mathbf{u}(n)$$
$$\mathbf{y}(n) = \mathbf{C}\mathbf{x}(n) + \mathbf{D}\mathbf{u}(n) \qquad (23)$$

where α is the fractional order and $\mathbf{A}_f = \mathbf{A} - \mathbf{I}$ is the difference between the discrete-time state-space system matrix $\mathbf{A}$ and the identity matrix $\mathbf{I}$. In (23), the fractional difference is defined as

$$\Delta^\alpha x(n) = \sum_{j=0}^{n} (-1)^j \binom{\alpha}{j} q^{-j} x(n) \qquad (24)$$

where q^{-1} is the backward shift operator.

Let us consider the fractional-order system represented by

$$\mathbf{A}_f = \begin{bmatrix} 0.6 & -1.45 \\ 1 & -1 \end{bmatrix}. \qquad (25)$$

For the system (23), the characteristic equation is given by

$$f(z) = det[z(1 - z^{-1})^\alpha \mathbf{I} - \mathbf{A}_f] = 0. \qquad (26)$$

By applying the transformation (7) and multiplying by w^2, one obtains

$$F(w) = w^2 f(w) = (1-w)^{2\alpha} + 0.4w(1-w)^\alpha + 0.85w^2. \qquad (27)$$

The stability test is executed for the parameter $\alpha = 1.1$. The algorithm accuracy is set to $\varepsilon = 10^{-5}$. The size of swarm populations is set to 200 particles. The final distribution of particles as well as Delaunay's triangulation applied to the

final swarm distribution are presented in Fig. 3. Two conjugate zeros are found inside the unit circle, i.e.

$$w = 0.529268 \pm 0.569170j. \qquad (28)$$

This stability test based on MPSO-WPA lasts about 0.17 second for the accuracy set to $\varepsilon = 10^{-5}$. The algorithm executes 9 iterations and generates 600 particles in total. The outer region of the unit circle is transformed into the inner one, thus, the system has zeros outside the unit circle on the complex z-plane. Therefore, the system is unstable.

It is worth mentioning that zero located very close to the unit circle represents a difficult case for the method, because the quadrant distance is calculated only between particles accumulated inside the defined region. In consequence, the function zero located close to the unit circle may be omitted. In order to avoid omitting such zeros, the swarm population may be increased to collect more phase samples of complex function, especially in points located close to the unit circle. Another solution is to increase the search area towards, e.g., $|w| < 1.1$. In our opinion, more reliable and less computationally demanding approach is to increase the w-plane search area.

D. High-Order Systems

High-order systems are computationally demanding because a large number of stable zeros complicates the phase portrait of complex function. It may lead to false detection of candidate regions and exploration of irrelevant regions by the algorithm. Let us consider the system whose the characteristic equation is given by

$$f(z) = (z-2)\prod_{i=1}^{L}[z-(-1)^{i}0.4(1+\frac{i}{L})]. \quad (29)$$

As one can notice, a single zero of this characteristic equation is located outside the unit circle on the complex z-plane. The stability test is executed for the accuracy set to $\varepsilon = 10^{-2}$. Unfortunately, higher accuracies require large computing times for the convergence of the algorithm. The initial number of particles in the swarm is increased with incrementation of the parameter L. This procedure is necessary because the phase portrait of the complex function becomes more and more complicated and false candidate detections occur. The parameter L, the number of generated particles, the number of iterations and time necessary for the algorithm convergence are presented in Table I.

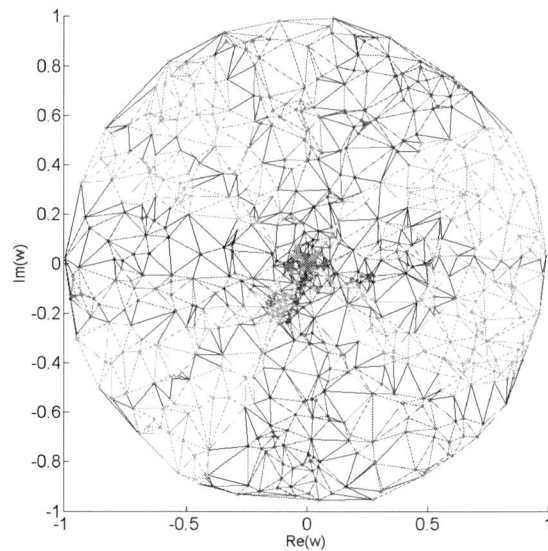

Fig. 4. Distribution of MPSO-WPA particles with Delaunay's triangulation for final iteration. Stability test of 4-th order system without removing poles is executed in transformed region $|w| < 1$. Phase quadrants of particles on the w-plane: $^{•}$Q=1, *Q=2, $^{•}$Q=3, $^{•}$Q=4.

TABLE I
MPSO-WPA Computing Time, Number of Generated Particles and Iterations for Testing Stability of Various Order Systems

L	no. of iterations	time to converge	no. of particles
1	3	0.18	400
5	3	0.20	400
10	4	0.20	400
15	4	0.21	400
20	9	0.23	600
25	17	0.36	1000
30	27	0.54	1200
35	41	0.82	1600
40	49	1.38	2600
45	91	2.80	3800
50	83	2.93	4200

MPSO-WPA is fast and efficient for low-order systems, i.e., for low L values. However, in the case of a complicated phase portrait of complex function, the algorithm requires high number of particles in the initial swarm to avoid convergence in regions that do not include zeros.

The significant advantage of MPSO-WPA is that removing singularities from the complex function is not necessary because it executes operations on phase values. Although it complicates function phase portrait, the algorithm is able to locate all zeros and singularities and return their values with multiplicities. Results of exemplary test of the system (29) for $L = 3$ without removing singularities is presented in Fig. 4. The stability test lasts about 0.44 second for the accuracy set to $\varepsilon = 10^{-2}$. A single zero and four poles are found inside the unit circle. The algorithm executes 22 iterations and generates 1400 particles in total. Subsequently, the stability test of the system (29) for $L = 5$ lasts about 0.94 second and generates 2200 particles. A single zero and six poles are found inside the unit circle. Tests for higher-order systems are not executed without removing poles due to long computing times.

V. CONCLUSION

The novel MPSO-WPA method is developed to evaluate the stability of integer- and fractional-order digital filters and systems. The algorithm merges the population based optimization technique with the phase quadrant analysis on the complex plane. The results obtained with the use of the MPSO-WPA algorithm are satisfactory. It is worth noticing that the algorithm prevents from omitting zeros due to the use of stochastic space exploration by subsequent swarms. Moreover, the algorithm does not require removing poles from the complex function. The MSPO-WPA method finds all zeros and poles within a defined region. However, testing stability with MPSO-WPA without removing poles, as well as testing very high-order systems, may be computationally inefficient for functions with complicated phase portraits.

REFERENCES

[1] K. Ogata, *Discrete-Time Control Systems*. Prentice-Hall International, 1995.
[2] J. R. B. A. Oppenheim, R. W. Schafer, *Discrete-Time Signal Processing*. Prentice-Hall International, 1999.
[3] E. I. Jury, *Theory and Application of the Z-Transform Method*. John Wiley and Sons, 1964.
[4] R. Stanislawski and K. Latawiec, "Stability analysis for discrete-time fractional-order lti state-space systems. part i: New necessary and sufficient conditions for the asymptotic stability," *Bull. Pol. Ac.: Tech.*, vol. 61, pp. 353–361, 2013.
[5] R. Stanislawski and K. Latawiec, "Stability analysis for discrete-time fractional-order lti state-space systems. part ii: New stability criterion for fd-based systems," *Bull. Pol. Ac.: Tech.*, vol. 61, pp. 363–370, 2013.
[6] L. Grzymkowski and T. P. Stefanski, "Numerical test for stability evaluation of discrete-time systems," in *2018 23rd International Conference on Methods Models in Automation Robotics (MMAR)*, Aug 2018, pp. 803–808.
[7] L. Grzymkowski and T. P. Stefanski. "A new approach to stability evaluation of digital filters," in *2018 25th International Conference "Mixed Design of Integrated Circuits and System" (MIXDES)*, June 2018, pp. 351–354.

Embedded Systems

Proceedings of the 27th International Conference *"Mixed Design of Integrated Circuits and Systems"*
June 25-27, 2020, Łódź, Poland

A Database Proposal for an Application Involving Industrial Networks for Industry 4.0 Concepts

Alexandre Baratella Lugli[1], Egidio Raimundo Neto[1], João Paulo Carvalho Henriques[1],
Maria Teresa de Carvalho Silva[1], Nayara Dias Pereira[1], Tales Cleber Pimenta[2]

[1] Department of Industrial Automation
National Institute of Telecommunications (INATEL)
CEP , 37.540-000, Santa Rita do Sapucaí, MG, Brazil

[2] Institute of Engineering System and Information Technology
Federal University of Itajubá (UNIFEI)
CEP 37.500-903, Itajubá, MG, Brazil

baratella@inatel.br, egidio.neto@inatel.br, joao.paulo@inatel.br, mariateresa@gea.inatel.br, nayaradias@gea.inatel.br
tales@unifei.edu.br

Abstract—**This paper proposes to develop an application with the PROFINET network applied in the context of the Industry 4.0. The proposal consists in collect data of an industry application, such as temperature and motor status, for instance, and then, by a tool developed in Python programming language, it is possible to monitor and control the magnitudes of the industrial process. The developed Database was used to monitor the variables in a PROFINET industrial network application.**

Keywords—**Database, Industrial Automation, Industry 4.0, PROFINET.**

I. INTRODUCTION

Nowadays the automation is essential in large industries, which the objective is to increase the speed of information processing, whereas they are increasingly complex, needing mechanisms to turn the decisions more agile, increasing productivity and efficiency of the production process. [1]

In industries where the environment is rugged and hostile, Ethernet standards are considered inappropriate for lacking of noise immunity and inadequate connectors. Moreover, in this environment, the most desired characteristic is determinism, which consists in having a maximum time. [2] [3]

PROFINET communication standard was developed to be used in Industry 4.0 concept. It consists in integrating the information technology solution applied in the factory floor and factory elements and it is possible to integrate them to the internet standard. [2] [3]

The purpose of this paper is to develop an application of this communication protocol, where industrial field elements will be able to access information stored in a database. Such application will be built especially for interaction between the PROFINET industrial communication network and the information technology network (IT), in real-time, being able to store the information of the industrial processes, accessed remotely and at any time.

II. CONCEPTS AND DEFINITIONS

In order to show the theoretical concepts and definitions of the paper, this chapter describes this concepts and definitions.

A. PROFINET Network

In order to connect electronic devices and controllers to control systems in the industrial environment, interferences such as moisture, abrasion and temperature changes, was necessary the creation of industrial protocols, such as PROFINET.

The PROFINET protocol is a network standardized by the PROFIBUS International Association, which operates in communication between a central station (controller/supervisor) and decentralized I/O devices, beyond to describe the complete data exchange performed in communication. The PROFINET protocol follows the provider/consumer model and is designed for real time data exchange between field equipment based in Ethernet standard. [4] [5]

Another significant advantage of this network is the use of Ethernet TCP/IP standard components, which makes it much simpler for customers to install and adopt PROFINET. [4]

1) Models of Operation and Network Types

The PROFINET protocol has two distinct forms of operation, one for real-time and another for the non-real time, illustrated in Fig. 1. [5] [6]

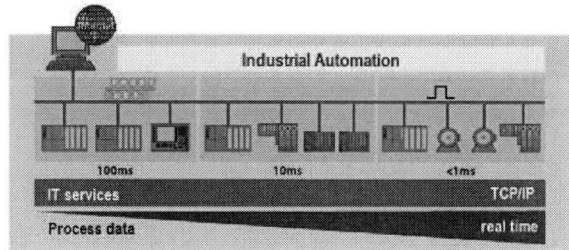

Fig. 1. Models of operation and their processing times. [6]

The SRT (Soft Real-Time) operating mode has a shorter transmission time because it connects the Ethernet layer to the application layer. Some protocol levels are eliminated and the length of the transmitted messages are reduced. [5] [6]

The other mode of operation is called IRT (Isochronous Real-Time Communication), or isochronous communication, which is a result of the need of some applications that require a planned communication for maximum integrated deterministic performance. Therefore, they are used in applications where the answer time should be less than 1ms, requiring dedicated hardware and point-to-point communication. [5]

The modes of operation mentioned above are used on the PROFINET I/O network that follows the Provider/Consumer model for the cyclic data exchange between the I / O Controller and an I/O Device. The assignment between provider and consumer is defined during the system settings. [5]

2) I/O Device Models

Fig. 2 illustrates the provider/consumer model of the PROFINET network.

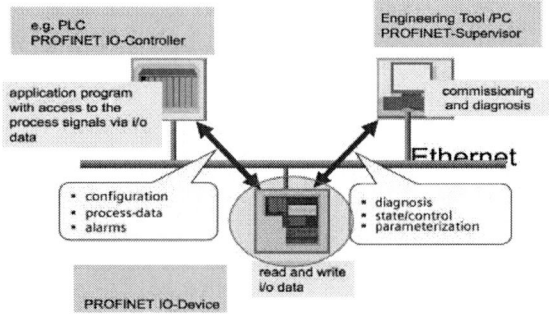

Fig. 2. Provider/consumer model. [6]

- IO-Controller: Intelligence central station, typically represented by the Programmable Logic Controller (PLC. It is responsible for providing output data to IO-Devices. [5] [6]

- IO-Device: Represented by field devices. It transmits process data, such as diagnostics and alarms, to one or more IO-Controllers. [5] [6]

- IO-Supervisor: Characterized by being a programming device, it realizes the configuration or diagnostics from the network devices and it can be a personal computer or an HMI (Human Machine Interface). [5] [6]

3) Network Topologies

Fig. 3 illustrates the possible topologies of the PROFINET network, cited below. [6]

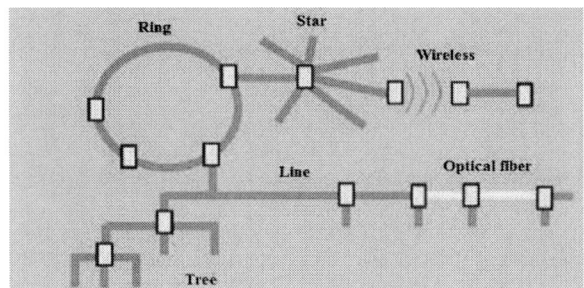

Fig. 3. Possible topologies in the PROFINET network. [6]

- **Ring:** Closed line communication.

- **Star:** A central switch connects the other devices.

- **Line:** Connects devices through integrated switches.

- **Tree:** It is a combination of the topologies mentioned above.

B. PLC - Programmable Logic Controller

The automation systems were based on electromechanical relays interconnected with many cables. The operators performed the control manually, reducing and increasing currents and voltages by controlling circuit breakers and valves through electromechanical pushbuttons and switches. [7]

The Programmable Logic Controller (PLC) was a breakthrough for industrial automation because it allows operations to be done with a high level of automation. The Fig. 4 illustrates a Programmable Logic Controller. [7]

Fig. 4. Programmable Logic Controller. [7]

Nowadays, the PLC is the most used industrial process control technology. It is a type of computer system that operates in real-time and can be programmed to execute control operations. It simplifies programming and installation because controllers reduce the number of wires associated with conventional relay control circuits. They have high-speed control, network compatibility, high reliability and fault checking. [6] [7]

The basic architecture of a PLC consists of a power supply, a central processing unit (CPU), a memory system and input and output circuits, as illustrated in Fig. 5. [7]

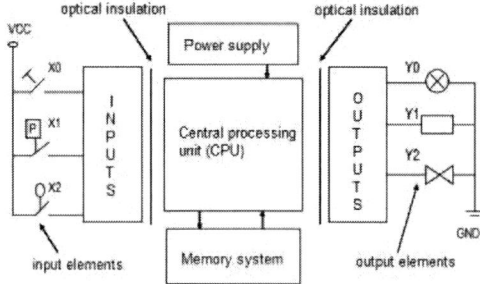

Fig. 5. The basic architecture of a PLC. [7]

C. Database

A DBMS (Database Management System) is a collection of interrelated data and a set of programs to access that data. The main objective of a DBMS is to retrieve and store relevant information in an efficient and secure way. [8]

The databases is useful because they ensure the management of a large volume of data, as well as to ensure that

240

the information stored, despite possible system failures or attempts at unauthorized access, is protected. It is also important to highlight that with the great advance of technology, people and companies need to use storage systems and recover compact and efficient data. [9]

The main advantages of using an SGDB are mentioned below:

- **Redundancy control:** It stores the same data many times generates a series of problems, such as the loss of consistency of information. [9]

- **Restrict unauthorized access:** It enables control of data considered confidential. [9]

- **Ensures Backup and Restore:** The backup and recovery subsystem of the DBMS subsystems are responsible for recovering from hardware or software failures. [9]

- **Ensures the storage of structures for efficient query processing:** It provides functionality for performing updates and queries efficiently. [9]

- **The data integrity is maintained.**

The database might be used to store business information such as sales, accounting, human resources, manufacturing and online retailers, as well as banking and financial information. This technology might also be used in universities, airlines and telecommunications applications, too. [8]

III. PRACTICAL APPLICATION

Fig. 6 illustrates the general block diagram of the practical application developed in this paper. The program application to store the data was developed in Python language (PyCharm) that it stores the information collected through the PROFINET network.

The goal of the paper is to keep the information generated in the industrial process stored in the database, such as temperature or any other industrial variable, in real time, and with remote access. A graphic visualization was also developed at the HMI, to monitor the information of the industrial process.

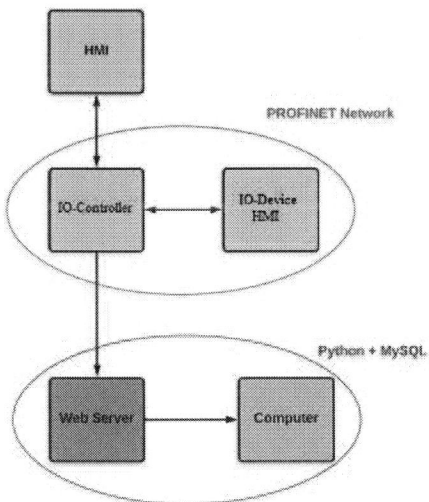

Fig. 6. Block diagram of the general project.

A. Web Server for PROFINET IO-Controller

The IO-Controller Web Server can be enabled through the IO-Controller configuration software and it allows the visualization of the data. This data is being monitored in the Programmable Logic Controller on a web page, which is accessed through the device's IP address.

It was needed to enable the PROFINET IO-Controller Web Server, as illustrated in Fig. 7.

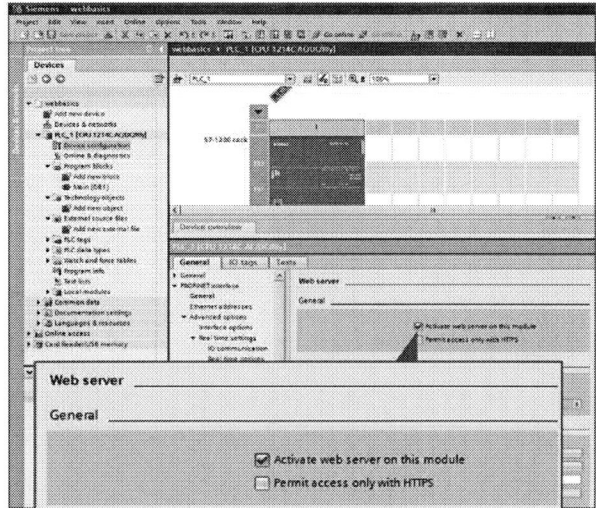

Fig. 7. Enable the device's Web Server.

After the enable the WebServer, it is possible to view on a web page, the variables that are being read by the program created in the IO-Controller, which can be accessed by the IP address configured in the IO-Controller, as illustrated in Fig. 8.

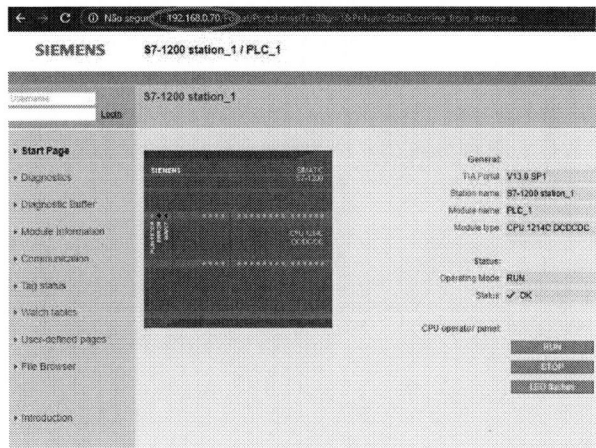

Fig. 8. Device's Web Server.

B. Integration with the Database

In order to integration with the Database was used Python programming language. Thereby, it is possible to develop a tool that collects data from the PROFINET network through the PROFINET IO-Controller WebServe. It establishes a connection with MySql, recording all the data collected in this Database. The Database, Webserver and IO-Controller must be connected at the same network.

Initially, the following libraries were imported, as illustrated in the code in Fig. 9.

```
import selenium
from selenium import webdriver
from datetime import date
from datetime import datetime
import time
import pyautogui
from datetime import datetime
import pymysql
```

Fig. 9. Source code of libraries used in development.

The Selenium library was used to collect data from the WebServer. Through this library, it is possible to manipulate and monitor elements present in the browser source code, such as ID (Identifier) and xpath, as illustrated in Fig. 10. Therefore, the library is used to: 1) open the browser and access the WebServer; 2) set the variables to be monitored and 3) read the values of the variables and store them in the program's memory.

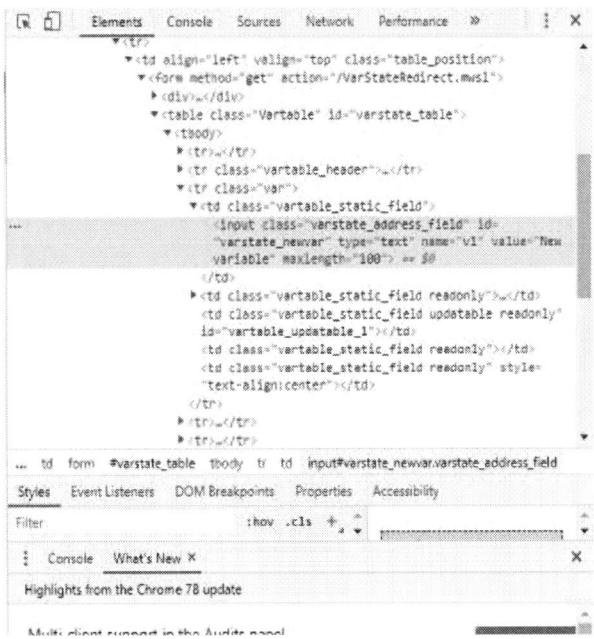

Fig. 10. Source code of the page with the elements.

The Pyautogui library was used to create a loop to update the Web page, getting more accurate monitoring of the analyzed variables.

The Pymsql library was used to establish the connection between Python and MySql and write the variables to the Database.

After importing the libraries, the connection with the MySQL database was established and variables were defined, as illustrated in Fig. 11.

```
connection = pymysql.connect(
    host = 'localhost',
    user = 'root',
    passwd = '',
    database = 'industry'
)
cursor = connection.cursor()
# cursor.execute("CREATE DATABASE industry")
# cursor.execute("SHOW DATABASES")
# cursor.execute("CREATE TABLE tabela(date
VARCHAR(255), tension VARCHAR(255), status
VARCHAR(255))")
# cursor.execute("SHOW TABLES")
tension = input('Enter with the Temperature variable: ')
variable = input('Enter with the digital variable: ')
```

Fig. 11. Code to establish a connection with the database and choose variables.

After that, the driver was developed and analyzed and the chosen variables were configured, as illustrated in Fig. 12.

```
driver = webdriver.Chrome()
driver.get("http://192.168.0.70/Portal/Portal.mwsl?PriNav=
Varstate")
driver.maximize_window()
driver.find_element_by_xpath('//*[@id="varstate_table"]/tbo
dy/tr[3]/td[1]/input').clear()
driver.find_element_by_xpath('//*[@id="varstate_table"]/tbo
dy/tr[3]/td[1]/input').send_keys(tension)
pyautogui.press('enter')
driver.find_element_by_xpath('//*[@id="varstate_newvar"]')
.clear()
driver.find_element_by_xpath('//*[@id="varstate_newvar"]')
.send_keys(variable)
pyautogui.press('enter')
k = 0
```

Fig. 12. Driver source code configured.

Finally, the variables were saved in the database, as illustrated in Fig. 13.

```
while True:
    driver.get(driver.current_url)
    tension =
driver.find_element_by_id("vartable_updatable_1").text
    booleano =
driver.find_element_by_id("vartable_updatable_2").text
    if str(booleano) == 'true':
        digital = 'ligado'
    elif str(booleano) =='false':
        digital = 'desligado'
    date = datetime.now().strftime('%Y-%m-%d %H:%M:%S')
    k = k + 1
    date = str(date)
    tension = str(tension)
    digital = str(digital)
    com_sql = "INSERT INTO tabela(date,tension, status)
VALUES(%s,%s,%s)"
    value = (date, tension, digital)
    cursor.execute(com_sql, value)
    connection.commit()
```

Fig. 13. Source code of variables saved in the project.

242

C. Results and Tests Conducted

Fig. 14 illustrates the assembly of the view tested in the laboratory, with all the instruments and components used in the development of this paper, such as: IO-Controller, IO-Device and HMI, connected in the PROFINET network.

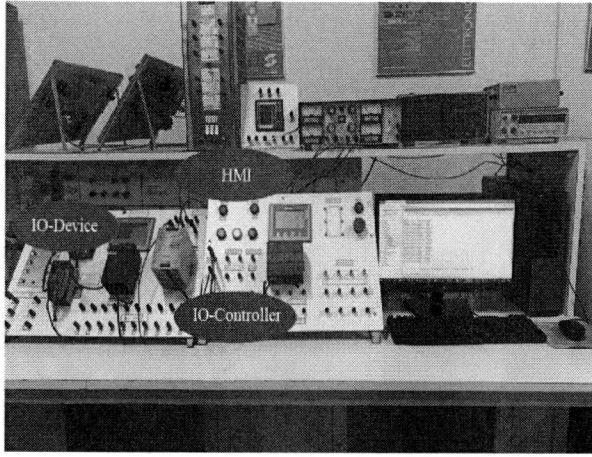

Fig. 14. View with all instruments connected.

Fig. 15 illustrates an executable of the tool developed in Python, where is possible to choose the variables of the PROFINET network that it will be analyzed and stored in the Database.

Enter with the Temperature variable: tension
Enter with the digital variable: Motor

Fig. 15. Executable of the tool developed in Python.

Fig. 16 illustrates the reading of the two developed tags in the IO-Controller.

"tension"	%MD20	Floating-poin...	▼	3.471785	
"Motor"	%I261.0	Bool		TRUE	
	Add new				

Fig. 16. Reading the tags on the IO-Controller.

Fig. 17 illustrates the reading of the two tags in the Web Server.

Address	Display Format	Monitor Value	Modify Value
tension	Floating_Point	▼ 3.474226	
Motor	BOOL	▼ true	
New variable		▼	

Refresh

Fig. 17. Reading the tags on the Web Server.

Fig. 18 illustrates the data, temperature (given in tension) and the motor status (ON / OFF), saved in the database. The temperature presented in the database is in the tension scale, between 0 to 10Vdc, which is the scale of the analog input used by the IO-Controller.

date		tension	status
2019-10-21	18:42:01	1.091036	on
2019-10-21	18:42:02	1.088595	on
2019-10-21	18:42:03	2.942289	on
2019-10-21	18:42:04	3.026215	on
2019-10-21	18:42:05	3.006683	on
2019-10-21	18:42:05	2.934964	off
2019-10-21	18:42:06	0.1556444	off
2019-10-21	18:42:07	0.01495406	off
2019-10-21	18:42:08	0.01007111	on
2019-10-21	18:42:08	0.007629615	off
2019-10-21	18:42:09	0.01007111	off
2019-10-21	18:42:10	0.01007111	on
2019-10-21	18:42:11	0.04455701	on
2019-10-21	18:42:11	0.6271552	off
2019-10-21	18:42:12	2.312997	off
2019-10-21	18:42:13	3.487959	on
2019-10-21	18:42:14	3.473005	on
2019-10-21	18:42:15	3.473005	on
2019-10-21	18:42:15	3.475447	on
2019-10-21	18:42:16	3.473005	on

Fig. 18. Data saved in real time in the database.

IV. CONCLUSION

The development of the practical system exemplifies the idea of monitoring and storing data from a PROFINET network. The purpose of this paper was to develop an application of this communication protocol, where industrial field elements accessed information stored in a database.

It is possible to conclude that the application developed for monitoring and data storage of a PROFINET network is completely viable and helps in the industrial process, because can store in a Database, including in the cloud and in real time, any desired network variables. It was also possible to conclude that the measure precision was good, with errors less than 0,5%.

In a real industrial application, it is very important to store industrial process data to monitor and control the process and in failure prevention and equipment maintenance.

A great advantage of use the tool developed is the cost-benefit, because it has a free license software, it is free of cost and its storage is limited only for the amount of data.

Possible suggestions for future work would be the insertion of this data in the cloud and the possibility of, in addition to monitoring, controlling and acting on the variables, using the tool developed in this paper.

243

REFERENCES

[1] Ralf, S.; Andreas, S. and Thilo, S. "Automatic Packing Mechanism for Simplification of the Scheduling in Profinet IRT", IEEE Transation on Industrial Informatics, Volume 12, Issue 5, 2015, pp. 1822-1831.

[2] Stephan, H.; Stefan, P. and Christian, D. "Design of communication systems for networked control system running on PROFINET", IEEE Workshop on Factory Communication Systems, WFCS, France, 2014, pp. 94-99.

[3] Sestito, G.S. "A methodological proposal for the prediction of throughput during the initialization of PROFINET networks through artificial neural networks." Master's Dissertation Presented to the Graduate Program in Electrical Engineering at São Carlos School of Engineering, USP/Brazil, Brazil, 2014, 116p.

[4] Dias, A.L.; Sestito, G.S.; Turcato, A.C. and Brandão, D. "Panorama, challenges and opportunities in PROFINET protocol research", IEEE International Conference on Industry Applications (INDUSCON), Brazil, 2018, pp55-60.

[5] Popp, M. "Industrial Communication with PROFINET." Book Translated by PI Brazil and revised by Alexandre Baratella Lugli, PROFIBUS Association, São Paulo/Brazil, 2018, 1st Ed., 326p.

[6] Lugli, A.B. and Santos, M.M.D. "Industrial Networks for Industrial Automation: AS-i, PROFIBUS and PROFINET." Saraiva publishing house, São Paulo/Brazil, 2019, 2st Ed., 189p.

[7] Van Hentenryck, P., "The PLC language CHIP: constraint solving and applications", IEEE Conference COMPCON, USA, 1991, pp. 64-76.

[8] Aliya S.; Aytasova N.A.S. and Pavel A.K. "Development the Risk Management System of Processes in the Enterprise", IEEE Conference of Russian Young Researchers in Electrical and Electronic Engineering (EIConRus), Russia, 2019, pp. 32-37.

[9] Zhijian Y.; Chengyang Y. and Ke Z. "A University Fixed Asset Database Information Management System Based on Internet of Things", IEEE Advanced Information Management,Communicates, Electronic and Automation Control Conference (IMCEC), China, 2018, pp. 40-45.

Consistency Preserving Development of Embedded Systems Using AADL

Tomasz Szmuc, Wojciech Szmuc

Department of Applied Computer Science
AGH University of Science and Technology
30-059 Kraków, Poland
tsz@agh.edu.pl, wszmuc@agh.edu.pl

Abstract—**Architecture Analysis and Design Language (AADL) supports consistence modeling and several analyses in designing of real-time systems. Additional features supporting modeling and analysis is proposed in the paper. The concept is based on automatic translation of AADL components into Colored Petri Net (CPN) models. The translated model may be verified using CPN tools, and also checking satisfability of requirements (described using temporal logic) in the model. The proposed extension supports detection of structural errors in early development stages.**

Keywords—**embedded systems; AADL; translation into CPN; temporal logic verification**

I. INTRODUCTION

Embedded systems are hardly depended on their environment, therefore multi-domain modeling is recommended to preserve consistency of application and reduce testing effort in the later phases of the development process. Software of embedded systems is deeply involved in hardware, hence design of software modules should even consider hardware features. Embedded systems in control domain should also care on dynamics of the control object and also time characteristics of sensors, actuators and other devices in the environment.

The necessity of the multi-domain approach is especially visible in the Cyber Physical Systems (CPS) domain. Architecture of these systems is often complicated structure which consists of microcontrolers, networks of sensors (IoT) and usually collections of data stored in nebula or cloud structure. Robot controllers, intelligent cars are other examples of the multi-domain approach. Correctness requirements (logics, time-response, safety, etc.) of Safety-Critical Systems magnifies the need for consistency preserving of modeling during the development.

SysML [2] and AADL (Architecture Analysis and Design Language) [1,3] are two main languages recommended for development of embedded systems. SysML is an extension of UML profile which adapts this language to embedded applications by additional diagrams: requirement, parametric, block definition and internal block diagrams. Requirement and parametric diagrams support consistency of developed models, while block definition and internal block diagrams are simpler and more precise than general class/object diagrams in UML. The total number of diagrams in SysML is also reduced. Besides these adaptations and extensions, SysML seems to be too heavy for modeling of embedded systems perhaps because of heritage

from "baroque" UML. It should be mentioned, however, that SysML is the most popular modeling language in the embedded world. AADL background is quite different. The language was built from scratch but basing on experience in development of real-time and embedded systems. The main AADL building constructs are process, thread, subprogram and data for software, and device, processor, memory and bus for hardware. The components are integrated in system block. The constructs are directly related to the corresponding ones in the embedded engineering space. Additional features dealing with grouping, abstractions and parametrizations support modeling of complex systems.

The goal of the paper is to present parallel modeling and verification path supporting development of AADL projects. The main idea is to apply automatic translation of AADL artifacts into the corresponding Colored Petri Nets (CPN) models. These models may be analyzed using existing tools [12, 13, 15] and provide qualitative feedback to AADL development process. The paper focuses on the translation AADL→ CPN provided by the developed translator [6], and describes verification lane of the models. An overview of main AADL features is presented to provide a background for the proposed approach.

AADL language reached some maturation in the area of tools supporting modeling and verification. The main tool is Eclipse plugin Osate providing several functions: graphical modeling, latency analysis, constraints checking and safety analysis. Ocarina extends Osate functions by code generation. Both tools are accessible under public license – EPL and GPL correspondingly. There are also commercial solutions AADL Inspector seems to be the most popular.

Systematic design and verification case study is presented in the paper. The study starts with definition of general system structure using AADL textual language. In the first step the description focuses on block structure, i.e. blocks with interfaces and connections. Devices which specify environment of the system are also defined. In the next steps the general structure is refined by definition of internal structures, hardware elements, time parameters and scheduling modes. Verification of time requirements (latency, time for specified paths etc.) are verified for the defined system. The next step is safety analysis which needs previous definition of error levels and error possible flows through architecture. The above mentioned activities allow to design system structure correct with regard to specified time and safety parameters. The system frames may be then injected by C/C++, Ada or Java code. The development may be also supported by the above mentioned formal modeling and verification using CPN modeling and verification using temporal logic.

II. AN OVERVIEW OF AADL

Basic AADL building blocks (components) are grouped within three sets:

- software: **process**, **thread**, **thread group**, **subprogram**, **subprogram group**, **data**;

- hardware: **device**, **processor**, **memory**, **bus**;

- integration: **system**.

The components are specified in textual way using the related keywords. The corresponding graphical representation is generated automatically. The both forms are interactively updated during design in development environment (e.g. Osate). Every component is specified using its declaration (type) and implementation being an (refined) instance of the related type. Communication interface for components (ports, bus access) is declared in **features** section. **Connections** clauses specify connection channels between ports or via bus. Additional specifications may be included in **properties** section. Time properties and safety related properties are especially applicable in design of Safety Critical and (Hard) Real-Time systems. Hardware and software components are integrated in the **system** component. The language provides also classical constructs supporting development of projects, i.e. packages, modes, abstracts, generics, prototypes, virtualization, etc. AADL tools support development of architectural model, analysis of the model and generation of the system. The generation may be targeted into predefined collection of platforms. Injections of code to the architectural frames is also possible (Ocarina plugin).

The above overview describes very general concept of AADL. More detailed description of the language may be found in [1,3]. Some illustrations of the language and its expressiveness are presented in the next sections.

III. SYSTEMATIC AADL DEVELOPMENT

Several development environments for AADL modeling are available. The most popular is Osate (Open Source AADL Tool Environment) [14] which is an eclipse plugin. The environment provides not only synchronized text and graphical editors but also several other features, i.e. translation of SCADE and Mathworks®Simulink blocks into AADL components, tools and processes supporting SAE ARP 4761 safety standard (Functional Hazard Assessment, Fault Tree Analysis, Depends Diagrams, Reliability Block Diagrams, etc.), structural analysis (Rockwell Collins project), requirements specification and analysis (ALISA). Several analysis are also possible using SEI Validation Plugins: end-to-end latency, functional integration, port connections consistency, weight analysis, computer resource budget, compositional verification. The code generation (C, Ada) is dedicated for RTOS used in avionic industry but also for WindRiver®VxWorks.

AADL supports model-based systematic design of embedded systems. The following steps of the approach may be defined.

1. Specification of the environment by declaration of devices and their implementations.

2. General specification of software structures declarations of processes, subsystems, etc. Integration with devices.

3. Refinements of the general descriptions related to devices and software components, sub-components, threads, communication channels, subprograms.

4. Deployment on hardware structure specification of processors, memory, bus, etc.

5. Setting of time parameters, scheduling polices, resource management polices. Checking time requirements using Cheddar (Osate component) tool.

6. Setting safety related parameters and safety analysis.

7. Injection of code to the AADL frames and code generation.

The presented above steps are illustrated on simplified cruise control example (Fig.1). The example is a modification of widely used cases. Selected parts of the design are presented to illustrate the concept.

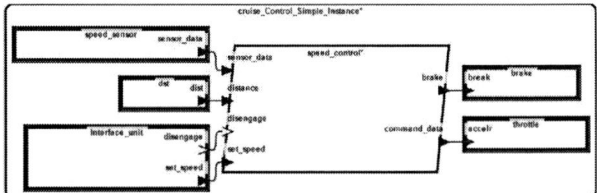

Fig. 1. AADL diagram of cruise control system

The system (i.e. controller + environment) consists of speed_control process, three input devices: speed_sensor, dst (distance to car in front), interface_unit (setting required speed and disabling the controller) and two output devices brake and throttle. Almost all ports are of data type, only disengage one is an event port. The diagram is generated from textual description using Osate.

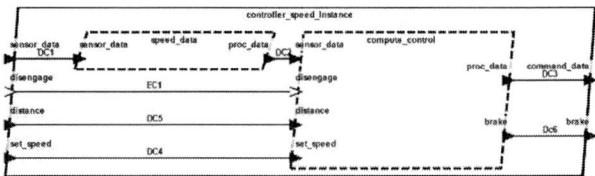

Fig. 2. Internal structure of controller_speed process

An instance of controller_speed process (Fig. 2) consists of two threads; speed_data transforming raw sensor data into input for compute_control, which implements control algorithm.

The two diagrams specify system after 3 steps of the above mentioned design process. Textual description of the controller_speed process is presented in Fig. 3

```
process controller
features
        command_data: out data port;
        sensor_data: in data port;
        set_speed: in data port;
        distance: in data port;
        brake: out data port;
        disengage: in event port;
end controller;

process implementation controller.speed
subcomponents
        speed_data: thread read_data.speed;
        compute_control: thread control.speed;
connections
        DC1: port sensor_data -> speed_data.sensor_data;
        DC2: port speed_data.proc_data ->
                 compute_control.sensor_data;
        DC3: port compute_control.proc_data -> command_data;
        DC4: port set_speed -> compute_control.set_speed;
        DC5: port distance -> compute_control.distance;
        DC6: port compute_control.brake -> brake;
        EC1: port disengage -> compute_control.disengage;
end controller.speed;
```

Fig. 3. Textual description of controller_speed process

IV. GENERATION OF PETRI NET MODELS

AADL is structural language therefore transformation of its components into CPN preserves the structures. This observation forms general mapping rule: every AADL component, i.e. process, thread, sub-program, device may be modeled using substituted transition where input and output socket places correspond to the related i/o ports or interface for transfer of parameters. General mapping rules presented in [7, 8] are refined and applied in building AADL to CPN translator [6].

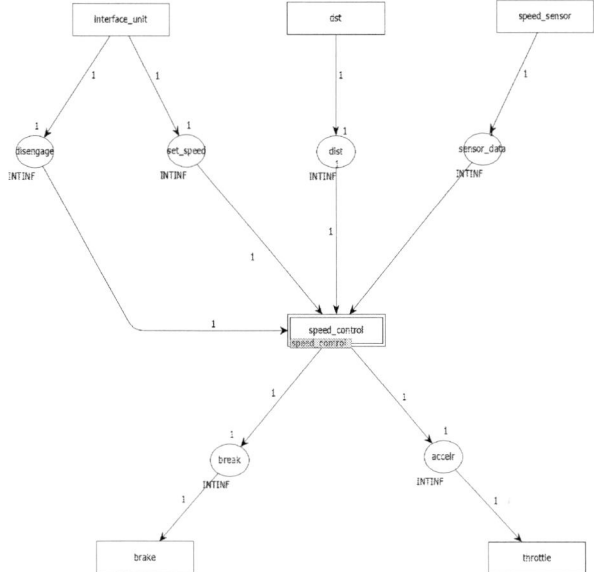

Fig. 4. Petri net model of cruiseControl system

Mapping of the cruiseControlSimple AAADL instance (Fig. 1) into CPN using the AADL to CPN translator [6] is presented in the next four diagrams. The diagrams are manually

tuned according graphical aesthetics to improve their visibility. It may be noticed that the CPN models of the system is represented by 3 levels hierarchical structure: system, controller, thread, and code.

CPN model in Fig. 4 specifies system structure (Fig. 1). Input and output devices are represented by the related transitions connected with speed_control by the corresponding places. The process name is taken by the algorithm from subcomponets section of AADL system definition. It is a reason of difference between names in the two diagrams. The devices are represented by simple transitions but will be replaced by the related substituted transitions when internal structure of the devices will be defined. Substituted transition speed_control is a gate to model of the process controller_speed from Fig. 2.

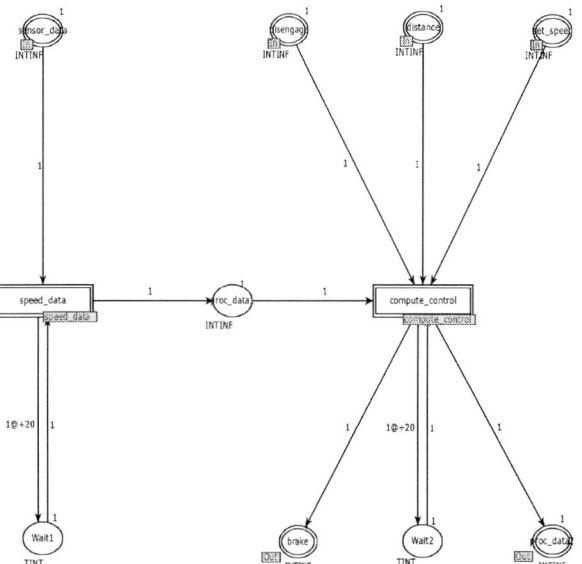

Fig. 5. Petri net model of the controller_speed process

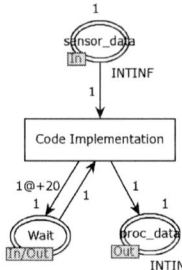

Fig. 6. Petri net model of speed_data thread

Petri net model of the controller_speed process instance (Fig. 5) consists of two substituted transitions (speed-data, compute_data) and several places. The transitions are gates to the next level, where internal structures of the related threads are defined. Two additional Wait places and related arcs model periodic scheduling policy declared in the AADL models in **properties** section by:

Dispatch_Protocol => Periodic;

Period => 20 ms;

Code_implementation transition in Fig. 6 represents the computation part. Substituted transition will be generated when internal structure will be specified in the AADL model.

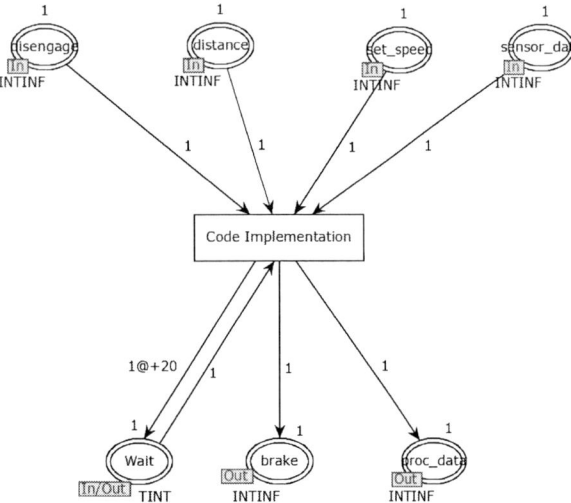

Fig. 7. Petri net model of the compute_control thread

Petri net model obtained after translation may be verified directly using CPN Tools [12]. Several properties classical for concurrent systems may be proved on the basis of generated reachability graph (coverability tree) and invariants - see [5] for details.

Verification scope may be extended by application of temporal logic for description of required properties and using of model checking for proving if the formal requirements are satisfied in the model generated by CPN Tools. The feature is not directly applicable due to different format of generated CPN output (LTS – Labeled Transition Tree) and data format for model checker – nuXmv [13] in this case. The intermediate mapping is carried out by PetriNet2ModelChecker [15] tool described in [11]. The complete formal path modeling and verification path is in Fig. 8.

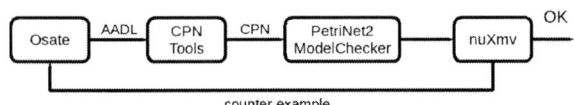

Fig. 8. Formal modeling and verification chain

The modeling verification sequence is similar to the related path for SysML language [10].

V. SUMMARY

Additional formal modeling and verification path to support consistency in model-based development is presented. AADL models are mapped into hierarchical colored Petri net which is then reformatted and transferred to temporal logic model checker. Required properties of the system are checked by proving if the related CTL formula is satisfied in the system formal CPN model. Negative result in the form of counter examples support searching places of errors in AADL design.

AADL to CPN translator is a simplified prototype which focuses on translation of the main structures. More advanced version is under development. It is related to extended features and also verification of time aspects. The later are implemented basing on the former research [9].

ACKNOWLEDGMENT

The research was supported by the Department Research Grant No 11.11.120.859: Methods, tools and environments for knowledge and software engineering. Topic: Modeling, analysis and optimisation of selected systems.

REFERENCES

[1] J. Delange, AADL in Practice. Design and Validate the Architecture of Critical Systems, Reblochon Development Company, 2017

[2] L. Delligatti, SysML Distilled. A Brief Guide to the Systems Modelling Language, Addison-Wesley, 2014.

[3] P.H. Feiler, J. Delange, L. Wrage. A requirement Specification Language for AADL, Software Engineering Institute, Carnegie Mellon University, Technical Report, 2016, https://resources.sei.cmu.edu/asset_files/TechnicalReport/2016_005_001_464378.pdf

[4] P. H. Feiler, and D. P. Gluch. Model-Based Engineering using AADL. An Introduction to the SAE Architecture Analysis & Design Language, Addison-Wesley, 2012.

[5] K. Jensen, and L. M. Kristensen. Coloured Petri Nets. Modelling and Valiation of Concurrent Systems, Springer, 2009

[6] Ł. Łukowicz. Modelling and analysis of software using AADL language and Petri nets (in Polish Modelowanie i analiza oprogramowania z zastosowaniem języka AADL i sieci Petriego) M.Sc. Thesis (Supervisor T. Szmuc), AGH University of Science and Technology, 2019

[7] X. RENAULT, F. KORDON, J. HUGUES. "From AADL architectural models to Petri Nets: Checking model viability". 2009 IEEE International Symposium on Object/Component/Service-Oriented Real-Time Distributed Computing. IEEE DOI 10.1109/ISORC.2009.11. pp. 313-320

[8] H. Reza, and A. Chatterjee. "Mapping AADL to Petri net Tool-Sets using PNML Framework, Journal of Software Engineering and Applications, 2014, 7, pp. 920-933, Published Online October 2014 in SciRes. http://www.scirp.org/journal/jsea, http://dx.doi.org/10.4236/jsea.2014.711082

[9] S. Samolej, and T. Szmuc. "HTCPN-based tool for web-server clusters development". Lecture Notes in Computer Science vol. 5082, Springer Verlag, 2008

[10] T. Szmuc, and W. Szmuc. "Rigorous development of embedded systems Supported by formal tools", submitted to Mixdes 2020

[11] M. Szpyrka, J. Biernacki, and A. Biernacka. "Tools and methods for RTCP-nets modelling and verification", Archives of Control Sciences, vol. 26, No 3, 2016, pp. 339-365, doi 10.1515/acsc-2016-0019

[12] CPN Tools: http://cpntools.org/

[13] nuXmv model checker: https://nuxmv.fbk.eu/

[14] Osate: https://osate.org/

[15] PetriNet2ModelChecker: http://alvis.kis.agh.edu.pl/wiki/software

Indoor Precise Infrared Navigation

Paweł Marzec, Andrzej Kos
AGH University of Science and Technology
Faculty of Computer Science, Electronics and Telecommunications
Krakow, Poland
pmarzec@agh.edu.pl, kos@agh.edu.pl

Abstract—The paper proposes an original solution enabling the navigation of a blind person in such buildings as shopping malls and offices as well as public and private institutions. A GPS system fitted with an infrared sensor in the form of a camera is used for navigation. The system detects the surrounding walls in buildings and enables a blind person to keep a safe distance from the walls while moving indoors. Navigational instructions and directions are transmitted to a blind person by means of vibration sensors placed on this person wrists.

Keywords—thermo-navigation, blind people, infrared sensor, indoor navigation

I. Introduction

Currently, the number of the blind is increasing, amounting to approx. 36 million, while roughly another 250 million people suffer from vision problems of some kind **Błąd! Nie można odnaleźć źródła odwołania.**. This problem is especially acute in underdeveloped countries and in those where the fast aging of the population is observed. The main problem affecting the blind is getting around in an urban environment. In an outdoor environment, what is mainly used for navigation is the GPS technology. This paper deals with mobility problems experienced by the blind moving inside such buildings as schools, shopping malls, offices, etc., where the GPS signal is not available, whereas a significant precision of navigation is required. At present, there are a number of solutions aimed at helping the blind to move indoors. An example of such solutions is an electronic stick for the blind that enables them to detect obstacles by means of a camera and other sensors coupled with the camera [2]. Another example is an acoustic solution allowing a dynamic detection of obstacles around a blind person [3]. Noteworthy is also the solution making use of a camera mounted inside a building [4], which enables identifying the position of a particular person or the solution enabling people to navigate by means of BLE sensors that have been appropriately deployed in a building [5]. An example of indoor navigation for visually impaired persons is the solution that helps them move efficiently and safely in indoor environments by mapping inputs from a depth camera to 3D-localized auditory cues [6].

Another solution is a system based on a video-camera and infrared sensor. The system detects obstacles by means of corner detection with a camera and distance measurement with a depth sensor [7]. Furthermore, solutions applying robotics can be used for navigating visually impaired persons indoors. An example is the method for detecting the distance and the boundary between a wall and a floor by extracting an ellipse projected from a robot-mounted projector and detecting this ellipse with a camera [8].

The proposed solution enables the indoor navigation of a blind person by using an infrared thermal camera for a precise determination of the distances between a blind person and a selected object, e.g. a building wall. This solution is a development of the method for determining the distance between a blind person and an object that was described by the authors in an earlier paper [6].

II. Indoor Navigation Methods for the Blind

Due to a very weak GPS signal, or even its complete absence, inside buildings and also to high precision requirements, indoor navigation of a visually impaired person is difficult. The proposed solution makes use of IR radiation to navigate a blind person. In contrast to the conventional methods of distance measurement, in the proposed method no measuring signal is generated. What is used instead is only the thermal energy radiated by all the ambient objects, thanks to which the device consumes significantly less energy. The occurrence of temperature and emissivity differences between particular objects allows to determine the distance between a blind person and an object, e.g. a wall [6]. Considering the variety of objects that can be found inside buildings, three different methods for determining the distance between a blind person and selected object are described below.

A. Method I – navigating by utilizing the temperature difference between the wall and the base surface.

This method makes use of the difference between the temperatures measured by a thermographic camera, which often results from different emissivities of, for example, the wall and the ground [6]. Fig. 1 shows a sample photograph taken inside a building with a conventional camera and a thermographic camera. The wall and the ground in the building are made of two different materials characterized by different colours and textures. Each material has a different emissivity, which makes it possible see different temperatures for the wall and the ground in the thermographic image. More information about distance measurement with this method is presented in [9]. In this example (Fig. 1), the floor is made of a rock material with an emissivity of 0.93, while the wall was painted with enamel paint and has an emissivity of 0.90.

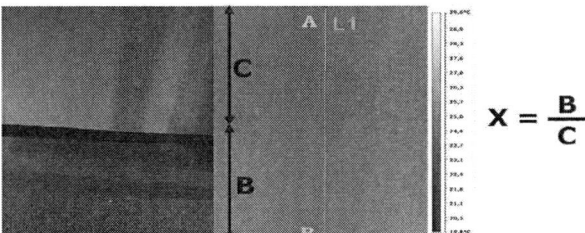

$$X = \frac{B}{C}$$

Fig. 1. A fragment of a wall and the ground inside a building (on the left - an image from a photo camera; on the right – from a thermographic camera)

In the thermographic image (Fig. 2), the line $L1$ was marked, for which a temperature diagram was next plotted, where the difference in the temperatures of the wall and the base surface of approx. 0.5°C can be observed.

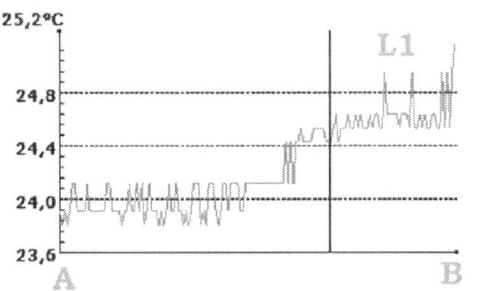

Fig. 2. The temperature diagram for the line $L1$ (from Fig.1)

To determine the distance D between a blind person and the inside wall, the method utilizing a thermographic camera, described in this paper [6], was applied.

$$D = \frac{H(\frac{1}{tg(\beta+\alpha)}+X)}{1+\frac{X}{tg(90-\beta)}} \quad (1)$$

where:

H – the sensor height,

X – the reading from the thermographic camera [6] Fig. 1,

α – the FOV angle of the infrared sensor,

β – the sensor inclination angle measured against the ground.

B. Method II – navigating by using one's own reflective heat

In this method, a moving person's own heat that is reflected from an object with a low emissivity coefficient (a high reflection coefficient) is used to determine the distance between a blind person and selected object. Examples of materials with a high reflection coefficient include: low-emission glass, aluminum, chromium, polished metals, shiny surfaces, etc. An example of a reflection of the heat generated by a person from a polished metal surface (lift doors) is presented in Fig. 3.

Fig. 4 and Fig. 5 show the position of a blind person being at a short ($D1$) and long ($D2$) distance, respectively, from a selected object. The IR sensor mounted on the arm detects the thermal radiation generated by the person which is reflected from the wall. From the thermographic image generated by a

thermal sensor we obtain the image S that results from the heat reflected from the object and generated by the human body. The size S of the image is proportional to the distance D. In Fig. 4, the blind person is at the distance $D1$ and generates the reflected image of the size $S1$. Similarly, for the distance $D2$, the image of the size $S2$ is generated.

Fig. 3. On the left – an object with a high reflection coefficient – doors made of a polished metal; on the right – a thermogram showing the reflection of the heat generated by the human body next to the object

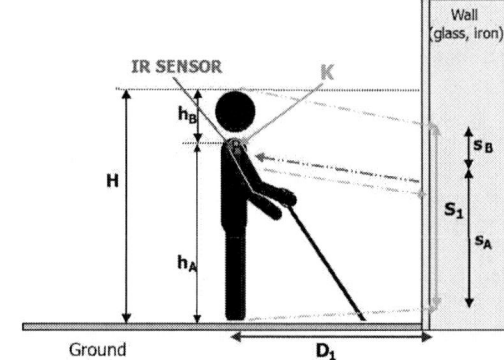

Fig. 4. The position of a blind person being at a short distance from the wall

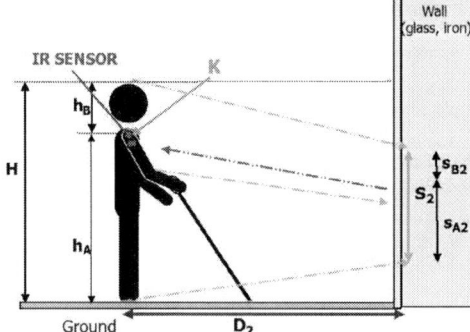

Fig. 5. The position of a blind person being at a long distance from the wall

Fig. 6 shows the dependence of the distance D of a blind person on the size S of the image obtained from a thermal sensor. The dependence of the distance D on the size S is linear and is mainly dependent on the parameters of the applied sensor and the person's height H. The value S is measured by the number of pixels. Before the first application, it is necessary to mark out a minimum of two points D in relation to the S parameter to be able to find the formula for the linear function describing this dependence. On the basis of the determined linear function formula and the reading of the value S from the thermographic camera, it is possible to identify the distance of the blind person from the selected object.

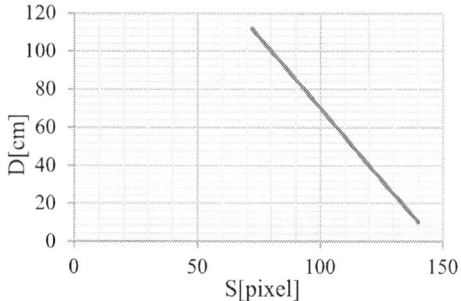

Fig. 6. The linear dependence of the distance D on the value S

Moreover, in order to correct measurement errors, an additional element with a higher temperature was mounted on the blind person's arm at the height h_A, thus dividing the overall height of the blind person at a ratio of h_A to h_B. This point is visible in the thermographic image (Fig. 7) as a brighter spot, so it divides the value S into the values s_A and s_B, respectively. The ratio of the value s_A to the value s_B is directly proportional to the ratio of h_A and h_B (2) and does not depend on the distance D.

$$Y = \frac{h_A}{h_B} = \frac{s_A}{s_B} \qquad (2)$$

Fig. 7. The thermographic image showing the reflection of the heat generated by a human from a glass wall; an additional element of a lower temperature is marked.

Knowing the value of the above mentioned ratio Y, we can correct the value of the distance. The measurement error of the distance can occur, for example, when an object made of a material characterized by a low reflection coefficient is located at the height of the person's eyes or legs, due to which the ratio of s_A to s_B will no longer be directly proportional to the ratio of h_A to h_B. Such a situation may occur, for example, when measuring the distance from a glass door with a frame made of a material with a low thermal reflection coefficient. The image

reflected from such an object will be partial and will not contain the fragment absorbed by the door frame.

In this case, the ratio of s_A to s_B in the generated image will be less than Y (3).

$$Y \geq \frac{s_A}{s_B} \qquad (3)$$

Therefore, the value s_A should be complemented with the value G so that the ratio of s_A to s_B is equal to Y (4).

$$\frac{s_A + G}{s_B} = Y => G = \frac{h_A \cdot s_B}{h_B} - s_A \qquad (4)$$

Next, knowing the value G, we can calculate the value S (5), which is linearly dependent on the value of the distance D.

$$s_A + s_B + G = S \sim D \qquad (5)$$

C. Method III – navigating with the use of an element with a low emissivity coefficient (a high reflection coefficient)

The suggested solution consists in the use of a thin element, e.g. in the form of a tape made from a material with a high thermal reflection coefficient, which is placed on an inner wall of a building at a specified height h over the base surface. This will allow to determine the distance between a blind person and the wall. The tape, having a high thermal reflection coefficient, will cause the heat generated by the person being next to the wall to be reflected from the tape and subsequently recorded by a thermal sensor mounted on the person's arm. This solution is presented in Fig. 8.

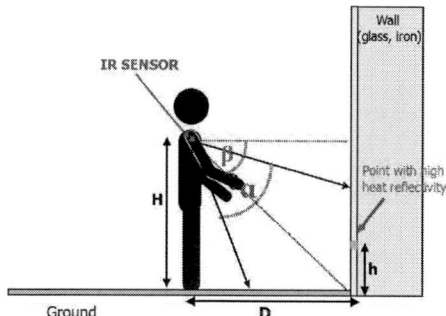

Fig. 8. The method for determining the distance with the use of an element having a high reflection coefficient

Using the formula (6), we can calculate the distance of the blind person from the wall,

$$D = \frac{tg\left(90° - \frac{1}{2}\alpha - \beta\right)HZ - Zh - h}{tg\left(\beta - \frac{1}{2}\alpha\right) + Z} \qquad (6)$$

where:

β – the inclination angle of the infrared sensor,

α – the fixed angle of the FOV sensor,

H – the sensor height over the ground,

Z – the value read from the image generated by a thermographic camera, which is equal to the ratio of number of pixels C to the number of pixels B, corresponding with the number of pixels over and below the horizontal line visible in the thermographic image, respectively.

251

Fig. 9 shows an image from a thermal sensor in which a horizontal line representing a higher temperature can be seen. The higher temperature of the horizontal line results from the reflection of the heat generated by the person located in front of the selected object from the thin and narrow aluminum tape mounted horizontally on the object.

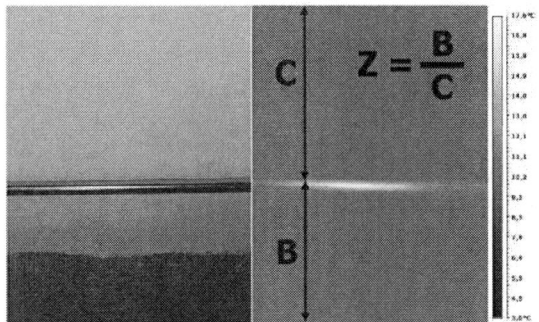

Fig. 9. On the left – a photograph showing a wall with a mounted aluminum tape; on the right – a thermographic image showing a horizontal line with a higher temperature, formed as a result of the reflection heat generated by the person located in front of the wall from the aluminum tape mounted on the wall. Z represents the number of pixels over and below the horizontal line (the ratio of B to C)

III. THE DESIGN OF THE IR NAVIGATION SYSTEM

The system consists of two main modules. The module gathering information from the surroundings, i.e. *Sensor module*, mounted on a blind person's arm, and a module sending information to the blind person, i.e. *Notice module*, mounted on the wrist of the blind person. Fig. 10 shows the block diagram of the system. The communication between the modules is effected via an energy-efficient interface Bluetooth Low Energy. Fig. 11 presents a photo of the device, i.e. the *Sensor module* and the *Notice module*.

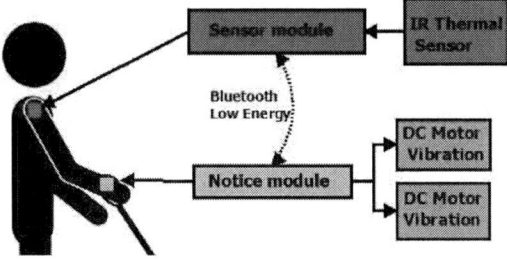

Fig. 10. The block diagram of the thermal navigation system

Data from the thermal sensor is transmitted to the *Sensor module*, and on the basis of this data the distance of the blind person from a selected object is calculated. Based on the data collected in the Sensor module, messages are generated, which are subsequently transmitted to the *Notice module* via the Bluetooth interface.

The *Notice module* converts the received messages into appropriate sequences of vibrations generated by vibration sensors mounted on a blind person's wrist.

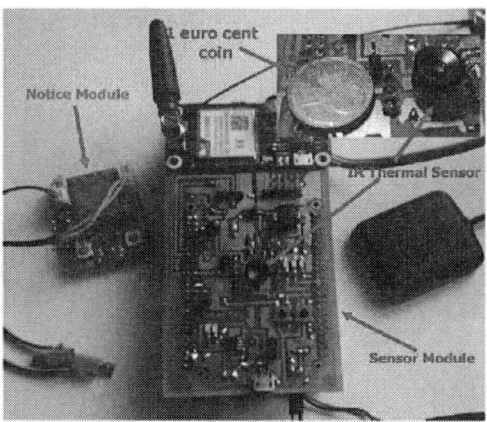

Fig. 11. IR Navigation System - *Sensor module* and *Notice module*

IV. MEASUREMENT RESULTS

A. Method I - navigating by utilizing the temperature difference between the wall and the ground.

In the measurement, a thermal sensor with a measurement angle $\alpha = 25°$ and a resolution of 160 pixels was used. The Sensor module together with the thermal sensor was attached to a blind person's arm and set at an angle $\beta = 50°$ with respect to the ground. Fig. 12 shows the results of the distance measurement as a function of the parameter X that was read from the thermal sensor image.

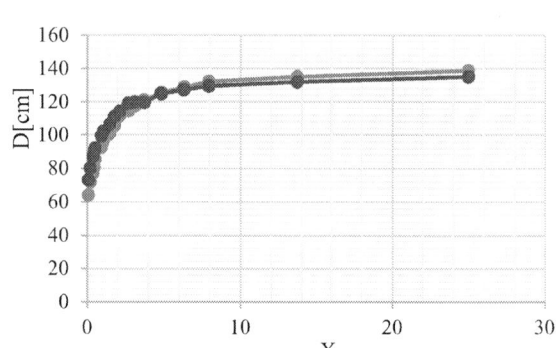

Fig. 12. The dependence of the distance D on the parameter X for the data obtained from the measurement and calculated from the formula (1)

During the measurement, the distance between the blind person and the selected object was both decreased and increased. The *Dm* graph shows the values measured with a measuring tape. These are the reference values. On the basis of the data obtained from the thermal sensor and formula (1), the distance of the blind person from the object was calculated and the *Dt* graph was plotted. The maximum measurement error of the distance obtained from the thermal sensor and formula (1) is approx. 5 cm relative to the reference measurement *Dm*.

B. Method II - navigating by using one's own reflective heat

During the measurement, a thermal sensor with a resolution of 160 pixels and a measurement angle $\alpha = 25°$, set at an angle $\beta = 30°$ was used. Next, for two points D, a linear function Dt was determined, which described the dependence of the distance D on the value S, formula (7).

$$Dt = -1,799S + 130,96 \text{ [cm]} \tag{7}$$

At the next stage, the person approached and walked away from a glass door, and on the basis of the data obtained from a thermal camera and the previously determined function (7), the *Dm* graph was plotted (Fig. 13).

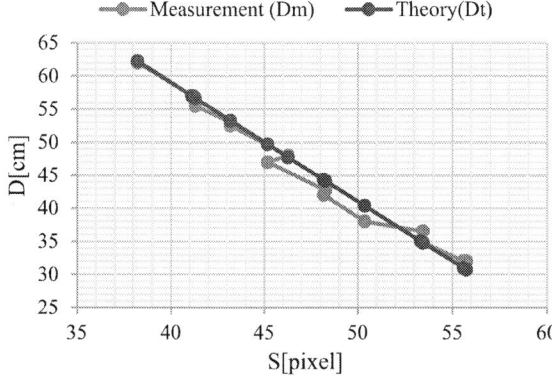

Fig. 13. The dependence of the distance on the value *S* (Method II)

The distance range depends primarily on the measurement angle α and the height *H* of the person whose heat is reflected from a particular object, e.g. a glass door. The measurement error is approx. 2 cm.

C. Method III - navigating with the use of an element with a low emissivity coefficient

In the measurement, a polished silvery aluminum tape with a low emissivity coefficient was used. The tape was mounted at a height h = 19 cm. A thermal sensor with a measurement angle $\alpha = 25°$ was attached at a height h = 52 cm and inclined by an angle $\beta = 35°$ with respect to the ground.

Fig. 14 presents the results of the measurements of the distance *D* with respect to the value *Z* for the data obtained from the thermographic image. The *Dt* graph shows the dependence, obtained from the formula (6), of the distance *D* with respect to the parameter *Z*. The *Dm* graph shows the dependence of the distance *D* measured with a measuring tape on the parameter *Z*. This is the reference measurement on the basis of which the accuracy of the distance calculated with the formula (6) was found.

For the obtained measurement values, the measurement error of the distance *D* with the use of the formula (6) is approx. 3 cm.

Table I. shows a comparison of different distance measurement methods (types of sensors).

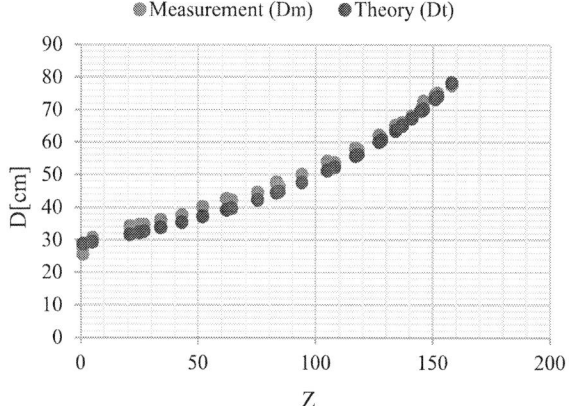

Fig. 14. The dependence of the distance *D* on the value *Z* (Method III)

TABLE I.
COMPARISON OF DIFFERENT TYPE OF SENSOR

Type of sensor	Average Power Consumption [mW]	Number of measurement points	Computational complexity
Ultrasonic distance sensor	75	1	small
Time-of-Flight	53	1	small
IR sensor	165	1	small
Single-point ranging LiDAR	300	1	small
Infrared array sensor	66	24[a]	small
VGA camera	80	depends of measurement method	large

[a] Depends on the horizontal resolution.

With the above comparative Table 1, it can be concluded that ToF sensors are the most energy efficient and are characterized by low computational complexity. Using an infrared array sensor and the method described in this presentation, the most energy efficient distance measurement system per one measurement point can be obtained. Additionally, the image obtained from an infrared sensor array is a filtered image, and thus it contains fewer details compared to the image obtained from a digital camera, allows for faster data analysis and the power consumption is lower.

V. CONCLUSIONS

This paper presents methods facilitating the mobility of the blind inside buildings like schools or shopping malls. All these methods are complementary and can be simultaneously used for navigating a blind person indoors. Methods I and II do not require any modifications in the existing infrastructure of a given building. In places where neither Method I nor II can be used, Method III can be applied, which requires only slight and inexpensive changes in the building infrastructure, i.e. mounting a tape with a high heat reflection coefficient at a particular height. In the conducted measurements, the lowest measurement error of approx. 2 cm was achieved for Method II. For Methods I and III, it was 5 cm and 3 cm, respectively.

ACKNOWLEDGEMENTS

The authors wish to thank for financial support from the subvention No. 16.16.230.434 payed for by the AGH University of Science and Technology, Krakow, Poland

REFERENCES

[1] World Health Organization. Visual Impairment and Blindness. Accessed: Jun. 2017. [Online]. Available: http://www.who.int/mediacentre/factsheets/fs282/en/

[2] C. T. Patel, V. J. Mistry, L. S. Desai and Y. K. Meghrajani, "Multisensor - Based Object Detection in Indoor Environment for Visually Impaired People," 2018 Second International Conference on Intelligent Computing and Control Systems (ICICCS), Madurai, India, 2018, pp. 1-4. doi: 10.1109/ICCONS.2018.8663016

[3] G. Yang and J. Saniie, "Indoor Navigation System Using Acoustic Cuing for Visually Impaired People," 2018 IEEE International Conference on Electro/Information Technology (EIT), Rochester, MI, 2018, pp. 0504-0508. doi: 10.1109/EIT.2018.8500192

[4] M. I. Islam, M. M. H. Raj, S. Nath, M. F. Rahman, S. Hossen and M. H. Imam, "An Indoor Navigation System for Visually Impaired People Using a Path Finding Algorithm and a Wearable Cap," 2018 3rd International Conference for Convergence in Technology (I2CT), Pune, 2018, pp. 1-6. doi: 10.1109/I2CT.2018.8529757

[5] S. Bilgi, O. Ozturk and A. G. Gulnerman, "Navigation system for blind, hearing and visually impaired people in ITU Ayazaga campus," 2017 International Conference on Computing Networking and Informatics (ICCNI), Lagos, 2017, pp. 1-5. doi: 10.1109/ICCNI.2017.8123814

[6] S. Blessenohl, C. Morrison, A. Criminisi and J. Shotton, "Improving Indoor Mobility of the Visually Impaired with Depth-Based Spatial Sound," 2015 IEEE International Conference on Computer Vision Workshop (ICCVW), Santiago, 2015, pp. 418-426, doi: 10.1109/ICCVW.2015.62.

[7] C. Stoll, R. Palluel-Germain, V. Fristot, D. Pellerin, D. Alleysson, and C. Graff. Navigating from a depth image converted into sound. Applied Bionics and Biomechanics, vol. 2015, 2015.

[8] M. Seki and Y. Sugaya, "Floor-wall boundary detection from projected ellipses for autonomous robot navigation," 2016 International Conference On Advanced Informatics: Concepts, Theory And Application (ICAICTA), George Town, 2016, pp. 1-4, doi: 10.1109/ICAICTA.2016.7803124.

[9] P. Marzec and A. Kos, "Low Energy Precise Navigation System for the Blind with Infrared Sensors," 2019 MIXDES - 26th International Conference "Mixed Design of Integrated Circuits and Systems", Rzeszów, Poland, 2019, pp. 394-397. doi: 10.23919/MIXDES.2019.8787093

Linux Kernel Driver for External Analog-to-Digital and Digital-to-Analog Converters

Paweł Skrzypiec, Zbigniew Marszałek
Department of Measurement and Electronics
AGH University of Science and Technology
Krakow, Poland
pawel.skrzypiec@agh.edu.pl, antic@agh.edu.pl

Abstract—**The paper presents the Linux kernel driver for analog-to-digital and digital-to-analog converters control. The proposed module is the main part of the distributed embedded system for impedance measuring applications. The designed prototype consists of three hardware components: Raspberry Pi 3, microcontroller and personal computer. Two experiments verified the functionality and parameters of the constructed system. The first one confirmed that the driver does not utilize processor time while the data processing application is not running. It also showed that the module consumes only 0.7 % and 3.3 % of two cores of a quad-core processor for data readout and its copying to the user space. The mentioned test was performed with the sampling frequency equal to 256 kHz. The second experiment determined the accuracy of impedance measurements done by the constructed prototype. Results were compared with ones realized with the commercially available measuring device usage. The determined measurements uncertainty did not exceed 1.5 %.**

Keywords—**Linux, kernel, driver, embedded system, microcontroller, impedance measurements.**

I. Introduction

The trend in the modern embedded systems design aims to increase the functionalities of devices working at the network edges [1]–[4]. One of the most difficult challenges related to that kind of system design is increasing its functionality without deterioration of its maintainability. The mentioned challenge can be met by open-source libraries or operating systems usage.

The embedded operating systems, such as Linux, allow programmers to focus their attention on problems much more abstract than particular registers accessing. It can solve problems related to data acquisition and processing, but it is not possible to develop one low-level library working on each hardware configuration. The way to improve embedded systems flexibility is an implementation of the hardware-specific logic as custom device drivers.

The proposed Linux kernel driver [5] implements functionalities related to the analog-to-digital (ADC) and digital-to-analog (DAC) converters control. Due to the lack of converters in the Raspberry Pi 3 device, it was adapted to work with external ones integrated with ARM Cortex-M7 core in the STM32F746ZG microcontroller, connected through the Serial Peripheral Interface (SPI) bus. The developed driver was used in the stability and performance requiring impedance measuring system.

Impedance is an electrical parameter that can be used to electronic circuits components characterization [6], metal objects detection [7], biological tissue analysis [8], and many others. It is generally defined as the total opposition of the device to alternating current for a given frequency and its measurements can be based on the auto-balancing bridge method [9], which was used in the constructed system.

Due to the necessity of the developed driver verification, the test environment was built. It contained two additional software components: data processing application executed in the Raspberry Pi user space and the data visualizer running on the personal computer. The system correctness was verified by the impedance measurements and the comparison of the results with ones done with the Agilent E4980A [10] multimeter usage.

In this paper authors propose usage of the Linux-powered embedded systems in measuring applications. Conducted experiments aimed to design a flexible and efficient device for analog signals processing, which allows access to collected data through the UDP network connection. The presented architecture can be an unconventional alternative for the commercial measuring devices.

The article is organized as follows: Section II shows the architecture of the constructed impedance measuring system, Section III provides details about the developed kernel driver, Section IV presents results of the conducted experiments, Section V contains conclusions and suggestions for the further researches.

II. Distributed Embedded System for Impedance Measurements

The constructed distributed system for impedance measurements consists of three hardware components:

- Raspberry Pi 3, executing developed kernel driver and data processing application,
- microcontroller, containing integrated DACs and ADCs,
- Personal Computer (PC), running application for the measurement results visualization.

In the designed system, conversions are performed in the microcontroller by its internal DAC and ADCs. After the buffer completion, the notification about data readiness is sent to the Raspberry Pi 3 through the general-purpose input/output (GPIO) digital line. After its reception, buffer readout is

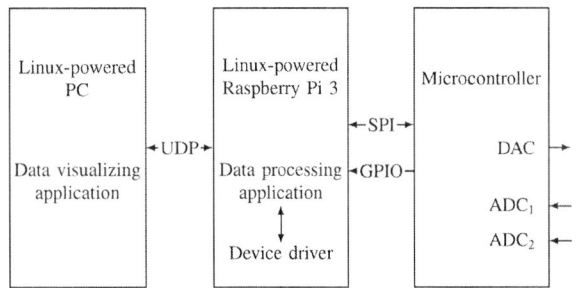

Fig. 1. Distributed system for impedance measurements architecture.

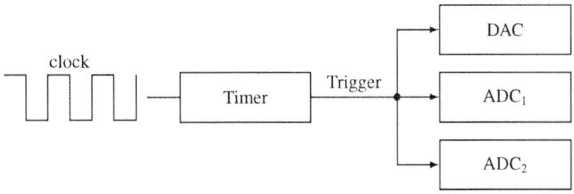

Fig. 2. D/A and A/D conversions synchronization.

performed, read data is processed and results are sent to the PC in the User Datagram Protocol (UDP) datagrams. The architecture of the constructed system is shown in Fig. 1.

A. Microcontroller

The lack of the ADCs and DAC in Raspberry Pi 3 was mitigated by the usage of the external ones integrated with an ARM Cortex-M7 core in the STM32F746ZG microcontroller. The chosen device contains three 12-bit ADCs, two 12-bits DACs and can operate with a frequency up to 216 MHz.

The impedance measurements require precisely synchronized DAC and ADCs. The timer integrated into the chosen microcontroller was configured to the parallel triggering of three converters (see Fig. 2). After the appropriate command reception, conversions with a frequency equal to 256 kHz are performed in a single DAC and two ADCs.

The conversions simultaneously executed by two ADCs and one DAC require execution of many processor instructions related to data copying between different memory addresses. During sampling performed with the frequency equal to 256 kHz, 768 thousand data transfers are executed in each second. The CPU utilization can be reduced by the usage of the direct memory access (DMA) controller.

During the microcontroller booting process, three DMA channels are configured: from the memory to the DAC buffer, from the ADCs buffer to the memory and from the memory to the SPI buffer. While conversions are performed, the processor is used only to DMA interrupts handling and notifying Raspberry Pi about data readiness. The high logical value on the dedicated digital line is set when the buffer is filled and the opposite one after the readout completion.

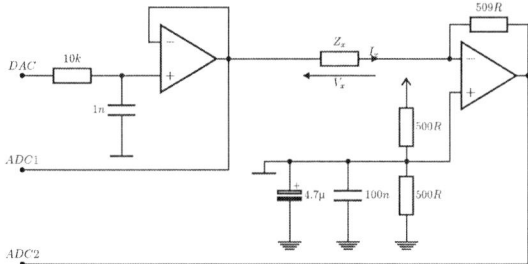

Fig. 3. The auto-balancing bridge as electronics frontend for impedance measurement.

B. Data processing application

The Linux kernel space is the area of virtual memory with increased privileges level. Applications running in it can directly access the hardware and kernel internal data. The role of the kernel applications (modules) should be reduced to the realization of only features that can not be realized in the user space.

The ADCs and DAC integrated into the microcontroller can be controlled by the developed application for data processing. The communication between the driver and the program is realized by the system calls allowing to reset the microcontroller, initialize sampling and read transferred converters buffers. The application contains two threads realizing data readout and processing, and UDP server implementation. The read buffers are averaged using 4096 samples before the impedance calculation. The calculated value is transferred in the UDP datagrams to the data visualizing application.

C. Data visualizing application

The constructed system for impedance measurements required the correctness verification method. For the acquired data visualization, the graphical user interface (GUI) based application was developed. The communication between the mentioned program and Raspberry Pi 3 was realized by the UDP datagrams exchange.

D. Electronics frontend for impedance measurement

The impedance measurements require measuring signals with amplitudes symmetrical about the reference voltage. Because the microcontroller converters operating voltages belong to the range ground - supply voltage, conditioning circuit was built. It sets the reference voltage to half of the supply one, as shown in Fig. 3.

In the constructed electronics frontend, the excitation is generated by the microcontroller DAC and it is smoothed by the buffered low pass filter. The impedance measurement requires measuring the voltage at the object and the current flowing through it. Since ADCs integrated with microcontroller are voltage measuring ones, the interesting current had to be converted into voltage. During experiments for the impedance calculations, voltages measured on the object and on the current-to-voltage converters were used.

III. KERNEL DRIVER FOR ADCs AND DAC

Linux is the open-source operating system allowing programmers to analyze its sources and adjust them to raise the performance of their applications. Installation of that operating system on an embedded device frees the user from the necessity of implementation of some kinds of libraries, such as related to file access ones or network stacks. It allows to divide complex applications into many smaller ones, implements its inter-process communication and provides mechanisms necessary to processes synchronization.

The important stage in the software architecture design is the separation of the hardware-interfacing features from the data processing ones. Linux provides separate areas of virtual memories for the drivers and applications - kernel and user spaces. Implementation of drivers in kernel space allows us to take access to hardware and implement event-based data acquisition, triggered, for example, by general-purpose input/output (GPIO) interrupts.

In the developed system, the communication between Raspberry Pi 3 and the microcontroller is initiated and controlled by the first one. The microcontroller, as the SPI bus slave, notifies Raspberry Pi about buffer readiness through the dedicated digital line. The occurrence of the rising signal edge at the mentioned line triggers an interrupt and initializes data readout and processing (see Fig. 4).

Due to the necessity of delegating time-consuming tasks from the interrupt service routines (ISRs) to the less privileged ones, the microcontroller buffer processing was partitioned. The code executed in the interrupt context places buffer readout in the dedicated work queue and informs the operating system about interrupt handling, as shown on Listing 1. The created work queue performs the microcontroller buffer readout in the kernel process context.

```
static irqreturn_t stm32ai_irq_handler(int irq, void *dev_id)
{
    (void)irq;
    struct stm32ai *stm32ai = (struct stm32ai *)dev_id;

    queue_work(stm32ai->workqueue, &stm32ai->work);
    return IRQ_HANDLED;
}
```

Listing 1. The GPIO interrupt service routine.

Because of the Linux kernel preemptive scheduling, the proper synchronization of driver threads had to be developed. The simultaneously executed microcontroller buffer readout and the previous buffer copying to the user space could cause data corruption. The risk of that was eliminated by wrapping out of the buffer readout with mutex locking before transmission and unlocking it after that. The definition of the work handler is shown on Listing 2.

```
static void stm32ai_work_handler(struct work_struct *ws)
{
    struct stm32ai *stm32ai = container_of(ws, struct stm32ai,
                                           work);

    mutex_lock(&stm32ai->mutex);
    spi_read(stm32ai->dev.spi, stm32ai->buffer, BUFFER_LENGTH);
    stm32ai->buffer_ready = true;
    mutex_unlock(&stm32ai->mutex);

    wake_up_interruptible(&stm32ai->waitqueue_head);
}
```

Listing 2. The microcontroller buffer reading work.

The communication between the kernel and user space is realized by the system calls executed on the character device file representing the developed driver. Due to the relatively long microcontroller buffer readout time, not properly defined read system call can cause processor time-wasting. The problem can occur when user space thread actively waits for data readiness. It was mitigated in the developed driver by putting user space thread in sleep while waiting for data. The thread is woken up in the readout work queue after the transfer completion.

IV. RESULTS

A. CPU usage measurements

The developed kernel module for the external converters control is adapted to be loaded during the operating system boot process. It was designed to not consume processor time while conversions are not performed. The first experiment confirmed that the CPU is utilized only while conversions are performed. The test was divided into two stages, during each the processor time utilization was measured:

1) the driver is loaded and sampling is not performed,
2) the driver is loaded and converters perform sampling.

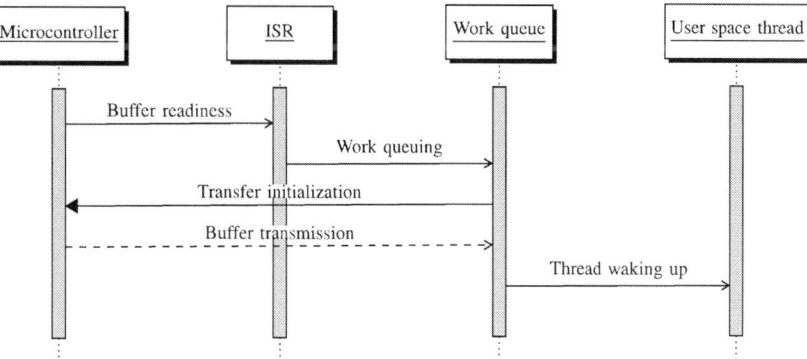

Fig. 4. The interrupt service routine processing sequence.

During the experiment, the microcontroller converters were sampling with a frequency equal to 256 kHz. The developed driver had to read through the SPI bus and copy to the user space about 1 MB of data each second. The results of the performed experiment are presented in Table I.

TABLE I
PROCESSOR TIME UTILIZED BY THE DEVELOPED DRIVER.

Stage	CPU usage [%]	
	Buffer readout	Data copying to user space
Driver loaded (1)	0.0	—
Sampling performed (2)	0.7	3.3

The experiment showed that the developed driver does not impact on the processor utilization while conversions are not performed. Because data processing is triggered by the GPIO interrupt, it can be safely loaded into the kernel during the device booting process. The CPU utilization measured while conversions were performed was relatively low (4 %) and it could be furtherly reduced by the mmap system call implementation.

B. Impedance measurements

The constructed electronics frontend (see Fig. 3) can be used for the Z_x impedance measurements. The data acquired by the designed system is sampled with a frequency equal to 256 kHz and it is averaged using 4096 samples before the impedance calculation. The constructed device can perform 62.5 measurements each second.

The second experiment confirmed that the developed driver can be used in the impedance measuring system. It was divided into three stages, during each impedance was measured:

1) measurement of the reference coil,
2) measurement of the reference coil with an object placed in its neighborhood,
3) measurement of the reference coil.

The third stage was the repetition of the first one done for better resistance (R) and reactance (X) changes visualization (see Fig. 5).

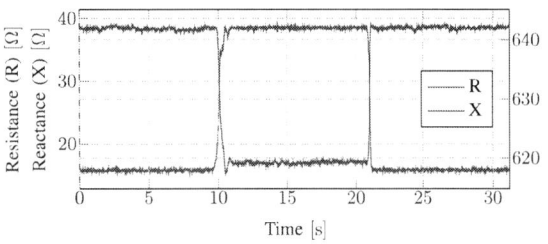

Fig. 5. The coil impedance during temporary presence of the steel sheet.

The impedance values measured during the temporary presence of the steel sheet were used for the system parameters analysis. The resistance and reactance values acquired in period 14 - 16 s were used for the noise root mean square (RMS) calculation. For the signal-to-noise ratio calculation, values measured in periods 4 - 6 s and 14 - 16 s were averaged and chosen as the signal lowest and highest values respectively. The results of the calculations are presented in Table II.

TABLE II
THE CALCULATED VALUES OF THE RMS NOISE AND SNR FOR THE COIL IMPEDANCE MEASUREMENTS

	Noise RMS [Ω]	Signal amplitude [Ω]	SNR
Resistance	0.18	22.75	129.48
Reactance	0.23	22.69	99.79

During the experiment results analysis, it was noticed that the measured values contained the constant offset. This observation was used for the measurements errors mitigation and can be easily applied to the constructed system by the calibration process modification. The impedances calculated after subtracting the mentioned offset are summarized with ones measured by the Agilent E4980A multimeter in Table III.

TABLE III
SUMMARISON OF THE REFERENCE COIL IMPEDANCE MEASUREMENTS.

Placed object type	Prototype		Agilent E4980A	
	R [Ω]	X [Ω]	R [Ω]	X [Ω]
—	12.7	630.8	12.7	630.8
Steel sheet	35.4	607.8	35.1	609.6
Aluminum sheet	16.7	568.5	16.8	568.9
Copper sheet	38.6	597.0	39.7	596.1

The impedances measured by the Agilent E4980A multimeter was used as references for the constructed system uncertainties estimation. The calculated absolute (Δ) and relative ($\delta_\%$) errors are presented in Table IV.

TABLE IV
ERRORS OF MEASUREMENTS MADE USING THE CONSTRUCTED SYSTEM.

Placed object type	Resistance		Reactance	
	Δ [Ω]	$\delta_\%$	Δ [Ω]	$\delta_\%$
—	0.00	0.00	0.00	0.00
Steel sheet	0.30	0.85	-1.83	-0.30
Aluminum sheet	-0.08	-0.48	-0.42	-0.07
Copper sheet	-1.08	-2.72	0.85	0.14
Mean of the absolute error values	0.49	1.35	1.03	0.17

The experiment confirmed that the developed driver can be used in the impedance measuring systems. The values of the noise RMS and SNR were calculated for the resistance and reactance separately. Measured values were summarized with measurements done with the Agilent E4980A multimeter usage.

V. CONCLUSIONS

In this work, the Linux kernel driver for external analog-to-digital and digital-to-analog converters was presented. The conducted experiments confirmed that it can be used in precision requiring applications like impedance measuring systems. The acquired data were compared with ones measured by the Agilent E4980A multimeter and the results were used for the proposed architecture uncertainty estimation. The observed differences were equal to 0.49 Ω and 1.03 Ω for resistance and reactance respectively, which suggests that the constructed system can be used in applications not requiring uncertainties smaller than 1 Ω, as a low-cost measuring device. The presented driver can be further improved by its adaptation to devices with integrated DAC and ADCs or by the mmap system call implementation.

REFERENCES

[1] M. Donno, K. Tange and N.Dragoni, *Foundations and Evolution of Modern Computing Paradigms: Cloud, IoT, Edge, and Fog*, IEEE Access, 10.1109/ACCESS.2019.2947652, Oct. 2019.

[2] C. Avasalcai, C. Tsigkanos and S.Dustdar, *Decentralized Resource Auctioning for Latency-Sensitive Edge Computing*, IEEE International Conference on Edge Computing (EDGE), 2019.

[3] J. Ren, G. Yu, Y. He and G.Li, *Collaborative Cloud and Edge Computing for Latency Minimization*, IEEE TRANSACTIONS ON VEHICULAR TECHNOLOGY, VOL. 68, NO. 5, MAY 2019.

[4] J. Sendorek, T. Szydlo, M. Windak and R.Brzoza-Woch, *Collaborative Cloud and Edge Computing for Latency MinimizationFogFlow - Computation Organization for Heterogeneous Fog Computing Environments*, International Conference on Computational Science, pp 634-647, 2019.

[5] P. Skrzypiec *Development and testing of the embedded operating system kernel module for handling fast A/D and D/A processing*, M.Sc. thesis. AGH 2019.

[6] G. Ramm and H. Moser. *New multifrequency method for the determination of the dissipation factor of capacitors and of the time constant of resistors*, IEEE Transaction on Instrumentation and Measurement, vol. 54, no. 2, 2005, pp. 521–524.

[7] Z. Marszalek, *Maxwell-Wien bridge with vector voltmeter system for measurement small and rapid changes in inductive-loop sensor impedance components*, Measurement, vol. 121, 2018, pp. 57-61.

[8] B. Sanchez and R. Bragos, *Multifrequency simultaneous bioimpedance measurements using multitone burst signals for dynamic tissue characterization*, Journal of Physics: Conference Series, vol. 224, 2010, pp. 1–4.

[9] Z. Marszalek, K. Duda, and P. Turcza. *Design and Experimental Validation of Multi-Frequency Impedance Measurement System*, International Conference on Signals and Electronic Systems, 2018.

[10] Keysight Technologies, *E4980A Precision LCR Meter*.

Multipoint Wireless Humidity and Temperature Monitoring Network for HVAC Systems Validation

Michał Zbieć, Dariusz Obrębski

ICs and Systems Design Department

Sieć Badawcza Łukasiewicz — Instytut Technologii Elektronowej

Warsaw, Poland

E-mails: zbiec@ite.waw.pl, obrebski@ite.waw.pl

Abstract—**This paper describes multipoint wireless humidity and temperature measurement network designed for validation of HVAC systems performance. It consists of main data logger module and multiple measurement nodes connected via wireless link in 868 MHz ISM band. It characterizes temperature and humidity distribution within measured residential or commercial spaces.**

Keywords—**Multipoint measurements, IoT, Embedded systems, Wireless communication**

I. INTRODUCTION

Heating, ventilation and air conditioning (HVAC) systems are not only widely used in big scale commercial buildings, but nowadays more and more often in residential ones too. Their proper operation influences both mood and health of people working and living in climate controlled environment, therefore it is crucial that HVAC system works properly, as designed, not only right after system launch, but through the whole exploitation period. Manual measurements, with ordinary thermometer, or hygrometer, are not sufficient to establish if temperature distribution and air humidity is consistent in large scale living or working space. They only provide local (from one point of measurement) data from very limited time frame. To properly characterise the HVAC system data from multiple points and long observation time are needed. Periodic measurements lasting few days can reveal instabilities in the system, chart influence of daily variations of external environment etc. They are also able to confirm the performance problems of HVAC installation, that can periodically appear due to the excessive sun operation during summer and low outside temperature in winter. In complicated big scale systems it is also important to measure climate conditions in multiple points. Overall performance of such a system is derived from proper behavior of many HVAC devices. Only by gathering data from multiple sensors located in different parts of controlled environment one can ensure proper behavior of every device involved in the system and, in case of problems, localize faulty ones. Such a wide HVAC system characterization can be useful both during system installation, when technicians need to validate if the system has been installed and configured properly, and during exploitation, when it can help user in detection where the problems with HVAC system are located.

Multipoint wireless humidity and temperature monitoring network was developed to answer the above needs. Key concepts in its designing were:

Data logger module - main device of the system. It is responsible for tasks like: measurement scheduling (triggering sensor nodes in correct time slots), data acquisition and logging (saving received data to the SD card), user interface, local measurements (measurements with sensor included in data logger module) and system wide settings management.

Sensor nodes - system measurement nodes. Their main role is to carry out measurements (when triggered) and sending them back to the data logger module. Sensor nodes are meant to be dispersed in monitored space and provide data from different locations within it.

Wireless communication - data logger module and sensor nodes are connected together via wireless link working in 868 MHz ISM band. Data logger uses it to schedule measurements, transfer data from sensor nodes, monitor state of their batteries and manage system wide settings.

User interface - data logger provides user interface with 240x128 graphic display and 4x4 keypad. It supports following features: file management (creating, deleting files and assigning them to sensors), sensor management (adding and deleting sensors within the network), current measurements visualization (timestamp, temperature, humidity and state of sensor battery), logged measurements visualization (as a list or graph), system settings (time and data, measurement period, node output power, display brightness).

Battery operation - sensor nodes should be easily settled in various temporary locations, where it may by impossible or impractical to connect them to the power grid. Having that in mind both data logger module and sensor nodes were designed to support battery operation. They are powered from single cell Li-Ion battery and utilize robust battery charging and monitoring system. Current battery state of the sensor nodes is available to data logger via wireless link.

Low power consumption - since the system supports battery operation mode it was crucial to decrease its power consumption to absolute minimum. The lower power consumption the longer devices can operate without recharging. Various software (RF module listen mode, microcontroller sleep mode) and hardware (SMPS burst mode) techniques were used to satisfy this simple statement.

II. HARDWARE

A. Data Logger Module

Data logger module was designed as a complete battery powered data acquisition and logging wireless device. Its architecture can be found in Figure 1. Whole device is presented in Figure 2. The data logger module bases on the ATMega128 (8-bits, Microchip AVR family) MCU, and features the RFM69 868MHz ISM transceiver from Hope-RF for bidirectional communication with slave modules. The RF module operates with on-board planar inverted-F antenna designed according to the guidelines from [1]. Since the data logger is a handheld instrument, a special attention was paid to locate the RF module and antenna as far as possible from the areas typically used to hold the device.

One of the main prerequisites for the data logger module was the large and removable data storage allowing the long lasting (few days) registration of environmental conditions in several points of the building. For that purpose, the microSD card was chosen as commonly used data carrier. This memory card formatted with typical file system enables easy transfer of gathered data to any computer for documentation and archiving purpose, however the data logger module also supports the standalone operation. The end-user - e.g. HVAC installer or serviceman - is able to browse and look through the gathered measurement series and even display them directly on that device in a form of a graph. For that purpose the relatively large graphical LCD - DOGL240-7 [2] from Electronic Assembly was deployed together with standard numeric and navigation membrane keypad. This highly integrated display device can operate within few variants of the SPI or I2C modes. In proposed application the LCD is connected to the I2C bus operating as a skeleton for internal communication of the data logger. In this mode system developer can configure two bits of LCD controller I2C slave address to avoid collision with another device. The developed module can be also deployed as a data source for PC or another type of controller supporting the USB connection. In this mode the communications is carried out via the standard FT232 UART to USB interface which is used to send periodically the values read from each particular sensor. The measurement cycle is synchronized by a standard PCF8583 RTC, featuring its own backup battery. The USB connection cable is also used for charging the internal data logger battery. For standalone operation data logger module features its own, buil-in integrated relative humidity and temperature sensor. The SHT25 from Sensirion was deployed, as already proven by the designers for construction of the multi-point wired sensor network coupled within the BMS structure, described in [3].

As a portable device, the data logger module was designed with a special attention payed to its power distribution and management system. The main power supply features the LTC3440 SMPS controller [4] form Linear Technology operating in buck-boots topology, very effective in applications when the output voltage is within the battery voltage range. Having its output stage designed in a bridge topology such a

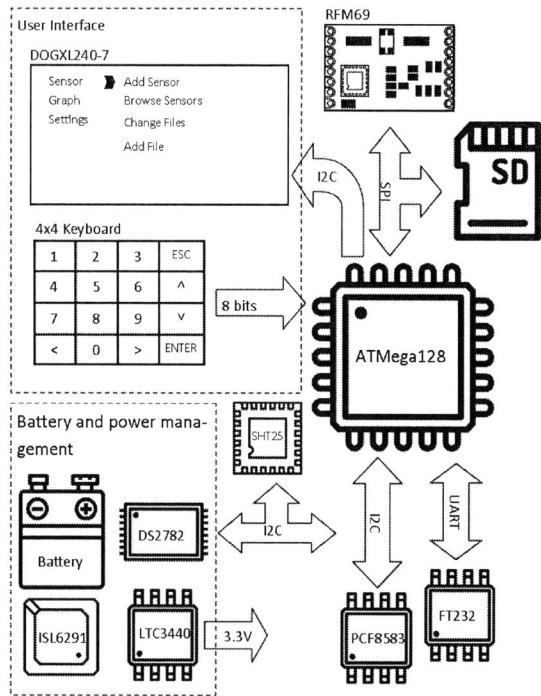

Fig. 1. The logger module block diagram. Icons from www.flaticon.com

converter can operate in buck mode whenever the input voltage is higher than the output one - i.e. in case of fully charged battery and can directly switch to the boost operating mode when the battery voltage drops below desired output level due to the discharging process. Deployment of the synchronous rectification technique guarantees the high energy efficiency, ranging to 96 %, the high operating frequency - up to 2 MHz allows to use smaller sizes inductors and filter capacitors. In proposed solution the converter based on LTC3440 produces the main supply of 3.3V for microcontroller, display, RTC and the RF communication module. Another switching-mode converter - CAT4238 [5] from ON Semiconductor, operating in boost topology is used as a high-efficiency power source for a chain of 9 white LEDs in the LCD backlight. The converter operation is directly controlled by the MCU - in that way the LCD backlight can be totally switched off to avoid its considerable contribution to overall power consumption in the idle state of the data logger, or its operation can be adjusted by means of PWM signal according to the ambient light intensity, measured by a photoresistor built-in the module and applied to one channel of the ADC embedded in the MCU. The main supply source of the system - a single cell Li-Ion battery is charged from the 5V USB bus voltage by a circuitry based on ISL6291 form Renesas [6]. This chip, dedicated to handling of a single cell with minimum external component count, supports safety features like testing the battery with low current for fast charge ability, as well as

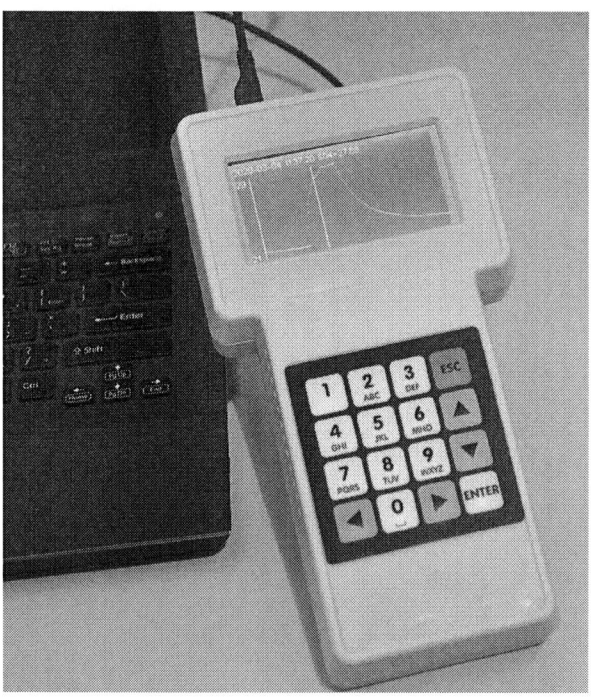

Fig. 2. Functioning data logger module

of the MCU. This allows to define multiple slave devices using only a single microcontroller port - the number of such nodes is limited by the ADC accuracy. The resistor network is switched on for a limited amount of time after power-up, to decrease the power consumption.

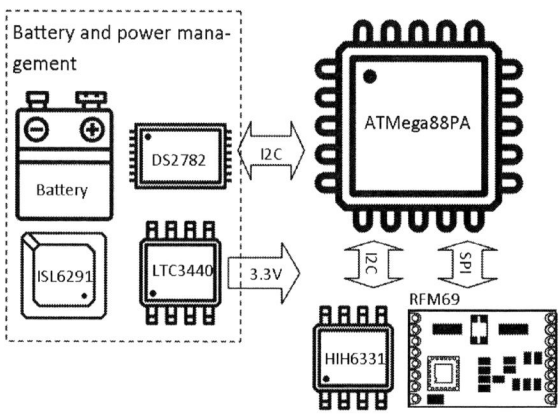

Fig. 3. The sensor node block diagram. Icons from www.flaticon.com

supervision of its power switch and battery temperature. Its important advantage is very low current consumption from the battery and elimination of reverse blocking diode from the application circuit. The battery charge and discharge process is monitored by DS2782 [7] - the fuel gauge chip from Maxim Semiconductors. Basing on the measurements of a current flow and cell voltage the chip provides very accurate estimates of the current capacity and remaining percentage of the full charge. The battery current measurement circuitry incorporated inside the DS2782 deploys the standard technique based on serial resistor connected within the ground path. The sensing resistor can be adjusted according to the application needs, its value and thermal coefficient are defined in specific registers of DS2782. The fuel gauge circuit is connected to the I2C bus skeleton. The switching mode converters were placed at the opposite site of the system PCB and as far as possible from the RF circuitry.

B. Sensor Node

Sensor node shares many design solutions with data logger module. Its architecture can be found in Figure 3. PCB of the sensor node is presented in Figure 4. The main difference in respect to the data logger module is its user interface limited to a dual-color LED only. The senor node is just expected to gather temperature and humidity readouts in specific location and send them via the short distance radio. The LED signals the beginning of measurement (blue) or battery checking cycle (yellow). The sensor node ID is programmed in hardware, by means of resistor network connected to the ADC channel

Fig. 4. The sensor node PCB.

This module also incorporates another type of sensor - the HIH6331 form Honeywell was deployed instead of SHT25. The sensor is assembled on a narrow part of the PCB separated from the rest by a molded slots (see Figure 4). This solution was used to overcome the thermal bridging problems between sensor and power circuits (battery loader and SMPS) of the device. The power supply and management system bases on the same components as in data logger module, however in the sensor node, the burst mode of LTC3440 was activated. In this mode the converter enables its power stage until the output voltage is regulated, then goes into a sleep mode when its power consumption is reduced to 25 μA. Although the peak efficiency of the converter in burst mode is lower than in fixed frequency one, this kind of operation maximizes the battery lifetime in those applications, when a device powered from the converter remains inactive for a majority of its working time, as it happens for mentioned sensor module. The SMPS is located as far as possible from the RF module and inverted F antenna.

III. SOFTWARE

Both types of devices, that together create the monitoring network, include a microcontroller (MCU) as a central control unit. With that in mind two versions of embedded software had to be designed. For data logger module the ATMega128 MCU [8] was chosen due to its large FLASH, and data memory. Larger control unit was necessary due to the wide functionality required from this module, but also due to large size of compiled library code used to interact with SD card (FatFs [9]). For sensor node the ATMega88PA MCU [10] was chosen. Since smaller FLASH size was not an issue (less functionality, no SD card) significantly smaller IC with lower power consumption was a better fit.

As described by NXP [11] "Sensor fusion is a process by which data from several different sensors are *fused* to compute something more than could be determined by any one sensor alone." It is a powerful technique employed when data from multiple sensors is used to drive some actuator devices, or fed to decision making algorithm. System described in this paper has, for now, only data logging and visualization capabilities. It is up to the user to draw conclusions from the gathered data. For that reason, currently, there is no need to use sensor fusion algorithms.

A. Sensor Node Software

Main role of the sensor node in the network is to carry out, when triggered by data logger module, measurements and send them back via the RF connection. Beside that embedded program supports also battery state monitoring feature and output power adjustment on demand. Due to the battery operating nature of the sensor node power consumption reduction was a key factor during development of this software. One of the steps towards that goal was implementation of the MCU power down between wireless trigger messages. However the most significant power consumption reduction was done by utilizing Listen Mode of the RFM69 [12]. During normal operation RF module waits for incoming communication in RX Mode, which requires 16 mA of supply current. With Listen Mode properly configured and enabled RFM69 spends most of the time in Idle Mode, during which only the RC oscillator runs. Periodically (each 16.3 ms) the module enters RX mode and listens for an RF signal. If power of incoming signal exceeds defined threshold the RX mode is kept on and the data is collected and processed. Otherwise, after 960 μs, the module returns to its Idle mode. The simplified timing diagram is illustrated in Figure 5.

Designed software follows flow chart presented in Figure 6.

After the power-up the initialization of node components (I2C, SPI, RFM69, HIH6331, node address) is carried out. Next, in order to reduce power consumption, RFM69 module

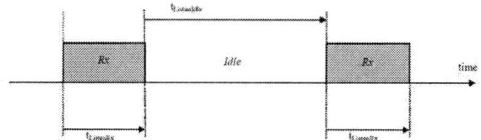

Fig. 5. Simplified timing diagram of Listen Mode [12]

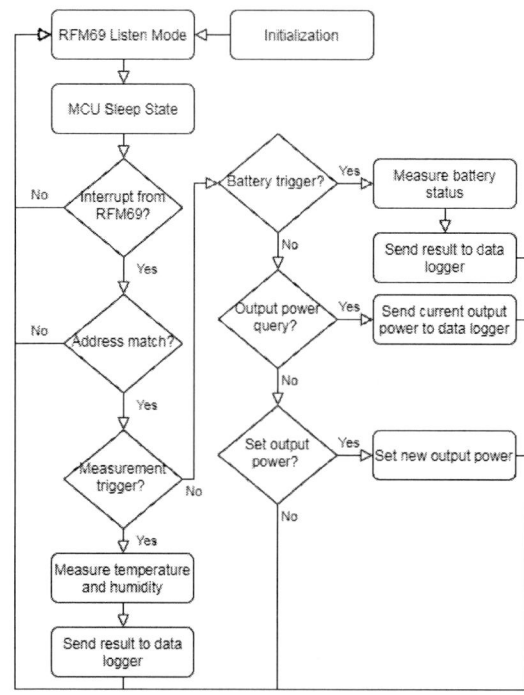

Fig. 6. The sensor node operation flow chart.

is put into listen mode and MCU to Sleep Mode. In that state sensor module waits for trigger message from data logger module. If any valid wireless communication occurs (input power on RFM69s receiver exceeds defined threshold) the RFM69 leaves the Listen Mode and compares first byte of incoming data with node address. Only in case of matching address rest of the received data can be handled. Otherwise communications is discarded, wireless module returns to Listen Mode and MCU to Sleep Mode. Four trigger messages (presented in Table I) has been defined.

TABLE I
VALID TRIGGER MESSAGES

Trigger Message	Additional Data	Description
Power	1 byte (new output power)	Output power (of RFM69) change trigger
Test	NONE	Output power(of RFM69) query
Trigg	NONE	Temperature and Humidity trigger message
Batt	NONE	Battery state measurement message

After successful trigger message reception proper action is carried out and, if necessary, data is sent back to data logger module. Lastly sensor node puts the RFM69 module in Listen Mode and MCU in Sleep Mode and waits for another trigger message.

B. Data Logger Module Software

Most important responsibilities of the data logger module, within this system, are: providing user interface, measurement scheduling and triggering, data acquisition and logging. After the power-up data logger undergoes initialization of: UART bus, DS2782 fuel gauge IC, battery control subsystem, MCU ports, PCF8583 RTC IC, RFM69 wireless module, keypad, LCD, microSD card, system variables from EEPROM, date and time, interrupt system, global flags. Most of functionality provided by this program results from cooperation of following software subsystems:

Micro SD card subsystem - as it was mentioned earlier, a microSD card subsystem utilizes FatFs library, which provides easy FAT filesystem access. It supports standard file manipulation operations like creating new one, editing, or deleting existing one and browsing content of current directory. Files on the microSD card have standardized name convention: YYMMDDAA.csv where YY, MM, DD stands for, respectively year (last two digits), month and day of file creation date. Finally AA stands for data source node address (00 for data logger local data). For example file created on 09.12.2019 that contains data from sensor node with address 01 would be called 19120901.csv. Measurement data is stored in comma separated format in order to make it easily importable into spreadsheet editors like MS Excel. Each measurement is recorded as 34 byte line in a file. For example three consecutive measurements taken in 5 second intervals would end up as following records:

2019-12-06 14:47:40,021.35,030.97
2019-12-06 14:47:45,021.36,030.97
2019-12-06 14:47:50,021.36,030.97

Standardized line length of each record simplifies file structure and data processing in embedded software environment.

Scheduling subsystem - scheduling subsystem controls event occurrence within the data logger module. It tracks date and time basing on interrupts triggered by 1Hz synchronization signal from PCF8583 RTC IC and synchronization procedure executed once a day. Based on a state of global variables it triggers system events (presented in Table II) that control workflow of the network.

RF subsystem - basic behaviour of this subsystem is complementary to that described in Sensor Node Software section. It also provides RF link and utilizes Listen Mode in order to reduce power consumption of whole device. When triggered by scheduling subsystem it issues one of described in Table I messages and expects one of defined in Table III responses (waits for it for number of seconds defined by TIMEOUT_SEC macro).

TABLE II
DEFINED SET OF EVENTS

Flag	Event
dateUpdate	Date change event. It triggers,once a day, date and time synchronization with PCF8583 IC
timeUpdate	Time change event. It triggers, each minute, time update on main screen
batteryUpdate	Battery check event. Triggers sequentially both for data logger and each of sensor nodes. Period of this events is controlled by BATTERY_UPDATE_PERIOD macro
txFlag	Humidity and Temperature measurement event. Triggers sequentially both in data logger and each of sensor nodes. Period is set in Settings menu and stored in EEPROM
timeoutFlag	Timeout event that happens after sensor node fails to answer with valid response within set by TIMEOUT_SEC macro time frame
backlightCounter	Backlight shutdown event. After set in BACKLIGHT_LIMIT macro nuber of seconds backlight is turned off to reduce modules power consumption

TABLE III
VALID RESPONSE MESSAGES

Response Message	Additional Data	Description
Data	4 bytes	Response message with measurement results from triggered sensor node
Batt	4 bytes	Response message with battery voltage and current
Alive	1 byte	Response message with current RF module output power

User interface subsystem - there are two parts to the user interface: 4x4 generic keypad (source of user inputs) and DOGXL240 graphic display (system main user output based on UC1611s [13] controller). Two sets of fixed and variable width fonts were defined, as well as set of utility functions for writing on the screen (from start of the line, with offset or in the center). DOGXL240 integrates RAM capable of storing 4 complete display contents. Whole RAM was used in order to reduce slow screen-wide clean operation. Four screen patterns were defined each in separate part of RAM memory: PATTERN_MENU (screen containing current state of menu), PATTERN_GRAPH (graphs of logged measure-ment data), PATTERN_MAIN (current time and measurement results), PATTERN_OTHER (multi purpose). After the data logger module finishes initialization the main screen (PAT-TERN_MAIN) is displayed. When any key is pressed user interface displays a menu screen. User menu was divided into three main items, each containing additional sub items. Detailed menu content is presented in Table IV. Sample user interface screen (with temperature graph) is presented in Figure 7. Graph contains visual representation of data (data points) and cursor with text measurement representation.

Described menu items are a result of the state machine implementation. Each menu screen is defined as a set of states implementing its functionalities. Complete set of menu states was presented in Table V.

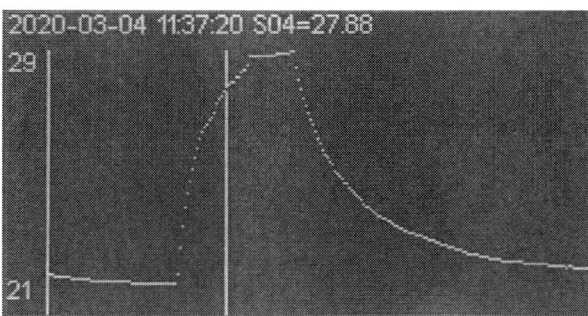

Fig. 7. Temperature graph generated by user interface subsystem on the display.

TABLE IV
AVAILABLE MENU ITEMS

Main menu	Sub menu	Description
SENSORS	Add Sensor	Adding new senor node to the system
	Browse Sensors	Browsing and managing system sensors
	Change Files	Matching sensor nodes with files
	New file	Creating new file for sensor node
GRAPH	Temperature	Single sensor node temperature graph
	Humidity	Single sensor node humidity graph
	Mixed Temp	Temperature graph for two data sets
	Mixed Hum	Humidity graph for two data sets
SETTINGS	Time	Setting of time and date
	Period	Measurement period setting
	Out Power	Data logger or sensor node output power setting
	Brightness	Display brightness setting

TABLE V
MENU STATES

State Name	Description	Pattern
IDLE	Main measurement state, menu turned off	PATTERN_MAIN
MENU0	Main menu state	PATTERN_MENU
MENU_SENSORS	Sensors sub menu state	PATTERN_MENU
ADD_SENSOR	Setting address of new sensor node state	PATTERN_OTHER
BROWSE _SENSORS	Selecting sensor node intermediate state	PATTERN_OTHER
DIRECTORY _MANAGEMENT	File selection state	PATTERN_OTHER
SAMPLES	Measurement data browsing state (in text)	PATTERN_OTHER
MENU _SETTINGS	Settings sub menu state	PATTERN_MENU
MENU_GRAPH	Graph sub menu state	PATTERN_MENU
GRAPH	Single data set graph state	PATTERN_GRAPH
GRAPH_MIX	Double data set graph state	PATTERN_GRAPH
SET_YEAR, SET_MONTH, SET_DAY, SET_HOUR, SET_MIN, SET_SEC	Date and Time setting state	PATTERN_OTHER
SET_PERIOD, SET_HOUR_P, SET_MIN_P, SET_SEC_P	Measurement period setting state	PATTERN_OTHER
SET_POWER	Output power of data logger and sensor nodes setting state	PATTERN_OTHER
SET_BRIGHTNESS	Display brightness setting state	PATTERN_OTHER

IV. CONCLUSION

The complete multipoint wireless humidity and temperature monitoring network for HVAC systems validation was developed. It provides easy way of monitoring proper temperature and humidity distribution over extended periods of time. Since it is fully battery operated (in addition to working from constant power supply) special attention was payed to lowering power consumption of each component, thus extending their battery life.

REFERENCES

[1] Fredrik Kervel, Texas Instruments. (2011) Design Note DN024 - 868 MHz, 915 MHz and 955 MHz Inverted F Antenna. [Online]. Available: http://www.ti.com/lit/an/swra228c/swra228c.pdf

[2] ELECTRONIC ASSEBLY. (2014) DOGXL240-7 GRAPHIC. [Online]. Available: https://www.lcd-module.com/fileadmin/eng/pdf/grafik/dogxl240-7e.pdf

[3] D. Obrebski and M. Zbiec, "Development of the sensor network for building technologies," in *24th International Conference Mixed Design of Integrated Circuits and Systems, MIXDES 2017, Bydgoszcz, Poland, June 22-24, 2017*. IEEE, 2017, pp. 558–563. [Online]. Available: https://doi.org/10.23919/MIXDES.2017.8005275

[4] Linear Technology Corporation. (2001) LTC3440 - Micropower Synchronous Buck-Boost DC/DC Converter. [Online]. Available: https://www.analog.com/media/en/technical-documentation/datasheets/3440fd.pdf

[5] ON Semiconductor. (2016) CAT4238 - High Efficiency 10 LED Boost Converter. [Online]. Available: https://www.onsemi.com/pub/Collateral/CAT4238-D.PDF

[6] Renesas. (2005) Li-ion/Li Polymer Linear Battery Charger. [Online]. Available: https://www.renesas.com/eu/en/www/doc/datasheet/isl6291.pdf

[7] Maxim Integrated Products. (2009) DS2782 Stand-Alone Fuel Gauge IC. [Online]. Available: https://datasheets.maximintegrated.com/en/ds/DS2782.pdf

[8] ATMEL. (2011) ATmega128(L) - Complete Datasheet. [Online]. Available: http://ww1.microchip.com/downloads/en/DeviceDoc/doc2467.pdf

[9] CHaN. (2017) FatFs - Generic FAT Filesystem Module. [Online]. Available: http://elm-chan.org/fsw/ff/00index_e.html

[10] ATMEL. (2014) ATmega48PA/ATmega88PA/ATmega168PA Automotive - Complete Datasheet. [Online]. Available: http://ww1.microchip.com/downloads/en/DeviceDoc/Atmel-9223-Automotive-Microcontrollers-ATmega48PA-ATmega88PA-ATmega168PA_Datasheet.pdf

[11] NXP. (2020) NXP Sensor Fusion. [Online]. Available: https://www.nxp.com/design/sensor-developer-resources/nxp-sensor-fusion:XTRSICSNSTLBOXX

[12] HOPERF. (2006) RFM69HW ISM TRANSCEIVER MODULE V1.3. [Online]. Available: https://www.hoperf.com/data/upload/portal/20190306/RFM69HW-V1.3%20Datasheet.pdf

[13] ULTRA CHIP. (2008) 160COM x 256SEG Matrix LCD Controller-Driver. [Online]. Available: https://www.lcd-module.com/fileadmin/eng/pdf/grafik/dogxl240-7e.pdf

Performance Analysis of Convolutional Neural Networks on Embedded Systems

Łukasz Grzymkowski, Tomasz P. Stefański
Faculty of Electronics, Telecommunications and Informatics
Gdańsk University of Technology, Gdańsk, Poland
emails: lukegrzym@gmail.com, tomstefa@pg.edu.pl

Abstract—**Machine learning is no longer confined to cloud and high-end server systems and has been successfully deployed on devices that are part of Internet of Things. This paper presents the analysis of performance of convolutional neural networks deployed on an ARM microcontroller. Inference time is measured for different core frequencies, with and without DSP instructions and disabled access to cache. Networks use both real-valued and complex-valued tensors and are tested using different inference engines. We conclude that the system must be tuned in a holistic way to achieve optimal efficiency.**

Keywords—**embedded systems, deep learning, edge computing, machine learning.**

I. INTRODUCTION

The number of applications, where machine learning and neural networks (NNs) have been successfully used, is constantly growing. Examples include computer vision for object classification [1], image segmentation [2] and object recognition [3],[4]. Other are focused on speech recognition [5], natural language processing [6], sensor data analysis [7] and predictive maintenance [8], [9].

Training NNs usually requires vast amount of data and takes place in cloud, data centers and servers with powerful computing resources, e.g., with graphics processing units (GPUs). For inference, however, the requirements on processing are not necessarily as high and NNs are often deployed on devices in Internet of Things (IoT). These are embedded systems with hardware and software built for a specific function (e.g., to drive a DC motor, collect sensory data), often with real-time response requirements. IoT systems are deployed in various areas, for example in cities, buildings, equipped with connectivity modules and are often battery-powered. Over the recent years, there have been significant improvements in hardware capabilities, software maturity and level of integration, while the cost has been going down. These have sparked interest in using IoT systems to run NN inference with computations run locally, without using a cloud.

Embedded systems have, however, specific requirements and constraints, especially when machine learning use cases are considered. One of the most important is the energy efficiency to prolong the battery life and maximize the duration for which the device can remain operational. Sophisticated power saving modes are supported with deep-sleep modes of operation, voltage and frequency scaling, dynamic disabling of peripherals and clocks. Many of such systems perform periodic operations, e.g., collect sensory data, process it and then switch

back to low-power mode for the majority of time. If the time to complete a task is long, however, then the duty cycle, i.e., percentage of time spent in high-power mode, is greater and so is the power consumption.

When using NN, the task may be to capture a frame from camera, run inference and send result with metadata to a cloud. To shorten inference time and duty cycle, it is necessary to either have high computing capabilities, e.g., higher frequency and system voltage, or to simplify the network using, e.g., fewer layers and lower depth. This can backfire with an increase in power consumption due to higher core frequency or produce a network with insufficient accuracy [10]. Methods of estimating energy consumption in machine learning and building efficient networks are actively investigated [21]. However, in the field of machine learning, researchers focus more often on model accuracy and the inference time over energy efficiency.

Other constraint of IoT systems is the dynamic memory. It is especially evident on microcontrollers that may only have SRAM available in hundreds of kilobytes inside the system-on-chip (SoC). This resource must be shared between all applications running on a target platform, including connectivity stack, operating system, drivers, data and I/O buffers and finally weights and intermediary buffers for NNs. An insufficient memory size is a hard-constraint on the depth and the size of used NNs. Additionally, memory access or cache latency have significant impact on the overall performance of the system. The cores or execution engines may become blocked until instructions are fetched or data is loaded.

Finally, the performance of the system is limited by its computing resources. These depend on the type and architecture of processing units used, e.g., a single or multi-core system, instruction set and parallelism, core frequency. IoT systems are, in contrast to a near infinitely scalable data center, always constrained and have a cap on maximum number of computations that can be executed. As mentioned before, embedded systems often run multiple applications simultaneously that share all resources. It is therefore important from the architectural perspective to view the system as a whole and understand the entire software stack when building such systems.

In this case, NN is simply one of the applications, hence, models should be designed with these constraints in mind. For instance, one must consider if there is enough memory available to run a large NN. With deeper networks, the capacity of NN to model data is higher leading to increased accuracy. However, is there enough processing power available to run these models without starving other tasks and draining the system battery too rapidly? The solution may be to build models with reduced number of operations and layers, so that the inference time is shorter. There are various on-going efforts to create such models, driven strongly by use cases from embedded and mobile platforms [11].

The purpose of this paper is to analyze how different constraints on computing resources, cache access (data, instruction) and the availability of DSP instructions operating on multiple data points at the same time, impact the performance of convolutional NNs (CNNs) deployed on an embedded system. We focus on relative comparisons between various CNN configurations on a single hardware platform. As demonstrated, the performance of NN depends on both processor core frequency and memory access, and these must be tuned together to obtain optimal results.

The paper has the following structure: We begin with a short introduction to real- and complex-valued CNNs, followed by a brief description of the benchmarking setup that outlines the hardware, inference engines and NNs used. Next, we present collected results for a series of benchmarks with analysis, and then move to conclusion and final remarks.

II. CNNs

A basic CNN structure is presented in Fig. 1. The convolution operation is used to extract features from input data, e.g., images in the considered case. A moving filter (kernel) is convolved with input data, i.e., the dot product of the kernel and input pixels is computed. All layers with parameters (weights of kernels for convolutional and fully connected layers) are followed by an activation function, i.e. a rectified linear unit (ReLU). Pooling layers are introduced after parametric layers that add spatial invariance by reducing the resolution of feature maps. Max and average pooling layers are the most commonly used.

NNs usually use real-valued input and kernel tensors. However, NNs can also be defined using complex arithmetic. Complex-valued NNs (CVNNs) use complex-valued tensors and can be considered as a generalization of real-valued NNs (RVNNs). The advantage of CVNN is that the phase structure of the input data is maintained, as well as circularity of signals [16]. As each complex number is represented by two real numbers, the number of parameters for CVNN is double that of RVNN for the same number of layers.

ReLU is a commonly used activation function for RVNNs [12] which is defined as

$$f(z) = max(0, z). \tag{1}$$

Initial approaches to CVNNs simply used real-valued activation functions separately on real and imaginary parts, which is referred to as a split activation function [18]. Another approach, used in this paper, is to only use the values when both real and imaginary parts are positive [15]

$$f(z) = \begin{cases} z & \text{if } \Re(z) \geq 0 \wedge \Im(z) \geq 0, \\ 0 & \text{otherwise} \end{cases}. \tag{2}$$

For a single patch of pixels, e.g., (2 x 2), the real-valued max pooling layer returns the maximum value [19]. Because the max operation is not defined for complex numbers, the max pooling operation must be generalized for CVNN. One approach is to take the maximum value of the magnitude $\max|z|$ from a patch of pixels. An alternative approach is to split the pooling operation by using the real-valued operations on real and imaginary parts separately. Another possibility is to use the softmax function [15], [17]. In this paper, we use the magnitude for complex-valued max pooling operation.

The convolution operation for RVNN is given by

$$X \cdot A = \sum_{ij} X_{ij} A_{ij} \tag{3}$$

where X and A are the input and kernel tensors, respectively. For CVNN, the arguments in the convolution operation are both defined in the complex domain. Hence, (3) takes the form

$$(X + jY) \cdot (A + jB) = X \cdot A - Y \cdot B + j(X \cdot B + Y \cdot A) \tag{4}$$

where pairs X, Y and A, B are real and imaginary parts of the input and kernel tensors, respectively.

III. BENCHMARKING SETUP

The purpose of benchmarks is to measure performance of CNNs on embedded systems under various conditions. In the presented benchmarks, a general-purpose central processing unit (CPU) is used with DSP instructions to accelerate execution of inference engines, i.e., software that implements NN kernels and operations. The benchmarks cover changing core frequency by switching to one of supported SoC run modes, disabling data and instruction cache, disabling DSP instructions support. On the software side, TensorFlow Lite and CMSIS NN are used as inference engines with CNN deployed with varying number of layers.

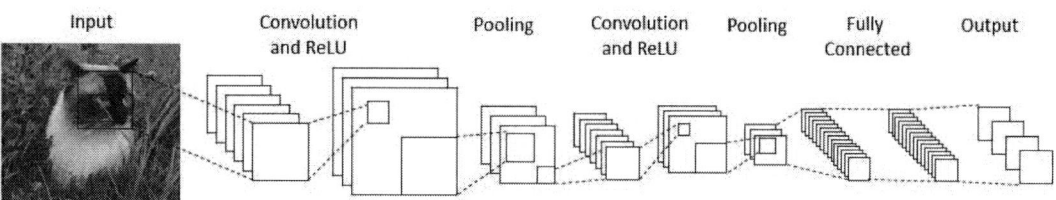

Fig. 1. Common topology of CNN.

A. System Under Test

Benchmarks are executed on an NXP i.MX RT1050 development board with ARM Cortex-M7 SoC with ARMv7E-M architecture (DSP instructions, floating-point unit). The SoC runs with up to 600MHz core frequency with 32kB instruction and data caches each. In addition, there are 512kB of low-latency on-chip TCM/SRAM and 32MB of slower, SPI-connected SDRAM memory. The latter is used for buffers for weights, intermediary results and images used as inputs to NNs. Code is compiled using GCC ARM.

i.MX RT1050 supports four run modes[20]:

- overdrive - core at 600MHz, SoC voltage 1.275V
- full speed - core at 528MHz, SoC voltage 1.15V
- low speed - core at 132MHz, SoC voltage 1.15V, system PLL powered up, other PLLs on request
- low power - core at 24MHz, SoC voltage 0.95V, all PLLs powered down, powered up on request.

In addition, SoC supports low-power modes used for standby and shutting down. Only run modes are tested.

B. CMSIS NN

CMSIS NN is an inference engine developed by ARM to accelerate NNs by utilizing DSP instructions on Cortex-M devices. The single-instruction multiple-data (SIMD) instructions, also referred to as the DSP instructions, are used to provide parallelism by operating on more than one data sample with each instruction. The system architecture is 32 bit. SIMD instructions can operate on either four 8-bit or two 16-bit samples simultaneously [13]. CMSIS NN is available as a library of NN kernels and operations. NNs layers are built by stacking function calls (convolution, pooling, etc.) and by passing as arguments pointers to buffers with I/O data and layer weights.

C. TensorFlow Lite

TensorFlow is an open-source, deep learning framework, developed primarily by Google. It supports building, training and using models for inference. TensorFlow Lite (TF Lite) is a stripped down version that supports only the forward-path, i.e., inferring on input data. TF Lite is used on mobile and embedded platforms, including Android phones. Recently, a version has been released that adds support for microcontrollers.

D. CNNs

A baseline CNN topology [13] used in benchmarks is outlined in Tab. I. In the benchmarks, the number following the CNN acronym, e.g., CNN3, CNN5, is the number of convolutional layers. Models are built by adding layers with indices 3-6 from the baseline model after the last convolutional layer and before the dense layer. The tests focus only on the inference time and the accuracy is not evaluated. The input image is a 32x32 colour image from the CIFAR-10 dataset [1].

Both RVNNs and CVNNs are tested. Most modern software frameworks focus primarily on RVNNs. For instance, CMSIS NN does not support CVNNs. To address this issue, kernels for RVNN are modified to support CVNNs as described in the previous section.

TABLE I.
BASELINE CNN3 WITH THREE CONVOLUTIONAL LAYERS.

Index	Layer Type	Filter Shape	Output Shape
1	Convolution1	5x5x3x32	32x32x32
2	MaxPooling1	N/A	16x16x32
3	Convolution2	5x5x32x32	16x16x32
4	MaxPooling2	N/A	8x8x32
5	Convolution3	5x5x32x64	8x8x64
6	MaxPooling3	N/A	4x4x64
7	Dense	4x4x64x10	10

TABLE II.
NON-VOLATILE (TEXT) AND VOLATILE MEMORY (BSS, DATA) SECTIONS SIZE IN FIRMWARE IMAGE WITH INFERENCE ENGINES ON EMBEDDED PLATFORM WITH A SAMPLE SETUP WITH RTOS, NNs AND MEMORY BUFFERS FOR CAMERA FRAMES.

Engine	Model	text	bss, data
CMSIS NN	CNN3	152.0kB	9.6MB
CMSIS NN	None	46.8kB	8.1MB
TensorFlow Lite	CNN3	1.07MB	9.6MB
TensorFlow Lite	None	0.97MB	8.1MB
RTOS-only	None	4.4kB	2.4kB

IV. RESULTS

In this section, we present and discuss the results of benchmarks.

A. Memory Footprint

The first test is simply building firmware image for the target with the inclusion of CMSIS NN and TF Lite to calculate the volatile (RAM) and non-volatile (flash) memory footprint. The results are presented in Tab. II. When compiled, CMSIS NN only requires 46.8kB of flash compared to 1.07MB for TF Lite. The size of the latter can be significantly reduced by removing unused operations from the library and only including operations used by the current model. However, even after further code optimization, TF Lite requires more flash memory than CMSIS NN. The total RAM usage depends largely on number of layers, parameters, size of buffers for intermediate results and input data. Both CMSIS NN and TF Lite show similar volatile memory footprint when using NN of the same size.

B. Inference Time

In the next test, the goal is to measure how inference time of CNN changes when the network is increasing in depth. The measurements are collected for CMSIS NN and TF Lite and are presented in Tab. III and Fig. 2. The inference time is proportional to the number of parameters and layers. The reference implementation in CMSIS NN is the slowest one, but enabling the DSP instruction set and switching to DSP

TABLE III.
INFERENCE TIME COMPARISON. CMSIS - REFERENCE IMPLEMENTATION WITHOUT DSP INSTRUCTIONS, C+DSP - CMSIS NN WITH DSP, TF LITE - TENSORFLOW LITE, TF+OPT - TENSORFLOW LITE WITH MODEL OPTIMIZATION (QUANTIZATION, INCLUDING ACTIVATIONS).

Model	Params	CMSIS	C+DSP	TF Lite	TF+Opt
CNN3	89688	0.84s	0.50s	374ms	103ms
CNN5	192184	1.58s	0.88s	407ms	116ms
CNN7	294680	2.31s	1.26s	479ms	123ms
CNN9	397176	3.04s	1.64s	506ms	131ms
CNN11	499672	3.77s	2.02s	519ms	155ms
CNN13	602168	4.50s	2.40s	632ms	168ms

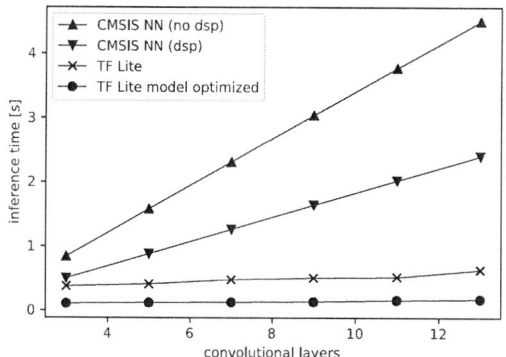

Fig. 2. Inference time as a function of number of convolutional layers when using different inference engines, DSP acceleration and model optimization.

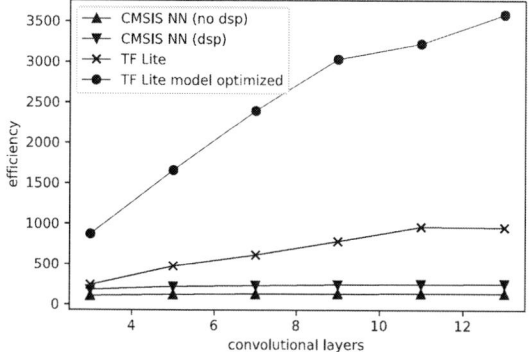

Fig. 3. Computation efficiency for each setup obtained by dividing the number of parameters by the time of inference (in milliseconds) as a function of the number of convolutional layers using different inference engines, DSP acceleration and model optimization.

kernels significantly reduces computation time. Further drop in inference time is achieved by switching to TF Lite. Finally, further gains are obtained by optimizing the model. The result is an over 26 times of reduction in the inference time.

When instantiating NN with TF Lite, an input FlatBuffer with model definition is used to build a constructed graph. CMSIS NN resembles eager execution where all kernels are functions and are evaluated when invoked with no graph built. Graph representation allows TF Lite to optimize and achieve higher performance in comparison to CMSIS NN.

With removed layers from CNN, TF Lite shows a fixed cost that is not present when using CMSIS kernels, as seen in Tab. III. For this reason, the efficiency of computations with TF Lite is higher with deeper NNs (Fig. 3). For CMSIS, it stays at the same level for all sizes of networks. When designing a system with TF Lite and deciding between using a deep network that is run infrequently or a shallow network, TF Lite is better suited for the former.

C. Computation Stages

Each computation stage of CNN3 is instrumented to measure execution time. Measurements are collected for models with real-valued and complex-valued tensors, both with the same total number of parameters. To ensure this, sizes of real and imaginary CVNN kernels are set to half of their real-

valued counterparts. The input image is real-valued and is convolved with the first layer using (4) to produce the first complex-valued intermediary output.

Execution measurements can be found in Tab. IV. The majority of computing time is spent performing convolutions with nearly 95% of the total. Optimizing convolutions should be the priority as it has the largest impact on the performance.

Introducing addition and subtraction operations to CVNN, required by (4), did not add much overhead. Despite that, the unoptimized implementation of CVNN is the slowest network. When using the DSP acceleration, the computation times for complex- and real-valued networks are similar. It indicates that CVNNs should be considered when complex-valued representation can be exploited to improve the accuracy without incurring any significant increase in the inference time.

TABLE IV.
EXECUTION TIME IN MILISECONDS FOR COMPUTATION STAGES OF CNN3, RVNN AND CVNN - IMPLEMENTATIONS WITHOUT DSP ACCELERATION FOR REAL-VALUED AND COMPLEX-VALUED NNs RESPECTIVELY, R+DSP AND C+DSP - CMSIS NN WITH DSP FOR REAL-VALUED AND COMPLEX-VALUED NNs, RESPECTIVELY.

Computation stage	RVNN	CVNN	R+DSP	C+DSP
Pre-processing	2	2	2	2
Convolution1	114	303	114	120
ReLU1	3	4	3	4
MaxPool1	10	9	2	9
Convolution2	513	541	249	255
AddSub2	-	6	-	6
ReLU2	2	2	2	2
MaxPool2	3	3	1	3
Convolution3	219	233	133	126
AddSub3	-	6	-	6
ReLU3	2	1	2	1
MaxPool3	3	3	2	3
Dense4	2	3	2	3
Softmax	1	1	1	1
Inference time [ms]	872	1117	513	541

D. Core and Memory Tuning

Benchmarks in this section are aimed at understanding the impact of different SoC core frequency, access to the low-latency cache and the DSP instructions on the inference time of NNs. The results for RVNN and CVNN are presented in Tabs. V and VI and Figs. 4 and 5.

The tables include records for each configuration - the measured inference time for all run modes and predictions based on full speed mode measurement that is then multiplied by the frequency ratio for other run modes. If the core is executing instructions as fast is it can and is not stalled by the access to memory or peripherals, then the prediction should closely match the measurement for a given configuration. For instance, if the inference time at full speed with 528MHz takes N ms, then after switching to low speed run mode with four times lower frequency, it should be four times longer, i.e., $4N$ ms. If the core is waiting for the memory access, the total time would be longer by this additional wait time.

In nearly all cases, both RVNN and CVNN inference times are similarly influenced by different configurations. As expected, computations are slower after disabling cache and reducing core frequency.

With full access to the cache (i+d cache), the predictions are close to measurements in almost each case. It means that the core is not blocked by the memory and is indeed the bottleneck. The exception is overdrive mode with the DSP instructions for RVNN, the fastest configuration, with the inference time shorter than predicted. The memory access is likely the limiting factor with the core running at its full capacity. There may be room for optimization, e.g., with larger data cache or lower memory latency.

After disabling the data cache, but with the instruction cache enabled (no dcache), the predicted time for high frequencies is close to the measured one. For low core frequencies, the actual inference time is significantly shorter than predicted in almost every case. It indicates that the data cache access is throttling the core at high frequencies. It does not occur to the same extent with lower frequencies, when the core is not able to process data quickly enough and so is not blocked as often.

With the instruction cache disabled and the data cache enabled (no icache), the overall inference time is significantly longer than in the previous case (no dcache). It shows that the instruction cache has greater impact and the core is unable to fetch instructions to process quickly enough. In almost all cases, predictions are close to measurements. The exception is the lower power mode when measured time is nearly half of the predicted. It means that the core is not stalled as often as its slow execution speed matches the high latency instruction fetching.

In the last configuration, both caches are disabled (no cache). Unsurprisingly, the inference time is the longest one. For the overdrive, the measured time is longer than predicted, while for lower frequencies measurements are much faster. It means that at lower core frequencies, the core is not stalled by the lack of data or instructions as much as it is with high frequencies. When cores run with low frequency, the cache size can be smaller as the core is not be able to utilize its full capacity.

TABLE V.
INFERENCE TIME FOR RVNN FOR DIFFERENT RUN MODES, CACHE AVAILABILITY, DSP INSTRUCTION ACCELERATION. ROW M IS MEASURED INFERENCE TIME, P IS A PREDICTION BASED ON MEASURED INFERENCE TIME FOR FULL SPEED MULTIPLIED BY FREQUENCY RATIO FOR A GIVEN RUN MODE.

Run mode Core freq.		overdrive 600MHz	full speed 528MHz	low speed 132MHz	low power 24MHz
Inference time, no DSP acceleration [s]					
no cache	M	11.35	11.67	26.41	126.96
	P	10.27	11.67	46.69	256.79
no icache	M	4.43	4.94	19.11	61.91
	P	4.35	4.94	19.76	108.66
no dcache	M	1.53	1.72	6.39	30.92
	P	1.51	1.72	6.87	37.79
i+d cache	M	0.85	0.97	3.89	21.33
	P	0.85	0.97	3.88	21.35
Inference time with DSP instructions acceleration [s]					
no cache	M	4.92	5.05	12.37	56.03
	P	4.90	5.05	20.19	111.03
no icache	M	2.45	2.66	9.83	31.92
	P	2.34	2.66	10.64	58.53
no dcache	M	0.82	0.92	3.07	16.33
	P	0.81	0.92	3.67	20.21
i+d cache	M	0.40	0.55	2.20	12.10
	P	0.49	0.55	2.21	12.17

TABLE VI.
INFERENCE TIME FOR CVNN FOR DIFFERENT RUN MODES, CACHE AVAILABILITY, DSP INSTRUCTION ACCELERATION. ROW M IS MEASURED INFERENCE TIME, P IS A PREDICTION BASED ON MEASURED INFERENCE TIME FOR FULL SPEED MULTIPLIED BY FREQUENCY RATIO FOR A GIVEN RUN MODE.

Run mode Core freq.		overdrive 600MHz	full speed 528MHz	low speed 132MHz	low power 24MHz
Inference time, no DSP acceleration [s]					
no cache	M	17.12	17.78	40.51	165.84
	P	15.65	17.78	71.13	391.19
no icache	M	5.35	6.08	24.25	78.39
	P	5.35	6.08	24.32	133.77
no dcache	M	1.65	1.87	7.44	35.66
	P	1.64	1.87	7.47	41.09
i+d cache	M	1.08	1.22	4.91	26.92
	P	1.08	1.22	4.90	26.94
Inference time with DSP instructions acceleration [s]					
no cache	M	5.05	5.18	12.73	57.28
	P	4.56	5.18	20.73	114.01
no icache	M	2.55	2.79	10.33	33.73
	P	2.46	2.79	11.16	61.41
no dcache	M	0.83	0.92	3.15	16.62
	P	0.81	0.92	3.67	20.21
i+d cache	M	0.50	0.57	2.29	12.55
	P	0.50	0.57	2.28	12.52

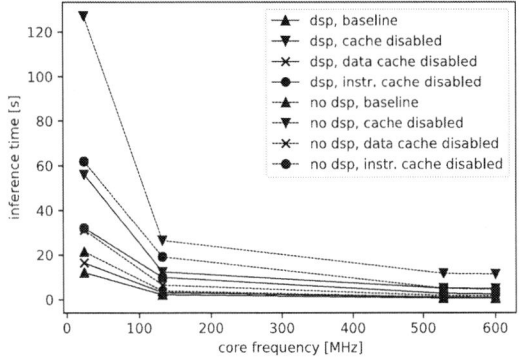

Fig. 4. Inference time for RVNN with different core frequencies, cache and DSP instruction acceleration availability. Points on the horizontal axis correspond to run modes - low power, low speed, full speed, overdrive.

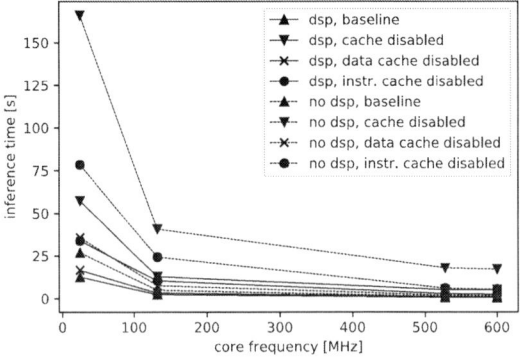

Fig. 5. Inference time for CVNN with different core frequencies, cache and DSP instruction acceleration availability. Points on the horizontal axis correspond to run modes - low power, low speed, full speed, overdrive.

The questions remain: What is the optimal size of the cache and what is the optimal frequency? The core frequency and the cache are intertwined and must be tuned together. The size of cache should be large enough so that the core is not stalled and is able to process data continuously. On the other hand, the core frequency should only be increased to match the memory latency to avoid waiting for data. There are multiple other factors that must be considered, e.g., type of workload (in our case limited to NNs), core and memory architecture, power efficiency. Therefore, to design an optimal system, it is necessary to consider all of its elements.

V. CONCLUSION

In this paper, the analysis of performance of CNNs deployed on an ARM microcontroller is presented. Analyzing performance requires a holistic approach as system components are intertwined. The higher the core frequency, the faster the computations. Introducing parallelism, for instance by using the DSP instructions, can further reduce the execution time for certain applications, especially deep learning. However, the theoretical increase in the number of instructions that can be executed in a fixed time, does not always produce the expected performance increase as higher throughput is hindered by the memory access or reading from I/Os. It is evident with deep learning applications that have high requirements on computing capabilities with low memory latency. It is therefore crucial to tune system components together and understand the system as a whole, including the software workloads and other requirements, , e.g. energy efficiency.

REFERENCES

[1] A. Krizhevsky, *Learning Multiple Layers of Features from Tiny Images*, University of Toronto, 2009.

[2] V. Badrinarayanan, A. Kendall, and R. Cipolla, "SegNet: A deep convolutional encoder-decoder architecture for image segmentation," arXiv preprint, arxiv:1511.00561, 2015.

[3] A. Krizhevsky, I. Sutskever, and G. E. Hinton, *ImageNet classification with deep convolutional neural networks*, Adv. Neural Inf. Process. Syst., vol. 25, 10.1145/3065386, 2012.

[4] C. Szegedyd, S. Ioffe, V. Vanhoucke, and A. A. Alemi, "Inception–v4, inception–ResNet and the impact of residual connections on learning," in Proceedings of the Thiry-First AAAI Conference on Artificial Intelligence, Feb. 2017, pp. 4278-4284.

[5] G. Hinton, et al., "Deep neural networks for acoustic modeling in speech recognition: the shared views of four research groups," IEEE Sig. Process. Mag., vol. 29, no. 6, pp. 82-97. 2012.

[6] T. Young, D. Hazarika, S. Poria, and E. Cambria, "Recent trends in deep learning based natural language processing," IEEE Comput. Intell. Mag., vol. 13. no. 3, pp. 55-75, 2018.

[7] D. Ravi, Ch. Wong, B. Lo, and G. Yang, "A deep learning approach to on-node sensor data analytics for mobile or wearable devices," IEEE J. Biomed. Health Inform., vol. 21, no. 1, pp. 56-64, 2017.

[8] T. A. Piedras Lopes and A. C. R. Troyman, "Neural networks on predictive maintenance of turbomachinery," IFAC Proc. Volumes, vol. 30, no. 18, pp 983-988, 1997.

[9] J. Krenek, K. Kuca, P. Blazek, O. Krejcar, and D. Jun, *Application of Artificial Neural Networks in Condition Based Predictive Maintenance*, in Recent Developments in Intelligent Information and Database Systems, Studies in Computational Intelligence, vol. 642, Springer, Cham, pp. 75-86, 2016.

[10] Y. Zhang, N. Suda, L. Lai, and V. Chandra, "Hello edge: keyword spotting on microcontrollers," arXiv preprint, arxiv:1711.07128, 2018.

[11] J. Redmon and A. Farhadi, "YOLOv3: an incremental improvement," arXiv preprint, arXiv:1804.02767, 2018.

[12] X. Glorot, A. Bordes, and Y. Bengio, "Deep sparse rectifier neural networks," in Proceedings of the Fourteenth International Conference on Artificial Intelligence and Statistics, vol. 15, Apr. 2011, pp. 315-323.

[13] L. Lai, N. Suda, and V. Chandra, "CMSIS-NN: Efficient Neural Network Kernels for Arm Cortex-M CPUs," arXiv preprint, arXiv:1801.06601, 2018.

[14] ARM Limited, *ARM Cortex-M7 Devices–Generic User Guide*, 2018.

[15] N. Guberman, "On complex valued convolutional neural networks," arXiv preprint, arXiv:1602.09046, 2016.

[16] Aks. Agrawal, et al., "TensorFlow eager: A multi-stage, Python-embedded DSL for machine learning," in Proceedings of the 2nd SysML Conference, 2019. pp. 1-11.

[17] S. Scardapane, S. Van Vaerenbergh. A. Hussain and A. Uncini, "Complex-valued neural networks with nonparametric activation functions," IEEE Trans. Emerg. Topics Comput. Intell., vol. 4, no. 2, pp. 140-150, 2020.

[18] T. Nitta, *An extension of the back-propagation algorithm to complex numbers*, Neural Netw., vol. 10, no. 8, pp. 1391–1415, 1997.

[19] D. Scherer, A. Muller, and S. Behnke, *Evaluation of pooling operations in convolutional architectures for object recognition*, in Proceedings of 20th International Conference on Artificial Neural Networks (ICANN), 2010, pp. 92-101.

[20] NXP Semiconductors, *Application Note AN12085 – How to use i.MX RT Low Power Feature*, 2019.

[21] E. García-Martín, C.F. Rodrigues, G. Riley, and H. Grahn, "Estimation of energy consumption in machine learning," Journal of Parallel and Distributed Computing, vol. 134, pp. 75–88, 2019.

Rigorous Development of Embedded Systems Supported by Formal Tools

Tomasz Szmuc, Wojciech Szmuc
Department of Applied Computer Science
AGH University of Science and Technology
30-059 Kraków, Poland
tsz@agh.edu.pl, wszmuc@agh.edu.pl

Abstract—**A rigorous approach to development of embedded systems is proposed in the paper. The concept is based on introduction of formal modeling branch in parallel to the classical V-development method. SysML is used for description of the developed components, and then these artifacts are translated into Colored Petri Nets (CPN) blocks. The correctness of the CPN models is described using temporal logic and finally verified using model checking tools. The proposed concept enables detection of structural errors in early development stages. The paper describes the next steps of research in this area. Translations of remaining SysML diagrams are included, and the modeling-verification chain is described.**

Keywords—**embedded systems; translation into formal models; temporal logic verification**

I. INTRODUCTION

Growing scope of embedded systems application domains and extending functionality of microcontrolers raise new challenges in their implementation. Microcontrolers embedded in complicated subsystems cooperating with different environments need careful approach in design and implementation to reduce risk of failures. Cyber Physical Systems (CPS) [15] are good examples of the complexity. Architecture of CPS systems is formed from microcontrolers, networks sensors (IoT) and usually collections of data stored in nebula or cloud structure. Complexity of system architecture enormously increases when Smart City applications are considered. Cyber Physical systems of systems architectures are applied in the development [16]. Integration and synchronization of several different subsystems are critical aspects in development of such systems. There are also other application domains where embedded systems are applied in different subsystems. Robot controllers, avionic controllers and autonomous (intelligent) cars should be mentioned here. The later domains are specific due to very hard safety requirements. Development of the all above described systems should be supported by systematic modeling and attentive analysis/verification of developed artifacts. Correctness and safety requirements should be hardly satisfied for Safety Critical Systems [1,10,11].

Development of embedded systems is involved both in hardware and software domains, therefore it is important to use modeling languages providing tools for merging these two domains. The most popular in this system engineering area are SysML [2,8] and AADL [3]. SysML is an extension of UML profile developed by OMG (Object Management Group) and currently is most popular modeling language in the area of Systems Engineering.

Increasing complexity of embedded systems and growing number of safety critical applications imply needs for systematic development methodologies supported by formal methods for analysis and verification during the development. The paper presents next steps in the research towards efficient application of formal methods to support development of embedded systems. Some results in this area are published in the previous publications. General rules for translation of main UML diagrams into CPN models are presented in [17]. Modeling of event handling primitives is described in [18]. Translations of state machine and modeling of class/objects behavior in CPN are presented in the next publication [19]. Results and experience gathered in these research stages are applied in mapping of selected SysML diagrams (block definition, internal block definition and activity diagrams) into CPN models [20].

The goal of the paper is to collect the earlier developed partial results and also to complete SysML → CPN translation algorithm by adding remaining modules, especially translating use case and sequence diagrams into CPN models. Complemented steps of the translation-verification chain are also described. The remaining links are focused on transformation of CPN output into nuXmv [25] input format and then use of the later system for verification (model checking) of the CPN models. The general approach is similar to the one described in [21], but related to the above mentioned AADL language. The two modeling languages are different, so translation rules and their implementations significantly differs.

The rest of the paper is organized in the following way. An overview of schemes for application of formal methods in system development is presented in the next section. Translations of remaining two (use case and sequential) SysML diagrams into CPN models is described. The translations are carried out using SysML2CPN [13]. The proposed verification chain including temporal logic model checker is shortly presented.

II. FORMAL METHODS FOR SYSTEM DEVELOPMENT. AN OVERVIEW

The proposed approach is based on creation of formal modeling and verification path in parallel to initial system/software development steps. Requirements and design (both general and detailed) phases are of our main interest here, but it seems that the approach may be also extended to other steps of development process, especially to integration phases.

The proposed formal modeling-verification concept is presented in Fig. 1.

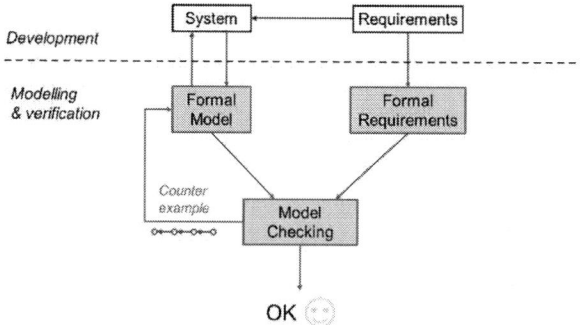

Fig. 1. General concept of formal methods applications

Dashed line (Fig. 1) demarcates the *Development* and *Modeling & verification* layers. Requirements specified in different ways in development part are translated into logical formulas and developed artifacts ("System" box) are translated into behavioral models. Model checking is used to verify if the logical formulas are satisfied within the behavioral models. OK result means that the formal system model is correct w.r.t. the formal requirements and consequently (if translations are correct) that system satisfies the requirements in the related development stage.

Quantitative (Fig. 2) and qualitative (Fig. 3) approaches are two instances of the general concept. The first approach is based on probabilistic descriptions while the second one applies deterministic modeling and temporal logic.

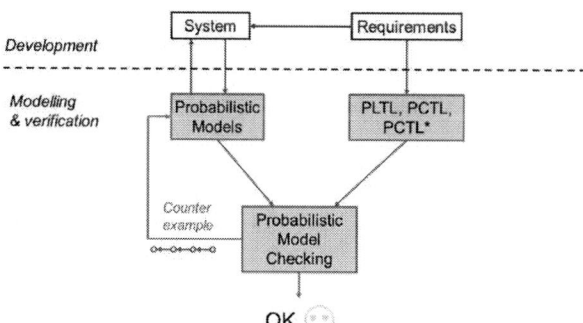

Fig. 2. Quantitative approach for using formal methods

Integrated (both modeling and verification) system has been developed at Oxford University by M. Kwiatkowska and coworkers [14]. The integrated tool PRISM [28] offers the following functionalities:

1. Modeling of processes: discrete-time Markov chains, continuous-time Markov chains, Markov decision processes, probabilistic automata, probabilistic timed automata, Stochastic Petri Nets.

2. Specification of properties (Formal Requirements) using (probabilistic) temporal logics: LTL, PCTL. PCTL*, CSL (continuous stochastic logic).

3. Verification of satisfability of the temporal formulas in the probabilistic models.

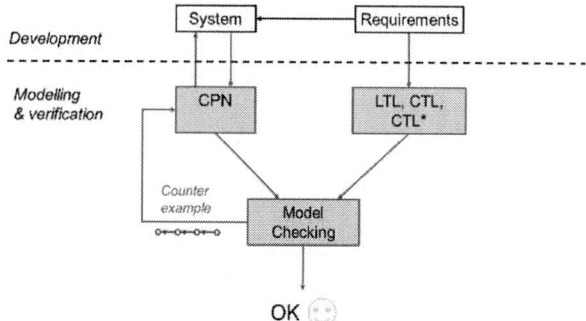

Fig. 3. Qualitative approach for using formal methods

The related languages and tools in the qualitative approach are given below.

1. Modeling of system behavior:

    a. (High level) Petri nets [12] (CPN Tools [24])

    b. Timed automata, hybrid automata (UPAAL [30])

    c. Process algebras - mainly LOTOS (CADP [23])

2. Description of requirements using temporal logics: LTL, CTL, CTL*, etc.

3. Verification of satisfability using model checkers or SAT Solvers [9,25].

The two modeling and verification approaches lead to different results. In the quantitative version positive verification results are ranged by values from the interval [0, 1]. It means that (system) model satisfies verification formula to some extent, for example 0.8. In the case of qualitative approach results are from the set {0, 1}, so the requirements are satisfied or not. The features determine application areas, i.e. quantitative approach may be recommended to soft verification (e.g time aspects in soft real-time systems), while the qualitative for exact (hard) satisfability (Safety Critical Systems).

In the paper (Hierarchical) Colored Petri nets are used in formal modeling domain and temporal logic is applied for description of required properties. This concept is illustrated in Fig.3.

III. TRANSLATION FROM SysML INTO CPN MODELS

SysML is a language supporting holistic approach for modeling of embedded systems. It means that SysML integrates models from different domains, and is mainly used in modeling of system architectures, while UML focuses on software models in development process.

SysML offers tools for integration of different modeling views described by different languages. The modeling tools are grouped in the free categories: requirements diagrams, behavior diagrams and structure diagrams. The current research stage focuses on structure design and analysis, therefore translations of requirements and parametric diagrams are not considered in the paper. The general level is an important first step in the development, especially in highly integrated systems. More detailed levels and dealing with the two remaining diagrams constitute the next research step.

Translation rules for use case and sequence diagrams are presented in simplified versions, which are implemented in the prototype of SysML into CPN (SysML2CPN) translator [13]. The following rules are specified:

1. Each actor and any use case are represented by the related places

2. Relations between actors and use cases are modeled by transitions

3. Include and extend relations between use cases are described by the corresponding transitions.

Lest us consider typical use case diagram for ATM example (Fig. 4).

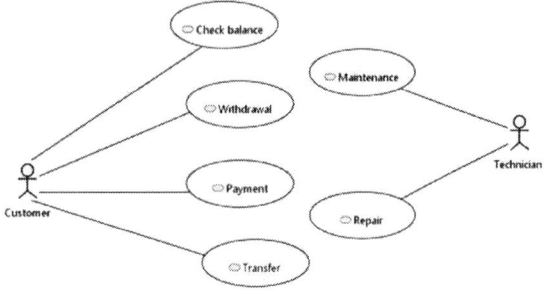

Fig. 4. Use case diagram for ATM

Left-hand part of the diagram specifies associations between Customer and the four main ATM functionalities: Check balance, Withdrawal, Payment, and Transfer. The right-hand part is related to technical (Maintenance, Repair) functions associated with Technician. The use case diagram is is translated into CPN using SysML2CPN tool [13], which generates the related Petri net in xml format of CPN Tools [24]. The resulted net after manual tuning (to improve its legibility) is presented in Fig. 5. Top part of the net is related to actor Customer and lower section specifies technical functionalities (Technician Actor). According to the defined translation rules use cases and actors are represented by the related places. Bidirectional association relations are modelled by two transitions (one for each direction) and the related arcs.

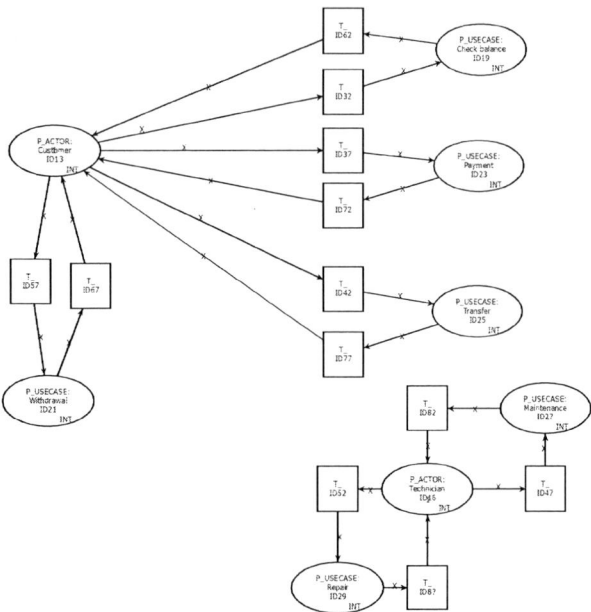

Fig. 5. Petri net model of use case diagram from Fig. 4

ATM communication with environment is presented in Fig. 5.

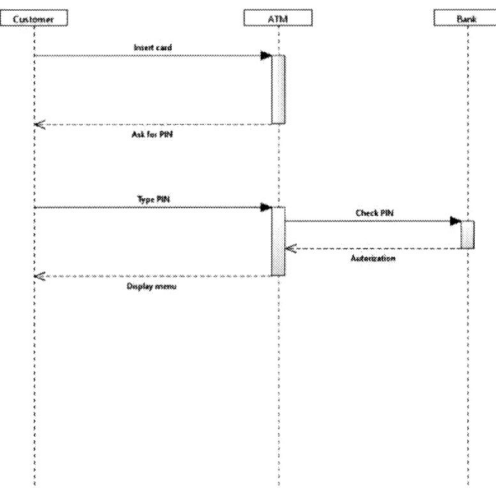

Fig. 6. Sequential diagram describing ATM communication

The diagram presents initial interactions between Customer, ATM and Bank leading to authorization of the Customer.

Translation rules for main sequence diagram elements are implemented in [13, 17] and described below.

1. Each head of lifeline specifying behavior of actor/object is denoted on the top (of the net) by place named as the related actor/object.

2. Sending message between two lifelines is modeled by two transitions placed on the related lifelines and place between them. The place represents the message.

3. Additional place between transitions sending or receiving messages in the same lifeline is inserted. These additional places preserve formal requirement of Petri nets and also model current point in the line.

4. Found/lost messages are modeled by input place-transition and transition output place correspondingly.

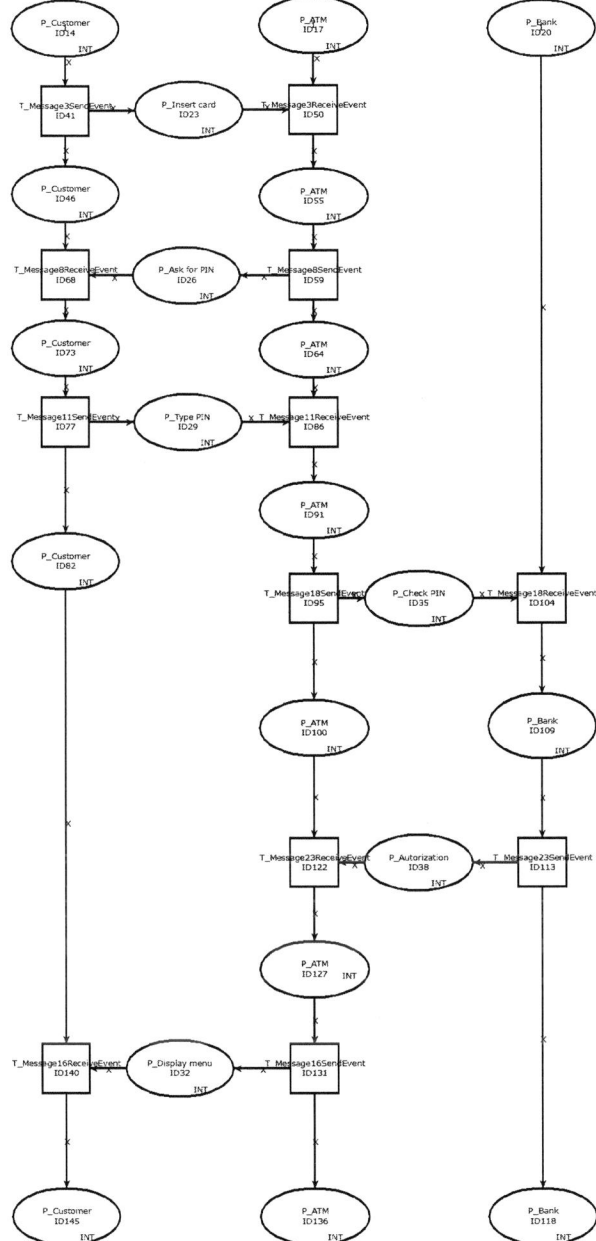

Fig. 7. Petri net model of sequence diagram (Fig. 6)

Petri net generated for the sequence diagram from Fig. 6 is presented in Fig. 7. The net is generated using the above mentioned rules for building the Petri net model. Places are put on the top (of the net) using the first rule. Transitions in separate lifelines and the connected places between them describe messages, while places in lifelines specify current points of the scenario. The translator generates identifiers for the additional places using places' names on the top of the related lifelines. The names are extended by ID with numbers generated by the system. Similar rules are used in composing identifiers of Petri net nodes in translations of other diagrams. The Petri net model is manually tuned similarly like in the previous (use case) diagram. The name convention supports finding relations to related elements in SysML diagrams and makes the tuning easier.

SysML2CPN translator is written in Java and may be installed as a plugin in eclipse environment. Use of eclipse or other IDE makes management of the translation process easier. SysML diagrams are created and edited using Papyrus [26] and Petri net models are displayed and analyzed using CPN Tools. The both systems use XML standard for specifications of the related graphs. Translation process builds graph of objects (of translated diagram) and then the corresponding scheme is applied for generation of input file for CPN Tools. The code is not generated from the sequence of characters but using the graph of objects, therefore its validation and modification is easier.

Prototype of SysML2CPN system enables automatic translation of six main SysML diagrams: requirement, use case, activity, block definition, internal block, and sequential. The translation is sometimes restricted to subset of diagrams (e.g. sequence diagrams), but the main goal was to build prototype covering relatively wide scope of SysML diagrams.

IV. VERIFICATION CHAIN

Generated Petri net models may be analyzed directly using CPN Tools. The following properties may be verified: reachability of specific (e.g. deadlock) states, home property, i.e. if the given state may be reached again in the future, safety, liveness and other features related to concurrent systems.

The verification scope may be extended by application of temporal logic for description of system requirements and then checking if the formulas are satisfied within Petri net models generated from SysML2CPN. The approach follows the verification chain presented in Fig. 3. This general chain must be refined due to data format used by available tools. CPN Tools generates LTS (Labeled Transition Tree) which has to be transformed into format acceptable by popular model checker nuXmv [25]. PetriNet2ModelChecker [27] tool described in [22] is used for mapping of LTS into nuXmv input format.

V. SUMMARY

Translations of two remaining SysML diagrams into Petri nets are presented in the paper. Simplified version of use case and sequence diagrams are considered to show possibilities of automatic translation and on the other hand to complete the set of translated diagrams presented in the previous papers. Prototype of SysML2CPN translator has been developed to prove that the automatic translation and the presented verification chain is feasible. It seems that it may be considered as a step on the way towards implementation of formal modeling and verification according to scheme presented in Fig. 3. Petri net models generated by SysML2CPN are more complicated than the related diagrams due to additional elements. Therefore, the proposed verification chain seems to be reasonable, because properties specified by temporal logic may be verified automatically without looking into complicated net structures. Tuning of the generated Petri net is not necessary, however the naming convention supports this action.

Further research will be carried out in the following main directions.

1. Design and implementation of translation rules for wider scope of SysML diagrams, i.e. mainly by reduction of restricting assumptions.

2. Integration of translation paths which now are performed independently for every type of diagrams.

The proposed approach is similar to general concept of TTool [29] developed in Telecom Paris. The similarity is in using SysML for systems modeling and additional tool for formal modeling/verification. TTool uses integrated UPAAL [30] which generates reachability graphs from SysML models. The graphs may be used for verification of safety, liveness, and user defined properties (using the so-called pragmas). In the proposed approach the reachability graphs are generated by CPN Tools and main properties may be verified also using this tool. A range of the verification is increased when properties are defined using temporal logic and model checker [25] as specified in the section IV.

ACKNOWLEDGMENT

The research was supported by the Department Research Grant No 11.11.120.859: Methods, tools and environments for knowledge and software engineering. Topic: Modeling, analysis and optimisation of selected systems.

REFERENCES

[1] J.-P. Blanquart, J.-M. Astruc, P. Baufreton, J.-L. Boulanger, H. Delseny, J.Gassino, et al., "Criticality categories across safety standards in different domains", ERTS2 Congress, Embedded Real Time Software and Systems, 2012, pp. 3–4.

[2] L. Delligatti, SysML Distilled. A Brief Guide to the Systems Modelling Language, Addison-Wesley, 2014.

[3] P. H. Feiler, and D. P. Gluch, Model-Based Engineering using AADL. An Introduction to the SAE Architecture Analysis & Design Language, Addison-Wesley, 2012.

[8] S. Friedental, A. Moore, and R. Steiner, A Practical Guide to SysML. The System Modeling Language, Morgan Kaufman OMG Press, 2010.

[9] R. Gore, J. Thomson, and F. Widmann, "An experimental comparison of theorem provers for CT", Proceedings of the 2011 International Symposium on Temporal Representation and Reasoning, IEEE Computer Society, 2011, pp. 49-56.

[10] K. Greb, A. Seely, Design of Microcontrollers for Safety Critical Operation (ISO 26262 Key Differences from IEC 61508), 2009.

[11] IEC 61508 Functional Safety of Electrical/Electronic/Programmable Electronic Safety-related Systems (E/E/PE, or E/E/PES).

[12] K. Jensen, and L. M. Kristensen, Coloured Petri Nets. Modelling and Validation of Concurrent Systems, Springer, 2009.

[13] M. Kałduś, Design and Implementation of System Mapping Selected SysML Diagrams into Petri Nets Models (in Polish: Projekt i implementacja systemu translującego wybrane diagramy języka SysML na modele sieci Petriego), MSc. Thesis (supervisor T. Szmuc), AGH University of Science and Technology, 2019

[14] M. Kwiatkowska, G. Norman, and D. Parker, PRISM 4.0: "Verification of Probabilistic Real-time Systems", in Proc. of International Conference on Computer Aided Verification (CAV'11), vol. 11, LNCS, vol. 6806, pp. 585-591, Springer, 2011

[15] H. Song, D. B. Rawat, S. Jeschke, and Christian Brecher, Cyber-Physical Systems. Foundations, Principles, and Applications, Elsevier, Intelligent Data Centric Systems, 2017

[16] H. Song, R. Srinsivasan, T. Sookkoor, and S. Jeschke: Smart Cities. Foundations, Principles, and Applications, Wiley, 2017

[17] W. Szmuc, Modelling of Selected UML 2.0 Diagrams with Coloured Petri Nets. PhD Report, Supervisor M. Szpyrka, AGH 2014

[18] W. Szmuc, and T. Szmuc, "Modeling UML object event handling with Petri Nets. Towards improvement of embedded systems analysis and design", 23rd International Conference on Mixed Design and Integrated Circuits and Systems, Mixdes 2016, June 2-25, 2016, Łódź, Poland, pp. 454-457.

[19] W.Szmuc, and T. Szmuc, "From UML object behavior description into Petri net models. Towards systematic development of embedded systems", International Journal of Microelectronics and Computer Science, Vol. 7, No 2, 2016, pp. 60-64.

[20] W. Szmuc, and T. Szmuc, "Towards embedded systems formal verification. Translation from SysML into Petri nets", Proceedings of the 25th International Conference "Mixed Design of Integrated Circuits and Systems, June 21-23, Gdynia , 2018, pp.420-423

[21] T. Szmuc, W. Szmuc. "Consistency preserving development of embedded systems using AADL", submitted to Mixdes 2020

[22] M. Szpyrka, J. Biernacki, A. Biernacka. "Tools and methods for RTCP-nets modelling and verification, Archives of Control Sciences", vol. 26, No 3, 2016, pp. 339-365, doi 10.1515/acsc-2016-0019

[23] CADP: https://cadp.inria.fr/

[24] CPN Tools: http://cpntools.org/

[25] nuXmv model checker: https://nuxmv.fbk.eu/

[26] Papyrus, https://www.isis-papyrus.com/software

[27] PetriNet2ModelChecker: http://alvis.kis.agh.edu.pl/wiki/software

[28] PRISM checker: https://www.prismmodelchecker.org

[29] TTools: https://ttool.telecom-paris.fr/index.html

[30] UPAAL: http://www.uppaal.org/

Proceedings of the 27th International Conference *"Mixed Design of Integrated Circuits and Systems"*
June 25-27, 2020, Łódź, Poland

Sensor Fusion Algorithm Implementation on Microchip PIC Microcontroller

Sergio Salas Arriarán*, Carlos Valdez*, Kalun Lau*, M. Hadi Amini†, Marek Kropidłowski‡, Paweł Śniatała‡

*Universidad Peruana de Ciencias Aplicadas (UPC), Prolongacion Primavera 2390, Monterrico, Surco Lima, Peru
†Florida International University, 11200 SW 8th Street Miami, FL 33199
‡Poznan University of Technology, Piotrowo 2, 60-965 Poznań, Poland
E-mail: pawel.sniatala@put.poznan.pl

Abstract—**The paper describes an implementation of a Brooks-Iyengar algorithm on the Microchip PIC18F4550 platform. The circuit is considered as a testing platform to check the algorithm concept simulated in MATLAB before the final implementation as ASIC IP core. The results confirm the correctness of the proposed approach, which will be used in the final IP Core design.**

Keywords—**Sensor Fusion, Microcontroller PIC18 family.**

I. INTRODUCTION

The world around us is an analog world, full of information useful to our civilization. We measure just about everything imaginable via sensors and convert the data into digital signals utilizing analog-to-digital converters (ADCs) [1], [2]. Sensors that were wired for many decades started communicating with no wire through existing protocols or by formulating new ones. This brought about a radical change in how sensors were used. It opened new applications and made processing possible through the concepts of distributed computing. It also promoted the sharing of data and resources through sensor nodes, which are configured in the form of networks. This led to the formation of Wireless Sensor Networks (WSN). A wireless sensor network (WSN) can be defined as a wireless network of sensors that are spatially distributed and have a specific function in the environment of deployment [2]. A WSN system that have been deployed in the environment are expected to communicate wither continuously or in bursts with the wired or distributed nodes where data aggregation and fusion occurs before the processing. WSNs have many advantages over the traditional sensors in terms of size, the ease of deployment, possibility of collaborative sensing, lower maintenance costs and self-organizing capabilities to name a few. The introduction of the IoT technology and the various smart environments open new areas of new applications that arise for these sensors and sensor networks [3]. Though WSN offers a lot on the plate, some issues have to be addressed. Some of these issues include the possibility of transmission of inaccurate or faulty data. Researchers have come up with multiple algorithms to address the fault tolerance. This work also also touches the issues related to the sensors' reliability. The signals taking from the sensors usually require some analog preprocessing in modules called analog-front-end (AFE) before ADC conversion [4], [5], [6]. In a modern wireless sensor network, this functionality often is implemented in a form of integrated circuit, which fulfils small size, low power

and low price requirements [5]. A problem with sensors that has always been bothering researchers is that of inaccuracy and lack of precision in the values collected by them, which could lead to faulty processing. Lots of research reveals that sensor fusion is a powerful method that can be used to mask and minimize the effects of such faulty data. Sensor fusion is a term that is associated with combining of sensory data or data derived from multiple sources such that the resulting information has less uncertainty than would be possible when these sources were used individually [2]. The term uncertainty reduction in this case can mean more accurate, more complete, or more dependable, or refer to the result of an emerging view, such as stereoscopic vision (calculation of depth information by combining two-dimensional images from two cameras at slightly different viewpoints) [7].

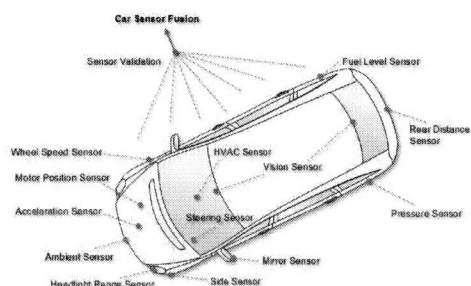

Fig. 1.   Example of sensors in a car.

The data sources for a fusion process are not specified to originate from identical sensors. One can distinguish direct fusion, indirect fusion, and fusion of the outputs of the former two. Direct fusion is the fusion of sensor data from a set of heterogeneous or homogeneous sensors, soft sensors, and history values of sensor data, while indirect fusion uses information sources like a priori knowledge about the environment and human input. We can find there lot sensors. Some of them are used to provide the same information (e.g. distance measure) and sensor fusion allows increasing the probability of the accurate enough result. An example of the sensor fusion application can be a distance measurement in a modern car, which is illustrated in Fig. 1. The information about the measure distance comes from a set of sensors. The system can have many sensors and/or they can be of different types.

277

As mentioned above the signals collected from the sensors can be inaccurate or even unavailable. The final decision about the measured distance should be calculated in a way that faulty sensors would not affect significantly the correctness of the result. It is the purpose of algorithms used in the sensor fusion block as depicted in Fig. 2.

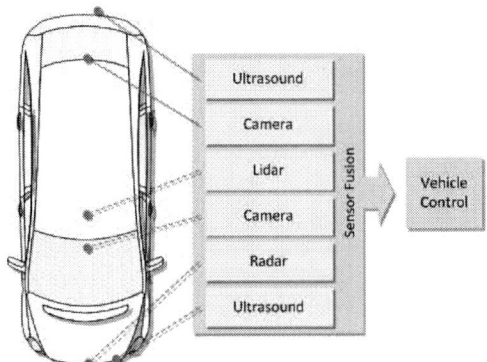

Fig. 2.   Example of a sensor fusion in car sensor system.

One of the possible solution to this problem is to use a sensor fusion algorithm.

II. BROOKS-IYENGAR ALGORITHM

Some commonly used sensor fusion algorithms are based on probability theories, which combines the values from different sensors into a single value that is more confident than each single sensor reading. Some widely used algorithms utilize the following concepts:

- Bayesian Filters
- Kalman Filters
- Particle filters or Sequential Monte Carlo (the term "particle filters" was first coined in 1996 by Del Moral about mean field interacting particle methods used in fluid mechanics since the beginning of the 1960s. The terminology "sequential Monte Carlo" was proposed by Liu and Chen in 1998.)
- Byzantine's Fault Tolerance
- Brooks-Iyengar Hybrid Algorithm [7], [8].

In this paper, we focus on the Brooks-Iyengar Hybrid Algorithm [9]. The Brooks-Iyengar hybrid algorithm is suitable for distributed systems that are working in the presence of noisy or inaccurate data. It combines the Byzantine agreement with sensor fusion and seamlessly bridges the gap between them. The algorithm can be touted to be the ideal mix of the famous Dolev's algorithm with the Mahaney and Schneider's Fast Convergence Algorithm (FCA). The algorithm is efficient and runs in $O(NlogN)$ time. Some of the many applications of this algorithm include distributed control, high performance computing, software reliability etc. The Brooks–Iyengar algorithm is executed in every processing element (PE) of a distributed sensor network. Each PE exchanges its measured interval with all other PEs in the network. The "fused" measurement is a weighted average of the midpoints of the regions found. Each PE performs the algorithm separately.

The input and output of the Brooks–Iyengar algorithm can be depicted as:

$$v, I = BI(s_1, s_2, s_3, \ldots, s_n)$$

where

$s_1, s_2, s_3, \ldots, s_n$ are the sensor readings in the specified interval;

v is the output value;

I is the output interval that defines the bounds of the output value.

The Brooks-Iyengar algorithm is a run in every processing element (PE) of a distributed sensor network. Every PE exchanges its measured interval value with all other PEs in the network. The received measurement data from all the PEs are fused to find the weighted average of the midpoints of the region [10].

The working of the Brooks-Iyengar algorithm is depicted below. The algorithm is run on each PE individually:

Input: The value sent by PE k to PE i is a closed interval $[l_{k,i}, h_{k,i}]$, $1 \leq k \leq N$.

Output: The output of PE i includes a point estimate and an interval estimate:

1) PE i receives measurements from all the other PEs.
2) Divide the union of collected measurements into mutually exclusive intervals based on the number of measurements that intersect, which is known as the weight of the interval.
3) Remove intervals with weight less than $N - \tau$ where τ is the number of faulty PEs.
4) If there are L intervals left, let A_i denote the set of the remaining intervals. We have $A_i = \{(I_1^i, w_1^i), \ldots, (I_L^i, w_L^i)\}$, where interval $I_L^i = [l_{I_l^i}, h_{I_l^i}]$ and w_l^i is the weight associated with interval I_l^i. We also assume $h_{I_l^i} \leq h_{I_{l+1}^i}$.
5) Calculate the point estimate v_i' of PE i as

$$v_i' = \frac{\sum_l \frac{l_{I_l^i} + h_{I_l^i}}{2}}{\sum_l w_l^i}$$

and the interval estimate is $[l_{I_l^i}, h_{I_l^i}]$.

Some of the characteristics of the algorithm are [11]:

- Faulty PEs tolerated $< N/3$
- Maximum faulty $PEs < 2N/3$
- Complexity $= O(N \log N)$
- Order of network bandwidth $= O(N)$
- Convergence $= 2t/N$
- Accuracy = limited by input
- Iterates for precision = often
- Precision over accuracy = no
- Accuracy over precision = no

The Brooks-Iyengar algorithm is highly flexible and could be implemented in any network environment. Accurate detection of sensor values is one of the fundamental requirements to successful tracking when working with multi-target and multi-sensor applications. Key factors affecting the multi-target detection include also sensor coverage, sensitivity, noise levels.

III. PIC18F IMPLEMENTATION OF THE TESTING ENVIRONMENT

The purpose of the implementation of the Brook-Iyengar algorithm on the microcontroller platform is to test the algorithm concept in a real circuit, before the final implementation dedicated for the ASIC realization. We have chosen the PIC18F4550 microcontroller from the Microchip PIC 8-bits family. MPLAB Integrated Development Environment was used to design and compile the program. Proteus simulator has allowed to emulate the circuit. The testing circuit consists of 5 temperature sensors connected to the PIC18F4550 microcontroler. Two analog sensors - LM35 - are connected to analog inputs, which measure the output voltage from the sensor. The ouput range, in the tested environment, is between $0V...5V$. There will be increase in $10mV$ for raise of every $1°C$. Next three sensors are digital: two DS1621 and one SHT11. The DS1621 digital thermometer and thermostat provides 9–bit temperature readings which indicate the temperature of the device. The SHT11 is a single chip relative humidity and temperature multi sensor module comprising a calibrated digital output. The screen-shot of the testing circuit simulated in Proteus simulator is presented in Fig. 3. We have assumed one out of five sensors was failed.

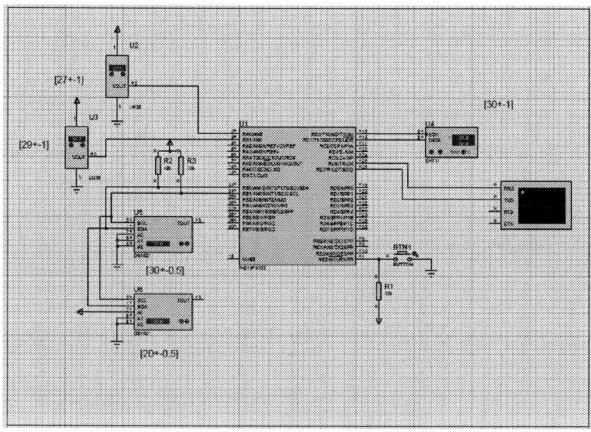

Fig. 3. Diagram of the tested circuit.

The flowchart of the implemented algorithm is depicted in Fig. 4 and Fig. 5. The core of the Brooks-Iyengar is presented in the flowchart presented in Fig.6.

IV. CONCLUSION AND RESULTS

The simluation and measurements were set to the condition presented in Table I.

For the testing purpose we have assumed the Sensor 5 to be the failed one. This sensor, depicted in the Table I in red, was set to generate different values transmitted to the

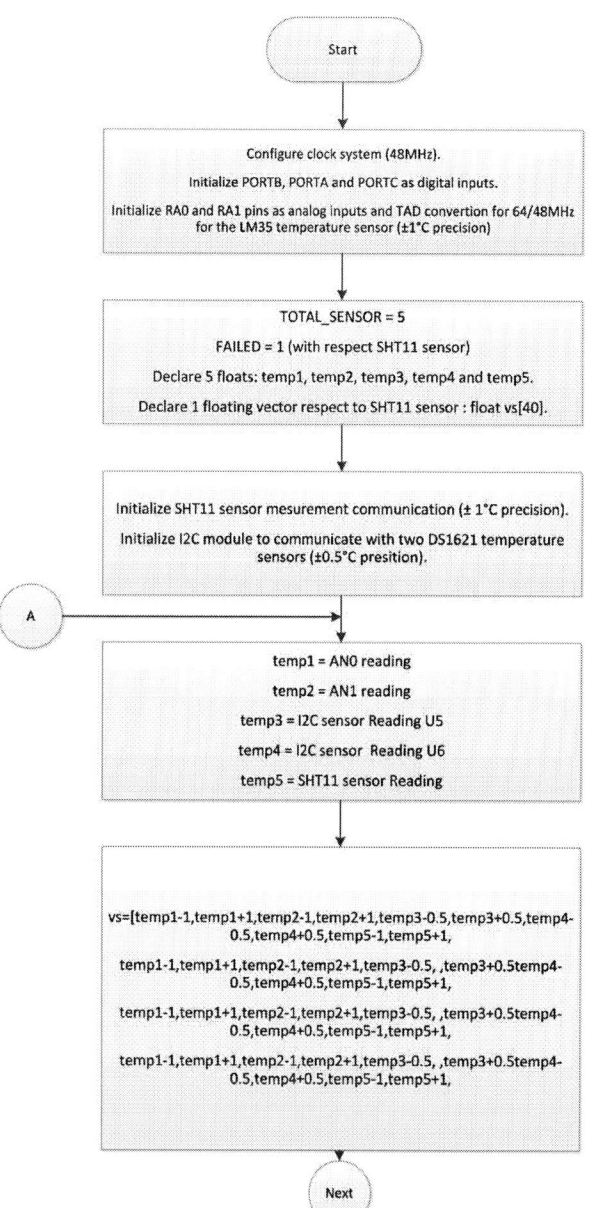

Fig. 4. Flowchart of the microcontroller program (part A)

other sensors. These inaccurate measurements, sent to the the other sensors, are listed in the fourth row of the Table I. The results calculated by the microcontroller, are presented in rows sixth (temperature output values) and seventh (calculated output intervals). The actual results, obtained based on the real circuit experiment, confirm the correctness of the proposed microcontroller based approach. Next, we would like to prepare IP core, which could be implemented as full hardware implementation. The testing platform described in this paper will be used to measure parameters needed as a references to be able to compare software and hardware implementation.

TABLE I
INPUT PARAMETERS AND RESULTS

	Sensor 1	Sensor 2	Sensor 3	Sensor 4	Sensor 5 (failed)
	Inputs				
Sensors' measured values	$28 \pm 1°C$	$29 \pm 1°C$	$29 \pm 0.5°C$	$29 \pm 0.5°C$	$30 \pm 1°C$
Values received by sensors from the Sensor 5	$30°C$	$28°C$	$29°C$	$27°C$	-
	Outputs calculated based on Brooks-Iyengar algorithm				
Sensors' calculated output values	$29.06°C$	$28.57°C$	$28.95°C$	$28.63°C$	-
Sensors' calculated output intervals	$28.5 \ldots 29.5$	$28.5 \ldots 29.0$	$28.5 \ldots 29.5$	$28.5 \ldots 29.0$	-

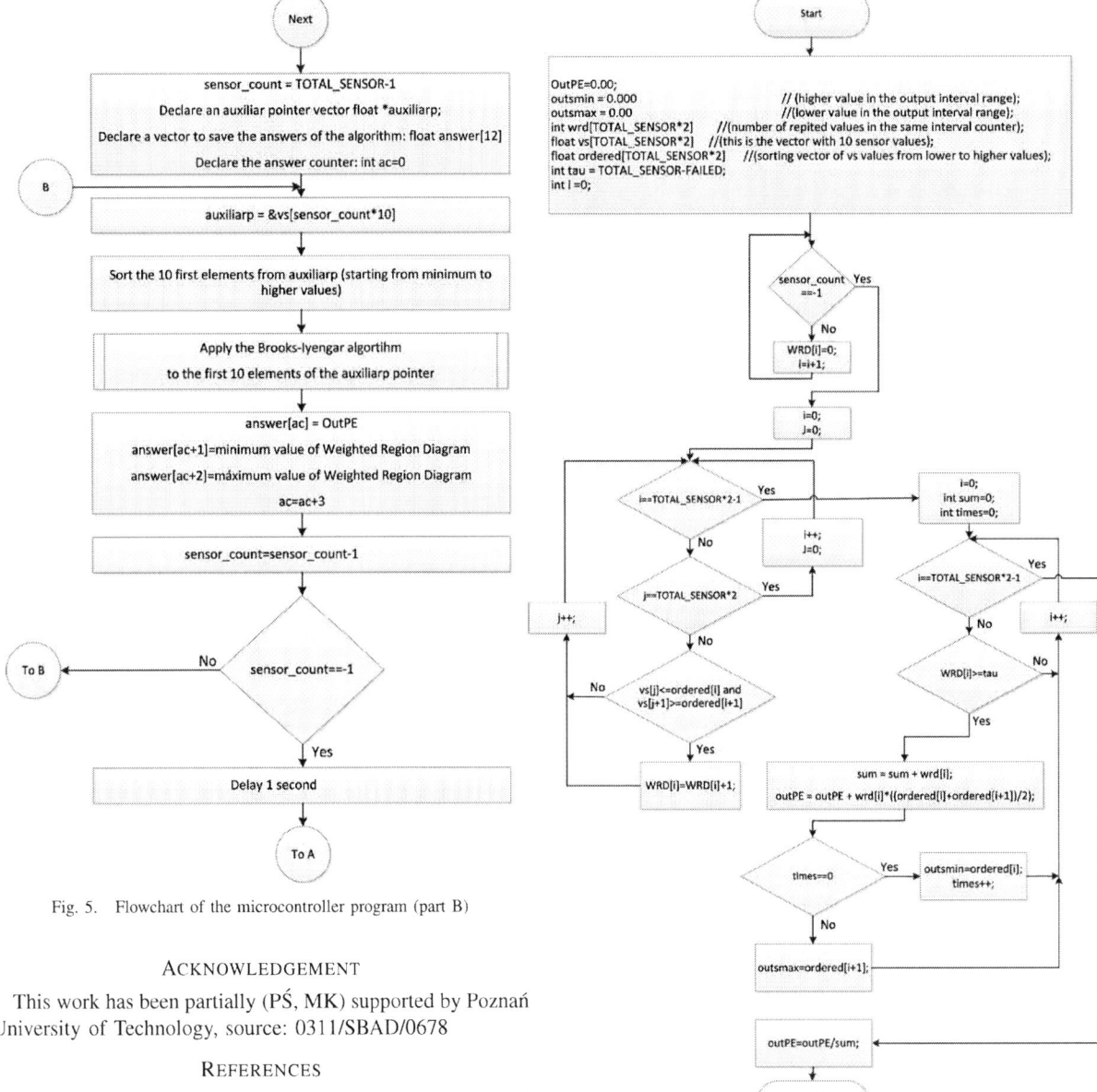

Fig. 5.   Flowchart of the microcontroller program (part B)

ACKNOWLEDGEMENT

This work has been partially (PŚ, MK) supported by Poznań University of Technology, source: 0311/SBAD/0678

REFERENCES

[1] Cleber M. Morais, Djamel Sadok, and Judith Kelner. An iot sensor and scenario survey for data researchers. *Journal of the Brazilian Computer Society*, 25(1):1–17, 2019.
[2] P. Arroyo, H. J. Luis, J. I. Suárez, and J. Lozano. Wireless sensor network combined with cloud computing for air quality monitoring. *Sensors (Basel, Switzerland)*, 19(3), 2019.

Fig. 6.   Flowchart of the core of the Brooks-Iyengar algorithm.

[3] B. P. L. Lau, S. H. Marakkalage, Y. Zhou, N. U. Hassan, Y. Chau, Z. Meng, and U. Tan. A survey of data fusion in smart city applications. *Information Fusion*, 52:357–374, 2019.

[4] P. Śniatała. *CMOS Current Mode Sigma-Delta Modulators*. Poznan Monographs in Computing and Its Applications, 2016.

[5] M. Wesolowska, A. Correia, P. Śniatała, and J. Goes. Speeding-up moderate-resolution dual-slope a/d converters by combining noise-shaping with efficient digital filtering. In *XVII KRAJOWA KONFERENCJA ELEKTRONIKI*, Krakow, Poland, Sep. 2018 2018. Polish Academy of Sciences, Polish Academy of Sciences.

[6] P. Śniatala, A. Handkiewicz, J. Goes, N. Paulino, and J.P. Oliveira. Fully differential sigma-delta modulator structure for current-mode sensors. In *Proc.International Conference on Signals and Electronic Systems (ICSES)*, pages 37–40, 2016.

[7] S. K. Ramani and S. S. Iyengar. Evolution of sensors leading to smart objects and security issues in iot. In *International Symposium on Sensor Networks, Systems and Security*, page 125–136, 2017.

[8] R.R Brooks and S.S Iyengar. Robust distributed computing and sensing algorithm. *Computer*, 29(6):53–60, 1996.

[9] S. S. Iyengar, S.K. Ramani, and A. Buke. Fusion of the brooks–iyengar algorithm and blockchain in decentralization of the data-source. *Journal of Sensor and Actuator Networks*, 8, no. 1:17, 2019.

[10] Sartaj Sahni and Xiaochun Xu. Algorithms for wireless sensor networks. University of Florida, Gainesville, September 2004.

[11] P Śniatała, M. Hadi Amini, and Kianoosh G. Boroojeni. *Fundamentals of Brooks-Iyengar Distributed Sensing Algorithm*. Springer, 2020.

Virtualization of an Aluminum Cans Production Line Using Virtual Reality

Lucas Sales de Oliveira Almeida[1], Alexandre Baratella Lugli[1], Tales Cleber Pimenta[2],
Matheus Vinícius Cirino e Silva[1], João Paulo Carvalho Henriques[1], Renzo Paranaíba Mesquita[1]

[1] Instituto Nacional de Telecomunicações - Inatel, Brasil
[2] Federal University of Itajubá - UNIFEI, Brasil
lucassales@gea.inatel.br, baratella@inatel.br, tales@unifei.edu.br,
matheuss@gec.inatel.br, joao.paulo@inatel.br, renzo@inatel.br

Abstract—In the context of the 4.0 industry, new technologies have been aggregated to the industries, which has been pushing production levels and through the use of Big Data keeping great control over its products. Connectivity has been the main focus of this new revolution highlighting the Internet of Things and Systems Integration. One pillar of this massive change industries is Augmented and Virtual reality, which shows itself as a promising area for training purposes and error diagnoses.

The focus of this paper is to develop a virtual reality system capable of showing a can production company operators the procedure for components exchange in a machine responsible for the extrusion of aluminum cups. For this end an HTC Vive device was employed for the immersion of the operator, the full modeling of the machine was also needed. The methods used to fulfill the project goals are presented as well as the tools used to develop the proposed system.

Keywords—**Autodesk Fusion, Industry 4.0, HTC Vive, Unity 3D, Virtual Reality.**

I. INTRODUCTION

The current context of industry is about resource optimization and swift procedures. As mentioned by Rojko [1], the current industrial process is driven by competition and a need to quickly adapt to an ever-changing market. Efforts are now directed to a new industrial revolution to help congregate the economic solutions to these modern problems with state-of-the-art technology. This new movement by which companies need to adapt is called industry 4.0.

According to Erboz [2], Industry 4.0 is based in the use and creation of technologies that are within nine different pillars, that are: Internet of Things (a network of multiple smart devices), Simulations (the evaluation of real-time data to reflect developments in the real world), Additive Manufacturing (the implementation of technologies such as 3D printing to rapid prototyping), System Integration (an interconnection of external and internal systems inside companies), Cloud Computing (online scalable storage and increased processing power), Autonomous Systems (further exploration of past discoveries and implementation in automation), Cybersecurity (the defense mechanisms against cyberattacks and minimization of any possible security breach), Big Data (the storage of massive analytical data sets) and finally, Augmented Reality (digital contents displayed on the physical world through a device). From the latter topic, it can also imply the use of

Virtual Reality (VR), a complete immersion of the user into a virtual environment.

The applications and use of VR are endless. It can be applied from digital games for pure entertainment to solutions for medical diagnostics or military applications [3]. The use of VR environments, specifically in training, has the potential to dramatically reduce its costs, as the users don't need to be physically present in a real environment to learn a new skill. Not to mention that, in a high-risk environment, such as a factory, lives could be spared and actions would not impact the real production process.

Based on the wide range of VR usage and aiming to create a solution to solve a real problem of a company, comes the proposal of this work: the creation of a VR production line machine with the goal of training employees of a can production company in how to operate, upkeep and overall understand the performance and workings of it. With this solution, factory employees are expected to know how to operate the equipment in advance before they get their hands on the physical equipment, preventing errors from impacting the production process and even causing financial losses for the company.

The remainder of this paper is organized as follows: section II provides a summary of related works that also tried to use VR technology to create new researches and solutions to the Industry 4.0. Section III highlights individual details of important tools and techniques that were used in this work in order to create the proposed solution. Section IV presents details of how these tools and techniques were integrated in order to build the virtual environment, and finally, in section V, a summary of conclusions and lessons learned are presented.

II. RELATED WORKS

Some authors have already researched about the advantages and use of virtual reality not only in the industry but in different application areas.

In Kovar et al. [4] a broad discussion of Industry 4.0 trends are brought. It highlights virtual reality technology applied for educational and research terms, although its possibilities for the industry are also argued. Focused on the visualization of microstructures of solar cells, it uses a technology called CAVE VR, which is a projection-based VR display [5], to create an immersive environment and show the resulting topology and measurements to the user.

A survey conducted by Berg et al. [3] aimed over the applicability of VR technology for product development and manufacturing. The survey occurred in the form of visits and conferences with 20 companies that uses VR in various forms. The objective was to determine how, and which VR technology has been enforced. It was esteemed that CAVE was the most common between the companies and Head Mounted Devices (HMD) [6] method appeared in third place, what revealed that is space for more research using these technologies when applied for industry.

Zhang Hui [7] describes a VR system developed for mining applications. The paper compares the applicability, regarding the immersion and easiness, of HMD devices, for VR's immersion, with the conventional screen-based system. The created environment allows the user to interact and utilize drilling machinery through the use of a Leap Motion device attached to the HMD. The results were collected by questionnaires, delivered to users, that addressed the following subjects: Immersive, Intuitive, Interactive, Easy to use, Easy to learn. In all subjects, the VR technology presented itself as superior.

A training system developed by Simone Borsci et al. [8] focused on car services maintenance. In the paper, a debate over the learn-by-doing training method is discussed, addressing the pros and highlighting as main con the cost issue of this approach. As a possible solution to the cost situation of a learn-by-doing training, a virtual environment was developed, in which, car's services maintenance proceeding could be performed. The presented results showed that users who were submitted to the virtual training were able to obtain higher scores in tests applied to non-virtual trained users and virtual trained users. As a side effect some users reported motion sickness due to the system .

Eschen et al. [9] conducted a study over the applicability of virtual and augmented reality (AR), highlighting the potential of each technology given its application and the level of interaction from the user and the real-world objects. The study shows that situations where no direct interaction between user and real-world object are needed, AR technology can be better applied, however, for the opposite situation, VR technology shows itself as promising. The paper discusses that VR can lead to a diminishing in the error rate of the user. Ultimately the author report that improvements in the system can be made, once some users reported that the method used for controlling was not suitable and through those improvements, VR technology can be better utilized.

III. EMPLOYED TECHNOLOGIES

In order to achieve this paper's goals, 3D modelling, game development and virtual reality tools were employed. These technologies are Autodesk Fusion 360, Unity 2019 and HTC Vive. These will be objectively discussed in the following subtopics.

A. Autodesk Fusion 360

Fusion 360 is a 3D modeling software from Autodesk Inc., a leader in 3D design, engineering, and entertainment software [10]. Some of important Fusion 360 features [11] are:

- 3D modeling: allows users to create 3D models with free form, direct, parametric, mesh and sheet metal designs;
- Simulation: users can test designs to ensure they can be applied to real-world applications. These simulations can be for static stress, thermal stress, events simulations and many others;
- Cloud-based: Designs and projects can be stored safely in the cloud, furthermore, projects can be shared and accessed from other computers and there is a cellphone application that gives access to the project;
- Collaborative environment: projects can be created cooperatively and can be edited and commented by the team.

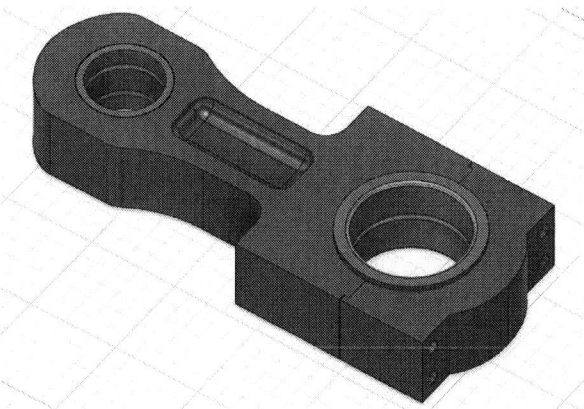

Fig. 1. Modelled bodymaker part.

A full bodymaker Standun B6 machine was modeled with Fusion software. This was possible thanks to the machine parts drawings provided by the factory. These drawings contain the dimensions for the machine's parts and how to assemble them. However, many visits were conducted to the factory to ensure the measurements and to have a better understanding of the modeled parts placements in the machine and how to assemble them. In Figure 1 an example of a modelled piece of the bodymaker using Fusion 360 is displayed.

Once the drawing is available, it is possible to import it into the Fusion environment, however, it is necessary to re-scale the image. This method is called Fusion's "Calibrate", which allows the user to make fine adjustments to the image size so that sketch's dimensions match the real part dimensions. This technique was done for most cases, however, not all pieces had their respective drawings and for these ones, visits to the factory were necessary to acquire the necessary measurements.

B. Unity

Unity is the most popular game development platform in the world. It facilitates most of the game development process supporting developers with a vast number of tools to create

283

2D, 3D, VR and AR projects. Furthermore, it also counts with a broad community, granting exchange of information among the users [12].

Unity Editor consists of many windows that can be visualized in Figure 2, each with its specific function. The most important [13] and relevant for this project were:

- Project Browser: it contains all project assets imported to the project, very similar to Windows File Explorer. It is a good practice to keep this environment well organized through the creation of folders that describes what kind of items it contains. Window "1" in Figure 2;
- Inspector: it is employed to monitor and customize GameObjects, Components, Assets, and others. This window is where Components are attached and also user-made scripts are stored. Window "2" in Figure 2;
- Game View: allows the user to test the project, it is a preview of what the final build will look like. It's a big time saver once the project doesn't have to be rebuilt every time a change is made.Window "3" in Figure 2;
- Scene View: where the game construction is made. The developer can drag and drop assets into the scene and position them accordingly. It consists of a grid area where functions such as Move, Rotate and Scale can be used. Window "4" in Figure 2;
- Hierarchy: contains all objects currently in the scene. It is automatically updated once a new object is inserted into the Scene View or directly in this window. Dragging an object on top of another will create parenting of the object. Window "5" in Figure 2;
- Asset Store: it's where community and company made assets can be downloaded to the project. The assets can be free or paid.

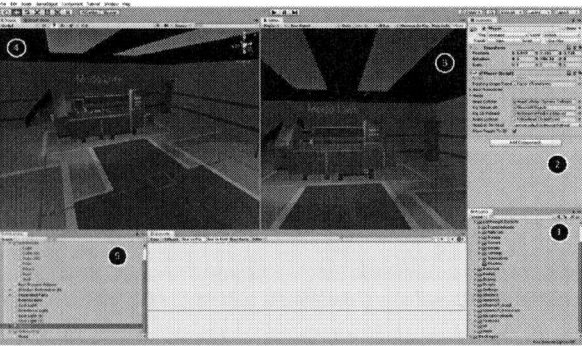

Fig. 2. Unity main screen.

To use Unity integration with HTC Vive, the Steam VR Plugin [14] was downloaded and added to the project using the Asset Store. This package works with Valve's SteamVR that allows configuration of the virtual space and equipment [15]. Its main purpose is to allow the HTC Vive input to be received by Unity and also allows Unity to display the Game View in the HTC's display. Moreover, all hand and head tracking are managed by SteamVR scripts. The package comes with several scripts and example scenes that allows the user to familiarize with the setup of a VR scene using the HTC Vive.

C. HTC Vive

The HTC Vive is a virtual reality system developed by HTC and Valve, composed of a head-mounted display (HMD), two wireless handheld controllers, a tracking mechanism and, on more advanced versions, headphones and a wireless adapter for ease of movement[16].

The Vive VR system used in this project was the Gamer edition, with the following specifications:

- Resolution of 1080 x 1200 pixels on each "eye display";
- Refresh rate of 90 frames per second;
- Field of view of 110 degrees;
- Integration to SteamVR tracking;
- Gyroscope;
- Proximity sensors;
- Multifunction trackpad with grip buttons.

Figure 3 illustrates the HTC Vive environment assembled to this project. The way the HTC Vive works within scope of this project is as a conduit between the user and the virtual environment created in Unity. The user will be able to move inside a virtual environment by selecting a point on the floor and pressing down on the trackpad, which will teleport their model to that location. The tracking system, head and hand movements are limited only by the user's physical boundaries, allowing them to look at any direction and move their virtual hands through and into any model inside the virtual environment.

The main focus of the project is the interaction and observation of the Bodymaker machine. Through the trigger (or grip) buttons, the user can perform a small number of actions on the machine, such as opening its hatches or initiating the can manufacturing operation. There are also spare parts of the machine on a table nearby the work station, so that the user may grab and manipulate them to their will in order to get a better understanding of how that portion of the machine works.

Fig. 3. Assembled environment.

IV. PROPOSED SOLUTION

In the context of Industry 4.0, the use of new technologies has been pushing production levels, however, maintenance and modifications continue to occur. Training proceedings may speed up tasks that require the stoppage of a machine operation.

It was suggested by the partner company a VR training system aiming to provide the operator closer to reality scenarios whilst in training sessions, also, this method would guarantee a form of interaction with the machinery without having to disrupt the operation of the production line. By these means, a system developed by the authors is expected to diminish machine downtime providing a good training experience prior to real operations with the equipment.

This partner company is specialized in the production of aluminum cans for beverage packaging. Given the variety of the can's sizes it's necessary to periodically execute alterations on the production line machinery, this involves adjustments and replacements of parts responsible to give the profile and finishing of products, this process is known as line conversion and allows one line to change the size/volume of the resulting can. This process consists of a series of tasks and can take hours or days depending on the alterations to be made. To improve the factory and to ingress in Industry 4.0 scenery an initiative to promote VR training sessions is put to test.

A. Project Planning

The first step for the project development was a meeting performed with the company personal. On this meeting the main idea of a VR training system was discussed, also, the main issues of the factory regarding the line conversion. In a second meeting, now with a technician of the factory plant, more detailed requirements were brought to the discussion, such as: how the flow of the training should occur and how the in-game interactions were to take place. By the end of the meeting a visit to the factory floor was conducted, this was essential to understand the environment that was needed to be reproduced in virtual means. Moreover, videos and sounds from inside the plant were recorded, as well as photos were taken. Those would be later used as a basis for modeling and scenery design.

It was defined that the development would be executed following the tasks:

- Study of proposed tools: in this task, the authors gathered information about which tools could be applied to achieve the project goal, in Section III those are highlighted;
- Gathering of information from real-world scenario: this stage is related to visits conducted to the company. Those were important to have a better understanding of the problem and to get acquainted with the machinery and production line;
- 3D modeling: it was clear that given the complexity of the machine, some time of the development would be used in modeling, furthermore, the programming of the mechanic, and movement of parts required a robust model to start with;

- Software development: consists of the actual making of a simulator, this involves: importing of 3D models, materials and shaders creation, project configuration, coding, creation of animations and debugging.
- Integration with VR equipment: by the end of the software development stage, the needed backbone of the software was done. The integration of Unity with HTC Vive was employed, requiring changes in how the interactions should take place.

B. Project Execution

The modeling of the Standun Bodymaker B6 [17] displayed in Figure 4, was accomplished via Fusion 360 software. As discussed in Section III-A, visits to the company were made ensuring the parts dimensions and their placements in the machine. The main concern of the team at this stage was the development of the parts separately so that later they could be grouped with other ready parts thus creating entire machine sections.

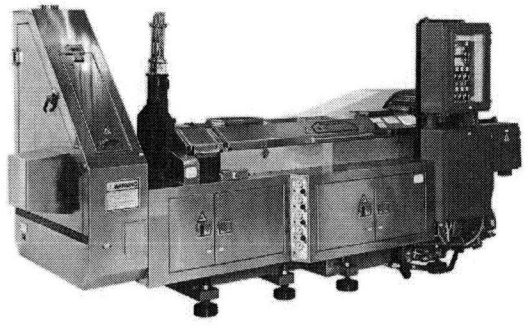

Fig. 4. Real Standun Bodymaker B6, Stolle Machinery.

In Figure 5 it's pointed that a part such as Part A, can have N subparts, this depends on its complexity. The subparts are then grouped using Fusion's "Joint" feature, whose types are Rigid, Revolute, Slider, Cylindrical, Pin-slot, Planar, and Ball. This will make one component to be "attached" to another and will give a motion that relies on the type of Joint applied. For most cases, rigid and revolute joints were types able to reproduce the expected motion. An example is to parts that should be fixed tight together, where a rigid joint is the most suitable, in contrast, for motions alike doors, a revolute joint will be the most suitable. By the end of the N attachments of Part A's subparts, the final Part A would be complete. The grouping of multiple parts would make a section and finally, grouping all sections would result in a full bodymaker.

The final model displayed in the Figure 6 was exported using Filmbox (fbx) file format, compatible with Unity Editor. Despite the fact that the exporting did not noticeably changed the objects dimensions, a large number of vertices, consequently triangles or "tris" were necessary to recreate the model built in the Fusion environment. This large amount of vertices

285

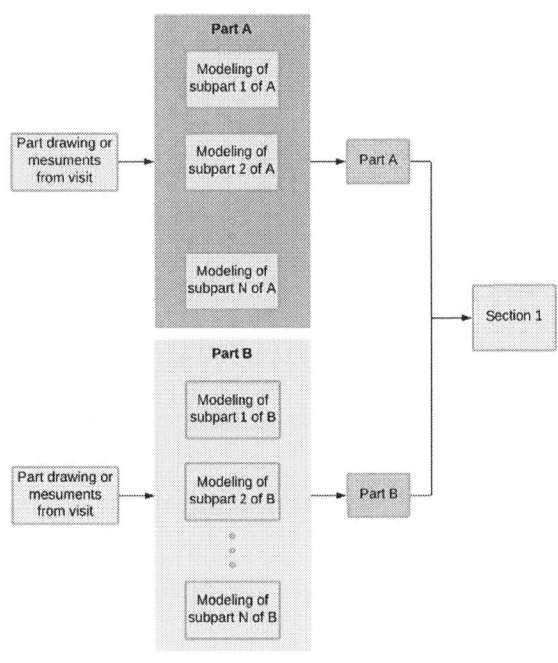

Fig. 5. Modeling procedure.

On the early stages of the software development, scenery design was the main focus. Photos, videos and visits served as basis for a modeling of a near to industrial environment. To improve graphics performance of Unity a common tool is to utilize the Lightweight Render Pipeline (LWRP) [19]. In Figure 7, one can notice on the left side image (Unity Standard Renderer) that the effect of lighting is very soft what can be visualized on the light source (Construction Light), note how the intensity is not represented as in the right side image (Lightweight Render Pipeline). Furthermore, aliasing and bloom effects are presented for the LWRP image.

Fig. 7. Comparison between renderers.

reaching two million could impair in the user experience, and in a VR scenario could lead to motion sickness. In order to ease those effects, Blender 2.8 [18] was used as a middle-ware from Fusion and Unity. Via Blender's "Limited Dissolve" tool, vertices closer than the user given value were merged together, this way only minor alterations to the model characteristics are made, in most cases the changes aren't noticeable and could lower from one million vertices to, approximately, 800 thousand.

C. Results

For test purposes, the first versions of the simulator could be controlled using a keyboard and mouse, for interactions. This idea was removed once the HTC Vive controller was added to the project. The integration of a VR device with Unity requires the user to apply the Virtual Reality Support checkbox in XR Settings, under Player Settings. On Section III-B Steam VR Plugin is discussed. This Asset is essential to guarantee the communication of the HTC Vive with Unity Editor and allows the developer to test the project without having to build it every time alterations are made. Several scripts are part of the Asset, including example scenes, SteamVR namespaces and objects.

The user movement inside the virtual environment can be done by regular walking inside the play area or by using a teleport function. Naturally, the play area is limited by sensors and available physical area, therefore, a teleport function allows the environment to be much bigger than the actual physical environment.

The procedures done by the user are briefly explained before he is able to perform activities on the machine. Inside the virtual environment tutorial stage is responsible to show the user where the activities are done and which parts have to be removed or replaced. This is done via a animated text explaining how to remove the part, including the specifications of the needed tools to perform the task and also highlighting the parts so the user can familiarize himself with the machine. In Figure 8 the tutorial stage is displayed.

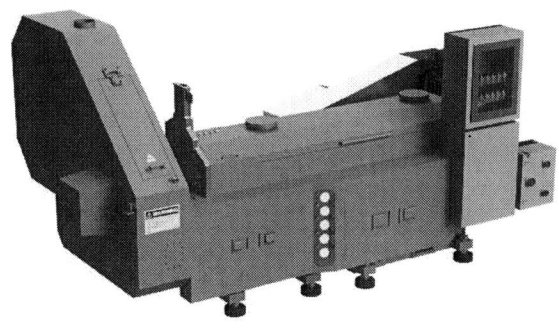

Fig. 6. Final model developed in Fusion 360.

Fig. 8. The tutorial stage.

The interaction with the virtual objects can be done using the trigger button on HTC Vive's controllers. One object can be picked and placed by the user just like one would in real life. This allows the training to be much more intuitive for the user. An example is the training for a part replacement on the machine. The user is able to, using his "hand", remove parts of the machine to get access to the part of interest, then remove the one and replace it for the new part as he would do in real life. This procedure is displayed in Figure 9.

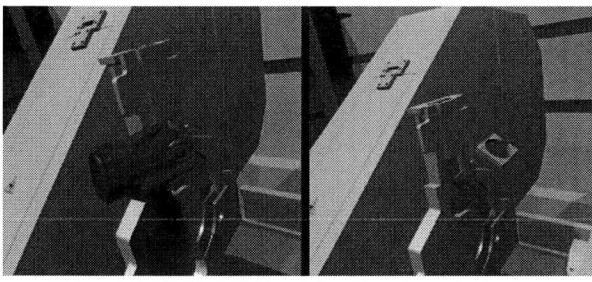

Fig. 9. An interaction of the user to remove a given part.

V. CONCLUSIONS

With the arrival of Industry 4.0, many innovative ways to solve old and new technical problems are being, or have been, invented, yet the use of Virtual Reality as means to train and specialize workers with difficult, heavy or elaborate machinery is still not as widespread as it could be. This project has shown that a function of VR, typically used on the gaming industry, could have compelling applications as a teaching device, with an interactive virtual environment designed to instruct an operator on how to maintain and repair a given machine. One of the obstacles on the making of this project was the complexity of certain parts of the Bodymaker machine, given that not all of them had their own detailed sketches, and required some guesswork and alternative methods to acquire their dimensions. And improvements could still be made, such as auditory feedback to the user, on both success and failure of a task, or the conversion procedures for other can sizes. Quality-of-life changes, such as the ability to grab objects from long-range or fixed text, and added polish to the machine's animations would also be substantial, but ultimately not as critical.

REFERENCES

[1] A. Rojko, "Industry 4.0 concept: background and overview," *International Journal of Interactive Mobile Technologies (iJIM)*, vol. 11, no. 5, pp. 77–90, 2017.

[2] G. Erboz, "How to define industry 4.0: main pillars of industry 4.0," *Szent Istvan University, Gödöllő*, pp. 1–9, 2017.

[3] L. P. Berg and J. M. Vance, "Industry use of virtual reality in product design and manufacturing: a survey," *Virtual reality*, vol. 21, no. 1, pp. 1–17, 2017.

[4] J. Kovar, K. Mouralova, F. Ksica, J. Kroupa, O. Andrs, and Z. Hadas, "Virtual reality in context of industry 4.0 proposed projects at brno university of technology," in *2016 17th International Conference on Mechatronics-Mechatronika (ME)*. IEEE, 2016, pp. 1–7.

[5] Visbox, Inc., "CAVE Automatic Virtual Environment," 2016, available at http://www.visbox.com/products/cave/. Acesso em Janeiro de 2020.

[6] A. Schindler and B. Rottenkolber, "Hmd device," Sep. 20 2005, uS Patent 6,945,648.

[7] H. Zhang, "Head-mounted display-based intuitive virtual reality training system for the mining industry," *International Journal of Mining Science and Technology*, vol. 27, no. 4, pp. 717–722, 2017.

[8] S. Borsci, G. Lawson, B. Jha, M. Burges, and D. Salanitri, "Effectiveness of a multidevice 3d virtual environment application to train car service maintenance procedures," *Virtual reality*, vol. 20, no. 1, pp. 41–55, 2016.

[9] H. Eschen, T. Kötter, R. Rodeck, M. Harnisch, and T. Schüppstuhl, "Augmented and virtual reality for inspection and maintenance processes in the aviation industry," *Procedia manufacturing*, vol. 19, pp. 156–163, 2018.

[10] Autodesk, "About Autodesk, Company," 2018, available in https://www.autodesk.com.br/company. Acessed in 01/2020.

[11] Autodesk, "Fusion 360 Features," 2020, available at https://www.autodesk.com/products/fusion-360/features. Acesso em Janeiro de 2020.

[12] Unity Public Relations, "Powering the real-time revolution," 2020, available at https://unity3d.com/public-relations?_ga=2.94387356.1 49083941.1579095568-579528162.1577723616. Acesso em Janeiro de 2020.

[13] J. K. Haas, "A history of the unity game engine," 2014.

[14] Valve, "Steam Vr Plugin," 2019, available at https://assetstore.unity.com/packages/tools/integration/steamvr-plugin-32647. Acesso em Janeiro de 2020.

[15] Valve, "Steam Vr," 2020, available at: https://www.steamvr.com/pt-br/. Acess in January 2020.

[16] VIVE, "HTC Product Comparison," https://www.vive.com/us/comparison/, 2020, accessed: 2020-02-08.

[17] S. Machinery, "Stolle Standun Bodymakers," 2019.

[18] Blender Fundation, "Blender home page," 2020, available at https://www.blender.org/. Acesso em Fevereiro de 2020.

[19] Unity Technologies, "About the lightweight render pipeline," 2020, available at https://docs.unity3d.com/Packages/com.unity.render-pipelines.lightweight@6.9/manual/index.html. Acesso em Fevereiro de 2020.

Index of Authors

ABBAS M. 107
ALJEHANI N. 107
AMINI A. 65
AMINI M.H. 277
AMROZIK P. 89
ANNUSEWICZ A. 223
BAJER K. 45
BANACH M. 101
BARATELLA LUGLI A. 282
BARZDENAS V. 59
BASCHIROTTO A. 199
BONAMENT A. 35
BRINSON M. 23, 50
BRZOZOWSKI I. 94
C. B MANNSFELD S. 40
CARVALHO HENRIQUES J.P. 282
CIRINO E SILVA M.V. 282
CLAUS M. 40
CLEBER PIMENTA T. 282
CORDIER Y. 181
CZABANSKI R. 213
DE CARVALHO SILVA M.T. 239
DEL CROCE P. 199
DELIGEORGIS G. 156
DIAS PEREIRA N. 239
DI LORENZO R. 199
DŁUGOSZ R. 101, 227
DŁUGOSZ Z. 101, 227
DONNHÄUSER S. 40
EL-SAYED A.-M. 168
FRAYSSINET E. 181
GAVELLE M. 181
GÓRECKI K. 127, 133, 204
GÓRECKI P. 133
GRZYMKOWSKI Ł. 266
GULGOWSKI J. 160, 164
HAASE K. 40
HALOUI C. 181
HELENIAK J. 204
HENRIQUES J.P. 239
ISOIRD K. 181
JAKUBOWSKI M. 112
JANICKI M. 127
JANKOWSKI M. 89
JANUS P. 112
JEZEWSKI M. 213
KALAMI A. 65
KAPPEL R. 147
KARAGIANNI E. 156
KEYHANAZAR M. 65
KŁECZEK R. 78
KMON P. 78
KOHUTKA L. 83
KOŁACIŃSKI C. 112
KOPEĆ M. 139
KOS A. 249
KOSTOPOULOS A. 156

KROPIDŁOWSKI M. 277
KUCHARSKI K. 112
KUCHARSKI M. 74
LALLEMENT C. 35
LAU K. 277
LESKI J.M. 213
LESSI C. 156
LUGLI A. 239
MADEC M. 35
MALINOWSKI A. 175
MARSZAŁEK Z. 255
MARTINEK R. 213
MARZEC P. 249
MATONIA A. 213
MISHRA S.K. 175
MORANCHO F. 181
MOTHES S. 40
MUDZA Z. 117
MUELLER M. 40
NAPIERALSKI A. 151
NAPIERALSKI J. 218
NAZDROWICZ J. 151
NETO E. 239
OBRĘBSKI D. 112, 260
OLIVEIRA J.P. 69
PACHECO-SANCHEZ A. 40
PANAGOPOULOS A. 156
PARANAIBA MESQUITA R. 282
PAUL S. 45
PETERS-DROLSHAGEN D. 45
PIDUTTI A. 199
PIEŃCZUK P. 112
PIESIEWICZ R. 74
PIETRALA D. 218
PIMENTA T.C. 239
PIRES L.M. 69
PREL A. 35
PTAK P. 127, 204
RAJEWSKI M. 101, 227
ROMANOVA A. 59
SALAS S. 277
SALES DE OLIVEIRA ALMEIDA L. 282
SALLESE J.-M. 35
SCHWARZ M. 13
SEJC I. 147
SELBERHERR S. 168
SHATARAH I.S.M. 192
SKRZYPIEC P. 255
STAVRINIDIS G. 156
STAWIŃSKI A. 151
STEFANSKI T. 160, 164, 232, 266
STOPJAKOVÁ V. 83
SVERDLOV V. 168
SZERMER M. 89
SZMUC T. 245, 272
SZMUC W. 245, 272
SZYMAŃSKI A. 112

ŚNIATAŁA P.	277
TALAŚKA T.	101, 227
TASSELLI J.	181
TOULON G.	181
TROFIMOWICZ D.	160, 164, 232
VALDEZ C.	277
VAZOURAS C.	156
WAWRYN K.	187
WĘGRZYN G.	78
WIDLOK M.	74
WIDULINSKI P.	187
WIĘCEK B.	139, 192
ZAJĄC P.	89
ZBIEĆ M.	112, 260
ZWIERZCHOWSKI J.	218, 223

IEEE
445 Hoes Lane
Piscataway, NJ 08854-4141

ISBN 978-1-7281-9781-4